Nonlinear Waves
in Integrable and Nonintegrable Systems

Mathematical Modeling and Computation

About the Series

The SIAM series on Mathematical Modeling and Computation draws attention to the wide range of important problems in the physical and life sciences and engineering that are addressed by mathematical modeling and computation; promotes the interdisciplinary culture required to meet these large-scale challenges; and encourages the education of the next generation of applied and computational mathematicians, physical and life scientists, and engineers.

The books cover analytical and computational techniques, describe significant mathematical developments, and introduce modern scientific and engineering applications. The series will publish lecture notes and texts for advanced undergraduate- or graduate-level courses in physical applied mathematics, biomathematics, and mathematical modeling, and volumes of interest to a wide segment of the community of applied mathematicians, computational scientists, and engineers.

Appropriate subject areas for future books in the series include fluids, dynamical systems and chaos, mathematical biology, neuroscience, mathematical physiology, epidemiology, morphogenesis, biomedical engineering, reaction-diffusion in chemistry, nonlinear science, interfacial problems, solidification, combustion, transport theory, solid mechanics, nonlinear vibrations, electromagnetic theory, nonlinear optics, wave propagation, coherent structures, scattering theory, earth science, solid-state physics, and plasma physics.

Jianke Yang, *Nonlinear Waves in Integrable and Nonintegrable Systems*

A. J. Roberts, *Elementary Calculus of Financial Mathematics*

James D. Meiss, *Differential Dynamical Systems*

E. van Groesen and Jaap Molenaar, *Continuum Modeling in the Physical Sciences*

Gerda de Vries, Thomas Hillen, Mark Lewis, Johannes Müller, and Birgitt Schönfisch, *A Course in Mathematical Biology: Quantitative Modeling with Mathematical and Computational Methods*

Ivan Markovsky, Jan C. Willems, Sabine Van Huffel, and Bart De Moor, *Exact and Approximate Modeling of Linear Systems: A Behavioral Approach*

R. M. M. Mattheij, S. W. Rienstra, and J. H. M. ten Thije Boonkkamp, *Partial Differential Equations: Modeling, Analysis, Computation*

Johnny T. Ottesen, Mette S. Olufsen, and Jesper K. Larsen, *Applied Mathematical Models in Human Physiology*

Ingemar Kaj, *Stochastic Modeling in Broadband Communications Systems*

Peter Salamon, Paolo Sibani, and Richard Frost, *Facts, Conjectures, and Improvements for Simulated Annealing*

Lyn C. Thomas, David B. Edelman, and Jonathan N. Crook, *Credit Scoring and Its Applications*

Frank Natterer and Frank Wübbeling, *Mathematical Methods in Image Reconstruction*

Per Christian Hansen, *Rank-Deficient and Discrete Ill-Posed Problems: Numerical Aspects of Linear Inversion*

Michael Griebel, Thomas Dornseifer, and Tilman Neunhoeffer, *Numerical Simulation in Fluid Dynamics: A Practical Introduction*

Khosrow Chadan, David Colton, Lassi Päivärinta, and William Rundell, *An Introduction to Inverse Scattering and Inverse Spectral Problems*

Charles K. Chui, *Wavelets: A Mathematical Tool for Signal Analysis*

Nonlinear Waves
in Integrable and Nonintegrable Systems

Jianke Yang
University of Vermont
Burlington, Vermont

siam.

Society for Industrial and Applied Mathematics
Philadelphia

10 9 8 7 6 5 4 3 2 1

Library of Congress Cataloging-in-Publication Data

Yang, Jianke.
Nonlinear waves in integrable and nonintegrable systems / Jianke Yang.
p. cm. – (Mathematical modeling and computation ; 16)
Includes bibliographical references and index.
ISBN 978-0-898717-05-1 (softcover)
1. Nonlinear waves. 2. Wave equation–Numerical solutions. I. Title.
QA927.Y365 2010
531′.1133–dc22

2010020210

Contents

List of Figures

Preface

Wave phenomena are abundant in nature. Familiar examples include water waves and optical waves. Low-amplitude waves are governed by linear partial differential equations. A main feature of linear wave phenomena is dispersion, i.e., different Fourier modes inside a disturbance travel at different speeds. This often leads to the spreading and decay of a local disturbance. For instance, if we throw a stone into a pond, concentric rings of ripples will spread out. Outer ripples generally have higher wavenumbers (shorter spatial periods) and travel faster. After a short time, all the ripples will disperse and disappear (even in the absence of dissipation). Similarly, when a low-intensity light beam passes through the air or a crystal, the beam broadens over distance. This phenomenon is called diffraction in optics, and it is the counterpart of dispersion in water waves.

In 1834, Scott Russell accidentally observed a type of water wave which did not disperse. Below is his original description:

"I was observing the motion of a boat which was rapidly drawn along a narrow channel by a pair of horses, when the boat suddenly stopped—not so the mass of water in the channel which it had put in motion; it accumulated round the prow of the vessel in a state of violent agitation, then suddenly leaving it behind, rolled forward with great velocity, assuming the form of a large solitary elevation, a rounded, smooth, and well-defined heap of water, which continued its course along the channel apparently without change of form or diminution of speed. I followed it on horseback, and overtook it still rolling on at a rate of some eight or nine miles an hour, preserving its original figure some thirty feet long and a foot to a foot and a half in height. Its height gradually diminished, and after a chase of one or two miles I lost it in the windings of the channel. Such, in the month of August 1834, was my first chance interview with that singular and beautiful phenomenon which I have called the Wave of Translation" (Russell (1844)).

This phenomenon was recreated on the Union Canal near Edinburgh in July 1995, and a photo is shown in Fig. 1.

Russell's observation puzzled physicists for a long time and caused much controversy, because it could not be explained by linear water wave theory. In 1895, Diederik

Figure 1. *Recreation of Russell's* 1834 *observation of solitary water waves in a canal. (Image courtesy of Mathematics Department, Heriot-Watt University, Edinburgh, Scotland.)*

Korteweg and Gustav de Vries considered this problem. They noticed that while dispersion causes a water wave to decay, nonlinear effects can cause it to steepen—a phenomenon we always see when waves run up a beach. When self-steepening balances dispersion, a solitary wave which moves without change of shape (i.e., the "wave of translation") can form. Performing detailed theoretical analysis, Korteweg and de Vries (1895) derived the following nondimensionalized wave equation which is now called the Korteweg–de Vries (KdV) equation:

$$\psi_t + \psi_{xxx} + 6\psi\psi_x = 0. \tag{1}$$

Here ψ is the elevation of the water surface. This equation admits traveling solitary waves

$$\psi(x,t) = \frac{1}{2} c \,\mathrm{sech}^2 \frac{1}{2}\sqrt{c}(x - ct), \tag{2}$$

where c is the wave speed. These solitary wave solutions correspond to the wave of translation in Russell's observation.

Not much progress was made on the KdV equation until the 1960s, when Zabusky and Kruskal (1965) numerically discovered the elastic collision between KdV solitary waves, and then Gardner, et al. (1967) invented the inverse scattering transform method and solved the KdV equation analytically. This pioneering work initiated an unprecedented burst of research activities on nonlinear waves. In subsequent years, many other nonlinear equations such as the nonlinear Schrödinger (NLS) equation, the sine-Gordon equation, and the Kadomtsev–Petviashvili (KP) equation were solved by this method, and such equations are now called integrable equations (see Ablowitz and Segur (1981), Zakharov et al. (1984), Newell (1985), Faddeev and Takhtadjan (1987), Ablowitz and Clarkson (1992) for reviews). In addition, the theory of integrable equations was also greatly expanded along many directions such as the Riemann–Hilbert formulation (Zakharov et al. (1984)) and the direct methods (Hirota (2004)). Interesting developments on integrable equations continue till this day.

While integrable equations constitute an important part of the nonlinear wave theory, most nonlinear wave equations encountered in physics and engineering are not integrable. The study of nonintegrable equations started as early as the 1970s, and intensified from the 1990s. These investigations reveal that solution dynamics in nonintegrable equations can be much richer and more complex. For instance, in nonintegrable equations, solitary waves can be unstable; their interactions can exhibit fractal scattering phenomena; solitary waves can be embedded inside the continuous spectrum and thus exhibit unique dynamical properties; solitary waves can suffer critical collapse in multidimensions, etc. Many exciting theories have also been developed, such as the Vakhitov–Kolokolov criterion for linear stability of solitary waves (Vakhitov and Kolokolov (1973)), the exponential asymptotics method for calculating nonlocal solitary waves (see Boyd (1998) for a review), the virial theorem and the wave collapse theory (see Sulem and Sulem (1999) for a review), and the theories for fractal scatterings in solitary wave interactions (Goodman and Haberman (2007), Zhu et al. (2008a)).

In nonlinear wave studies, numerical computations play an important role. This is particularly so for nonintegrable equations. Numerical computations include evolution simulation of initial value problems, computation of solitary wave solutions, computation of stability spectra of solitary waves, etc. In early computations, finite difference methods were often used. In recent years, a wide range of spectrally accurate numerical methods were developed. Examples include the split-step method for evolution simulations, the Newton-conjugate-gradient method for solitary wave computations, and many others. Such methods deliver superior performance while maintaining easy implementation. These numerical methods provide powerful tools for nonlinear wave studies.

Experiments have been closely associated with nonlinear wave studies since the early days of Russell's observation in 1834. In the 1960s and 1970s, much progress on nonlinear waves was made in the context of water waves. Since 1980s, optical experiments have become very active due to technological advances and exciting commercial applications (such as fiber communications). The high sophistication of optical experiments makes it possible to compare experimental results with nonlinear wave theories not only qualitatively but also quantitatively. This has become one of the main driving forces behind nonlinear wave studies. In recent years, optical experiments in periodic or quasi-periodic media (such as photonic crystal fibers and optically induced photonic lattices) have been attracting a lot of interest, and this motivates many theoretical investigations in this area. Another different but closely related physical subject is the Bose–Einstein condensates. The nonlinear atom-atom interaction in Bose–Einstein condensates gives rise to collective wave behaviors which are intimately related to nonlinear optical phenomena. The first experimental observation of Bose–Einstein condensates in 1995 and the subsequent award of Nobel prize to its observers in 2001 greatly stimulated theoretical and experimental research in this area. The dynamic interplay between theory and experiments in these physical subjects has been a signature of current nonlinear wave research, and this trend will continue in the days to come.

The aim of this book is to cover nonlinear waves from integrable to nonintegrable equations, from analysis to numerics, and from theory to experiments. These different aspects of nonlinear waves, which in the past have generally been covered in separate books, are combined in this single book. Thus this book is not only valuable for preparing graduate students of applied mathematics to enter the nonlinear wave field, but also useful for scientists in other disciplines (such as optics, fluid mechanics, and Bose–Einstein condensates) to learn how recent progress in the nonlinear wave theory can help them in their own research. Most materials in this book are self-contained, with detailed calculations carefully explained. Thus the reader should be able to learn these materials from this book without much help from other sources. Although we often use familiar wave equations to develop the analysis, the treatments we present are largely general and can be readily extended to other related equations.

The first chapter derives various nonlinear wave equations from generic model equations or from concrete physical systems. These wave equations will be heavily studied in later chapters. The second chapter develops the integrable theory for the NLS equation using the Riemann–Hilbert formulation. Various aspects of the integrable theory, such as the inverse scattering method, infinite conservation laws, the integrable hierarchy, and the connection between squared eigenfunctions and the linearized integrable equations, will be presented. The third chapter extends the integrable theory of the second chapter to higher orders (such as the vector NLS systems). Much of this extension is straightforward, thus the universal aspects of the integrable theory in Chapter 2 become evident. But some new features do appear in higher-order systems. Such new features will be highlighted in this chapter. The fourth chapter develops the soliton perturbation theory for weakly perturbed integrable equations, and then applies it to study various physical problems. This soliton perturbation theory is based on multiscale perturbation expansions, and its key component is to directly solve the linearized integrable equations using squared eigenfunctions. This perturbation theory not only can yield dynamical equations for soliton parameters, but can also obtain radiation fields. It will be shown that the calculation of radiation is critical for certain types of perturbed solitons (such as embedded solitons). The fifth chapter covers various analytical theories for nonintegrable equations. Topics in this chapter include the Vakhitov–Kolokolov stability criterion and its generalizations, the exponential asymptotics technique for nonlocal waves, dynamical theories for embedded solitons, theoretical analysis of fractal scatterings in solitary wave interactions, transverse stability analysis of solitary waves, and theories for wave collapse in multidimensions. These theories highlight the drastic differences between solution dynamics in integrable and nonintegrable equations. The sixth chapter covers a particular but important branch of nonintegrable theories, which is the nonlinear wave dynamics in periodic media. This subject exhibits rich and interesting phenomena which have no counterpart in homogeneous media. Both the analytical theories and experiments in optics and Bose–Einstein condensates will be presented, and the interplay between theory and experiment will be demonstrated. The last chapter presents miscellaneous numerical methods for various aspects of nonlinear wave computations. Simple but efficient numerical schemes for evolution simulations, computations of solitary waves and their stability spectra will be developed. The accuracy, numerical stability, and convergence rates of these numerical schemes will also be analyzed. In addition, sample MATLAB codes for all these numerical methods will be displayed.

It should be mentioned that the field of nonlinear waves has become huge. Numerous directions have been pursued, and rich results have been obtained. The topics covered in this book only represent a small fraction of this vast field. Many interesting results and methods have been left out. For instance, on the theories of integrable equations, we have completely omitted the Hirota method, the Hamiltonian and bi-Hamiltonian structures, Darboux and Bäcklund transformations, the Painlevé properties, inverse scattering in multidimensions, and so on. Other omitted topics include semiclassical limits of integrable equations, the normal-form perturbation theory for integrable equations, analysis of discrete systems, etc. In addition, an important area of application for nonlinear waves, namely nonlinear water waves, is not covered either. Nonetheless, the topics we do cover in this book form a coherent and self-contained body of knowledge on nonlinear waves. More importantly, many of these covered topics witnessed remarkable growth in recent years, and they could show further new developments in the near future.

The only software used in this book is MATLAB. In the last chapter on numerical methods, sample MATLAB codes are included for all numerical schemes which are presented, so that readers can directly use them or modify them to their purposes. These codes can also be found on the SIAM Web page associated with this book, www.siam.org/books/mm16, or on the author's Web page, www.cems.uvm.edu/~jyang. These MATLAB codes can be readily converted into other software such as *Mathematica* or C if the reader prefers.

This book can be used as a textbook for a graduate-level course on nonlinear waves. It can also be used as a reference book for researchers working in various fields such as applied mathematics, physics, and engineering, where nonlinear wave phenomenon is a common theme.

I am indebted to Prof. David Benney for introducing me to the area of nonlinear waves, and to Prof. C.C. Lin for constant encouragement over the years. I also thank all my collaborators, postdocs and students, since collaborations with them have significantly broadened my view on this subject. In addition, I would like to thank T.R. Akylas, X.J. Chen, E.V. Doktorov, R. Grimshaw, R. Haberman, J.D. Wang, D.J. Zhang, and Y. Zhu for proofreading certain parts of this book and providing valuable comments. Much of the author's research in this area was partially funded by the Air Force Office of Scientific Research and the National Science Foundation. Their support is gratefully appreciated. I also thank my department for offering one-course teaching release, which made my book writing move faster. Last but not least, I am grateful to my family for great support during the time of my book writing.

Jianke Yang
Burlington, Vermont, USA

Chapter 1

Derivation of Nonlinear Wave Equations

In this chapter, we show how nonlinear wave equations can be derived from physical systems under certain asymptotic limits.

1.1 The Nonlinear Schrödinger Equation for Weakly Nonlinear Wave Packets

The nonlinear Schrödinger (NLS) equation governs weakly nonlinear and dispersive wave packets in one-dimensional (1D) physical systems. It was first derived by Benney and Newell (1967) in a general setting. Zakharov (1968) also derived this equation in his study of modulational stability of deep water waves. Then Hasegawa and Tappert (1973) showed that the same equation governs light pulse propagation in optical fibers. Below, we illustrate the derivation of the NLS equation for wave packets in a general KdV-type model equation:

$$\psi_t + \psi_{xxx} + f(\psi)\psi_x = 0, \tag{1.1}$$

where ψ is a real variable, and $f(\psi)$ is a general real function of ψ. For the convenience of analysis, we expand $f(\psi)$ into a Taylor series:

$$f(\psi) = c_1\psi + c_2\psi^2 + c_3\psi^3 + \cdots. \tag{1.2}$$

Here the constant term c_0 in this Taylor expansion is omitted, because this term can be eliminated from Eq. (1.1) by a trivial coordinate transform. We also assume that c_1 and c_2 are not zero simultaneously, which is the generic case.

Now we consider the evolution of a low-amplitude slowly varying wave packet in Eq. (1.1). This wave packet is illustrated in Fig. 1.1. The leading-order approximation for this wave packet is

$$\psi \sim \varepsilon[U(x,t)e^{i(k_0x-\omega_0t)} + c.c.]. \tag{1.3}$$

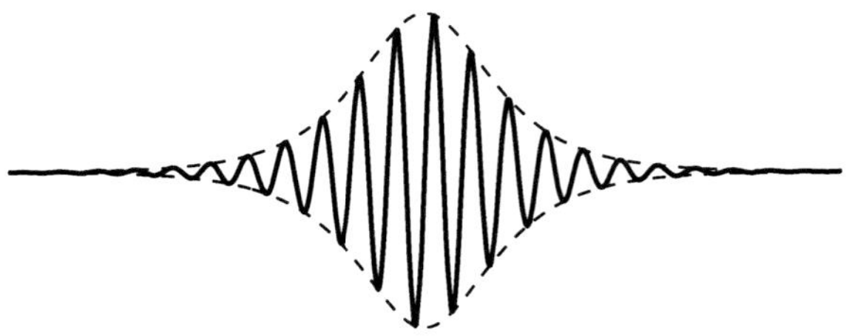

Figure 1.1. *A wave packet in the model equation* (1.1).

Here $\varepsilon \ll 1$ is a small amplitude parameter, k_0 is the dominant wavenumber of the packet,

$$\omega = -k^3 \tag{1.4}$$

is the linear dispersion relation of Eq. (1.1), $\omega_0 = \omega(k_0)$, U is the envelope function which is slowly varying in space and time, and "*c.c.*" stands for complex conjugates, which is $U^* e^{-i(k_0 x - \omega_0 t)}$ in the present context. According to the linear dispersive wave theory (see Whitham (1974) for instance), we expect the envelope of this wave packet to move at the group velocity

$$v_0 = \omega'(k_0) = -3k_0^2. \tag{1.5}$$

In the frame moving at this group velocity, the envelope function U then depends on the slow spatial variable ξ and slow temporal variable τ with

$$\xi = \varepsilon(x - v_0 t), \quad \tau = \varepsilon^2 t, \tag{1.6}$$

as we will see below.

To derive the evolution of this wave packet, we use the multiscale perturbation method. For this purpose, we expand the solution ψ as a multiscale perturbation series in ε:

$$\psi = \varepsilon[\psi_0 + \varepsilon\psi_1 + \varepsilon^2\psi_2 + \cdots], \tag{1.7}$$

where

$$\psi_0 = U(\xi, \tau)e^{i\theta} + c.c. \tag{1.8}$$

is a wave packet, and $\theta \equiv k_0 x - \omega_0 t$. Notice that this wave packet contains both fast variables (x,t) and slow variables (ξ, τ). Inserting the above expressions into the model equation (1.1), we get

$$\begin{aligned}
&\varepsilon^2(\psi_{1t} + \psi_{1xxx}) + \varepsilon^3(\psi_{2t} + \psi_{2xxx}) + O(\varepsilon^4) \\
&= -\varepsilon^2 c_1 \left(ik_0 U^2 e^{2i\theta} + c.c.\right) - \varepsilon^3\left[(U_\tau + 3ik_0 U_{\xi\xi})e^{i\theta} + c.c.\right] \\
&\quad -\varepsilon^3 c_1 \left[UU_\xi e^{2i\theta} + c.c. + (|U|^2)_\xi\right] - \varepsilon^3 c_1 \left[(Ue^{i\theta} + c.c.)\psi_1\right]_x \\
&\quad -\varepsilon^3 c_2 (Ue^{i\theta} + c.c.)^2 (ik_0 Ue^{i\theta} + c.c.) + O(\varepsilon^4).
\end{aligned} \tag{1.9}$$

Here, the dispersion relation (1.4) and the group-velocity formula (1.5) have been utilized to cancel certain terms. In this equation, the partial derivatives with respect to space x and time t are still total derivatives; i.e., they contain differentiations with respect to both fast variables (x,t) and slow variables (ξ,τ). At $O(\varepsilon^2)$, this equation gives

$$\psi_{1t}+\psi_{1xxx}=-ic_1k_0U^2e^{2i\theta}+c.c. \tag{1.10}$$

Here, the partial derivatives in ψ_{1t} and ψ_{1xxx} are with respect to fast variables x and t only. The solution ψ_1 to this equation is

$$\psi_1=U_{12}(\xi,\tau)e^{2i\theta}+c.c.+U_{10}(\xi,\tau), \tag{1.11}$$

where

$$U_{12}=\frac{c_1k_0}{2\omega_0+8k_0^3}U^2. \tag{1.12}$$

When this ψ_1 solution is substituted into Eq. (1.9), at $O(\varepsilon^3)$ we get

$$\begin{aligned}\psi_{2t}+\psi_{2xxx}=&-e^{i\theta}\left\{U_\tau+3ik_0U_{\xi\xi}+\frac{ic_1^2k_0^2}{2\omega_0+8k_0^3}|U|^2U+ic_1k_0U_{10}U+ic_2k_0|U|^2U\right\}\\&+c.c.+v_0U_{10,\xi}-c_1(|U|^2)_\xi+O\left\{e^{\pm 2i\theta},e^{\pm 3i\theta}\right\}.\end{aligned} \tag{1.13}$$

Here, the term $v_0U_{10,\xi}$ comes from $\varepsilon^2\psi_{1t}$ on the left side of Eq. (1.9) when time differentiation of the $U_{10}(\xi,\tau)$ term in ψ_1 is taken with respect to the slow variable ξ. In order to suppress linear (secular) growth of the e^0 term in the solution ψ_2, the e^0 term on the right-hand side of the above equation must vanish. Thus

$$U_{10}=\frac{c_1}{v_0}|U|^2. \tag{1.14}$$

In order to suppress secular growth of the $e^{i\theta}$ term in the solution ψ_2, the coefficient of the $e^{i\theta}$ term on the right-hand side of (1.13) must also vanish. Recalling the expressions for ω_0 and U_{10} above, we obtain the following NLS equation for the envelope U of the wave packet:

$$iU_\tau-3k_0U_{\xi\xi}+\alpha|U|^2U=0, \tag{1.15}$$

where the nonlinear coefficient α is

$$\alpha=\frac{c_1^2}{6k_0}-c_2k_0. \tag{1.16}$$

Notice that only the quadratic and cubic nonlinearities in Eq. (1.1), i.e., the c_1 and c_2 coefficients in (1.2), contribute to this envelope equation. The reason is that the wave packet (1.3) has small amplitude, thus quartic and higher nonlinearities in this wave packet are very weak and hence do not contribute to the envelope equation (1.15). In the two special cases where $f(\psi) = \psi$ (the KdV equation) or $f(\psi) = \psi^2$ (the modified KdV equation), the above results reproduce those in Boyd and Chen (2001) and Grimshaw et al. (2005), respectively.

The NLS equation (1.15) was derived for the model equation (1.1) where the linear dispersion relation was $\omega = -k^3$. For a more general model equation

$$\psi_t + \mathcal{L}\psi + f(\psi)\psi_x = 0, \tag{1.17}$$

where $\mathcal{L}$ is a linear differential operator with constant coefficients (or a linear pseudodifferential operator in general), and $f(\psi)$ is given by the Taylor series (1.2), assuming that the linear dispersion relation for Fourier waves $e^{i(kx-\omega t)}$ is $\omega = \omega(k)$, then repeating the above derivation, one can find that the evolution of the wave envelope U in the wave packet (1.3) is governed by the following NLS equation:

$$iU_\tau + \frac{1}{2}\omega_0'' U_{\xi\xi} + \alpha|U|^2U = 0, \tag{1.18}$$

where $U = U(\xi,\tau)$, ξ and τ are as given in Eq. (1.6) with $v_0 = \omega'(k_0)$, the coefficient $\omega_0'' = \omega''(k_0)$ is the second-order dispersion at the carrier wavenumber k_0, and

$$\alpha = -c_1^2\left[\frac{k_0^2}{2\omega(k_0) - \omega(2k_0)} + \frac{k_0}{\omega'(k_0)}\right] - c_2k_0 \tag{1.19}$$

is the nonlinear coefficient. For the model equation (1.1) where $\omega = -k^3$, the general envelope equation (1.18) reproduces (1.15) as a special case.

Envelope equation (1.18) is written as a temporal evolution equation. Sometimes it is more convenient to write it as a spatial evolution equation (such as for light transmission in optical fibers). To do so, we define new slow variables

$$\zeta = \varepsilon\left(t - \frac{x}{v_0}\right), \qquad \eta = \varepsilon^2 x. \tag{1.20}$$

It can be seen that these new slow variables are related to the previous slow variables (ξ,τ) as $\zeta = -\xi/v_0$, $\eta = v_0\tau + \varepsilon\xi$. Then for $\xi,\tau = O(1)$, η is asymptotically equal to $v_0\tau$. Thus the derivatives to τ and ξ in Eq. (1.18) can be converted to derivatives to η and ζ, and we get

$$iU_\eta - \frac{1}{2}k_0'' U_{\zeta\zeta} + k_0'\alpha|U|^2U = 0. \tag{1.21}$$

Here $k_0' \equiv k'(\omega_0) = 1/\omega'(k_0) = 1/v_0$, and $k_0'' \equiv k''(\omega_0)$. Notice that after this temporal and spatial swapping, the nonlinear coefficient in the new NLS equation (1.21) has changed. In addition, the sign of the second-order "dispersion" coefficient in (1.21) has also changed.

In optical fibers, the electromagnetic field of the light signal is a slowly varying wave packet. Its carrier wave is quickly oscillating, and its envelope which codes signal information is slowly varying. The light signal also experiences weak nonlinearity due to the Kerr nonlinear effect. The Kerr nonlinearity means that the effective refractive index n light experiences in the fiber depends on the light intensity $|E|^2$ as $n = n_0 + n_2|E|^2$, where n_0 is the refractive index of the fiber at infinitesimal light intensities, and n_2 is the Kerr coefficient which is usually quite small (hence the fiber is weakly nonlinear). In this case, the leading-order approximation for the electric field $\mathbf{E}(x,y,z,t)$ of the light is a slowly varying wave envelope:

$$\mathbf{E}(x,y,z,t) = \frac{1}{2}\hat{\mathbf{x}}\,[U(z,t)F(x,y)e^{i(\beta_0 z - \omega_0 t)} + c.c.], \tag{1.22}$$

where ω_0 is the carrier frequency, β_0 is the wavenumber of light in the fiber at the carrier frequency, $\beta_0 = \beta(\omega_0)$, $\beta = \beta(\omega)$ is the linear dispersion relation, $\hat{\mathbf{x}}$ is the unit polarization vector, $F(x,y)$ is the transverse electric field function (assuming the fiber is single mode), z is the propagation distance along the fiber, t is time, and $U(z,t)$ is the envelope function of the light signal. Note that the wavenumber β is related to the refractive index n by $\beta = \omega n/c$, where c is the speed of light in the vacuum.

Starting from the Maxwell equations and assuming that the Kerr coefficient n_2 is small, a calculation similar in spirit to the one performed above shows that the light amplitude function U satisfies the following NLS equation (see Hasegawa and Tappert (1973), Hasegawa and Kodama (1995), Agrawal (2007)):

$$iU_z - \frac{1}{2}\beta_2 U_{TT} + \gamma|U|^2U = 0, \tag{1.23}$$

where

$$T = t - \beta'(\omega_0)z, \quad \beta_2 = \beta''(\omega_0), \tag{1.24}$$

$$\gamma = \frac{2\pi n_2}{\lambda_0}\frac{\int_{-\infty}^{\infty}\int_{-\infty}^{\infty}|F(x,y)|^4 dxdy}{\int_{-\infty}^{\infty}\int_{-\infty}^{\infty}|F(x,y)|^2 dxdy}, \tag{1.25}$$

and λ_0 is the carrier's wavelength in the vacuum. Equation (1.23) is the same as (1.21) except that the expression for the nonlinear coefficient is different.

The NLS equations (1.18) and (1.23) can be solved exactly by the inverse scattering transform method; see Zakharov and Shabat (1971). This will be done in the next chapter.

1.2 The Generalized Nonlinear Schrödinger Equation for Light Beam Propagation

In this section, we derive the generalized NLS equation in spatial optics, where the light source is a monochromatic continuous-wave beam whose intensity is stationary in time but evolves along the propagation direction in a nontrivial way.

In an isotropic charge-free medium, the Maxwell equations take the form

$$\nabla \times \mathbf{E} = -\frac{\partial \mathbf{B}}{\partial t}, \tag{1.26}$$

$$\nabla \times \mathbf{H} = \frac{\partial \mathbf{D}}{\partial t}, \tag{1.27}$$

$$\nabla \cdot \mathbf{D} = 0, \tag{1.28}$$

$$\nabla \cdot \mathbf{B} = 0, \tag{1.29}$$

and the corresponding medium equations are

$$\mathbf{D} = \epsilon \mathbf{E}, \tag{1.30}$$

$$\mathbf{B} = \mu_0 \mathbf{H}. \tag{1.31}$$

Here $\mathbf{E}$ and $\mathbf{H}$ are the electric and magnetic field vectors, $\mathbf{D}$ and $\mathbf{B}$ are the electric and magnetic displacement vectors, and ϵ and μ_0 are the dielectric constant and the medium's permeability, respectively. Assuming that the fields are monochromatic and harmonic, i.e.,

$$\mathbf{E}(x,y,z,t) = \mathbf{E}(x,y,z)e^{-i\omega t}, \quad \mathbf{B}(x,y,z,t) = \mathbf{B}(x,y,z)e^{-i\omega t}, \tag{1.32}$$

taking the curl of both sides of (1.26) and substituting (1.27) and (1.31) into the resulting equation, we get

$$\nabla \times (\nabla \times \mathbf{E}) = \nabla \times (i\omega\mu_0 \mathbf{H}) = \omega^2 \mu_0 \epsilon \mathbf{E}. \tag{1.33}$$

The left side of (1.33) can be further expanded by using the following vector identity:

$$\nabla \times (\nabla \times \mathbf{E}) = \nabla(\nabla \cdot \mathbf{E}) - \nabla^2 \mathbf{E}. \tag{1.34}$$

Moreover, using (1.28) we can write

$$\nabla \cdot \mathbf{D} = \nabla \cdot (\epsilon \mathbf{E}) = \nabla\epsilon \cdot \mathbf{E} + \epsilon \nabla \cdot \mathbf{E} = 0. \tag{1.35}$$

Replacing the left-hand side of Eq. (1.33) by (1.34), and using (1.35), we then find

$$\nabla^2 \mathbf{E} + \omega^2 \mu_0 \epsilon \mathbf{E} = -\nabla \left(\frac{1}{\epsilon} \nabla\epsilon \cdot \mathbf{E} \right). \tag{1.36}$$

If the dielectric constant ϵ changes little over a single optical wavelength, the right side of (1.36) can be neglected (Yariv (1989)). This is known as a scalar approximation which reduces (1.36) to the following Helmholtz equation:

$$\nabla^2 \mathbf{E} + k_0^2 n^2 \mathbf{E} = 0, \tag{1.37}$$

where $k_0 = \omega/c$, $n = \sqrt{\mu_0 \epsilon}\, c$ is a refractive index of the medium, and c is the speed of light in a vacuum. Note that if the electric field $\mathbf{E}$ is linearly polarized along one transverse axis, then the Helmholtz equation (1.37) becomes scalar (hence the name of the scalar approximation). Therefore, in what follows we will use a scalar notation for the electric field E in place of a vector one.

Scalar approximation is usually satisfied if the refractive index n does not vary strongly in the media, which is the case in planar etched waveguide arrays (Eisenberg et al. (1998)), photorefractive crystals imprinted with photonic lattices (Fleischer et al. (2003b)), low-refractive-index-contrast fibers (Argyros et al. (2005)), laser-written waveguides (Szameit et al. (2006)), as well as many nonlinear optics problems. In this case, the refractive index n can be considered as a constant background index n_b plus a weak variation Δn,

$$n = n_b + \Delta n, \qquad \Delta n \ll n_b. \tag{1.38}$$

Variation Δn generally depends on the spatial coordinates of the medium and/or the light intensity $|E|^2$. Now we consider the electric field E in the form of a slowly varying wave envelope along the propagation direction $\hat{\mathbf{z}}$,

$$E = U(x,y,z)e^{ikz}, \tag{1.39}$$

where $k = k_0 n_b$, and $U(x,y,z)$ changes slowly with z (hence $|U_{zz}| \ll |U_z|$). This slowly varying envelope assumption is equivalent to the paraxial approximation in geometric optics, which assumes that the ray of light makes a small angle to the central propagation direction $\hat{\mathbf{z}}$. Substituting (1.39) into (1.37) and recalling the assumption (1.38), the Helmholtz equation (1.37) then becomes

$$i\frac{\partial U}{\partial z} + \frac{1}{2k}\nabla_\perp^2 U + \frac{k\Delta n}{n_b}U = 0. \tag{1.40}$$

Here the higher-order terms of Δn as well as U_{zz} are small and thus omitted, and $\nabla_\perp^2 = \partial_{xx} + \partial_{yy}$ is the transverse Laplace operator. Equation (1.40) is a Schrödinger equation. When describing light beam propagation, the first term on the left side of (1.40) represents envelope propagation, the second term represents transverse diffraction of light, and the last term represents the effect of refractive index change on the light propagation.

In a homogeneous Kerr nonlinear medium, $\Delta n = n_2|U|^2$, where n_2 is the Kerr coefficient. In this case, Eq. (1.40) becomes

$$i\frac{\partial U}{\partial z} + \frac{1}{2k}\nabla_\perp^2 U + \frac{kn_2}{n_b}|U|^2 U = 0, \tag{1.41}$$

which is a two-dimensional (2D) NLS equation. In a non-Kerr nonlinear medium, $\Delta n = f(|U|^2)$ in general. In this case, (1.40) becomes

$$i\frac{\partial U}{\partial z} + \frac{1}{2k}\nabla_\perp^2 U + \frac{k}{n_b}f(|U|^2)U = 0, \tag{1.42}$$

which is a generalized 2D NLS equation. When the light is confined in one transverse direction by a waveguide, then Eqs. (1.41)–(1.42) reduce from two dimensions to one dimension. In photonic crystal fibers, planar etched waveguide arrays, and laser-written waveguides where a refractive index variation $\Delta n_0(x,y)$ is present on the transverse plane, then $\Delta n = \Delta n_0(x,y) + n_2|U|^2$. In this case, Eq. (1.40) becomes

$$i\frac{\partial U}{\partial z} + \frac{1}{2k}\nabla_\perp^2 U + \frac{k}{n_b}\Delta n_0(x,y)U + \frac{kn_2}{n_b}|U|^2U = 0, \tag{1.43}$$

which is a 2D NLS equation with an external potential.

In photorefractive crystals, the expression for Δn is somewhat different. Suppose the crystal line as well as the direction of an external DC electric field (with voltage differential V across the crystal) are both chosen along the $\hat{\mathbf{x}}$-axis. In this configuration, the refractive index along $\hat{\mathbf{x}}$ direction is n_e, and the refractive index along $\hat{\mathbf{y}}$ direction is n_o, with the corresponding electro-optic coefficients $r_{33} \gg r_{13}$. When an extraordinarily polarized beam (along $\hat{\mathbf{x}}$ direction) or an ordinarily polarized beam (along $\hat{\mathbf{y}}$ direction) is launched into the crystal, the respective change in the refractive index is given by (see Singh et al. (1995), Segev et al. (1996), Efremidis et al. (2002))

$$\Delta\left(\frac{1}{n_e^2}\right) = r_{33}E_{SC} \tag{1.44}$$

and

$$\Delta\left(\frac{1}{n_o^2}\right) = r_{13}E_{SC}, \tag{1.45}$$

where $E_{SC} = E_0/(1+I)$ is a steady-state space-charge field, I is the total beam intensity normalized to the background illumination, and E_0 is the value of a biased DC field. Hence, if an ordinarily polarized lattice beam with transverse intensity distribution $I_L(x,y)$ is launched into the crystal and made uniform along the propagation direction $\hat{\mathbf{z}}$ (Fleischer et al. (2003b), Chen and Yang (2007)), then for an extraordinarily polarized probe beam, since $\Delta n_e \ll n_e$, we can approximate

$$\Delta\left(\frac{1}{n_e^2}\right) = -2\frac{1}{n_e^3}\Delta n_e. \tag{1.46}$$

Inserting this equation into (1.44), we get

$$\Delta n_e = -\frac{n_e^3}{2}r_{33}E_{SC}. \tag{1.47}$$

After substitution of this equation into (1.40), the guiding equation for the probe beam then becomes

$$i\frac{\partial U}{\partial z} + \frac{1}{2k}\nabla_\perp^2 U - \frac{kn_e^2 r_{33}}{2}\frac{E_0}{1+I_L(x,y)+|U|^2}U = 0. \tag{1.48}$$

This is a generalized NLS equation with saturable nonlinearity and a lattice potential. It is noted that this model neglected the anisotropic property of photorefractive nonlinearity, where the nonlinearity experienced by an optical beam depends not only on its polarization, but also on its orientation relative to the direction of the applied electric field. A more elaborate model which incorporates this anisotropic nonlinearity of photorefractive crystals can be found in Zozulya and Anderson (1995).

The generalized NLS equations (1.42), (1.43), and (1.48) cannot be solved exactly by the inverse scattering transform method. For such nonintegrable equations, alternative analytical methods need to be used. In addition, numerical methods are often needed as well. These analytical and numerical methods for nonintegrable equations will be developed in Chaps. 5–7.

1.3 Other Nonlinear Wave Equations in Physical Systems

In addition to the nonlinear wave equations derived in the previous sections, many other equations can also be derived in different physical settings. Below we list a few of them.

The NLS Equation with Higher-Order Corrections

For a wave packet in the model equation (1.17), the envelope of the wave packet is governed to the leading order by the NLS equation (1.18). When that perturbative derivation is pursued to higher orders, corrections to the NLS equation (1.18) will be obtained. The corrected equation takes the form (see Kano (1989), Boyd and Chen (2001))

$$iU_\tau + \frac{1}{2}\omega_0'' U_{\xi\xi} + \alpha |U|^2 U = i\varepsilon \left[\frac{1}{6}\omega_0''' U_{\xi\xi\xi} + a_1 U^2 U_\xi^* + a_2 |U|^2 U_\xi \right] + O(\varepsilon^2), \tag{1.49}$$

where $\omega = \omega(k)$ is the linear dispersion relation for Fourier waves $e^{i(kx-\omega t)}$, $\omega_0''' = \omega'''(k_0)$, and a_1, a_2 are constant coefficients. The first correction term on the right-hand side of (1.49) is the third-order dispersion term, and the second and third correction terms are called nonlinear dispersion (or self-steepening) terms.

In fiber optics, the higher-order corrections to the NLS equation (1.23) for optical pulse transmission can be derived as well, and the corrected equation is (see Hasegawa and Kodama (1995), Agrawal (2007))

$$iU_z - \frac{1}{2}\beta_2 U_{TT} + \gamma |U|^2 U = \frac{1}{6} i\beta_3 U_{TTT} + \gamma \left[T_R U(|U|^2)_T - \frac{i}{\omega_0}(|U|^2 U)_T \right], \tag{1.50}$$

where $\beta = \beta(\omega)$ is the linear dispersion relation for Fourier waves $e^{i(\beta z - \omega t)}$, β_2 is the second-order dispersion coefficient as defined in Eq. (1.24), $\beta_3 = \beta'''(\omega_0)$ is the third-order dispersion coefficient, and T_R is the Raman coefficient which is related to the slope of the

Raman gain spectrum. The fiber loss term is not included here since it may be compensated by erbium doped fiber amplifiers (EDFA).

Unlike the integrable NLS equations (1.18) and (1.23), the above NLS equations with higher-order corrections (1.49) and (1.50) are no longer integrable in general. But if these higher-order corrections are small, then these equations can be treated by a perturbation theory. Such perturbation theories for perturbed integrable equations will be covered in Chapter 4.

The Coupled NLS Equations

When several low-amplitude wavepackets propagate simultaneously in a nonlinear dispersive medium, these wavepackets can interact with each other. In such cases, the evolution of these wavepackets is often governed by the coupled NLS equations. This was first shown in surface water waves by Benney and Newell (1967) and Roskes (1976). In nonlinear fiber optics, these equations have also been derived for nonlinear pulse propagation in birefringent fibers and wavelength-division-multiplexed fiber systems (Menyuk (1987), Agrawal (2007)). In spatial optics, interaction of several incoherent beams propagating in crystals is governed by these equations as well. Below we briefly describe these equations in fiber optics.

In a birefringent fiber, there are two orthogonal principal axes along which light travels at different speeds. The electric field in such a fiber can be written as

$$\mathbf{E}(x,y,z,t) = \frac{1}{2}F(x,y)\left[\hat{\mathbf{x}}A(z,t)e^{i\beta_{0x}z} + \hat{\mathbf{y}}B(z,t)e^{i\beta_{0y}z}\right]e^{-i\omega_0 t} + c.c., \tag{1.51}$$

where $\hat{\mathbf{x}}$ and $\hat{\mathbf{y}}$ are the unit polarization vectors of the two principal axes, $F(x,y)$ is the transverse spatial distribution of the fiber mode, ω_0 is the carrier frequency, β_{0x} and β_{0y} are the wavenumbers of light along the two principal axes at the carrier frequency ω_0, and $A(z,t)$, $B(z,t)$ are the envelope functions of light along the principal axes. When the fiber refringence is relatively large, these envelope functions can be found to satisfy the following evolution equations (Menyuk (1987), Agrawal (2007))

$$iA_z + i\beta_{1x}A_t - \frac{1}{2}\beta_2 A_{tt} + \gamma(|A|^2 + \delta|B|^2)A = 0, \tag{1.52}$$

$$iB_z + i\beta_{1y}B_t - \frac{1}{2}\beta_2 B_{tt} + \gamma(|B|^2 + \delta|A|^2)B = 0, \tag{1.53}$$

where $\beta_{1x} \neq \beta_{1y}$ are the inverse group velocities of light along the two principal axes at the carrier frequency ω_0, β_2 is the second-order dispersion at the carrier frequency, γ is the nonlinear coefficient given in Eq. (1.25), and δ is the cross-phase-modulation coefficient. In linearly birefringent fibers, $\delta = 2/3$, while in circularly birefringent fibers, $\delta = 2$. Through a gauge transformation

$$A(z,t) = \widehat{A}(z,t)e^{i(\beta_{1x}/\beta_2)t - i(\beta_{1x}^2/2\beta_2)z}, \tag{1.54}$$

$$B(z,t) = \widehat{B}(z,t)e^{i(\beta_{1y}/\beta_2)t - i(\beta_{1y}^2/2\beta_2)z}, \tag{1.55}$$

and with the hats dropped, the above evolution equations reduce to the following coupled NLS equations:

$$iA_z - \frac{1}{2}\beta_2 A_{tt} + \gamma(|A|^2 + \delta|B|^2)A = 0, \tag{1.56}$$

$$iB_z - \frac{1}{2}\beta_2 B_{tt} + \gamma(|B|^2 + \delta|A|^2)B = 0. \tag{1.57}$$

When $\delta = 1$, these equations are the Manakov equations which can be solved exactly (see Manakov (1973)). When $\delta = 0$, these equations become two decoupled NLS equations which are exactly integrable too; see Zakharov and Shabat (1971). Solution methods for these two special cases will be covered in Chapters 2 and 3. For other δ values, however, these equations are nonintegrable. The analytical and numerical techniques to treat such nonintegrable cases will be covered in Chapters 5 and 7.

When two light waves at different wavelengths copropagate along one of the principal axes of a birefringent single-mode fiber, the interaction between these two light waves is governed by equations of the form (1.52)–(1.53) as well. In this case, the electric field can be written as

$$\mathbf{E}(x,y,z,t) = \frac{1}{2}\hat{\mathbf{x}}[E_1 e^{-i\omega_1 t} + E_2 e^{-i\omega_2 t}] + c.c., \tag{1.58}$$

where ω_1 and ω_2 are the carrier frequencies of these two light waves, $\hat{\mathbf{x}}$ is the unit polarization vector,

$$E_1 = A(z,t)F_1(x,y)e^{i\beta_{01}z}, \quad E_2 = B(z,t)F_2(x,y)e^{i\beta_{02}z}, \tag{1.59}$$

$F_{1,2}(x,y)$ are the transverse distributions of the fiber mode at the two frequencies $\omega_{1,2}$, $A(z,t)$ and $B(z,t)$ are the envelope functions of the two waves, and β_{01}, β_{02} are the corresponding wavenumbers at the two carrier frequencies. The envelope functions A and B can be found to satisfy the following coupled equations (Agrawal (2007)):

$$iA_z + i\beta_{11}A_t - \frac{1}{2}\beta_{21}A_{tt} + \gamma_1(|A|^2 + 2|B|^2)A = 0, \tag{1.60}$$

$$iB_z + i\beta_{12}B_t - \frac{1}{2}\beta_{22}B_{tt} + \gamma_2(|B|^2 + 2|A|^2)B = 0, \tag{1.61}$$

where β_{1j} is the inverse group velocity at frequency ω_j, β_{2j} is the second-order dispersion at frequency ω_j, and γ_j is the nonlinear coefficient for each frequency whose expression is analogous to that in Eq. (1.25); see Agrawal (2007). These evolution equations are very similar to (1.52)–(1.53), except that the coefficients are different. These equations are not integrable, thus need to be treated by nonintegrable theories and numerical techniques as described in Chapters 5 and 7.

The Gross–Pitaevskii Equation

A Bose–Einstein condensate (BEC) is a dilute gas of weakly interacting boson particles confined in an external potential and cooled to temperatures very close to absolute zero. Under such conditions, a large fraction of the bosons collapse into the lowest quantum state of the external potential, and all wave functions overlap each other, at which point quantum effects become apparent on a macroscopic scale. This state of matter was first predicted by S.N. Bose and A. Einstein in 1924–1925, and then experimentally observed by E. Cornell, W. Ketterle, and C. Wieman in 1995 (Anderson et al. (1995), Davis et al. (1995)). For their work, Cornell, Ketterle, and Wieman jointly won the Nobel Prize in Physics in 2001.

In quantum mechanics, the motion of a boson in a potential well is described by a wave function satisfying a linear Schrödinger equation. But since many bosons in a BEC are in the same lowest quantum state and their wave functions overlap, the state of BEC can be described by a collective wave function $\psi(\mathbf{x},t)$, where $\mathbf{x}$ is the three-dimensional (3D) spatial coordinate, $|\psi|^2$ is interpreted as the particle density, and the total number of atoms is $\int |\psi|^2 d\mathbf{x}$. In addition, due to interactions between bosons which give rise to nonlinear effects, this collective wave function $\psi(\mathbf{x},t)$ satisfies a *nonlinear* Schrödinger equation with an external potential:

$$i\hbar\psi_t = \left[-\frac{\hbar^2}{2m}\nabla^2 + V(\mathbf{x}) + g|\psi|^2 \right]\psi, \tag{1.62}$$

where $\hbar$ is the Planck constant, m is the mass of the boson, $V(\mathbf{x})$ is the external potential, $g = 4\pi\hbar^2 a_s/m$ is the nonlinear coefficient, and a_s is the scattering length of two interacting bosons. This equation was first derived by Gross (1961) and Pitaevskii (1961), and is now called the Gross–Pitaevskii equation. It provides a good description of the collective behavior of BECs and is thus often used for theoretical analysis. This equation closely resembles Eq. (1.43) for spatial light propagation in index-varying media. It cannot be solved exactly in general, and thus it needs to be treated by nonintegrable theories and numerical methods as described in Chapters 5–7.

In addition to the nonlinear wave equations mentioned above, many other equations arise in different physical situations as well. For instance, the KdV equation

$$\psi_t + \psi_{xxx} + 6\psi\psi_x = 0 \tag{1.63}$$

is a generic model for propagation of unidirectional waves in weakly nonlinear and weakly dispersive systems. Indeed, the derivation of this equation in shallow water by Korteweg and de Vries (1895) explained for the first time Russell's famous observation of solitary waves in a water canal in 1834. Other well-known nonlinear wave equations include the sine-Gordon equation for differential geometry and superconductivity in Josephson junctions, the Kadomtsev–Petviashvili equation for long wave propagation in two spatial dimensions, the Ginzburg–Landau equation for superconductivity and pattern formation, and so on. Despite the great diversities of these nonlinear wave equations, these equations often allow unified treatments, such as by the inverse scattering transform method, by the soliton perturbation theory, by the solitary-wave stability theory, and by numerical methods. These unified theories and methods will be covered in the following chapters.

Chapter 2

Integrable Theory for the Nonlinear Schrödinger Equation

Integrable equations are an important class of nonlinear wave equations. Notable examples include the KdV equation, the NLS equation, the sine-Gordon equation, the Kadomtsev–Petviashvili (KP) equation, and many others. These equations have the remarkable property that their solitary waves collide elastically; i.e., they pass through each other without change of shape or speed. In addition, these equations can be integrated exactly by the inverse scattering transform method. This method was developed first by Gardner et al. (1967) for the KdV equation. Lax (1968) realized that if a nonlinear wave equation is the compatibility condition between two linear operators (now called the Lax pair), then this equation would be integrable by the inverse scattering transform method. Based on this idea, Zakharov and Shabat (1971) showed that the NLS equation is also integrable. Subsequently, Ablowitz et al. (1974) developed a method to derive a large class of integrable equations such as an integrable hierarchy. From that time on, numerous advances have been made on integrable equations, and this field is still very active today.

In this and the next chapters, we cover the inverse scattering transform method and related theories on integrable equations. The original inverse scattering transform method was based on the Gel'fand–Levitan–Marchenko integral equations (Gardner et al. (1967), Ablowitz and Segur (1981)). Later on, a Riemann–Hilbert formulation was developed which streamlines and simplifies the inverse scattering transform method considerably (Zakharov et al. (1984)). Our presentation will be based on the Riemann–Hilbert formulation. In this chapter, we develop the theory for the (focusing) NLS equation and its integrable hierarchy. In the next chapter, theories for other higher-order integrable equations will be developed.

2.1 Inverse Scattering Transform Method

We consider the NLS equation of the form

$$iu_t + u_{xx} + 2|u|^2 u = 0. \tag{2.1}$$

This nonlinear equation is the compatibility condition of the following two linear equations:

$$Y_x = \begin{pmatrix} -i\zeta & u \\ -u^* & i\zeta \end{pmatrix} Y, \tag{2.2}$$

$$Y_t = \begin{pmatrix} -2i\zeta^2 + i|u|^2 & iu_x + 2\zeta u \\ iu_x^* - 2\zeta u^* & 2i\zeta^2 - i|u|^2 \end{pmatrix} Y, \tag{2.3}$$

where ζ is a spectral parameter, $Y(x,t,\zeta)$ is a vector or matrix function, and the superscript "*" represents complex conjugation. Indeed, when u satisfies the NLS equation (2.1), then by differentiating (2.2) and (2.3) with respect to t and x, respectively, we find that their right-hand sides are indeed equal to each other; i.e., Eqs. (2.2) and (2.3) are compatible with each other. The spatial linear operator (2.2) and the temporal linear operator (2.3) are the Lax pair of the NLS equation (2.1), and they were discovered by Zakharov and Shabat (1971). The spatial equation (2.2) is now called the Zakharov–Shabat scattering problem. In our analysis, we always assume that u decays to zero sufficiently fast as $x \to \pm\infty$. For a prescribed initial condition $u(x,0)$, we seek the solution $u(x,t)$ at any later time t. That is, we solve an initial value problem for the NLS equation (2.1).

In the following analysis, we treat Y in (2.2)–(2.3) as a fundamental matrix of those linear equations. In addition, as $x \to \pm\infty$, since $u \to 0$, we see from (2.2)–(2.3) that $Y \propto e^{-i\zeta\Lambda x - 2i\zeta^2\Lambda t}$, where Λ is a diagonal constant matrix,

$$\Lambda = \operatorname{diag}(1, -1). \tag{2.4}$$

Hence it will be convenient to express Y as

$$Y = Je^{-i\zeta\Lambda x - 2i\zeta^2\Lambda t}, \tag{2.5}$$

so that the new matrix function J is (x,t)-independent at infinity. Inserting (2.5) into (2.2)–(2.3), we find that the Lax pair (2.2)–(2.3) becomes

$$J_x = -i\zeta[\Lambda, J] + QJ \tag{2.6}$$

and

$$J_t = -2i\zeta^2[\Lambda, J] + VJ, \tag{2.7}$$

where

$$Q = \begin{pmatrix} 0 & u \\ -u^* & 0 \end{pmatrix}, \quad V = 2\zeta Q + \begin{pmatrix} i|u|^2 & iu_x \\ iu_x^* & -i|u|^2 \end{pmatrix}, \tag{2.8}$$

and $[\Lambda, J] \equiv \Lambda J - J\Lambda$ is the commutator. Notice that both matrices Q and V are anti-Hermitian, i.e.,

$$Q^\dagger = -Q, \quad V^\dagger = -V, \tag{2.9}$$

where the superscript "†" represents the Hermitian of a matrix (here, the spectral parameter ζ in the expression of V is assumed to be real). In addition, their traces are both equal to zero, i.e., $\mathrm{tr}Q = \mathrm{tr}V = 0$.

The basic idea of the inverse scattering transform method for solving the initial value problem of the NLS equation (2.1) is the following. We first treat Eq. (2.6) as a spatial scattering problem at a fixed time t. In this context, u plays the role of a potential function. From this scattering problem, we can get scattering data. We will show that the temporal dependence of the scattering data is very simple, thus from the initial scattering data (at time $t = 0$), we can obtain the scattering data at any later time t. A key feature of this method is that the potential u can be reconstructed from the scattering data (hence the name "inverse scattering"), thus the solution u can be obtained at any later time.

The above inverse scattering transform method for solving nonlinear partial differential equations (PDEs) is analogous to the Fourier transform method for solving linear PDEs. In the Fourier transform method, one first obtains the Fourier coefficients for the initial solution through the Fourier transform, then determines the temporal evolution of the Fourier coefficients, and lastly determines the solution at any later time from the Fourier coefficients through the inverse Fourier transform. In the inverse scattering transform method, the scattering data plays the role of the Fourier coefficients. In this connection, the inverse scattering transform method is sometimes referred to as the Fourier analysis for nonlinear equations (Ablowitz et al. (1974)).

Below, we implement the above procedure of the inverse scattering transform method. In Secs. 2.1.1 and 2.1.2, the scattering and inverse scattering problems of the Zakharov–Shabat system (2.6) will be presented using the Riemann–Hilbert formulation. In Sec. 2.1.3, time evolution of the scattering data will be determined.

2.1.1 Riemann–Hilbert Formulation

In this subsection, we consider the scattering problem of the Zakharov–Shabat system (2.6). In this consideration, the time t is fixed and is a dummy variable, and thus it will be suppressed in our notation.

In the scattering problem, we first introduce matrix Jost solutions $J_{\pm}(x,\zeta)$ of Eq. (2.6) with the following asymptotics at large distances:

$$J_{\pm}(x,\zeta) \to I, \qquad x \to \pm\infty. \tag{2.10}$$

Here, I is the 2×2 unit matrix, and the subscripts in $J_{\pm}$ refer to which end of the x-axis the boundary conditions are set.

For a general linear system of differential equations

$$Y_x = A(x)Y, \tag{2.11}$$

Abel's identity says that

$$\det Y(x) = \det Y(x_0)\, e^{\int_{x_0}^{x} \operatorname{tr} A(\xi)d\xi}, \tag{2.12}$$

where $Y(x)$ is a fundamental matrix of Eq. (2.11), tr$A$ is the trace of the coefficient matrix A, and x_0 is an arbitrary constant. Applying this identity to the Zakharov–Shabat system (2.2) and recalling the relation (2.5), we find that $\det J(x,\zeta)$ is a constant for all x. Then using the boundary conditions (2.10), we see that

$$\det J_{\pm}(x,\zeta) = 1 \tag{2.13}$$

for all (x,ζ). Introducing the notation

$$E(x,\zeta) \equiv e^{-i\zeta\Lambda x} \tag{2.14}$$

and

$$\Phi \equiv J_{-}E, \qquad \Psi \equiv J_{+}E, \tag{2.15}$$

then since $\Phi(x,\zeta)$ and $\Psi(x,\zeta)$ are both solutions of the linear equation (2.2), they are not independent and are linearly related by a matrix $S(\zeta)$:

$$\Phi(x,\zeta) = \Psi(x,\zeta)\,S(\zeta), \quad \zeta \in \mathbb{R}, \tag{2.16}$$

i.e.,

$$J_{-} = J_{+}ESE^{-1}, \quad \zeta \in \mathbb{R}. \tag{2.17}$$

Here $\mathbb{R}$ is the set of real numbers (for nonreal ζ, certain elements in S may not be well defined). This matrix $S(\zeta)$ is called the scattering matrix. Clearly,

$$\det S(\zeta) = 1 \tag{2.18}$$

in view of Eqs. (2.13) and (2.17).

The scattering matrix S contains important scattering data from which we can reconstruct the potential u. This reconstruction will rely heavily on the analytical properties of Jost solutions $J_\pm(x,\zeta)$ and the scattering matrix $S(\zeta)$ in the complex ζ plane, and thus we will delineate these analytical properties first. Notice that (Φ,Ψ) satisfy the scattering equation (2.2), i.e.,

$$Y_x + i\zeta\Lambda Y = QY. \tag{2.19}$$

If we treat the QY term in the above equation as an inhomogeneous term and notice that the solution to the homogeneous equation on its left side is E, then using the method of variation of parameters as well as the boundary conditions (2.10), we can turn (2.19) into Volterra integral equations for (Φ,Ψ). These equations can be cast in terms of $J_\pm$ as

$$J_-(x,\zeta) = I + \int_{-\infty}^{x} e^{i\zeta\Lambda(y-x)} Q(y) J_-(y,\zeta) e^{i\zeta\Lambda(x-y)} dy, \tag{2.20}$$

$$J_+(x,\zeta) = I - \int_{x}^{\infty} e^{i\zeta\Lambda(y-x)} Q(y) J_+(y,\zeta) e^{i\zeta\Lambda(x-y)} dy. \tag{2.21}$$

Thus $J_\pm(x,\zeta)$ allow analytical continuations off the real axis $\zeta\in\mathbb{R}$ as long as the integrals on the right sides of the above Volterra equations converge. Due to the structure (2.8) of the potential Q, we can easily see that the first column of J_- and the second column of J_+ can be analytically continued to the upper half plane $\zeta\in\mathbb{C}_+$, while the second column of J_- and the first column of J_+ can be analytically continued to the lower half plane $\mathbb{C}_-$. For instance, the integral equation (2.20) for the first column of J_-, say $(\varphi_1,\varphi_2)^T$, is

$$\varphi_1(x,\zeta) = 1 + \int_{-\infty}^{x} u(y)\varphi_2(y,\zeta)dy, \tag{2.22}$$

$$\varphi_2(x,\zeta) = -\int_{-\infty}^{x} u^*(y)\varphi_1(y,\zeta)e^{2i\zeta(x-y)}dy. \tag{2.23}$$

When $\zeta\in\mathbb{C}_+$, since $e^{2i\zeta(x-y)}$ in (2.23) is bounded, and $u(x)$ decays to zero sufficiently fast at large distances, both integrals in the above two equations converge. Thus the Jost solution $(\varphi_1,\varphi_2)^T$ can be analytically extended to $\mathbb{C}_+$. The analytic properties of the other Jost solutions can be obtained similarly. If we express (Φ,Ψ) as a collection of columns,

$$\Phi = (\phi_1,\phi_2), \quad \Psi = (\psi_1,\psi_2), \tag{2.24}$$

then the Jost solutions

$$P^+ = (\phi_1,\psi_2)e^{i\zeta\Lambda x} = J_-H_1 + J_+H_2 \tag{2.25}$$

are analytic in $\zeta\in\mathbb{C}_+$, and the Jost solutions

$$(\psi_1,\phi_2)e^{i\zeta\Lambda x} = J_+H_1 + J_-H_2 \tag{2.26}$$

are analytic in $\zeta\in\mathbb{C}_-$. Here

$$H_1 \equiv \mathrm{diag}(1,0), \qquad H_2 \equiv \mathrm{diag}(0,1). \tag{2.27}$$

In addition, from the Volterra integral equations (2.20)–(2.21), we see that the large-ζ asymptotics of these analytical functions are

$$P^+(x,\zeta) \to I, \qquad \zeta \in \mathbb{C}_+ \to \infty, \tag{2.28}$$

and

$$(\psi_1, \phi_2)e^{i\zeta\Lambda x} \to I, \qquad \zeta \in \mathbb{C}_- \to \infty. \tag{2.29}$$

To obtain the analytic counterpart of P^+ in $\mathbb{C}_-$, we consider the adjoint scattering equation of (2.6):

$$K_x = -i\zeta[\Lambda, K] - KQ. \tag{2.30}$$

The inverse matrices $J_\pm^{-1}$ satisfy this adjoint equation. Indeed, by utilizing the relation

$$0 = \left(JJ^{-1}\right)_x = J_x J^{-1} + J\left(J^{-1}\right)_x \tag{2.31}$$

as well as the scattering equation (2.6), we get

$$\left(J^{-1}\right)_x = -i\zeta[\Lambda, J^{-1}] - J^{-1}Q; \tag{2.32}$$

thus $J_\pm^{-1}(x,\zeta)$ satisfies the adjoint equation (2.30). If we express Φ^{-1} and Ψ^{-1} as a collection of rows,

$$\Phi^{-1} = \begin{pmatrix} \widehat{\phi}_1 \\ \widehat{\phi}_2 \end{pmatrix}, \qquad \Psi^{-1} = \begin{pmatrix} \widehat{\psi}_1 \\ \widehat{\psi}_2 \end{pmatrix}, \tag{2.33}$$

then by techniques similar to those used above, we can show that the adjoint Jost solutions

$$P^- = e^{-i\zeta\Lambda x}\begin{pmatrix} \widehat{\phi}_1 \\ \widehat{\psi}_2 \end{pmatrix} = H_1 J_-^{-1} + H_2 J_+^{-1} \tag{2.34}$$

are analytic in $\zeta \in \mathbb{C}_-$, and the adjoint Jost solutions

$$e^{-i\zeta\Lambda x}\begin{pmatrix} \widehat{\psi}_1 \\ \widehat{\phi}_2 \end{pmatrix} = H_1 J_+^{-1} + H_2 J_-^{-1} \tag{2.35}$$

are analytic in $\zeta \in \mathbb{C}_+$. In addition,

$$P^-(x,\zeta) \to I, \qquad \zeta \in \mathbb{C}_- \to \infty, \tag{2.36}$$

and

$$e^{-i\zeta\Lambda x}\begin{pmatrix} \widehat{\psi}_1 \\ \widehat{\phi}_2 \end{pmatrix} \to I, \qquad \zeta \in \mathbb{C}_+ \to \infty. \tag{2.37}$$

The above analytic properties of Jost solutions can be summarized as

$$\Phi = (\phi_1^+, \phi_2^-), \qquad \Psi = (\psi_1^-, \psi_2^+), \tag{2.38}$$

$$\Phi^{-1} = \begin{pmatrix} \widehat{\phi}_1^- \\ \widehat{\phi}_2^+ \end{pmatrix}, \qquad \Psi^{-1} = \begin{pmatrix} \widehat{\psi}_1^+ \\ \widehat{\psi}_2^- \end{pmatrix}, \tag{2.39}$$

where the superscripts "±" indicate the half plane of analyticity for the underlying quantities. These analytic properties of Jost solutions have immediate implications on the analytic properties of the scattering matrix $S(\zeta)$. Indeed, since

$$S = \Psi^{-1}\Phi = \begin{pmatrix} \widehat{\psi}_1^+ \\ \widehat{\psi}_2^- \end{pmatrix} (\phi_1^+, \phi_2^-) \tag{2.40}$$

and

$$S^{-1} = \Phi^{-1}\Psi = \begin{pmatrix} \widehat{\phi}_1^- \\ \widehat{\phi}_2^+ \end{pmatrix} (\psi_1^-, \psi_2^+), \tag{2.41}$$

we see immediately that scattering matrices S and S^{-1} have the following analyticity structures:

$$S = \begin{pmatrix} s_{11}^+ & s_{12} \\ s_{21} & s_{22}^- \end{pmatrix}, \qquad S^{-1} = \begin{pmatrix} \hat{s}_{11}^- & \hat{s}_{12} \\ \hat{s}_{21} & \hat{s}_{22}^+ \end{pmatrix}, \tag{2.42}$$

where the superscripts "±" here also refer to which half of the ζ plane the scattering elements are analytic in. Elements without superscripts indicate that such elements do not allow analytical extensions to $\mathbb{C}_\pm$ in general. Note that in the present case, S is a 2×2 matrix with unit determinant (see (2.18)), and thus elements in S^{-1} are simply related to elements in S as

$$\hat{s}_{11} = s_{22}, \quad \hat{s}_{22} = s_{11}, \quad \hat{s}_{12} = -s_{12}, \quad \hat{s}_{21} = -s_{21}. \tag{2.43}$$

Hence analytic properties of S^{-1} can be directly read off from analytic properties of S. But in higher-order scattering problems (see the next chapter), the matrix S is of higher order, and hence elements in S^{-1} are related to elements in S in a more complicated way. In such cases, analytic properties of elements in S^{-1} have to be derived by the general method above which is based on the adjoint scattering problem (2.30).

Hence we have constructed two matrix functions $P^\pm(x,\zeta)$ which are analytic for ζ in $\mathbb{C}_\pm$, respectively. On the real line, using Eqs. (2.17), (2.25), and (2.34), we easily get

$$P^-(x,\zeta)P^+(x,\zeta) = G(x,\zeta), \qquad \zeta \in \mathbb{R}, \tag{2.44}$$

where

$$G = E(H_1 + H_2 S)(H_1 + S^{-1} H_2) E^{-1} = E \begin{pmatrix} 1 & \hat{s}_{12} \\ s_{21} & 1 \end{pmatrix} E^{-1}. \tag{2.45}$$

Equation (2.44) forms a matrix Riemann–Hilbert problem. The normalization condition for this Riemann–Hilbert problem can be obtained from (2.28) and (2.36) as

$$P^{\pm}(x,\zeta) \to I, \quad \zeta \to \infty, \tag{2.46}$$

which is the canonical normalization condition. If this Riemann–Hilbert problem can be solved from the given scattering data $(s_{21}, \hat{s}_{12})$, then the potential Q can be reconstructed from the asymptotic expansion of its solution at large ζ. Indeed, recall that P^+ and P^- are solutions of the scattering problem (2.6) and its adjoint problem (2.30), respectively. Thus if we expand $P^{\pm}$ at large ζ as

$$P^{\pm}(x,\zeta) = I + \zeta^{-1} P_1^{\pm}(x) + O(\zeta^{-2}), \quad \zeta \to \infty, \tag{2.47}$$

and insert them into (2.6) and (2.30), then by comparing terms of the same power in ζ^{-1}, we find at $O(1)$ that

$$Q = i[\Lambda, P_1^+] = -i[\Lambda, P_1^-]. \tag{2.48}$$

Hence the solution u can be reconstructed by

$$u = 2i \left(P_1^+\right)_{12} = -2i \left(P_1^-\right)_{12}. \tag{2.49}$$

This completes the inverse scattering process. How to solve the matrix Riemann–Hilbert problem (2.44) will be discussed in the next subsection.

Continuing the above calculations, at $O(\zeta^{-1})$ in Eq. (2.6), we get

$$\text{diag}(P_1^+)_x = \text{diag}(Q P_1^+). \tag{2.50}$$

From (2.48), (2.50) as well as the large-x asymptotics of $P^+(x,\zeta)$ from Eq. (2.10), we see that the full matrix $P_1^+(x)$ is

$$P_1^+(x) = \frac{1}{2i} \begin{pmatrix} \int_{-\infty}^{x} |u(y)|^2 dy & u(x) \\ u^*(x) & \int_x^{\infty} |u(y)|^2 dy \end{pmatrix}. \tag{2.51}$$

This matrix, together with Eq. (2.47), gives the asymptotic expansion for the analytical function P^+ up to $O(\zeta^{-1})$. The asymptotic expansion for P^- can be obtained similarly.

Using the asymptotic expansions of $P^{\pm}$, we can obtain the asymptotic expansions for the elements of the scattering matrices S and S^{-1}. Recalling the definitions (2.25) and (2.34) of $P^{\pm}$ as well as the scattering relation (2.17), we see that

$$\det P^+ = \hat{s}_{22} = s_{11}, \quad \det P^- = s_{22} = \hat{s}_{11}. \tag{2.52}$$

Thus using the asymptotics (2.47) of $P^{\pm}$, we get

$$\hat{s}_{22}(\zeta) \rightarrow 1 + \frac{1}{2i\zeta}\int_{-\infty}^{\infty}|u|^2 dx + O\left(\frac{1}{\zeta^2}\right), \quad \zeta \in \mathbb{C}_+ \rightarrow \infty, \tag{2.53}$$

and

$$s_{22}(\zeta) \rightarrow 1 - \frac{1}{2i\zeta}\int_{-\infty}^{\infty}|u|^2 dx + O\left(\frac{1}{\zeta^2}\right), \quad \zeta \in \mathbb{C}_- \rightarrow \infty. \tag{2.54}$$

The anti-Hermitian property (2.9) of the potential matrix Q gives rise to involution properties in the scattering matrix as well as in the Jost solutions. Indeed, by taking the Hermitian of the scattering equation (2.6) and utilizing the anti-Hermitian property of the potential matrix $Q^{\dagger} = -Q$, we get

$$J_x^{\dagger} = -i\zeta^*[\Lambda, J^{\dagger}] - J^{\dagger}Q. \tag{2.55}$$

Thus $J^{\dagger}(x,\zeta^*)$, or $J_{\pm}^{\dagger}(x,\zeta^*)$ in particular, satisfies the adjoint scattering equation (2.30). However, $J_{\pm}^{-1}(x,\zeta)$ satisfies this adjoint equation as well (see Eq. (2.32)). Consequently, $J_{\pm}^{\dagger}(x,\zeta^*)$ and $J_{\pm}^{-1}(x,\zeta)$ must be linearly dependent on each other. Recalling the boundary conditions (2.10) of Jost solutions $J_{\pm}$, we further see that $J_{\pm}^{\dagger}(x,\zeta^*)$ and $J_{\pm}^{-1}(x,\zeta)$ have the same boundary conditions at $x \rightarrow \pm\infty$, and hence they must be the same solutions of the adjoint equation (2.30), i.e.,

$$J_{\pm}^{\dagger}(x,\zeta^*) = J_{\pm}^{-1}(x,\zeta). \tag{2.56}$$

From this involution property as well as the definitions (2.25) and (2.34) for $P^{\pm}$, we see that the analytic solutions $P^{\pm}$ satisfy the involution property as well:

$$(P^+)^{\dagger}(\zeta^*) = P^-(\zeta). \tag{2.57}$$

In addition, in view of the scattering relation (2.17) between J_+ and J_-, we see that S also satisfies the involution property:

$$S^{\dagger}(\zeta^*) = S^{-1}(\zeta). \tag{2.58}$$

2.1.2 Solution of the Riemann–Hilbert Problem

In this subsection, we discuss how to solve the matrix Riemann–Hilbert problem (2.44) in the complex ζ plane. In most of these discussions, x is a dummy variable, hence will be suppressed in our notations.

Solution to the Regular Riemann–Hilbert Problem

If $\det P^{\pm}(\zeta) \neq 0$ in their respective planes of analyticity, the Riemann–Hilbert problem (2.44) is called regular. Under the canonical normalization condition (2.46), the solution to this regular Riemann–Hilbert problem is unique. This can be easily proved as follows. Suppose (2.44) has two sets of solutions $P^{\pm}$ and $\widetilde{P}^{\pm}$. Then $P^{-}(\zeta)P^{+}(\zeta) = \widetilde{P}^{-}(\zeta)\widetilde{P}^{+}(\zeta)$, and thus

$$P^{+}(\zeta)(\widetilde{P}^{+})^{-1}(\zeta) = (P^{-})^{-1}(\zeta)\widetilde{P}^{-}(\zeta), \quad \zeta \in \mathbb{R}. \tag{2.59}$$

Since the Riemann–Hilbert problem is regular, $\det P^{\pm}$ and $\det \widetilde{P}^{\pm}$ are nonzero in their respective planes of analyticity. Thus $P^{+}(\widetilde{P}^{+})^{-1}$ and $(P^{-})^{-1}\widetilde{P}^{-}$ are analytic in $\mathbb{C}_{+}$ and $\mathbb{C}_{-}$, respectively. On the real axis $\mathbb{R}$, they are equal to each other; see (2.59). Thus they together define a matrix function which is analytic in the whole $\mathbb{C}$ plane. Due to the canonical normalization condition (2.46), this analytic matrix function approaches the unit matrix I as $\zeta \to \infty$. In complex analysis, Liouville's theorem says that if a function is analytic and bounded in the entire complex plane, then this function must be a constant. Using this theorem, we see that the analytic matrix function mentioned above is equal to the unit matrix I everywhere, i.e.,

$$P^{+}(\zeta)(\widetilde{P}^{+})^{-1}(\zeta) = (P^{-})^{-1}(\zeta)\widetilde{P}^{-}(\zeta) = I, \quad \zeta \in \mathbb{C}. \tag{2.60}$$

Hence $P^{\pm}(\zeta) = \widetilde{P}^{\pm}(\zeta)$, which means that the solution to the regular Riemann–Hilbert problem (2.44) is unique.

This unique solution to the regular matrix Riemann–Hilbert problem (2.44) defies explicit expressions. But its formal solution can be given in terms of a Fredholm integral equation. To do so, we first recall an important formula in the theory of Riemann–Hilbert problems—the Plemelj formula. Assume that γ is a simple, smooth, and closed contour or a line dividing the complex ζ plane into two regions D_{+} and D_{-}, and $f(\xi)$ is a continuous function on the contour γ. Suppose a function $\phi(\zeta)$ is sectionally analytic in D_{+} and D_{-}, vanishing at infinity, and on γ,

$$\phi_{+}(\xi) - \phi_{-}(\xi) = f(\xi), \quad \xi \in \gamma, \tag{2.61}$$

where $\phi_{\pm}(\xi)$ is the limit of $\phi(\zeta)$ as ζ approaches $\xi \in \gamma$ in $D_{\pm}$. Then the Plemelj formula for the solution to this Riemann–Hilbert problem is

$$\phi(\zeta) = \frac{1}{2\pi i}\int_{\gamma} \frac{f(\xi)}{\xi - \zeta}\, d\xi. \tag{2.62}$$

This formula can be proved by complex analysis. To use this Plemelj formula on the regular Riemann–Hilbert problem (2.44), we first rewrite (2.44) as

$$(P^{+})^{-1}(\zeta) - P^{-}(\zeta) = \widehat{G}(\zeta)(P^{+})^{-1}(\zeta), \quad \zeta \in \mathbb{R}, \tag{2.63}$$

where

$$\widehat{G} = I - G = -E \begin{pmatrix} 0 & \hat{s}_{12} \\ s_{21} & 0 \end{pmatrix} E^{-1}, \tag{2.64}$$

$(P^+)^{-1}(\zeta)$ is analytic in $\mathbb{C}_+$, and $P^-(\zeta)$ is analytic in $\mathbb{C}_-$. Applying the Plemelj formula (2.62) to (2.63) and utilizing the canonical boundary conditions (2.46), the solution to the regular Riemann–Hilbert problem (2.44) then is provided by the following integral equation:

$$(P^+)^{-1}(\zeta) = I + \frac{1}{2\pi i}\int_{-\infty}^{\infty} \frac{\widehat{G}(\xi)(P^+)^{-1}(\xi)}{\xi - \zeta}\,d\xi, \quad \zeta \in \mathbb{C}_+. \tag{2.65}$$

Solution to the Nonregular Riemann–Hilbert Problem

In the more general case, the Riemann–Hilbert problem (2.44) is not regular; i.e., $\det P^+(\zeta)$ and $\det P^-(\zeta)$ can be zero at certain discrete locations $\zeta_k \in \mathbb{C}_+$ and $\bar{\zeta}_k \in \mathbb{C}_-$, $1 \le k \le N$, where N is the number of these zeros. In view of Eq. (2.52), we see that $(\zeta_k, \bar{\zeta}_k)$ are zeros of the scattering coefficients $\hat{s}_{22}(\zeta)$ and $s_{22}(\zeta)$. Due to the involution property (2.58), we have the involution relation

$$\bar{\zeta}_k = \zeta_k^*. \tag{2.66}$$

For simplicity, we assume that all zeros $\{(\zeta_k, \bar{\zeta}_k), k = 1, \dots, N\}$ are simple zeros of $(\hat{s}_{22}, s_{22})$, which is the generic case. In this case, the kernels of $P^+(\zeta_k)$ and $P^-(\bar{\zeta}_k)$ contain only a single column vector v_k and row vector $\bar{v}_k$, respectively,

$$P^+(\zeta_k)\,v_k = 0, \quad \bar{v}_k\,P^-(\bar{\zeta}_k) = 0, \quad 1 \le k \le N. \tag{2.67}$$

Taking the Hermitian of the first equation in (2.67) and utilizing the involution properties (2.57) and (2.66), we get

$$v_k^\dagger\,P^-(\bar{\zeta}_k) = 0. \tag{2.68}$$

Then comparing it with the second equation in (2.67), we see that eigenvectors $(v_k, \bar{v}_k)$ satisfy the involution property

$$\bar{v}_k = v_k^\dagger. \tag{2.69}$$

The zeros $\{(\zeta_k, \bar{\zeta}_k)\}$ of $\det P^{\pm}(\zeta)$ as well as vectors $\{v_k, \bar{v}_k\}$ in the kernels of $P^+(\zeta_k)$ and $P^-(\bar{\zeta}_k)$ constitute the discrete scattering data which is also needed to solve the general Riemann–Hilbert problem (2.44).

Regarding this nonregular Riemann–Hilbert problem (2.44) under the canonical normalization condition (2.46), its solution is also unique and is given by the following theorem.

Theorem 2.1. *The solution to the nonregular Riemann–Hilbert problem* (2.44) *with zeros* (2.67) *under the canonical normalization condition* (2.46) *is*

$$P^+(\zeta) = \widehat{P}^+(\zeta)\,\Gamma(\zeta), \tag{2.70}$$

$$P^-(\zeta) = \Gamma^{-1}(\zeta)\,\widehat{P}^-(\zeta), \tag{2.71}$$

where

$$\Gamma(\zeta) = I + \sum_{j,k=1}^{N} \frac{v_j \left(M^{-1}\right)_{jk} \bar{v}_k}{\zeta - \bar{\zeta}_k}, \tag{2.72}$$

$$\Gamma^{-1}(\zeta) = I - \sum_{j,k=1}^{N} \frac{v_j \left(M^{-1}\right)_{jk} \bar{v}_k}{\zeta - \zeta_j}, \tag{2.73}$$

M is an $N \times N$ matrix with its (j,k)th element given by

$$M_{jk} = \frac{\bar{v}_j v_k}{\bar{\zeta}_j - \zeta_k}, \quad 1 \le j,k \le N, \tag{2.74}$$

$$\det \Gamma(\zeta) = \prod_{k=1}^{N} \frac{\zeta - \zeta_k}{\zeta - \bar{\zeta}_k}, \tag{2.75}$$

and $\widehat{P}^{\pm}(\zeta)$ is the unique solution to the following regular Riemann–Hilbert problem:

$$\widehat{P}^{-}(\zeta)\widehat{P}^{+}(\zeta) = \Gamma(\zeta) G(\zeta) \Gamma^{-1}(\zeta), \quad \zeta \in \mathbb{R}, \tag{2.76}$$

where $\widehat{P}^{\pm}(\zeta)$ are analytic in $\mathbb{C}_{\pm}$, respectively, and $\widehat{P}^{\pm}(\zeta) \to I$ as $\zeta \to \infty$.

This theorem reduces a nonregular Riemann–Hilbert problem to a regular one, and it was obtained by Zakharov and Shabat (1979) (see also Zakharov et al. (1984) and Kawata (1984)).

Proof. We will use the construction method to prove this theorem. First, from the pair of zeros $(\zeta_1, \bar{\zeta}_1)$ and vectors $\{|v_1\rangle, \langle \bar{v}_1|\}$ in Eq. (2.67), we construct the matrix

$$\Gamma_1(\zeta) = I + \frac{\bar{\zeta}_1 - \zeta_1}{\zeta - \bar{\zeta}_1} \frac{|v_1\rangle\langle \bar{v}_1|}{\langle \bar{v}_1 | v_1\rangle}, \tag{2.77}$$

which is meromorphic with a simple-pole singularity at $\zeta = \bar{\zeta}_j \in \mathbb{C}_-$. Here, to distinguish between column vectors v_j and row vectors $\bar{v}_j$, we have used the Dirac bra-ket notation $|v_j\rangle$ and $\langle \bar{v}_j|$. It is easy to check that

$$\Gamma_1^{-1}(\zeta) = I - \frac{\bar{\zeta}_1 - \zeta_1}{\zeta - \zeta_1} \frac{|v_1\rangle\langle \bar{v}_1|}{\langle \bar{v}_1 | v_1\rangle}, \tag{2.78}$$

$$\Gamma_1(\zeta_1)|v_1\rangle = 0, \quad \langle \bar{v}_1|\Gamma_1^{-1}(\bar{\zeta}_1) = 0, \tag{2.79}$$

and

$$\det \Gamma_1(\zeta) = (\zeta - \zeta_1)/(\zeta - \bar{\zeta}_1). \tag{2.80}$$

Next we define matrix functions

$$R_1^+(\zeta) = P^+(\zeta)\Gamma_1^{-1}(\zeta), \quad R_1^-(\zeta) = \Gamma_1(\zeta)P^-(\zeta). \tag{2.81}$$

Using the relation (2.67), it is easy to see that the residues of $R_1^+(\zeta)$ at ζ_1 and $R_1^-(\zeta)$ at $\bar{\zeta}_1$ are both zero, and thus $R_1^+(\zeta)$ and $R_1^-(\zeta)$ are analytic in $\mathbb{C}_+$ and $\mathbb{C}_-$, respectively. In addition, $\det R_1^+(\zeta_1)$ and $\det R_1^-(\bar{\zeta}_1)$ are no longer zero. Notice from (2.67) that

$$R_1^+(\zeta_2)|w_2\rangle = 0, \quad \langle \bar{w}_2|R_1^-(\bar{\zeta}_2) = 0, \tag{2.82}$$

where vectors $\{|w_2\rangle, \langle \bar{w}_2|\}$ are related to $\{|v_2\rangle, \langle \bar{v}_2|\}$ as

$$|v_2\rangle = \Gamma_1^{-1}(\zeta_2)|w_2\rangle, \quad \langle \bar{v}_2| = \langle \bar{w}_2|\Gamma_1(\bar{\zeta}_2). \tag{2.83}$$

Thus we can construct the matrix $\Gamma_2(\zeta)$ as in (2.77), except that vectors $\{|v_1\rangle, \langle \bar{v}_1|\}$ are replaced by $\{|w_2\rangle, \langle \bar{w}_2|\}$, and zeros $(\zeta_1, \bar{\zeta}_1)$ replaced by $(\zeta_2, \bar{\zeta}_2)$. Subsequently, we can define matrix functions $R_2^+(\zeta)$ and $R_2^-(\zeta)$ as in (2.81) but with $P^\pm$ on the right sides replaced by $R_1^\pm$. This process can be repeated to the last pair of zeros $(\zeta_N, \bar{\zeta}_N)$. Putting these results together, we find that functions $P^\pm(\zeta)$ can be decomposed as (2.70)–(2.71), where

$$\Gamma(\zeta) = \Gamma_N(\zeta)\Gamma_{N-1}(\zeta)\cdots\Gamma_1(\zeta), \tag{2.84}$$

$$\Gamma_j(\zeta) = I + \frac{\bar{\zeta}_j - \zeta_j}{\zeta - \bar{\zeta}_j}\frac{|w_j\rangle\langle \bar{w}_j|}{\langle \bar{w}_j|w_j\rangle}, \quad j = 2, \dots, N, \tag{2.85}$$

and vectors $\{|w_j\rangle, \langle \bar{w}_j|\}$ are related to $\{|v_j\rangle, \langle \bar{v}_j|\}$ as

$$|v_j\rangle = \Gamma_1^{-1}(\zeta_j)\Gamma_2^{-1}(\zeta_j)\cdots\Gamma_{j-1}^{-1}(\zeta_j)|w_j\rangle \tag{2.86}$$

and

$$\langle \bar{v}_j| = \langle \bar{w}_j|\Gamma_{j-1}(\bar{\zeta}_j)\Gamma_{j-2}(\bar{\zeta}_j)\cdots\Gamma_1(\bar{\zeta}_j). \tag{2.87}$$

In addition, functions $\widehat{P}^\pm(\zeta)$ in the decompositions (2.70)–(2.71) are analytic in $\mathbb{C}_\pm$, respectively, $\det \widehat{P}^\pm(\zeta)$ are nonzero in their planes of analyticity, and $\widehat{P}^\pm(\zeta) \to I$ as $\zeta \to \infty$.

Equation (2.84) is the product formula for the decomposition matrix $\Gamma(\zeta)$. From this formula and the determinant of each constituent matrix such as (2.80), we easily find that $\det \Gamma(\zeta)$ is given by Eq. (2.75). The product formula for $\Gamma^{-1}(\zeta)$ can be easily obtained from (2.84) as

$$\Gamma^{-1}(\zeta) = \Gamma_1^{-1}(\zeta)\Gamma_2^{-1}(\zeta)\cdots\Gamma_N^{-1}(\zeta), \tag{2.88}$$

where each constituent inverse matrix $\Gamma_j^{-1}(\zeta)$ is

$$\Gamma_j^{-1}(\zeta) = I - \frac{\bar{\zeta}_j - \zeta_j}{\zeta - \zeta_j}\frac{|w_j\rangle\langle \bar{w}_j|}{\langle \bar{w}_j|w_j\rangle}, \quad j = 2, \dots, N. \tag{2.89}$$

The function $\Gamma(\zeta)$ is meromorphic with simple poles at $\{\bar{\zeta}_j \in \mathbb{C}_-, 1 \le j \le N\}$, and $\Gamma^{-1}(\zeta)$ is meromorphic with simple poles at $\{\zeta_j \in \mathbb{C}_+, 1 \le j \le N\}$.

The product representations (2.84) and (2.88) for $\Gamma(\zeta)$ and $\Gamma^{-1}(\zeta)$ are not convenient to use for later calculations in the inverse scattering transform method, but they can lead us to the more useful fraction representations (2.72)–(2.73) as we will show below. First, we see from (2.84)–(2.85) that $\Gamma(\zeta)$ has simple-pole singularities at each $\bar{\zeta}_j$. In view of (2.87), the residue of $\Gamma(\zeta)$ at $\zeta = \bar{\zeta}_j$ can be found to be in the form $|z_j\rangle\langle\bar{v}_j|$, where $|z_j\rangle$ is a certain column vector. Thus $\Gamma(\zeta)$ can be expanded into the following fractional form,

$$\Gamma(\zeta) = 1 + \sum_{j=1}^{N} \frac{1}{\zeta - \bar{\zeta}_j} |z_j\rangle\langle\bar{v}_j|. \tag{2.90}$$

Similarly, we see from (2.86) and (2.88)–(2.89) that $\Gamma^{-1}(\zeta)$ can be expanded into a similar fractional form,

$$\Gamma^{-1}(\zeta) = 1 - \sum_{j=1}^{N} \frac{1}{\zeta - \zeta_j} |v_j\rangle\langle\bar{z}_j|, \tag{2.91}$$

where $\langle\bar{z}_j|$ is a certain row vector. In these two fractional representations, vectors $\{|z_j\rangle, \langle\bar{z}_j|\}$ are related to the kernel vectors $\{|v_j\rangle, \langle\bar{v}_j|\}$. This dependence can be determined from the relation $\Gamma(\zeta)\Gamma^{-1}(\zeta) = I$. Inserting the expansion (2.91) of $\Gamma^{-1}(\zeta)$ into the function $\Gamma(\zeta)\Gamma^{-1}(\zeta)$ and equating its residue at each ζ_j to zero, we get

$$\Gamma(\zeta_j)|v_j\rangle = 0, \quad 1 \le j \le N. \tag{2.92}$$

Then inserting $\Gamma(\zeta)$'s expansion (2.90) into this equation, we obtain a linear system of equations for column vectors $\{|z_j\rangle, 1 \le j \le N\}$ as

$$M\left(|z_1\rangle, |z_2\rangle, \ldots, |z_N\rangle\right) = \left(|v_1\rangle, |v_2\rangle, \ldots, |v_N\rangle\right), \tag{2.93}$$

where the matrix M is given in Eq. (2.74). Solving this linear system for vectors $\{|z_j\rangle\}$ and substituting the result into (2.90), the fractional representation (2.72) for $\Gamma(\zeta)$ is then obtained. If we insert $\Gamma(\zeta)$'s expansion (2.90) into $\Gamma(\zeta)\Gamma^{-1}(\zeta)$ and equating its residue at each $\bar{\zeta}_j$ to zero, we would get

$$\langle\bar{v}_j|\Gamma^{-1}(\bar{\zeta}_j) = 0, \quad 1 \le j \le N. \tag{2.94}$$

Inserting the expansion (2.91) of $\Gamma^{-1}(\zeta)$ into the above equation, we would get a linear system for row vectors $\{\langle\bar{z}_j|\}$. Solving this linear system and substituting the result into (2.91), the fractional representation (2.73) for $\Gamma^{-1}(\zeta)$ would then be obtained. This completes the proof of Theorem 2.1. □

In Theorem 2.1, the zeros $\{(\zeta_k, \bar{\zeta}_k), k = 1, \ldots, N\}$ were assumed to be simple zeros of $(\hat{s}_{22}, s_{22})$, which is the simplest case. If some or all of these zeros are multiple zeros of $(\hat{s}_{22}, s_{22})$, the corresponding matrices $\Gamma(\zeta)$ and $\Gamma^{-1}(\zeta)$ can be similarly constructed, but their expressions will be different from those in (2.72)–(2.73) (see Shchesnovich and Yang (2003a, 2003b) for details).

From the solutions of the nonregular Riemann–Hilbert problem (2.44) as given in Theorem 2.1, we see that the scattering data needed to solve this nonregular Riemann–Hilbert problem is

$$\left\{s_{21}(\xi), \hat{s}_{12}(\xi), \xi \in \mathbb{R}; \quad \zeta_k, \bar{\zeta}_k, v_k, \bar{v}_k, 1 \le k \le N\right\}. \tag{2.95}$$

This is called the minimal scattering data. Of this scattering data, vectors v_k and $\bar{v}_k$ are x-dependent, while the others are not. Below we determine the x-dependence of v_k and $\bar{v}_k$. In this calculation, the x-dependence of relevant variables will be restored in the notations.

Our starting point is Eq. (2.67) for v_k and $\bar{v}_k$. Taking the x-derivative to the v_k equation and recalling that P^+ satisfies the scattering equation (2.6), we get

$$P^+(\zeta_k; x)\left(\frac{dv_k}{dx} + i\zeta_k \Lambda v_k\right) = 0. \tag{2.96}$$

Due to our assumption, the only vector in the kernel of $P^+(\zeta_k; x)$ is v_k. Thus

$$\frac{dv_k}{dx} + i\zeta_k \Lambda v_k = \alpha_k(x) v_k, \tag{2.97}$$

where $\alpha_k(x)$ is a scalar function. The solution to the above equation is

$$v_k(x) = e^{-i\zeta_k \Lambda x} v_{k0}\, e^{\int_{x_0}^{x} \alpha_k(y)dy}, \tag{2.98}$$

where $v_{k0} = v_k|_{x=0}$. When this function is substituted into the formulae (2.72)–(2.74), it is found that the $e^{\int_{x_0}^{x} \alpha_k(y)dy}$ term cancels out and does not affect the results. Thus, without loss of generality, we take $\alpha_k = 0$ and write the solution $v_k(x)$ as

$$v_k(x) = e^{-i\zeta_k \Lambda x} v_{k0}. \tag{2.99}$$

Following similar calculations for $\bar{v}_k$, we readily get

$$\bar{v}_k(x) = \bar{v}_{k0}\, e^{i\bar{\zeta}_k \Lambda x}. \tag{2.100}$$

These two equations give the simple x-dependence of vectors v_k and $\bar{v}_k$.

2.1.3 Time Evolution of Scattering Data

In this subsection, we determine the time evolution of the scattering data (2.95).

First we determine the time evolution of the scattering matrices S and S^{-1}. Our starting point is the definition (2.17) for the scattering matrix, which can be rewritten as

$$J_- E = J_+ E S, \quad \zeta \in \mathbb{R}. \tag{2.101}$$

Since $J_\pm$ satisfies the temporal equation (2.7) of the Lax pair, then multiplying (2.7) by the time-independent diagonal matrix $E = e^{-i\zeta\Lambda x}$, we see that $J_- E$, i.e., J_+ES, satisfies the same temporal equation (2.7) as well. Thus, by inserting J_+ES into (2.7), taking the limit $x \to +\infty$, and recalling the boundary condition (2.10) for J_+ as well as the fact that $V \to 0$ as $x \to \pm\infty$, we get

$$S_t = -2i\zeta^2[\Lambda, S]. \tag{2.102}$$

Similarly, by inserting J_-ES^{-1} into (2.7), taking the limit $x \to -\infty$, and recalling the asymptotics (2.10) for J_-, we get

$$(S^{-1})_t = -2i\zeta^2[\Lambda, S^{-1}]. \tag{2.103}$$

From these two equations, we get

$$\frac{\partial \hat{s}_{22}}{\partial t} = \frac{\partial s_{22}}{\partial t} = 0 \tag{2.104}$$

and

$$\frac{\partial \hat{s}_{12}}{\partial t} = -4i\zeta^2\hat{s}_{12}, \quad \frac{\partial s_{21}}{\partial t} = 4i\zeta^2 s_{21}. \tag{2.105}$$

The two equations in (2.104) show that $\hat{s}_{22}$ and s_{22} are time independent. Recall that ζ_k and $\bar{\zeta}_k$ are zeros of det $P^\pm(\zeta)$; i.e., they are zeros of $\hat{s}_{22}(\zeta)$ and $s_{22}(\zeta)$ in view of (2.52). Thus ζ_k and $\bar{\zeta}_k$ are also time independent. The two equations in (2.105) give the time evolution for the scattering data $\hat{s}_{12}$ and s_{21}, which is

$$\hat{s}_{12}(t;\zeta) = \hat{s}_{12}(0;\zeta)e^{-4i\zeta^2 t}, \quad s_{21}(t;\zeta) = s_{21}(0;\zeta)e^{4i\zeta^2 t}. \tag{2.106}$$

Next we determine the time dependence of the scattering data v_k and $\bar{v}_k$. This determination is similar to that for the x-dependence of v_k and $\bar{v}_k$ at the end of the previous subsection. We also start with Eq. (2.67) for v_k and $\bar{v}_k$. Taking the time derivative to the v_k equation and recalling that P^+ satisfies the temporal equation (2.7), we get

$$P^+(\zeta_k;x,t)\left(\frac{\partial v_k}{\partial t} + 2i\zeta_k^2\Lambda v_k\right) = 0; \tag{2.107}$$

thus

$$\frac{\partial v_k}{\partial t} + 2i\zeta_k^2\Lambda v_k = 0. \tag{2.108}$$

Combining it with the spatial dependence (2.99), we get the temporal and spatial dependence for the vector v_k as

$$v_k(x,t) = e^{-i\zeta_k\Lambda x - 2i\zeta_k^2\Lambda t} v_{k0}, \tag{2.109}$$

where v_{k0} is a constant. Similar calculations for $\bar{v}_k$ give

$$\bar{v}_k(x,t) = \bar{v}_{k0} e^{i\bar{\zeta}_k\Lambda x + 2i\bar{\zeta}_k^2\Lambda t}. \tag{2.110}$$

By now, the time evolution for the minimal scattering data (2.95) is all obtained. From this scattering data at any later time, we can solve the nonregular Riemann–Hilbert problem (2.44) with zeros (2.67), and thus reconstruct the solution $u(x,t)$ at any later time from the formula (2.49). This completes the inverse scattering transform process for solving the NLS equation (2.1).

2.1.4 Long-Time Behavior of the Solution

For a general localized initial condition, the explicit expression for the NLS solution at a later time is not available since the regular matrix Riemann–Hilbert problem (2.76) is not explicitly solvable. However, the long-time asymptotic state of the NLS solution can still be obtained. According to the formal solution (2.65) to a regular Riemann–Hilbert problem (2.44), the solution to Eq. (2.76) can be expressed as an integral equation:

$$(\widehat{P}^{+})^{-1}(\zeta) = I + \frac{1}{2\pi i}\int_{-\infty}^{\infty} \frac{\Gamma(\xi)\widehat{G}(\xi)\Gamma^{-1}(\xi)(\widehat{P}^{+})^{-1}(\xi)}{\xi - \zeta}\, d\xi, \qquad \zeta \in \mathbb{C}_{+}, \tag{2.111}$$

where $\widehat{G}$ is given in Eq. (2.64). As $\zeta \to \infty$,

$$(\widehat{P}^{+})^{-1}(\zeta) \to I - \frac{1}{2\pi i\zeta}\int_{-\infty}^{\infty} \Gamma(\xi)\widehat{G}(\xi)\Gamma^{-1}(\xi)(\widehat{P}^{+})^{-1}(\xi)\, d\xi, \tag{2.112}$$

and thus

$$\widehat{P}^{+}(\zeta) \to I + \frac{1}{2\pi i\zeta}\int_{-\infty}^{\infty} \Gamma(\xi)\widehat{G}(\xi)\Gamma^{-1}(\xi)(\widehat{P}^{+})^{-1}(\xi)\, d\xi. \tag{2.113}$$

From the expression (2.72), we see that as $\zeta \to \infty$,

$$\Gamma(\zeta) \to I + \frac{1}{\zeta}\sum_{j,k=1}^{N} v_j \left(M^{-1}\right)_{jk} \bar{v}_k. \tag{2.114}$$

Thus from formula (2.70) and the above two equations, we find that P_1^+ in the expansion (2.47) of $P^+(\zeta)$ is

$$P_1^{+}(x,t) = \sum_{j,k=1}^{N} v_j \left(M^{-1}\right)_{jk} \bar{v}_k + \frac{1}{2\pi i}\int_{-\infty}^{\infty} \Gamma(\xi)\widehat{G}(\xi)\Gamma^{-1}(\xi)(\widehat{P}^{+})^{-1}(\xi)\, d\xi. \tag{2.115}$$

In view of the time evolution equations (2.106) of the scattering data as well as the definition of $E = e^{-i\zeta\Lambda x}$, we see from (2.64) that

$$\widehat{G}(\xi;x,t) = \begin{pmatrix} 0 & -\hat{s}_{12}(0;\xi)e^{-2i\xi x - 4i\xi^2 t} \\ -s_{21}(0;\xi)e^{2i\xi x + 4i\xi^2 t} & 0 \end{pmatrix}. \tag{2.116}$$

Notice that the temporal part of $\widehat{G}$ is in the form of $e^{\pm 4i\xi^2 t}$. Then in the integral of (2.115), by making the substitution $\hat{\xi} = \xi\sqrt{t}$, it is easy to see that as $t \to \infty$, this integral goes to zero (at the rate of $t^{-1/2}$). Thus

$$P_1^+(x,t) \to \sum_{j,k=1}^{N} v_j \left(M^{-1}\right)_{jk} \bar{v}_k, \qquad t \to \infty. \tag{2.117}$$

The corresponding asymptotic solution $u(x,t)$ from (2.48) is then

$$u(x,t) \to 2i \left(\sum_{j,k=1}^{N} v_j \left(M^{-1}\right)_{jk} \bar{v}_k \right)_{12}, \qquad t \to \infty, \tag{2.118}$$

where vectors $\{v_k(x,t)\}$ are given by the formula (2.109), $\bar{v}_k = v_k^{\dagger}$, and matrix M is given in (2.74). From this asymptotics, we see that the integral term in (2.115) corresponds to the radiation part of the solution, while the summation term in (2.115) corresponds to the nonradiative part of the solution. This nonradiative solution is very important for the NLS equation and will be examined in more detail in the next section.

2.2 N-Soliton Solutions

The nonradiative solution in (2.118) corresponds to the case where $G = I$, whence the integral in (2.115) disappears in view of (2.64). In this case, the continuous scattering data $(\hat{s}_{12}, s_{21})$ is zero, and the corresponding scattering problem (2.6) (or (2.2)) is called reflectionless. The corresponding solution u is thus called a reflectionless potential. This solution can be written out explicitly. As we will show below, this solution possesses very special properties, and is called an N-soliton solution in the literature.

When $G = I$, $\widehat{G} = 0$. Thus from (2.48) and (2.115) we have

$$P_1^+(x,t) = \sum_{j,k=1}^{N} v_j \left(M^{-1}\right)_{jk} \bar{v}_k \tag{2.119}$$

and

$$u(x,t) = 2i \left(\sum_{j,k=1}^{N} v_j \left(M^{-1}\right)_{jk} \bar{v}_k \right)_{12}. \tag{2.120}$$

Here vectors v_j are given by (2.109), $\bar{v}_k = v_k^{\dagger}$, and matrix M is given by (2.74). Without loss of generality, we let $v_{k0} = (c_k, 1)^T$. In addition, we introduce the notation

$$\theta_k = -i\zeta_k x - 2i\zeta_k^2 t. \tag{2.121}$$

Then the above solution u can be written out explicitly as

$$u(x,t) = 2i \sum_{j,k=1}^{N} c_j \, e^{\theta_j - \theta_k^*} \left(M^{-1}\right)_{jk}, \tag{2.122}$$

where the elements of the $N \times N$ matrix M are given by

$$M_{jk} = \frac{1}{\zeta_j^* - \zeta_k} \left[e^{-\left(\theta_k + \theta_j^*\right)} + c_j^* c_k e^{\theta_k + \theta_j^*} \right]. \tag{2.123}$$

Notice that M^{-1} can be expressed as the transpose of M's cofactor matrix divided by $\det M$. Also recall that the determinant of a matrix can be expressed as the sum of its elements along a row or column multiplying their corresponding cofactors. Hence the solution (2.122) can be rewritten as

$$u(x,t) = -2i \frac{\det F}{\det M}, \tag{2.124}$$

where F is the following $(N+1) \times (N+1)$ matrix:

$$F = \begin{pmatrix} 0 & e^{-\theta_1^*} & \ldots & e^{-\theta_N^*} \\ c_1 e^{\theta_1} & M_{11} & \ldots & M_{1N} \\ \vdots & \vdots & \vdots & \vdots \\ c_N e^{\theta_N} & M_{N1} & \ldots & M_{NN} \end{pmatrix}. \tag{2.125}$$

This form of the N-soliton solution can be seen in Faddeev and Takhtadjan (1987). It is noted that N-soliton solutions in the NLS equation can also be written as the ratio between two determinants of $2N \times 2N$ matrices. These matrices contain four $N \times N$ submatrices whose determinants are Wronskian (see Nimmo (1983)).

Note that in the above Riemann–Hilbert derivation of the N-soliton solutions, only three pieces of information were used. One is the asymptotic forms of the Lax pair (2.2) and (2.3) as $|x| \to \infty$ (where $u \to 0$), which give the spatial and temporal dependence (2.109) of the discrete scattering data $v_k(x,t)$. Another piece of information is the involution properties (2.66) and (2.69) between $(\bar{\zeta}_k, \bar{v}_k)$ and (ζ_k, v_k), which are induced by the anti-Hermitian property of the potential matrix Q. The third piece of information is the formula (2.48) for the reconstruction of the potential Q, which is derived from the scattering problem (2.6). For a general integrable equation, the above fact holds as well (see the next chapter, and Sec. 3.5 in particular).

We examine properties of these N-soliton solutions next.

1. Single-soliton solutions

When $N = 1$, the solution (2.124) is

$$u(x,t) = 2i(\zeta_1^* - \zeta_1)\frac{c_1 e^{\theta_1 - \theta_1^*}}{e^{-(\theta_1+\theta_1^*)} + |c_1|^2 e^{\theta_1+\theta_1^*}}. \tag{2.126}$$

Letting

$$\zeta_1 = \xi + i\eta, \quad c_1 = e^{-2\eta x_0 + i\sigma_0}, \tag{2.127}$$

where ξ, η are the real and imaginary parts of ζ_1, and x_0, σ_0 are real parameters, then the above solution can be rewritten as

$$u(x,t) = 2\eta \operatorname{sech}[2\eta(x + 4\xi t - x_0)] \exp\left\{-2i\xi x - 4i\left(\xi^2 - \eta^2\right)t + i\sigma_0\right\}. \tag{2.128}$$

This solution is a solitary wave in the NLS equation (2.1). Its amplitude function $|u|$ has the shape of a hyperbolic secant with peak amplitude 2η, and its velocity is -4ξ. The phase of this solution depends linearly on both space x and time t. The spatial gradient of the phase is proportional to the speed of the wave. Parameters x_0 and σ_0 are the initial location and phase of this solitary wave. This solution is called a single-soliton solution in the NLS equation (2.1).

2. Two-soliton solutions

When $N = 2$, the solution (2.124) can also be written out explicitly. Here instead of giving that complicated expression, we show two typical solution behaviors in Fig. 2.1(a, b), one for the case of $\xi_1 \neq \xi_2$, and the other one for the case of $\xi_1 = \xi_2$. Here

$$\zeta_k = \xi_k + i\eta_k, \quad k = 1, 2, \tag{2.129}$$

i.e., ξ_k and η_k are the real and imaginary parts of ζ_k. In Fig. 2.1(a), the solution parameters in the formulae (2.122)–(2.125) are

$$\zeta_1 = 0.1 + 0.7i, \quad \zeta_2 = -0.1 + 0.4i, \quad c_1 = c_2 = 1, \tag{2.130}$$

while in Fig. 2.1(b), the solution parameters are

$$\zeta_1 = 0.7i, \quad \zeta_2 = 0.4i, \quad c_1 = c_2 = 1. \tag{2.131}$$

In the first case ($\xi_1 \neq \xi_2$), we see from Fig. 2.1(a) that as $t \to -\infty$, the solution consists of two single-solitons which are far apart and moving toward each other. When they collide, they interact strongly. But when $t \to \infty$, these solitons re-emerge out of interactions without any change of shape and velocity, and there is no energy radiation emitted to the far field. Thus the interaction of these solitons is elastic. This elastic interaction is a remarkable

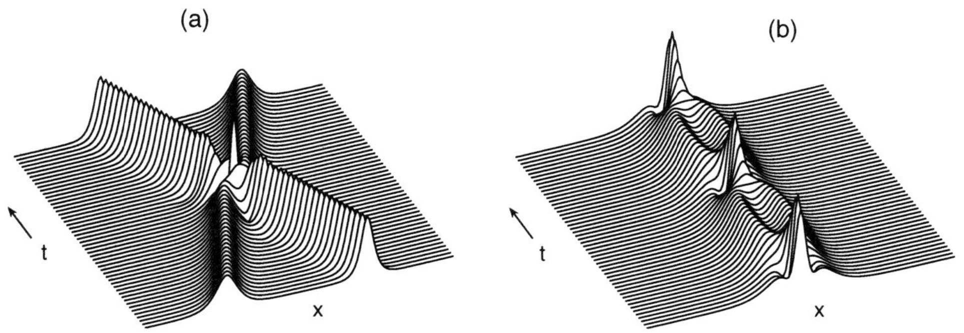

Figure 2.1. *Two-soliton solutions* $|u(x,t)|$ *in the NLS equation* (2.1)*:* (a) *collision, where* $\mathrm{Re}(\zeta_1) \neq \mathrm{Re}(\zeta_2)$*;* (b) *bound state, where* $\mathrm{Re}(\zeta_1) = \mathrm{Re}(\zeta_2)$.

property which signals that the NLS equation (2.1) is integrable. There is still some trace of the interaction however. Indeed, after the interaction, each soliton acquires a position shift (visible in Fig. 2.1(a)) and a phase shift (not visible in this $|u|$ figure). The position of each soliton is always shifted forward (toward the direction of propagation), as if the soliton accelerates during interactions.

Fig. 2.1(a) is typical of all two-soliton solutions (2.124) with $\xi_1 \neq \xi_2$. To show this fact, we analyze the asymptotic states of the solution (2.124) as $t \to \pm\infty$. Without loss of generality, let us assume that $\xi_1 > \xi_2$. This means that at $t = -\infty$, soliton-1 is on the right side of soliton-2 and moves slower. Note also that $\eta_k > 0$ since $\zeta_k \in \mathbb{C}_+$. In the moving frame with velocity $-4\xi_1$,

$$\mathrm{Re}(\theta_1) = \eta_1(x + 4\xi_1 t) = O(1). \tag{2.132}$$

Since

$$\mathrm{Re}(\theta_2) = \eta_2(x + 4\xi_2 t) = \eta_2(x + 4\xi_1 t) + 4\eta_2(\xi_2 - \xi_1)t, \tag{2.133}$$

when $t \to -\infty$, $\mathrm{Re}(\theta_2) \to +\infty$. In this case, simple calculations show that the asymptotic state of the solution (2.124) is

$$u(x,t) \to 2i(\zeta_1^* - \zeta_1)\frac{c_1^- e^{\theta_1 - \theta_1^*}}{e^{-(\theta_1+\theta_1^*)} + |c_1^-|^2 e^{\theta_1+\theta_1^*}}, \quad t \to -\infty, \tag{2.134}$$

where $c_1^- = c_1(\zeta_1 - \zeta_2)/(\zeta_1 - \zeta_2^*)$. Comparing this expression with (2.126), we see that this asymptotic solution is a single-soliton solution with peak amplitude $2\eta_1$ and velocity $-4\xi_1$. When $t \to +\infty$, $\mathrm{Re}(\theta_2) \to -\infty$. In this case, the asymptotic state of the solution (2.124) is

$$u(x,t) \to 2i(\zeta_1^* - \zeta_1)\frac{c_1^+ e^{\theta_1 - \theta_1^*}}{e^{-(\theta_1+\theta_1^*)} + |c_1^+|^2 e^{\theta_1+\theta_1^*}}, \quad t \to +\infty, \tag{2.135}$$

where $c_1^+ = c_1\left(\zeta_1 - \zeta_2^*\right)/(\zeta_1 - \zeta_2)$. This is also a single-soliton solution with peak amplitude $2\eta_1$ and velocity $-4\xi_1$. This indicates that this soliton does not change its shape and velocity after collision. Its position and phase have shifted, however, as Fig. 2.1(a) has shown. The position shift is

$$\Delta x_{01} = -\frac{1}{2\eta_1}\left(\ln|c_1^+| - \ln|c_1^-|\right) = \frac{1}{\eta_1}\ln\left|\frac{\zeta_1 - \zeta_2}{\zeta_1 - \zeta_2^*}\right|, \tag{2.136}$$

and the phase shift is

$$\Delta\sigma_{01} = \arg(c_1^+) - \arg(c_1^-) = -2\arg\left(\frac{\zeta_1 - \zeta_2}{\zeta_1 - \zeta_2^*}\right). \tag{2.137}$$

Notice that $\Delta x_{01} < 0$ since $\zeta_k \in \mathbb{C}_+$, and thus the (slower) soliton-1 acquires a negative position shift.

Following similar calculations, we find that in the moving frame with velocity $-4\xi_2$, as $t \to \pm\infty$, the asymptotic solutions are both single solitons with the same peak amplitude $2\eta_2$ and velocity $-4\xi_2$, and the soliton constants $c_2^\pm$ before and after collision are related as

$$c_2^+ = c_2^-(\zeta_2 - \zeta_1)^2/(\zeta_2 - \zeta_1^*)^2. \tag{2.138}$$

Thus, after collision, this second soliton acquires a position shift

$$\Delta x_{02} = -\frac{1}{\eta_2}\ln\left|\frac{\zeta_2 - \zeta_1}{\zeta_2 - \zeta_1^*}\right|, \tag{2.139}$$

and a phase shift

$$\Delta\sigma_{02} = 2\arg\left(\frac{\zeta_2 - \zeta_1}{\zeta_2 - \zeta_1^*}\right). \tag{2.140}$$

Notice that $\Delta x_{02} > 0$, indicating that the (faster) soliton-2 acquires a positive position shift. In addition,

$$-\frac{\Delta x_{02}}{\Delta x_{01}} = \frac{\eta_1}{\eta_2}; \tag{2.141}$$

thus the amount of each soliton's position shift is inversely proportional to its amplitude.

Next we consider the case where $\xi_1 = \xi_2$. In this case, the two constituent solitons have equal velocities, thus they will stay together and form a bound state which moves at the common speed $-4\xi_1$. In a frame moving at this speed, which is equivalent to a Galilean transformation on the NLS equation (2.1), this bound state will be spatially localized, and its amplitude function $|u(x,t)|$ will oscillate periodically with time. Such a bound state is illustrated in Fig. 2.1(b). It can be seen that the "width" of this solution changes periodically with time, thus this solution is called a "breather" in the literature. The temporal period of this breather is $\pi/|\eta_1^2 - \eta_2^2|$.

2.3 Infinite Number of Conservation Laws

An important property of the NLS equation is that it possesses an infinite number of conservation laws. This is clear from the expansion (2.53) of $\hat{s}_{22}(\zeta)$ at large ζ. For the NLS equation, we have known that $\hat{s}_{22}(\zeta)$ is time independent; thus the coefficient of each ζ^{-n} term in the expansion (2.53) is also time independent. These coefficients are integrals of u and its spatial derivatives, and hence they give an infinite number of conserved quantities in the NLS equation. For instance, (2.53) shows that $\int_{-\infty}^{\infty}|u|^2dx$ is the first conserved quantity. To derive higher conserved quantities, one could pursue the expansion (2.53) to higher orders. But that is awkward to do. A more efficient method was developed by Zakharov and Shabat (1971), which utilizes Jost solutions and their connection to the time-independent scattering coefficient $\hat{s}_{22}(\zeta)$. Below we use a method due to Wadati et al. (1975). This method is a variation of the above Zakharov–Shabat method, but has become purely algebraic and does not require any knowledge of the inverse scattering transform method.

Let us consider a solution $Y=(y_1,y_2)^T$ to the Lax pair (2.2)–(2.3). Defining $\mu = y_1/y_2$, then it is easy to find from this Lax pair that

$$(\ln y_2)_x = i\zeta - u^*\mu, \tag{2.142}$$

$$(\ln y_2)_t = (2i\zeta^2 - i|u|^2)+(iu_x^* - 2\zeta u^*)\mu. \tag{2.143}$$

Cross-differentiating these two equations with respect to t and x, respectively, we get

$$\left(i\zeta - u^*\mu\right)_t = \left[(2i\zeta^2 - i|u|^2)+(iu_x^* - 2\zeta u^*)\mu\right]_x. \tag{2.144}$$

Thus, if we expand μ into a power series in ζ^{-1},

$$\mu(x,t,\zeta) = -\sum_{n=1}^{\infty}\frac{\mu_n(x,t)}{(-2i\zeta)^n}, \tag{2.145}$$

and insert it into Eq. (2.144), then by comparing the coefficients at various orders of ζ^{-1}, we get

$$(u^*\mu_n)_t = i\left(u^*\mu_{n+1} - u_x^*\mu_n\right)_x, \quad n=1,2,\dots. \tag{2.146}$$

These equations give an infinite number of local conservation laws for the NLS equation, with $u^*\mu_n$ being the density, and $u^*\mu_{n+1} - u_x^*\mu_n$ being the flux. The infinite number of conserved quantities for the NLS equation is then

$$I_n = \int_{-\infty}^{\infty} u^*\mu_n\,dx, \quad n=1,2,\dots. \tag{2.147}$$

To determine the expansion coefficients μ_n, we use the Zakharov–Shabat scattering equation (2.2) in the Lax pair. After simple calculations, we see from (2.2) that μ satisfies the following Riccati equation:

$$\mu_x = u^* \mu^2 - 2i\zeta\mu + u. \tag{2.148}$$

Then by inserting μ's expansion (2.145) into this Riccati equation, we readily find that μ_n is given as

$$\mu_1 = u, \quad \mu_2 = u_x, \tag{2.149}$$

$$\mu_{n+1} = \mu_{n,x} + u^* \sum_{k=1}^{n-1} \mu_k \mu_{n-k}, \quad n \geq 2. \tag{2.150}$$

Then the corresponding local conservation laws (2.146) and the conserved quantities (2.147) can be quickly obtained. The first four conserved quantities of the NLS equation are

$$I_1 = \int_{-\infty}^{\infty} |u|^2 \, dx, \tag{2.151}$$

$$I_2 = \int_{-\infty}^{\infty} u^* u_x \, dx, \tag{2.152}$$

$$I_3 = \int_{-\infty}^{\infty} \left(u^* u_{xx} + |u|^4 \right) dx, \tag{2.153}$$

$$I_4 = \int_{-\infty}^{\infty} u^* \left\{ u_{xxx} + (u^2 u^*)_x + 2u^* u u_x \right\} dx. \tag{2.154}$$

The first three quantities are conservations of mass (also called power), momentum, and energy. The fourth and higher conserved quantities do not have physical correspondence.

In the above derivation, the local conservation laws (2.146) were derived using both equations of the Lax pair, but the conserved quantities (2.147) were obtained only from the Zakharov–Shabat scattering equation of the Lax pair. Thus any integrable equation whose scattering operator is the Zakharov–Shabat system (2.2) will share the same conserved quantities as above.

2.4 Discrete Eigenvalues in the Zakharov–Shabat System

When the initial condition is not an N-soliton, the previous two sections show that the solution will evolve into an N-soliton plus energy radiation. The radiation component will disperse away over time, thus the soliton component will dominate the long-time solution. In this section, we determine this soliton component of the solution for a general initial condition.

Recall that the soliton part of the solution is determined by the discrete scattering data $\{\zeta_k, \bar{\zeta}_k, v_k, \bar{v}_k, 1 \leq j \leq N\}$. For the NLS equation, $\bar{\zeta}_k = \zeta_k^*$, and $\bar{v}_k = v_k^\dagger$ due to the involution property. Below we first show that this discrete scattering data simply corresponds to the discrete eigenmodes of the Zakharov–Shabat system (2.2), with $\{\zeta_k, \bar{\zeta}_k, 1 \le j \le N\}$ being the discrete eigenvalues. Then we discuss various ways to determine or estimate these discrete eigenvalues and eigenfunctions.

To show that (ζ_k, v_k) corresponds to a discrete eigenmode of the Zakharov–Shabat system (2.2), we start from the kernel relation (2.67). Recalling the spatial dependence (2.99) of v_k, we get from (2.67) that

$$P^+(x;\zeta_k)e^{-i\zeta_k\Lambda x}v_{k0} = 0. \tag{2.155}$$

Without loss of generality, we take $v_{k0} = (1,\alpha)^T$. We also denote

$$[\phi_d(x), \psi_d(x)] = P^+(x;\zeta_k)e^{-i\zeta_k\Lambda x}. \tag{2.156}$$

Then (2.155) becomes

$$\phi_d(x) = -\alpha\,\psi_d(x). \tag{2.157}$$

By virtue of the definition (2.25) of P^+ and the large-x asymptotics (2.10) of Jost solutions $J_\pm$, we see that for $\zeta_k \in \mathbb{C}_+$, $\phi_d(x) \to 0$ as $x \to -\infty$, and $\psi_d(x) \to 0$ as $x \to \infty$. Then (2.157) shows that $\phi_d(x)$ is a localized function in space. From (2.5) and (2.156), we also see that $\phi_d(x)$ satisfies the original Zakharov–Shabat equation (2.2) at $\zeta = \zeta_k$, thus ζ_k and $\phi_d(x)$ are the discrete eigenvalue and eigenfunction of the Zakharov–Shabat system (2.2). Thus the calculation of the discrete scattering data is equivalent to the calculation of the discrete eigenmodes in the Zakharov–Shabat scattering problem (2.2). In particular, the large-x asymptotics of the discrete eigenfunction ϕ_d can be found from (2.156) and (2.157) as

$$\phi_d(x) \longrightarrow \begin{cases} \begin{pmatrix} e^{-i\zeta_k x} \\ 0 \end{pmatrix}, & x \to -\infty, \\ \begin{pmatrix} 0 \\ -\alpha\, e^{i\zeta_k x} \end{pmatrix}, & x \to +\infty. \end{cases} \tag{2.158}$$

Hence if the discrete eigenfunction of the Zakharov–Shabat system (2.2) has been obtained, its large-x asymptotics will yield the corresponding discrete scattering data $v_{k0} = (1,\alpha)^T$.

Next we determine discrete eigenvalues of the Zakharov–Shabat system (2.2) for various initial conditions $u(x,0)$.

2.4.1 Eigenvalue Formulae for Special Initial Conditions

For certain initial conditions, formulae for discrete eigenvalues of the Zakharov–Shabat system (2.2) can be obtained explicitly. We consider two such initial conditions in this subsection.

1. The Satsuma–Yajima initial condition

The first initial condition we consider is

$$u(x,0) = A\,\mathrm{sech}(x), \tag{2.159}$$

which has a sech profile and a varying amplitude A. If A is a complex number, its phase term only causes the phase of the solution $u(x,t)$ to shift by the same amount, and thus we will assume $A > 0$ below. For this initial condition, Satsuma and Yajima (1974) explicitly solved the Zakharov–Shabat system (2.2) and obtained its whole spectra. Their result is the following. Let $Y = (y_1, y_2)^T$ be an eigenfunction of Eq. (2.2). Eliminating y_2 from these equations and using (2.159), we get

$$s(1-s)\frac{d^2y_1}{ds^2} + \left(\frac{1}{2}-s\right)\frac{dy_1}{ds} + \left[A^2 + \frac{\zeta^2 + i\zeta(1-2s)}{4s(1-s)}\right]y_1 = 0, \tag{2.160}$$

where

$$s \equiv (1-\tanh x)/2 = 1/(1+e^{2x}). \tag{2.161}$$

Note that the boundary $x = +\infty$ corresponds to $s = 0$ and $x = -\infty$ to $s = 1$. Further transformation of the dependent variable y_1 into $s^{i\zeta/2}(1-s)^{-i\zeta/2}w_1$ reduces (2.160) to the hypergeometric equation

$$s(1-s)\frac{d^2w_1}{ds^2} + \left(i\zeta + \frac{1}{2} - s\right)\frac{dw_1}{ds} + A^2 w_1 = 0, \tag{2.162}$$

from which the two linearly independent solutions for y_1 are obtained as

$$y_1^{(1)}(s;\zeta) = s^{i\zeta/2}(1-s)^{-i\zeta/2}F\left(-A, A, i\zeta + \frac{1}{2}; s\right) \tag{2.163}$$

and

$$y_1^{(2)}(s;\zeta) = s^{1/2-i\zeta/2}(1-s)^{-i\zeta/2}F\left(\frac{1}{2} - i\zeta + A, \frac{1}{2} - i\zeta - A, \frac{3}{2} - i\zeta; s\right), \tag{2.164}$$

where $F(\alpha,\beta,\gamma;s)$ is the hypergeometric function. In view of the Zakharov–Shabat equations (2.2) and the initial condition (2.159), the two linearly independent solutions for y_2 can be obtained by replacing ζ with $-\zeta$ in (2.163)–(2.164). Using these results as well as properties of hypergeometric functions, it is easy to check that the function

$$\psi_1(x;\zeta) = \begin{pmatrix} y_1^{(1)}(s;\zeta) \\ -A\left(\zeta - \frac{1}{2}i\right)^{-1} y_1^{(2)}(s;-\zeta) \end{pmatrix} \tag{2.165}$$

is a Jost solution, i.e., it satisfies the Zakharov–Shabat equations (2.2). In addition, by taking the $x \to \pm\infty$ asymptotics of this solution and utilizing the following properties of hypergeometric functions:

$$F(\alpha,\beta,\gamma;0) = 1,$$

$$\begin{aligned}F(\alpha,\beta,\gamma;s) = {} & \frac{\Gamma(\gamma)\Gamma(\gamma-\alpha-\beta)}{\Gamma(\gamma-\alpha)\Gamma(\gamma-\beta)}F(\alpha,\beta,\alpha+\beta+1-\gamma;1-s) \\ & + \frac{\Gamma(\gamma)\Gamma(\alpha+\beta-\gamma)}{\Gamma(\alpha)\Gamma(\beta)}(1-s)^{\gamma-\alpha-\beta}F(\gamma-\alpha,\gamma-\beta,1+\gamma-\alpha-\beta;1-s),\end{aligned}$$

we find that the Jost solution (2.165) has the following large-x asymptotics:

$$\psi_1(x;\zeta) \longrightarrow \begin{cases} \begin{pmatrix} 1 \\ 0 \end{pmatrix} e^{-i\zeta x}, & x \to \infty, \\ \begin{pmatrix} \frac{\Gamma^2(i\zeta+\frac{1}{2})}{\Gamma(i\zeta+\frac{1}{2}+A)\Gamma(i\zeta+\frac{1}{2}-A)}e^{-i\zeta x} \\ -i\sin(\pi A)\operatorname{sech}(\pi\zeta)e^{i\zeta x} \end{pmatrix}, & x \to -\infty. \end{cases} \tag{2.166}$$

Notice from (2.10), (2.15), and (2.16) that when $x \to -\infty$, $\psi_1 \to (\hat{s}_{11}e^{-i\zeta x}, \hat{s}_{21}e^{i\zeta x})^T$. Thus, the scattering coefficients $\hat{s}_{11}$ and $\hat{s}_{21}$ can be obtained from the above asymptotics (2.166). Then in view of the involution property (2.58) as well as relations (2.43) on scattering coefficients, we find that

$$s_{11}(\zeta) = \hat{s}_{22}(\zeta) = \frac{\Gamma^2(-i\zeta+\frac{1}{2})}{\Gamma(-i\zeta+\frac{1}{2}+A)\Gamma(-i\zeta+\frac{1}{2}-A)}, \tag{2.167}$$

$$s_{12}(\zeta) = s_{21}(\zeta) = i\sin(\pi A)\operatorname{sech}(\pi\zeta). \tag{2.168}$$

These explicit expressions show that $s_{11}(\zeta)$ and $\hat{s}_{22}(\zeta)$ are indeed analytic everywhere in $\mathbb{C}_+$, but (s_{12}, s_{21}) have pole singularities at $\zeta = i(n-\frac{1}{2})$ for all integers of $n = 0, \pm1, \pm2, \ldots$. These singularities lie in both $\mathbb{C}_+$ and $\mathbb{C}_-$, and thus (s_{12}, s_{21}) are not analytic in either $\mathbb{C}_+$ or $\mathbb{C}_-$. These facts are consistent with the general analytic properties of the scattering matrix as established in Sec. 2.1.1.

Discrete eigenvalues of the Zakharov–Shabat system (2.2) in $\mathbb{C}_+$ are zeros of $s_{11}(\zeta)$. In view of the expression (2.167), we find that such discrete eigenvalues have the following simple expressions:

$$\zeta_n = i\left(A + \frac{1}{2} - n\right), \tag{2.169}$$

where n is a positive integer satisfying $n < A + 1/2$ so that $\zeta_n \in \mathbb{C}_+$. The first eigenvalue appears at $A = \frac{1}{2}$, a second one at $A = \frac{3}{2}$, and so on. As A increases further, more discrete eigenvalues will arise. For a given A value, the number of discrete eigenvalues is equal to the largest integer n which satisfies $n-\frac{1}{2}<A$. Complex conjugates of these eigenvalues ζ_n^* in the lower half plane $\mathbb{C}_-$ are also discrete eigenvalues of the Zakharov–Shabat system (2.2) due to the involution property (2.58).

When $A=N$ is a positive integer, the initial condition (2.159) is quite special. In this case, it follows from (2.169) that there exist N discrete eigenvalues. More importantly, it is seen from (2.168) that $s_{12}(\zeta)=s_{21}(\zeta)=0$; thus this initial condition is a reflectionless potential which corresponds to a pure N-soliton solution (i.e., there is no energy radiation). This N-soliton solution forms a bound state since all the eigenvalues (2.169) are purely imaginary. In this case, s_{11} and $\hat{s}_{22}$ reduce from (2.167) to

$$s_{11}(\zeta)=\hat{s}_{22}(\zeta)=\prod_{n=1}^{N}\frac{\zeta-\zeta_n}{\zeta-\zeta_n^*}. \tag{2.170}$$

When $N-1/2<A<N+1/2$ for some integer N but $A\neq N$, then there also exist N discrete eigenvalues. But in this case, $s_{12}(\zeta)$ and $s_{21}(\zeta)$ are nonzero, and thus energy radiation will be present. The initial condition (2.159) in this case will evolve into an N-soliton bound state, plus energy radiation which disperses away over time.

2. The box initial condition

The second initial condition we consider is piecewise constant and has a rectangular profile with a varying height; i.e.,

$$u(x,0)=\begin{cases} A, & 0\le x\le 1, \\ 0 & \text{elsewhere.}\end{cases} \tag{2.171}$$

Here A is a constant which is assumed to be real positive without loss of generality. For this initial condition, simple calculations show that

$$s_{11}(\zeta)=e^{i\zeta}\left[\cos\sqrt{\zeta^2+A^2}-\frac{i\zeta}{\sqrt{\zeta^2+A^2}}\sin\sqrt{\zeta^2+A^2}\right]. \tag{2.172}$$

Thus discrete eigenvalues $\zeta\in\mathbb{C}_+$ are given by the following equation:

$$\cot\sqrt{\zeta^2+A^2}=\frac{i\zeta}{\sqrt{\zeta^2+A^2}}. \tag{2.173}$$

It can be seen that this equation admits only purely imaginary eigenvalues. At the birth of each eigenvalue, $\zeta=0$. Thus the above equation reduces to $\cot A=0$. This indicates that new eigenvalues appear whenever $A=(n-\frac{1}{2})\pi$, where n is a positive integer. The first eigenvalue appears at $A_1=\frac{1}{2}\pi$, a second one at $A_2=\frac{3}{2}\pi$, etc. At higher values of A, additional eigenvalues will arise. For a given A value, the number of discrete eigenvalues is equal to the largest integer n which satisfies $(n-\frac{1}{2})\pi < A$. For this box initial condition, $s_{12}(\zeta)$ and $s_{21}(\zeta)$ are never zero and thus a box profile can never be a pure soliton or multisoliton solution.

The above box potential is on a compact support. For any potential on a compact support, it can be shown that all Jost solutions $J_{\pm}(x,\zeta)$ of the Zakharov–Shabat system are entire (i.e., analytic in the whole ζ plane). The reason is that in this case, the Volterra integral equations (2.20) for $J_{\pm}(x,\zeta)$ are on a finite interval, thus they always have absolutely convergent Neumann series solutions (Pogorzelski (1966), Ablowitz and Segur (1981)). As a consequence, all elements in the scattering matrix S are entire as well. However, the formula (2.172) for s_{11} contains quantities $\sqrt{\zeta^2+A^2}$ (so do the formulae for other scattering coefficients and Jost solutions). This may create the impression that these scattering coefficients and Jost solutions are not entire due to the branch cuts in $\sqrt{\zeta^2+A^2}$. But closer examination shows that branch cuts in $\sqrt{\zeta^2+A^2}$ actually are not branch cuts for the scattering coefficients and Jost functions. For instance, the function $\cos\sqrt{\zeta^2+A^2}$ in (2.172) actually has no branch cuts and is entire in fact. Thus for the above box potential (2.171), the scattering coefficients and Jost functions are indeed entire, in agreement with the general theory.

The above two initial conditions exhibit some common and important features on discrete eigenvalues. First, eigenvalues appear only when the L_1-norm of the initial condition $\|u(x,0)\|_{L_1}$ is above a certain threshold. Here the L_1-norm is defined as $\|f(x)\|_{L_1} \equiv \int_{-\infty}^{\infty}|f(x)|dx$. Second, at this threshold of discrete eigenvalues, the L_1-norm of the initial condition is $\|u(x,0)\|_{L_1} = \pi/2$. Third, the number of discrete eigenvalues is the largest positive integer n such that $(n-\frac{1}{2})\pi < \|u(x,0)\|_{L_1}$. Fourth, all the eigenvalues are purely imaginary. All of these features have important extensions to more general initial conditions. This will be discussed in the next subsection.

2.4.2 General Criteria for Discrete Eigenvalues

Many of the results for the above two special initial conditions can be extended to more general initial conditions. First, we show that for a general initial condition, if its L_1-norm is too small, then discrete eigenvalues cannot exist. A nonsharp result on this problem is that, if $\|u(x,0)\|_{L_1} < 1.317$, then the Zakharov–Shabat system cannot have discrete eigenvalues (Zakharov et al. (1984), see also Ablowitz et al. (1974)). To prove it, we consider the first column of the Jost matrix $J_-(x,\zeta)$ at a discrete eigenvalue $\zeta=\zeta_k \in \mathbb{C}_+$, and denote this Jost solution as $(\varphi_1,\varphi_2)^T$. Eqs. (2.155)–(2.157) tell us that $(\varphi_1,\varphi_2)^T e^{-i\zeta_k x}$ is a localized discrete eigenfunction. Thus $(\varphi_1,\varphi_2)^T \to 0$ as $x\to+\infty$. Recall that $(\varphi_1,\varphi_2)^T$ satisfies the integral equations (2.22)–(2.23), thus

$$\int_{-\infty}^{\infty} u(x)\varphi_2(x,\zeta_k)dx = -1. \tag{2.174}$$

Now we estimate the left integral in the above equation. When $\zeta \in \mathbb{C}_+$, $|e^{2i\zeta(x-y)}| \le 1$ for $y \le x$, thus from (2.23) we get

$$|\varphi_2(x,\zeta)| \le \int_{-\infty}^{x} |u(x_1)||\varphi_1(x_1,\zeta)|\, dx_1. \tag{2.175}$$

Inserting (2.22) into the above equation, we obtain

$$|\varphi_2(x,\zeta)| \le \int_{-\infty}^{x} |u(x_1)| dx_1 + \int_{-\infty}^{x} \int_{-\infty}^{x_1} |u(x_1)||u(x_2)||\varphi_2(x_2,\zeta)|\, dx_2 dx_1.$$

Repeatedly iterating the above inequality and performing simple integrations, we find that

$$|\varphi_2(x,\zeta)| \le M(x) + \frac{1}{3!}M^3(x) + \frac{1}{5!}M^5(x) + \cdots = \sinh[M(x)], \tag{2.176}$$

where $M(x) = \int_{-\infty}^{x} |u(x_1)| dx_1$. Thus

$$\left| \int_{-\infty}^{\infty} u(x)\varphi_2(x,\zeta_k)\, dx \right| \le \int_{-\infty}^{\infty} |u(x)| \sinh[M(x)]\, dx = \cosh[M(\infty)] - 1. \tag{2.177}$$

If $\cosh[M(\infty)] < 2$, i.e., $M(\infty) = \|u(x)\|_{L_1} < 1.317$, then (2.174) and (2.177) would contradict each other, and hence discrete eigenvalues cannot exist.

The above condition $\|u(x,0)\|_{L_1} < 1.317$ for the nonexistence of discrete eigenvalues is not sharp. Recently, a sharp condition for the nonexistence of discrete eigenvalues was obtained by Klaus and Shaw (2003). It says that for a complex initial condition $u(x,0)$, if

$$\|u(x,0)\|_{L_1} \le \pi/2, \tag{2.178}$$

then the Zakharov–Shabat system (2.2) has no discrete eigenvalues. This result, together with the two special examples considered in the previous subsection, establishes that this condition (2.178) is sharp.

Next we consider under what general conditions discrete eigenvalues of the Zakharov–Shabat system are all purely imaginary as in the two examples of the previous subsection. On this question, Klaus and Shaw (2002) obtained the following "single-lobe" theorem.

> *If the initial condition $u(x,0)$ is real, nonnegative, piecewise smooth, having a finite L_1-norm, and having a single lobe (i.e., the function is nondecreasing on the left of a certain x-value and nonincreasing on the right of this x-value), then all discrete eigenvalues in the Zakharov–Shabat system* (2.2) *must be purely imaginary.*

The two special initial conditions (2.159) and (2.171) considered in the previous subsection satisfy the conditions of this theorem, and thus all their eigenvalues are purely imaginary.

Note that it was once believed that if an initial condition were real and symmetric, then all its eigenvalues would be purely imaginary. This is false. Indeed there are examples of initial conditions which are positive, symmetric but with two humps, and their eigenvalues are not purely imaginary (Klaus and Shaw (2002)). One such example will be shown in the next subsection (see Fig. 2.3(a)). Thus symmetry (evenness) of the initial condition is not essential, but the single-lobe property is critical.

In addition to the above results, Klaus and Shaw (2003) have also obtained a number of other important results. For instance, for a real potential $u(x,0)$ with a finite L_1-norm, there are at least n purely imaginary eigenvalues, where n is the largest nonnegative integer such that $(n-\frac{1}{2})\pi < \|u(x,0)\|_{L_1}$. In particular, a real potential has at least one purely imaginary eigenvalue if $\|u(x,0)\|_{L_1} > \pi/2$. For single-lobe potentials as described in the above single-lobe theorem, the eigenvalues are all simple. In addition, the number of (purely imaginary) eigenvalues is exactly equal to the largest integer n such that $(n-\frac{1}{2})\pi < \|u(x,0)\|_{L_1}$. For real and single-signed potentials, the largest magnitude of an eigenvalue on the imaginary axis strictly dominates the imaginary part of any other eigenvalue in the complex plane. In particular, if there are no imaginary eigenvalues, then there are no eigenvalues at all. Moreover, for a complex potential $u(x,0)$, if $|u(x,0)|$ has no eigenvalues on the imaginary axis, then $u(x,0)$ has no eigenvalues at all. These results significantly extended the ones observed for the two special initial conditions in the previous subsection.

Most of the above results and examples are for real-valued initial conditions. When the initial condition is a real-valued function $f(x)$ multiplied by $e^{i\alpha x}$, where α is a real constant, then its eigenvalues are simply those of $f(x)$ minus $\alpha/2$. Thus a linear phase gradient in the initial condition does not impose any difficulty. However, if the phase is not a linear function of x, then its eigenvalues can be more exotic. One such example will be displayed in the next subsection (see Fig. 2.3(b)).

2.4.3 Numerical Computations of Eigenvalues

For most initial conditions, discrete eigenvalues of the Zakharov–Shabat system (2.2) cannot be obtained analytically. Thus numerical computations of these eigenvalues are necessary. Equation (2.2) can be rewritten as a standard linear eigenvalue problem:

$$\begin{pmatrix} -\partial_x & u \\ u^* & \partial_x \end{pmatrix} Y = i\zeta Y. \tag{2.179}$$

One of the methods to solve it is to use finite-difference discretization. In this method, one first truncates the infinite x-axis into a finite interval and discretizes it into grid points. Then one approximates the spatial derivatives in (2.179) by a finite-difference scheme on the grid points (for instance, by the central difference formula). This way, the eigenvalue problem (2.179) becomes a matrix eigenvalue problem which can be solved by standard methods such as the Arnoldi algorithm (Arnoldi (1951), Sorensen (1992)). However, the accuracy of this finite-difference method is quite low. In addition, our numerical testing shows that this method can give spurious eigenfunctions even if the eigenvalues it obtains are approximately correct. Thus this method will not be elaborated.

Below we describe another method—the Fourier collocation method (Boyd (2001)), which is much more accurate and reliable in both eigenvalues and eigenfunctions. In this method, one expands the eigenfunction Y into a Fourier series and turns the Zakharov–Shabat system (2.179) into a matrix eigenvalue problem for the Fourier coefficients of the eigenfunction. Another way to look at it is that, one first takes the Fourier transform of the Zakharov–Shabat equations and then discretizes the transformed equation in Fourier space. For this purpose, we first truncate the infinite x-axis into a finite interval $[-L/2, L/2]$, where L is the length of the interval. On this interval, we expand the eigenfunction $Y = (y_1, y_2)^T$ as well as the potential function $u(x,0)$ into Fourier series with $2N+1$ modes:

$$y_1(x) = \sum_{n=-N}^{N} a_{1,n} e^{ink_0x}, \quad y_2(x) = \sum_{n=-N}^{N} a_{2,n} e^{ink_0x}, \quad u(x,0) = \sum_{n=-N}^{N} c_n e^{ink_0x},$$

where $k_0 = 2\pi/L$. Substituting these expansions into the Zakharov–Shabat equations (2.179) and equating the coefficients of the same Fourier modes, the following eigenvalue system for the coefficients $\{a_{1,k}, a_{2,k}\}$ will be obtained:

$$\begin{pmatrix} -\mathcal{B}_1 & \mathcal{B}_2 \\ \mathcal{B}_2^\dagger & \mathcal{B}_1 \end{pmatrix} \begin{pmatrix} A_1 \\ A_2 \end{pmatrix} = i\zeta \begin{pmatrix} A_1 \\ A_2 \end{pmatrix}. \tag{2.180}$$

Here

$$\mathcal{B}_1 = ik_0 \operatorname{diag}(-N,\ -N+1,\ \ldots,\ N-1,\ N),$$

$$\mathcal{B}_2 = \begin{pmatrix} c_0 & c_{-1} & \cdots & c_{-N} & & & \\ c_1 & c_0 & c_{-1} & \ddots & \ddots & & \\ \vdots & c_1 & c_0 & \ddots & \ddots & \ddots & \\ c_N & \ddots & \ddots & \ddots & \ddots & \ddots & c_{-N} \\ & c_N & \ddots & \ddots & \ddots & \ddots & \vdots \\ & & \ddots & \ddots & \ddots & \ddots & c_{-1} \\ & & & c_N & \cdots & c_1 & c_0 \end{pmatrix},$$

$$A_1 = (a_{1,-N},\ a_{1,-N+1},\ \ldots,\ a_{1,N})^T, \quad A_2 = (a_{2,-N},\ a_{2,-N+1},\ \ldots,\ a_{2,N})^T,$$

and $\mathcal{B}_2^\dagger$ is the Hermitian of $\mathcal{B}_2$. Note that in the Fourier space, $\mathcal{B}_1$ is the differentiation matrix for ∂_x, and $\mathcal{B}_2$ is the convolution matrix for the functional multiplication by $u(x,0)$. Thus (2.180) is the discretization of (2.179) in the Fourier space. The matrix in (2.180) is non-Hermitian. Its dimension is $2(2N+1)$, and its computer memory usage is $O(N^2)$. When N is small or moderate, this eigenvalue problem can be solved by the QR algorithm which gives all the eigenvalues of the matrix (Golub and Van Loan (1996)). If N is large, it can be solved by the Arnoldi algorithm which can give a few of the eigenvalues very efficiently (Arnoldi (1951), Sorensen (1992)). Both algorithms are built in software such as MATLAB.

This Fourier collocation method is simple to implement. To illustrate, we take the Satsuma–Yajima initial condition (2.159) with $A = 1.8$. For this initial condition, the MATLAB code of this method is displayed below. This code can also be downloaded directly from the SIAM Web page, www.siam.org/books/mm16, or from the author's Web page, www.cems.uvm.edu/~jyang.

Program 1

```
% p1.m:  Fourier collocation method for the eigenvalue spectrum of
% the Zakharov-Shabat system, output shown in Fig. 2.2(a)

  N=100; L=100; Nx=2*N;       % number of Fourier modes is 2N+1
  dx=L/Nx;   x=-L/2:dx:L/2-dx;   k0=2*pi/L;
  u=1.8*sech(x);              % potential function
  for n=-N:1:N
     C(n+N+1)=dx*sum(u.*exp(-i*k0*n*x))/L;
  end
  B1=i*k0*diag([-N:N]);
  B2=toeplitz([C(N+1:2*N+1) zeros(1,N)],[C(N+1:-1:1) zeros(1,N)]);
  M=[ -B1        B2
      B2'        B1];
  eigvalues=eig(-i*M);        % calculate eigenvalues
  plot(eigvalues, '.')        % plotting spectrum
  axis([-2 2 -1.7 1.7])
```

In this code, the length of the x-interval is taken as $L = 100$, the number of Fourier modes is taken as $2N+1 = 201$ (i.e., $N = 100$), and eigenvalues of the resulting matrix are computed by the QR algorithm (its MATLAB command is "eig"). On a personal computer (PC) with 2 GHz processor, this code takes about 2 seconds to finish. Running this code, the spectrum of this Satsuma–Yajima initial condition is obtained and displayed in Fig. 2.2(a). It is seen that this numerically obtained spectrum contains two pairs of discrete eigenvalues $\pm 1.3i$ and $\pm 0.3i$, which agree with the analytical formula (2.169).

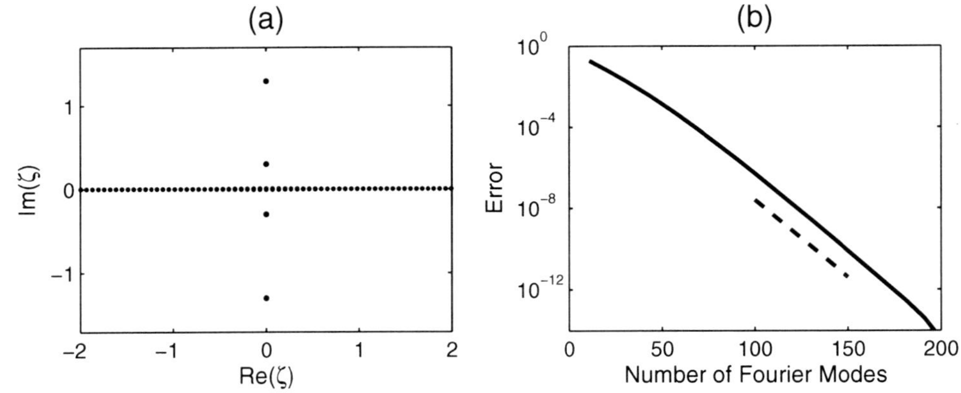

Figure 2.2. (a) *Output of Program* 1, *which gives the eigenvalue spectrum of the Zakharov–Shabat system* (2.2) *for the Satsuma–Yajima initial condition* (2.159) *with* $A = 1.8$ *by the Fourier collocation method. Here* $L = 100$, *and* $N = 100$; (b) *error diagram versus the number of Fourier modes* $(2N+1)$ *for the eigenvalue* $\zeta = 1.3i$. *In this diagram,* $L = 50$ *is used. The dashed line is the benchmark exponentially decaying function* $e^{-0.175(2N+1)}$.

Regarding the accuracy of this method, for smooth potential functions $u(x,0)$, its accuracy is spectral (Trefethen (2000), Boyd (2001)). That is, its error decays faster than any power of $(2N+1)^{-1}$. Here $2N+1$ is the number of Fourier modes. This spectral accuracy is due to the Fourier series approximation of a function, whose error is spectral for smooth functions. To demonstrate this spectral accuracy, we again take the Satsuma–Yajima initial condition (2.159) with $A = 1.8$, and we monitor the numerical error for the discrete eigenvalue $\zeta = 1.3i$. For this eigenvalue, the interval length $L = 50$ is sufficient and is thus used. On this interval, Program 1 is run with various numbers of Fourier modes. The error diagram versus the number of Fourier modes $2N+1$ for this eigenvalue is displayed in Fig. 2.2(b). It is seen that the error decays exponentially with the number of Fourier modes, roughly in proportion to $e^{-\beta(2N+1)}$, where $\beta = 0.175$. Thus the spectral accuracy of this method is confirmed.

One may notice that the Fourier coefficients c_n in Program 1 were computed by simple Riemann summation, and thus may wonder why this "primitive" calculation of the Fourier coefficients did not corrupt the spectral accuracy of numerical eigenvalues. The reason is that the accuracy of this "primitive" Riemann summation for Fourier coefficients is actually also spectral rather than first order in the grid spacing Δx. Indeed, for a smooth periodic function $u(x)$, evaluation of its integral over one period by simple Riemann summation is spectrally accurate, because this integral is equal to the constant term of this function's

Fourier components multiplying the period of this function, and this constant term is given discretely by the simple Riemann summation of this function with spectral accuracy (Boyd (2001, page 457)). If the function $u(x)$ is not periodic but rather decays to zero as $x \to \pm\infty$, as long as the truncated interval $[-L/2, L/2]$ is large enough, the boundary effect then is insignificant, and this localized function can be treated as a periodic function. Hence the accuracy of the Fourier coefficients c_n in the above code is also spectral and consequently does not affect the spectral accuracy of the above Fourier collocation method.

Using the above numerical method, we can compute the spectrum of the Zakharov–Shabat system (2.179) for any localized or periodic initial condition $u(x,0)$. Below we present numerically obtained spectra for two interesting initial conditions.

The first initial condition is

$$u(x,0) = 2\left(e^{-4(x-1)^2} + e^{-4(x+1)^2}\right), \tag{2.181}$$

which is positive, symmetric but with two lobes. For this initial condition, the numerically computed spectrum (with $L = 400, N = 400$) is displayed in Fig. 2.3(a). This spectrum contains one pair of purely imaginary eigenvalues and two pairs of complex eigenvalues. Thus it disproves the belief that real symmetric potentials could only have purely imaginary eigenvalues. Other examples of real symmetric potentials giving nonimaginary eigenvalues can be found in Klaus and Shaw (2002) and Desaix et al. (2003).

The second initial condition is

$$u(x,0) = A(\epsilon x)e^{iS(\epsilon x)/\epsilon}, \quad A(\epsilon x) = S(\epsilon x) = \operatorname{sech} 2\epsilon x. \tag{2.182}$$

This type of initial condition arises in the consideration of the semiclassical limit of the NLS equation where the dispersion coefficient goes to zero. Here the phase function S/ϵ is nonlinear in x, and thus most theoretical results in the previous subsection do not apply. At $\epsilon = 0.1$, Bronski (1996) computed the spectrum of this initial condition and found that its discrete eigenvalues exhibited a Y-shape. We computed this spectrum by the Fourier collocation method above (with $L = 400, N = 200$), and the result is displayed in Fig. 2.3(b). It is seen that this spectrum contains two pairs of purely imaginary eigenvalues and two quadruples of complex eigenvalues which form Y-shapes. When $\epsilon \to 0$, Bronski (1996) showed numerically that the number of discrete eigenvalues scales as ϵ^{-1}, and these eigenvalues approach a limiting curve with a Y-shape.

Lastly, we would like to mention another spectrally accurate numerical method for eigenvalue computations. This is the spectral differentiation method. Here one discretizes a truncated x-interval and works with the values of the eigenfunction on the spatial grid. The derivative ∂_x of this eigenfunction is computed by the discrete Fourier transform method, and it is implemented by multiplying a spectral-differentiation matrix on the grid values of the eigenfunction (see Trefethen (2000)). The accuracy of this spectral differentiation

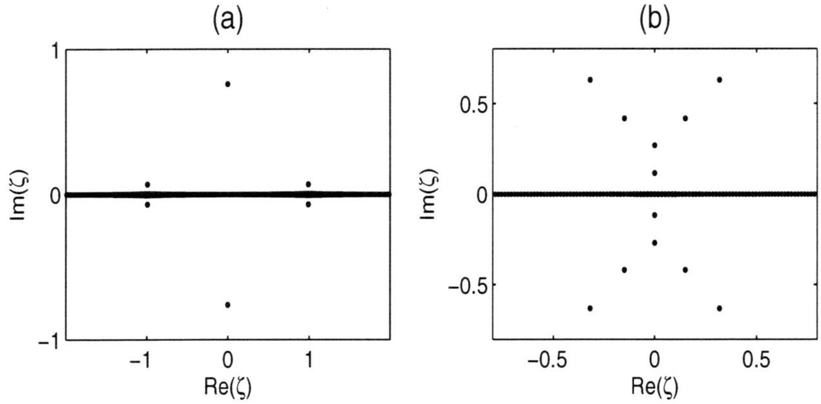

Figure 2.3. *Eigenvalue spectra of the Zakharov–Shabat system for two potentials by the Fourier collocation method:* (a) *a two-lobe potential* (2.181)*, and* (b) *a complex-valued potential* (2.182) *with* $\epsilon = 0.1$.

method is also spectral like the Fourier collocation method. But our numerical testing on the Zakharov–Shabat system shows that this method is less accurate than the Fourier collocation method for the same amount of computations, thus it will not be elaborated.

2.5 Closure of Zakharov–Shabat Eigenstates

The closure (i.e., completeness) of eigenfunctions for a linear operator in the squared integrable functional space is an important problem, since this closure is critical for many applications (see Yang and Kaup (2000) for instance). If the linear operator is Hermitian (such as the linear Schrödinger operator), then the closure of its eigenfunctions is well known. However, for non-Hermitian operators, the closure is much harder to prove, and only limited progress has been made. The Zakharov–Shabat system (2.2) is non-Hermitian. Fortunately, since analyticity properties of its eigenfunctions have been elaborated earlier in this chapter, these analyticity properties will enable us to prove the closure of Zakharov–Shabat eigenstates. This closure was first established by Ablowitz et al. (1974) by using the Marchenko equations. Our proof below uses an alternative method, which is based on contour integrations. The idea of this alternative method was first used by Gerdjikov and Khristov (1980) (see also Gerdjikov (2005)) in their analysis of an integrodifferential operator (which is the recursion operator of the Zakharov–Shabat system; see Sec. 2.8). Here, we will further develop this method and apply it to prove the closure of Zakharov–Shabat eigenstates. This contour-integration method is easier. In addition, this method can be readily generalized to higher-order scattering operators (see Sec. 3.7).

First we define the matrix functions

$$R^+(x,y,\zeta) = \chi^+(x,\zeta)\,\mathrm{diag}\,[\theta(y-x),\, -\theta(x-y)]\,(\chi^+)^{-1}(y,\zeta), \tag{2.183}$$

$$R^-(x,y,\zeta) = \chi^-(x,\zeta)\,\mathrm{diag}\,[\theta(x-y),\, -\theta(y-x)]\,(\chi^-)^{-1}(y,\zeta), \tag{2.184}$$

where

$$\chi^+ = (\phi_1, \psi_2), \quad \chi^- = (\psi_1, \phi_2), \tag{2.185}$$

$$\Phi = (\phi_1, \phi_2) = J_- E, \quad \Psi = (\psi_1, \psi_2) = J_+ E, \tag{2.186}$$

and $\theta(x)$ is the standard step function,

$$\theta(x) = \begin{cases} 1, & x \geq 0, \\ 0, & x < 0. \end{cases} \tag{2.187}$$

These functions $R^\pm$ are meromorphic in the upper and lower halves of the ζ plane, respectively. From definitions (2.185), we see that

$$\chi^+ = \Phi H_1 + \Psi H_2 = \Psi(S H_1 + H_2) = \Phi(H_1 + S^{-1} H_2) \tag{2.188}$$

and

$$\chi^- = \Psi H_1 + \Phi H_2 = \Psi(H_1 + S H_2) = \Phi(S^{-1} H_1 + H_2). \tag{2.189}$$

Thus

$$\det \chi^+ = s_{11} = \hat{s}_{22} \tag{2.190}$$

and

$$\det \chi^- = s_{22} = \hat{s}_{11}. \tag{2.191}$$

Hence $R^\pm(x,y,\zeta)$ have pole singularities at the zeros of $\hat{s}_{22}(\zeta)$ and $s_{22}(\zeta)$, respectively. The step functions are introduced in $R^\pm$ so that they remain bounded as $\zeta \to \infty$. Indeed, we see from the asymptotics (2.10) of Jost solutions $J_\pm$ that as $\zeta \to \infty$ in the respective half plane of analyticity,

$$\chi^\pm(x,\zeta) \to e^{-i\zeta\Lambda x}; \tag{2.192}$$

thus

$$R^+(x,y,\zeta) \to \mathrm{diag}\left[\theta(y-x)e^{i\zeta(y-x)},\, -\theta(x-y)e^{i\zeta(x-y)}\right], \tag{2.193}$$

$$R^-(x,y,\zeta) \to \mathrm{diag}\left[\theta(x-y)e^{i\zeta(y-x)},\, -\theta(y-x)e^{i\zeta(x-y)}\right], \tag{2.194}$$

which are both bounded. Next, we define two contours in the complex ζ plane. One is γ_+, which starts from $\zeta = -\infty + i0^+$, passes over all zeros of $\hat{s}_{22}(\zeta)$ in $\mathbb{C}_+$, and ends at $\zeta = \infty + i0^+$. The other one is γ_-, which starts from $\zeta = -\infty + i0^-$, passes under all zeros of $s_{22}(\zeta)$ in $\mathbb{C}_-$, and ends at $\zeta = \infty + i0^-$. Then using the above large-ζ asymptotics of $R^{\pm}$, we see that

$$\int_{\gamma_+} R^+(x,y,\zeta)d\zeta = \int_{\gamma_+} \text{diag}\left[\theta(y-x)e^{i\zeta(y-x)}, -\theta(x-y)e^{i\zeta(x-y)}\right]d\zeta. \tag{2.195}$$

The integrand on the right side of the above equation is analytic, and thus its integration contour γ_+ can be brought down to the real axis, i.e.,

$$\int_{\gamma_+} R^+(x,y,\zeta)d\zeta = \int_{-\infty}^{\infty} \text{diag}\left[\theta(y-x)e^{i\xi(y-x)}, -\theta(x-y)e^{i\xi(x-y)}\right]d\xi.$$

Similarly, we have

$$\int_{\gamma_-} R^-(x,y,\zeta)d\zeta = \int_{-\infty}^{\infty} \text{diag}\left[\theta(x-y)e^{i\xi(y-x)}, -\theta(y-x)e^{i\xi(x-y)}\right]d\xi.$$

Adding the above two equations, we get

$$\begin{aligned}\int_{\gamma_+} R^+(x,y,\zeta)d\zeta + \int_{\gamma_-} R^-(x,y,\zeta)d\zeta &= \int_{-\infty}^{\infty} \text{diag}\left[e^{i\xi(y-x)}, -e^{i\xi(x-y)}\right]d\xi \\ &= 2\pi\delta(x-y)\Lambda,\end{aligned} \tag{2.196}$$

where Λ is the diagonal matrix diag$(1,-1)$; see (2.4). Now we calculate the left side of the above equation. The goal is to get rid of the step functions in it and reduce it to integrals of products between eigenfunctions and adjoint eigenfunctions of the Zakharov–Shabat system. First, using the residue theorem we get

$$\begin{aligned}\int_{\gamma_+} R^+(x,y,\zeta)d\zeta + \int_{\gamma_-} R^-(x,y,\zeta)d\zeta &= \int_{-\infty}^{\infty} \left[R^+(x,y,\xi) + R^-(x,y,\xi)\right]d\xi \\ &+ 2\pi i \sum_j \left\{\text{Res}\left[R^-(x,y,\zeta),\bar{\zeta}_j\right] - \text{Res}\left[R^+(x,y,\zeta),\zeta_j\right]\right\}.\end{aligned} \tag{2.197}$$

Here ζ_j and $\bar{\zeta}_j$ are zeros of $\hat{s}_{22}(\zeta)$ and $s_{22}(\zeta)$. From the definitions of $R^{\pm}$, we see that

$$R^+(x,y,\zeta) = \chi^+(x,\zeta)\text{diag}(1,0)(\chi^+)^{-1}(y,\zeta) - \theta(x-y)\chi^+(x,\zeta)(\chi^+)^{-1}(y,\zeta) \tag{2.198}$$

and

$$R^-(x,y,\zeta) = -\chi^-(x,\zeta)\text{diag}(0,1)(\chi^-)^{-1}(y,\zeta) + \theta(x-y)\chi^-(x,\zeta)(\chi^-)^{-1}(y,\zeta). \tag{2.199}$$

On the real axis, $\xi \in \mathbb{R}$, since both $\chi^+(x,\xi)$ and $\chi^-(x,\xi)$ are fundamental matrices of the Zakharov–Shabat system, they must be related by a constant matrix $A(\xi)$ as $\chi^+(x,\xi) = \chi^-(x,\xi)A(\xi)$. Because of this,

$$\chi^+(x,\xi)(\chi^+)^{-1}(y,\xi) = \chi^-(x,\xi)(\chi^-)^{-1}(y,\xi), \quad \xi \in \mathbb{R}. \tag{2.200}$$

Thus,

$$\int_{-\infty}^{\infty} \{R^+(x,y,\xi) + R^-(x,y,\xi)\}\, d\xi = \int_{-\infty}^{\infty} \Big\{\chi^+(x,\xi)\,\mathrm{diag}(1,0)(\chi^+)^{-1}(y,\xi) \\ - \chi^-(x,\xi)\,\mathrm{diag}(0,1)(\chi^-)^{-1}(y,\xi)\Big\}\, d\xi, \tag{2.201}$$

where the step functions have disappeared.

Next we simplify the residue terms in (2.197) and show that

$$\mathrm{Res}\left[R^+(x,y,\zeta),\zeta_j\right] = \mathrm{Res}\left[\chi^+(x,\zeta)\,\mathrm{diag}(1,0)(\chi^+)^{-1}(y,\zeta),\zeta_j\right] \tag{2.202}$$

and

$$\mathrm{Res}\left[R^-(x,y,\zeta),\bar{\zeta}_j\right] = -\mathrm{Res}\left[\chi^-(x,\zeta)\,\mathrm{diag}(0,1)(\chi^-)^{-1}(y,\zeta),\bar{\zeta}_j\right], \tag{2.203}$$

where the step functions have disappeared as well. In view of the expressions (2.198)–(2.199) for $R^{\pm}$, to prove (2.202)–(2.203), we only need to show that

$$\mathrm{Res}\left[\chi^+(x,\zeta)(\chi^+)^{-1}(y,\zeta),\zeta_j\right] = 0 \tag{2.204}$$

and

$$\mathrm{Res}\left[\chi^-(x,\zeta)(\chi^-)^{-1}(y,\zeta),\bar{\zeta}_j\right] = 0. \tag{2.205}$$

The proofs for these two equations are similar, so we will prove only (2.204) below. Equation (2.190) shows that at the zero ζ_j of $\hat{s}_{22}(\zeta)$, the meromorphic function $(\chi^+)^{-1}(y,\zeta)$ has a pole singularity. Let us first suppose that ζ_j is a simple pole, and we denote

$$B(x,y) \equiv \mathrm{Res}\left[\chi^+(x,\zeta)(\chi^+)^{-1}(y,\zeta),\zeta_j\right]. \tag{2.206}$$

Then taking the x-derivative of this equation and recalling that χ^+ satisfies the Zakharov–Shabat equations (2.2), we get

$$B_x = \mathrm{Res}\left\{(Q(x) - i\zeta\Lambda)\chi^+(x,\zeta)(\chi^+)^{-1}(y,\zeta),\zeta_j\right\} \\ = (Q(x) - i\zeta_j\Lambda)B - i\Lambda\,\mathrm{Res}\left[(\zeta-\zeta_j)\chi^+(x,\zeta)(\chi^+)^{-1}(y,\zeta),\zeta_j\right]. \tag{2.207}$$

Since ζ_j is a simple pole in $(\chi^+)^{-1}(y,\zeta)$, the function $(\zeta-\zeta_j)(\chi^+)^{-1}(y,\zeta)$ then is analytic at ζ_j; thus the residue term on the right-hand side of the above equation is zero. So we get

$$B_x = (Q(x) - i\zeta_j\Lambda)B. \tag{2.208}$$

The initial condition to this linear homogeneous differential equation is that $B=0$ at $x=y$, which can be easily seen from the definition (2.206). Thus the solution to this equation is obviously

$$B(x,y)=0 \quad \text{for all } (x,y), \tag{2.209}$$

which proves Eq. (2.204). In the general case where ζ_j is a multiple pole, by repeating the above calculations, we will get a system of linear homogeneous differential equations for the residue $B(x,y)$ together with other residue functions such as the one on the right-hand side of (2.207), subject to the zero initial condition at $x=y$. Its solution is also zero, and thus $B(x,y)=0$ for all (x,y) as well. Another way to prove (2.204) in this general case of multiple poles is to treat the multiple pole as a limit of several simple poles approaching each other. Since (2.204) holds for each simple pole, it would hold for the multiple pole under this limit as well.

Lastly, we further simplify (2.201)–(2.203) and derive the final closure relation. For this purpose, recall that

$$(\chi^+)^{-1}=(\phi_1,\psi_2)^{-1}=(\Phi H_1+\Psi H_2)^{-1}, \tag{2.210}$$

and they are the adjoint solutions of the Zakharov–Shabat system (2.2). So are

$$\begin{pmatrix} \widehat{\psi}_1 \\ \widehat{\phi}_2 \end{pmatrix} = H_1\Psi^{-1}+H_2\Phi^{-1} \tag{2.211}$$

in view of Eq. (2.35). Thus these adjoint solutions must be linearly related. Using Eqs. (2.210)–(2.211) as well as the scattering relation $\Phi=\Psi S$, it is easy to see that

$$(\chi^+)^{-1}=\begin{pmatrix} 1/s_{11} & \\ & 1/\hat{s}_{22} \end{pmatrix}\begin{pmatrix} \widehat{\psi}_1 \\ \widehat{\phi}_2 \end{pmatrix}. \tag{2.212}$$

Similarly,

$$(\chi^-)^{-1}=\begin{pmatrix} 1/\hat{s}_{11} & \\ & 1/s_{22} \end{pmatrix}\begin{pmatrix} \widehat{\phi}_1 \\ \widehat{\psi}_2 \end{pmatrix}. \tag{2.213}$$

Substituting the above two equations into (2.201)–(2.203), the final closure relation will then be obtained from (2.196)–(2.197). Assuming the zeros $(\zeta_j,\bar{\zeta}_j)$ are all simple, the closure relation reads

$$\begin{aligned}\delta(x-y)\Lambda = &\frac{1}{2\pi}\int_{-\infty}^{\infty}\left[\frac{1}{s_{11}(\xi)}\phi_1(x,\xi)\widehat{\psi}_1(y,\xi)-\frac{1}{s_{22}(\xi)}\phi_2(x,\xi)\widehat{\psi}_2(y,\xi)\right]d\xi \\ &-i\sum_j\left[\frac{1}{s_{11}'(\zeta_j)}\phi_1(x,\zeta_j)\widehat{\psi}_1(y,\zeta_j)+\frac{1}{s_{22}'(\bar{\zeta}_j)}\phi_2(x,\bar{\zeta}_j)\widehat{\psi}_2(y,\bar{\zeta}_j)\right].\end{aligned} \tag{2.214}$$

Note that ζ_j is the zero of $s_{11}(\zeta)$ since $s_{11} = \hat{s}_{22}$ (see (2.43)). This closure relation shows that the discrete and continuous Zakharov–Shabat eigenstates

$$\left\{\phi_1(x,\xi),\phi_2(x,\xi),\xi \in \mathbb{R};\quad \phi_1(x,\zeta_j),\phi_2(x,\bar{\zeta}_j), 1 \le j \le N\right\} \tag{2.215}$$

form a complete set. So do the discrete and continuous adjoint Zakharov–Shabat eigenstates

$$\left\{\widehat{\psi}_1(x,\xi),\widehat{\psi}_2(x,\xi),\xi \in \mathbb{R};\quad \widehat{\psi}_1(x,\zeta_j),\widehat{\psi}_2(x,\bar{\zeta}_j), 1 \le j \le N\right\}. \tag{2.216}$$

The inner products between these Zakharov–Shabat eigenstates and their adjoint states can be read off from the closure relation (2.214) as well. For multiple zeros of $(\zeta_j, \bar{\zeta}_j)$, the closure relation is similar, except that the residues (2.202)–(2.203) need to be calculated differently due to the multiple poles involved.

2.6 Squared Eigenfunctions of the Zakharov–Shabat System

The concept of squared eigenfunctions arises when one considers the problem of what would happen when the potential is slightly changed. In this case, one would like to determine how the scattering data varies due to the change of the potential. The functions which provide this mapping are called "adjoint squared eigenfunctions." The inverse of this mapping provides how the potential changes when the scattering data is slightly varied. The functions which provide this mapping are called "squared eigenfunctions." Squared eigenfunctions and their adjoints are always in the form of quadratic combinations between Jost solutions and adjoint Jost solutions. For 2×2 scattering problems such as the Zakharov–Shabat system (2.2), adjoint Jost solutions are linearly related to the Jost solutions, and thus squared eigenfunctions become "squares" of Jost solutions themselves (hence the name "squared eigenfunctions"). Squared eigenfunctions and their adjoints are intimately related to various aspects of the integrable theory. For instance, these squared eigenfunctions are also eigenfunctions of the recursion operator of the integrable equation (see Sec. 2.9). Squared eigenfunctions also appear as self-consistent sources of integrable equations; see Zeng et al. (2000), Wang and Hu (2008) for instance. More importantly, squared eigenfunctions are essential to any studies of perturbed integrable equations (such as the soliton perturbation theory in Chapter 4). In these studies, the expansion of the perturbed solution over the set of squared eigenfunctions forms the basis of analysis.

The original approach for deriving squared eigenfunctions is the following (see Ablowitz et al. (1974), Kaup (1976a,1976b), Newell (1980)). First, one finds variations of the scattering data in terms of perturbations of the potential. The coefficients in these formulae are then the adjoint squared eigenfunctions. Next, one shows that these adjoint squared eigenfunctions are also eigenfunctions of an integrodifferential operator. Then one finds the adjoint of that operator, which will be the eigenvalue operator for the squared eigenfunctions (this operator is the recursion operator of the integrable equation). Then by trial and error, one finds eigenfunctions of this recursion operator, which are then the squared eigenfunctions. Next, the inner products between these squared eigenfunctions and their adjoints are calculated explicitly from the asymptotics of Jost solutions. After these inner products are known, one can construct what should be the closure relation. Lastly, one proves this closure relation either by the use of the Marchenko equations (Kaup (1976b)), or by contour integration (Gerdjikov and Khristov (1980), Gerdjikov (2005)).

This original way of deriving squared eigenfunctions can run into difficulties if structures of the squared eigenfunctions become complicated. In such cases, the integrodifferential operator for adjoint squared eigenfunctions may be hard to find. Determining squared eigenfunctions from the adjoint of that integrodifferential operator (the recursion operator) by trial and error could be even harder. In addition, calculating the inner products and proving the closure relation of these squared eigenfunctions also involve nontrivial effort. Below we derive squared eigenfunctions by a different and simpler method, which was used by Kaup and Lakoba (1996), Kaup et al. (1999), and Yang and Kaup (2009) for various integrable equations. In this method, one first calculates variations of the scattering data via variation of the potential, which yields the adjoint squared eigenfunctions. This step is the same as in the original method. Next, one calculates the variation of the potential via variations of the scattering data by the Riemann–Hilbert method. This step is new, and it directly yields the squared eigenfunctions. These two steps together will quickly give the inner products as well as the closure relation of squared eigenfunctions. This method simplifies and streamlines the derivation of squared eigenfunctions. In addition, it can treat complicated problems quite easily (see Sec. 3.8).

In this section, we use this method to derive squared eigenfunctions and their adjoints for the NLS equation. For the ease of presentation, we first assume that $\hat{s}_{22}$ and s_{22} have no zeros in their respective planes of analyticity, i.e., the scattering problem (2.2) has no discrete eigenvalues. This facilitates the derivation of squared eigenfunctions. The results for the general case of $\hat{s}_{22}$ and s_{22} having zeros will be given in the end of this section.

2.6.1 Variations of Scattering Data via Variation of Potential

In this subsection, we calculate variations of the scattering data in terms of variation of the potential, which will yield adjoint squared eigenfunctions. This is the easier step in this method and thus will be done first.

The basic idea is to take variation to the scattering equation (2.2) for Jost solutions, and then solve the variation equation. Let us first consider the scattering equation (2.2) for

Jost solutions Φ, which is

$$\Phi_x = -i\zeta\Lambda\Phi + Q\Phi. \tag{2.217}$$

Taking a variation to this equation, we get

$$\delta\Phi_x = -i\zeta\Lambda\,\delta\Phi + Q\,\delta\Phi + \delta Q\,\Phi. \tag{2.218}$$

Recalling that $\Phi \to E$ as $x \to -\infty$ (see Eqs. (2.10) and (2.15)), thus $\delta\Phi \to 0$ as $x \to -\infty$. Using this boundary condition, the solution to the above inhomogeneous equation (2.218) can be found by the method of variation of parameters as

$$\delta\Phi(x,\zeta) = \Phi(x,\zeta)\int_{-\infty}^{x} \Phi^{-1}(y,\zeta)\,\delta Q(y)\,\Phi(y,\zeta)\,dy. \tag{2.219}$$

Here

$$\delta Q = \begin{pmatrix} 0 & \delta u \\ -\delta u^* & 0 \end{pmatrix} \tag{2.220}$$

is a variation of the potential. Now we take the limit of $x \to +\infty$ in the above equation. Recalling that $\Phi = \Psi S$ and the asymptotics of $\Psi \to E$ as $x \to +\infty$, we see that $\Phi \to ES$, $\delta\Phi \to E\,\delta S$ as $x \to +\infty$. Thus in this limit, (2.219) becomes

$$\delta S(\xi) = \int_{-\infty}^{\infty} \Psi^{-1}(x,\xi)\,\delta Q(x)\,\Phi(x,\xi)\,dx, \qquad \xi \in \mathbb{R}, \tag{2.221}$$

i.e.,

$$\delta s_{ij}(\xi) = \int_{-\infty}^{\infty} \widehat{\psi}_i(x,\xi)\,\delta Q(x)\,\phi_j(x,\xi)\,dx, \qquad \xi \in \mathbb{R}. \tag{2.222}$$

Similarly, by working with the scattering equation (2.2) for Jost solutions Ψ, we can get

$$\delta\left(S^{-1}\right)(\xi) = -\int_{-\infty}^{\infty} \Phi^{-1}(x,\xi)\,\delta Q(x)\,\Psi(x,\xi)\,dx, \qquad \xi \in \mathbb{R}, \tag{2.223}$$

i.e.,

$$\delta\hat{s}_{ij}(\xi) = -\int_{-\infty}^{\infty} \widehat{\phi}_i(x,\xi)\,\delta Q(x)\,\psi_j(x,\xi)\,dx, \qquad \xi \in \mathbb{R}. \tag{2.224}$$

These formulae give variations of the scattering coefficients δs_{ij} and $\delta\hat{s}_{ij}$ via variation of the potential δQ. It is important to notice that these variation formulae are in terms of products between Jost solutions (Φ,Ψ) and their adjoints (Φ^{-1},Ψ^{-1}). Also notice that these variation formulae for the off-diagonal scattering coefficients δs_{21} and $\delta\hat{s}_{12}$ involve products of Jost solutions and their adjoints which are not simultaneously analytic in $\mathbb{C}_+$ or $\mathbb{C}_-$. For instance, the formula for δs_{21} involves the product between ϕ_1 and $\widehat{\psi}_2$, but ϕ_1 is analytic in $\mathbb{C}_+$, while $\widehat{\psi}_2$ is analytic in $\mathbb{C}_-$, thus their product is not analytic in either $\mathbb{C}_+$ or $\mathbb{C}_-$. To avoid this undesirable nonanalytic behavior, we define new scattering coefficients,

$$\rho \equiv \frac{s_{21}}{s_{22}}, \quad \hat{\rho} \equiv \frac{\hat{s}_{12}}{\hat{s}_{22}}. \tag{2.225}$$

Then using relations (2.221)–(2.223), variations for ρ and $\hat{\rho}$ can be obtained as

$$\delta\rho(\xi) = \frac{1}{s_{22}^2(\xi)} \int_{-\infty}^{\infty} \widehat{\psi}_2(x,\xi)\, \delta Q(x)\, [s_{22}(\xi)\phi_1(x,\xi) - s_{21}(\xi)\phi_2(x,\xi)]\, dx \tag{2.226}$$

and

$$\delta\hat{\rho}(\xi) = -\frac{1}{\hat{s}_{22}^2(\xi)} \int_{-\infty}^{\infty} \left[\hat{s}_{22}(\xi)\widehat{\phi}_1(x,\xi) - \hat{s}_{12}(\xi)\widehat{\phi}_2(x,\xi)\right] \delta Q(x)\, \psi_2(x,\xi)\, dx. \tag{2.227}$$

When ϕ_j and $\widehat{\phi}_j$ in these expressions are eliminated through the relation $\Phi = \Psi S$ and the facts $\det S = \det S^{-1} = 1$ utilized, the above variation relations then simplify to

$$\delta\rho(\xi) = \frac{1}{s_{22}^2(\xi)} \int_{-\infty}^{\infty} \widehat{\psi}_2(x,\xi)\, \delta Q(x)\, \psi_1(x,\xi)\, dx \tag{2.228}$$

and

$$\delta\hat{\rho}(\xi) = -\frac{1}{\hat{s}_{22}^2(\xi)} \int_{-\infty}^{\infty} \widehat{\psi}_1(x,\xi)\, \delta Q(x)\, \psi_2(x,\xi)\, dx. \tag{2.229}$$

Notice that functions $(\widehat{\psi}_2, \psi_1)$ are simultaneously analytic in $\mathbb{C}_-$, while $(\widehat{\psi}_1, \psi_2)$ are simultaneously analytic in $\mathbb{C}_+$; thus their products in the above variation equations are analytic in $\mathbb{C}_-$ and $\mathbb{C}_+$, respectively. In addition, these variation equations involve only Jost solutions Ψ and their adjoints Ψ^{-1} whose boundary conditions are all set at $x = +\infty$. These variation equations can be rewritten in the following more convenient form:

$$\delta\rho(\xi) = \frac{1}{s_{22}^2(\xi)} \left\langle \Omega^-(x,\xi), \begin{pmatrix} \delta u(x) \\ \delta u^*(x) \end{pmatrix} \right\rangle, \tag{2.230}$$

$$\delta\hat{\rho}(\xi) = \frac{1}{\hat{s}_{22}^2(\xi)} \left\langle \Omega^+(x,\xi), \begin{pmatrix} \delta u(x) \\ \delta u^*(x) \end{pmatrix} \right\rangle, \tag{2.231}$$

where $\xi \in \mathbb{R}$,

$$\Omega^- \equiv \begin{pmatrix} \psi_{21}\widehat{\psi}_{21} \\ -\psi_{11}\widehat{\psi}_{22} \end{pmatrix}, \quad \Omega^+ \equiv \begin{pmatrix} -\psi_{22}\widehat{\psi}_{11} \\ \psi_{12}\widehat{\psi}_{12} \end{pmatrix}, \tag{2.232}$$

the inner product is defined as

$$\langle \mathbf{f}, \mathbf{g} \rangle = \int_{-\infty}^{\infty} \mathbf{f}^T(x)\, \mathbf{g}(x)\, dx, \tag{2.233}$$

and

$$\Psi = (\psi_1, \psi_2) \equiv \begin{pmatrix} \psi_{11} & \psi_{12} \\ \psi_{21} & \psi_{22} \end{pmatrix}, \quad \Psi^{-1} = \begin{pmatrix} \widehat{\psi}_1 \\ \widehat{\psi}_2 \end{pmatrix} \equiv \begin{pmatrix} \widehat{\psi}_{11} & \widehat{\psi}_{12} \\ \widehat{\psi}_{21} & \widehat{\psi}_{22} \end{pmatrix}. \tag{2.234}$$

Formulae (2.230)–(2.231) give variations of the scattering data via variation of the potential, and the functions Ω^- and Ω^+ in these expansions are the adjoint squared eigenfunctions of the Zakharov–Shabat system. These adjoint squared eigenfunctions are products between Jost solutions and their adjoints. More importantly, they are analytic in the lower and upper half planes $\mathbb{C}_\mp$, respectively (this analyticity is indicated by their superscripts "$\mp$"). These analytic properties will be valuable when we derive their closure relations in the presence of discrete scattering data (i.e., when $\hat{s}_{22}$ and s_{22} have zeros).

It is noted that since the Zakharov–Shabat system is a 2×2 system, the adjoint Jost solution Ψ^{-1} is linearly related to Ψ itself. Specifically,

$$\begin{pmatrix} \widehat{\psi}_{11} & \widehat{\psi}_{12} \\ \widehat{\psi}_{21} & \widehat{\psi}_{22} \end{pmatrix} = \begin{pmatrix} \psi_{22} & -\psi_{12} \\ -\psi_{21} & \psi_{11} \end{pmatrix}. \tag{2.235}$$

In this case, the adjoint squared eigenfunctions Ω^- and Ω^+ become

$$\Omega^- = -\begin{pmatrix} \psi_{21}^2 \\ \psi_{11}^2 \end{pmatrix}, \quad \Omega^+ = -\begin{pmatrix} \psi_{22}^2 \\ \psi_{12}^2 \end{pmatrix}, \tag{2.236}$$

which are simple squares of Jost solutions Ψ.

2.6.2 Variation of Potential via Variations of Scattering Data

In this subsection, we derive variation of the potential via variations of the scattering data, which will yield squared eigenfunctions. This is the second and harder step of our two-step procedure and will be done using the Riemann–Hilbert method.

Our basic idea is to take a variation to the Riemann–Hilbert problem (2.44) and then solve it. This will yield the formula for the variation of the potential, because the potential can be reconstructed from the Riemann–Hilbert solutions $P^\pm$ (which is the key idea of the inverse scattering transform method; see Eq. (2.48)). However, our previous variation formulae (2.230)–(2.231) are for the new scattering coefficients $(\rho, \hat{\rho})$, while the scattering data in the Riemann–Hilbert problem (2.44) is $(s_{21}, \hat{s}_{12})$, which are inconsistent. To fix this problem, we modify the Riemann–Hilbert problem (2.44) so that the scattering data in it becomes $(\rho, \hat{\rho})$. For this purpose, we define the following matrix functions:

$$F^+ = P^+ \operatorname{diag}\left(1, \frac{1}{\hat{s}_{22}}\right), \qquad F^- = (P^-)^{-1} \operatorname{diag}(1, s_{22}). \tag{2.237}$$

When $\hat{s}_{22}$ and s_{22} have no zeros in $\mathbb{C}_+$ and $\mathbb{C}_-$, then $F^{\pm}$ as well as $(F^{\pm})^{-1}$ are still analytic in $\mathbb{C}_{\pm}$. On the real line, since $P^{\pm}$ are related by (2.44), $F^{\pm}$ are then related by

$$F^{+}(\zeta)=F^{-}(\zeta)\widetilde{G}(\zeta), \qquad \zeta\in\mathbb{R}, \tag{2.238}$$

where

$$\widetilde{G}=\operatorname{diag}\left(1,\frac{1}{s_{22}}\right)G\operatorname{diag}\left(1,\frac{1}{\hat{s}_{22}}\right)=E\begin{pmatrix}1 & \hat{\rho}\\ \rho & 1+\rho\hat{\rho}\end{pmatrix}E^{-1}. \tag{2.239}$$

In deriving (2.239), the fact that $\det S=1$ has been used. Equation (2.238) defines a new Riemann–Hilbert problem which contains only the new scattering coefficients $(\rho,\hat{\rho})$ now. The normalization condition for this Riemann–Hilbert problem is still canonical, i.e.,

$$F^{\pm}(\zeta)\to I, \qquad \zeta\to\infty, \tag{2.240}$$

since $P^{\pm}\to I$ and $s_{22},\hat{s}_{22}\to 1$ as $\zeta\to\infty$ (see Eqs. (2.46) and (2.53)–(2.54)). Notice that for this new Riemann–Hilbert problem (2.238), $\det\widetilde{G}=1$. This contrasts the original Riemann–Hilbert problem (2.44) for $P^{\pm}$, where $\det G\neq 1$ in general.

Next we take variation of this new Riemann–Hilbert problem (2.238) and get

$$\delta F^{+}=\delta F^{-}\widetilde{G}+F^{-}\delta\widetilde{G}, \qquad \zeta\in\mathbb{R}. \tag{2.241}$$

Utilizing (2.238), we can rewrite this equation as

$$\delta F^{+}(F^{+})^{-1}-\delta F^{-}(F^{-})^{-1}=F^{-}\delta\widetilde{G}(F^{+})^{-1}, \qquad \zeta\in\mathbb{R}, \tag{2.242}$$

which defines yet another Riemann–Hilbert problem for $\delta F F^{-1}$. In view of the canonical normalization condition (2.240) for $F^{\pm}$, and utilizing the Plemelj formula (2.62), the general solution to this Riemann–Hilbert problem (2.242) is

$$\left(\delta F F^{-1}\right)(x,\zeta)=\frac{1}{2\pi i}\int_{-\infty}^{\infty}\frac{\Pi(x,\xi)}{\xi-\zeta}\,d\xi, \tag{2.243}$$

where

$$\Pi(x,\xi)\equiv F^{-}(x,\xi)\delta\widetilde{G}(x,\xi)(F^{+})^{-1}(x,\xi), \qquad \xi\in\mathbb{R}. \tag{2.244}$$

Now we consider the large-ζ asymptotics of this solution. Notice from expansions (2.47), (2.51), (2.54) and definition (2.237) that as $\zeta\to\infty$,

$$F^{+}(x,\zeta)\to\begin{pmatrix}1+O(\frac{1}{\zeta}) & \frac{u(x)}{2i\zeta}\\ \frac{u^{*}(x)}{2i\zeta} & 1+O(\frac{1}{\zeta})\end{pmatrix} \tag{2.245}$$

and

$$\delta F^{+}(x,\zeta) \rightarrow \begin{pmatrix} O(\frac{1}{\zeta}) & \frac{\delta u(x)}{2i\zeta} \\ \frac{\delta u^{*}(x)}{2i\zeta} & O(\frac{1}{\zeta}) \end{pmatrix}. \tag{2.246}$$

In addition, it is easy to see that

$$\int_{-\infty}^{\infty} \frac{\Pi(x,\xi)}{\xi-\zeta} d\xi \longrightarrow -\frac{1}{\zeta} \int_{-\infty}^{\infty} \Pi(x,\xi)\, d\xi, \qquad \zeta \to \infty. \tag{2.247}$$

When these large-ζ expansions are substituted into Eq. (2.243), at $O(\zeta^{-1})$ we get

$$\delta u(x) = -\frac{1}{\pi} \int_{-\infty}^{\infty} \Pi_{12}(x,\xi)\, d\xi \tag{2.248}$$

and

$$\delta u^{*}(x) = -\frac{1}{\pi} \int_{-\infty}^{\infty} \Pi_{21}(x,\xi)\, d\xi. \tag{2.249}$$

Now we calculate the elements Π_{12} and Π_{21}. From Eq. (2.239), we see that

$$\delta \widetilde{G} = E \begin{pmatrix} 0 & \delta\hat{\rho} \\ \delta\rho & \rho\,\delta\hat{\rho} + \hat{\rho}\,\delta\rho \end{pmatrix} E^{-1}. \tag{2.250}$$

Then inserting the $F^{\pm}$ definitions (2.237), $P^{\pm}$ expressions (2.25), (2.34), the scattering relation (2.17), as well as the definition $\Phi = J_{-}E$ into (2.244), we get

$$\Pi = \Phi(H_1 + H_2 S)^{-1} \mathrm{diag}(1, s_{22}) E^{-1}\, \delta\widetilde{G}\, E\, \mathrm{diag}(1, \hat{s}_{22})(H_1 + S^{-1}H_2)^{-1}\Phi^{-1},$$

which simplifies to

$$\Pi = \Phi \begin{pmatrix} 0 & \delta\hat{\rho} \\ \delta\rho & 0 \end{pmatrix} \Phi^{-1}, \tag{2.251}$$

i.e.,

$$\Pi = \phi_2 \widehat{\phi}_1\, \delta\rho + \phi_1 \widehat{\phi}_2\, \delta\hat{\rho}. \tag{2.252}$$

Relation (2.252), together with Eqs. (2.248)–(2.249), shows that variation of the potential $(\delta u, \delta u^{*})^{T}$ can be expanded into products between Jost solutions Φ and their adjoints Φ^{-1} whose boundary conditions are both set at $x = -\infty$. Notice that these products in (2.252) are analytic in either $\mathbb{C}_{+}$ or $\mathbb{C}_{-}$.

Now we calculate explicit expressions for $(\delta u, \delta u^{*})^{T}$. Let

$$\Phi = (\phi_1, \phi_2) \equiv \begin{pmatrix} \phi_{11} & \phi_{12} \\ \phi_{21} & \phi_{22} \end{pmatrix}, \quad \Phi^{-1} = \begin{pmatrix} \widehat{\phi}_1 \\ \widehat{\phi}_2 \end{pmatrix} \equiv \begin{pmatrix} \widehat{\phi}_{11} & \widehat{\phi}_{12} \\ \widehat{\phi}_{21} & \widehat{\phi}_{22} \end{pmatrix}. \tag{2.253}$$

Inserting (2.252)–(2.253) into (2.248)–(2.249), we readily find that variation of the potential is related to variations of the scattering data as

$$\begin{pmatrix} \delta u(x) \\ \delta u^*(x) \end{pmatrix} = -\frac{1}{\pi}\int_{-\infty}^{\infty} \left[Z^-(x,\xi)\,\delta\rho(\xi) + Z^+(x,\xi)\,\delta\hat{\rho}(\xi)\right] d\xi, \tag{2.254}$$

where column vectors

$$Z^- \equiv \begin{pmatrix} \phi_{12}\widehat{\phi}_{12} \\ \phi_{22}\widehat{\phi}_{11} \end{pmatrix}, \quad Z^+ \equiv \begin{pmatrix} \phi_{11}\widehat{\phi}_{22} \\ \phi_{21}\widehat{\phi}_{21} \end{pmatrix} \tag{2.255}$$

are the squared eigenfunctions we are seeking. Note that Z^- and Z^+ are analytic in $\mathbb{C}_-$ and $\mathbb{C}_+$, respectively. Note also that the two component equations in (2.254) for δu and its complex conjugate δu^* are consistent with each other due to the involution properties (2.56) and (2.58). For the 2×2 Zakharov–Shabat system, Φ^{-1} is linearly related to Φ, and thus these squared eigenfunctions reduce to

$$Z^- = \begin{pmatrix} -\phi_{12}^2 \\ \phi_{22}^2 \end{pmatrix}, \quad Z^+ = \begin{pmatrix} \phi_{11}^2 \\ -\phi_{21}^2 \end{pmatrix}, \tag{2.256}$$

which are simple squares of the Jost solutions Φ.

2.6.3 Inner Products and Closure Relation of Squared Eigenfunctions

After variations of the scattering data via variation of the potential (2.230)–(2.231) and its converse (2.254) have been obtained, we can then read off the inner products between squared eigenfunctions and establish their closure relation without any extra effort. This is one of the many advantages of the present method. To get the closure relation, we insert Eqs. (2.230)–(2.231) into the potential expansion (2.254). Exchanging the order of integration, we get

$$\begin{pmatrix} \delta u(x) \\ \delta u^*(x) \end{pmatrix} = \int_{-\infty}^{\infty} A(x,y) \begin{pmatrix} \delta u(y) \\ \delta u^*(y) \end{pmatrix} dy, \tag{2.257}$$

where

$$A(x,y) = -\frac{1}{\pi}\int_{-\infty}^{\infty} \left[\frac{1}{s_{22}^2(\xi)} Z^-(x,\xi)\,\Omega^{-T}(y,\xi) + \frac{1}{\hat{s}_{22}^2(\xi)} Z^+(x,\xi)\,\Omega^{+T}(y,\xi)\right] d\xi.$$

Since $(\delta u, \delta u^*)^T$ are arbitrary localized functions, in order for (2.257) to always hold, $A(x,y)$ must be the delta function $\delta(x-y)I$. Thus we have

$$-\frac{1}{\pi}\int_{-\infty}^{\infty}\left[\frac{1}{s_{22}^2(\xi)}Z^-(x,\xi)\Omega^{-T}(y,\xi)+\frac{1}{\hat{s}_{22}^2(\xi)}Z^+(x,\xi)\Omega^{+T}(y,\xi)\right]d\xi=\delta(x-y)I, \tag{2.258}$$

where $\delta(x)$ is the Dirac delta function. This equation is the closure relation. Here, discrete-spectrum contributions are absent due to our assumption that $\hat{s}_{22}$, s_{22} have no zeros.

To obtain inner products between squared eigenfunctions and their adjoints, we do the opposite; i.e., we insert the potential expansion (2.254) into the scattering coefficient formulae (2.230)–(2.231). Exchanging the order of integration, we get

$$\begin{pmatrix}\delta\rho(\xi)\\ \delta\hat{\rho}(\xi)\end{pmatrix}=\int_{-\infty}^{\infty}B(\xi,\xi')\begin{pmatrix}\delta\rho(\xi')\\ \delta\hat{\rho}(\xi')\end{pmatrix}d\xi', \tag{2.259}$$

where

$$B(\xi,\xi')=\begin{pmatrix}-\frac{1}{\pi s_{22}^2(\xi)}\langle\Omega^-(x,\xi),Z^-(x,\xi')\rangle & -\frac{1}{\pi s_{22}^2(\xi)}\langle\Omega^-(x,\xi),Z^+(x,\xi')\rangle\\ -\frac{1}{\pi \hat{s}_{22}^2(\xi)}\langle\Omega^+(x,\xi),Z^-(x,\xi')\rangle & -\frac{1}{\pi \hat{s}_{22}^2(\xi)}\langle\Omega^+(x,\xi),Z^+(x,\xi')\rangle\end{pmatrix}.$$

Since the relation (2.259) is valid for arbitrary functions of $\delta\rho(\xi)$ and $\delta\hat{\rho}(\xi)$, and thus

$$B(\xi,\xi')=\delta(\xi-\xi')I. \tag{2.260}$$

Consequently, we get the inner products and orthogonality relations between squared eigenfunctions and their adjoints as

$$\langle\Omega^-(x,\xi),Z^-(x,\xi')\rangle=-\pi s_{22}^2(\xi)\,\delta(\xi-\xi'), \tag{2.261}$$

$$\langle\Omega^+(x,\xi),Z^+(x,\xi')\rangle=-\pi \hat{s}_{22}^2(\xi)\,\delta(\xi-\xi'), \tag{2.262}$$

$$\langle\Omega^-(x,\xi),Z^+(x,\xi')\rangle=\langle\Omega^+(x,\xi),Z^-(x,\xi')\rangle=0 \tag{2.263}$$

for any $\xi,\xi'\in\mathbb{R}$.

2.6.4 Extension to the General Case

In the previous subsections, the squared eigenfunctions and their closure relation were established under the assumption that $\hat{s}_{22}$ and s_{22} have no zeros, i.e., the scattering equation (2.2) has no discrete eigenvalues. In this subsection, we extend those results to the general case where $\hat{s}_{22}$ and s_{22} have zeros in their respective planes of analyticity $\mathbb{C}_\pm$. Suppose all the zeros of $\hat{s}_{22}$ and s_{22} are $\zeta_j \in \mathbb{C}_+$ and $\bar{\zeta}_j \in \mathbb{C}_-$, $j = 1, 2, \ldots, N$, respectively. For simplicity, we also assume that all these zeros are simple.

In this general case with zeros, the sets of (continuous) squared eigenfunctions $\{Z^{\pm}(x,\xi), \xi \in \mathbb{R}\}$ and adjoint squared eigenfunctions $\{\Omega^{\pm}(x,\xi), \xi \in \mathbb{R}\}$ are no longer complete, i.e., the closure relation (2.258) does not hold any more, and contributions from the discrete spectrum must be included now. These discrete-spectrum contributions turn out to be nothing but the residues of integrand functions in (2.258) at poles ζ_j and $\bar{\zeta}_j$ (notice that zeros of $\hat{s}_{22}$ and s_{22} are poles of those integrand functions). This behavior is the same as the closure of Zakharov–Shabat eigenstates in Sec. 2.5. At simple zeros $(\zeta_j, \bar{\zeta}_j)$, the integrand functions in (2.258) have double poles. After their residues are calculated, the closure relation with discrete-spectrum contributions is then

$$
\begin{aligned}
&-\frac{1}{\pi}\int_{-\infty}^{\infty}\left[\frac{1}{s_{22}^2(\xi)}Z^-(x,\xi)\,\Omega^{-T}(y,\xi)+\frac{1}{\hat{s}_{22}^2(\xi)}Z^+(x,\xi)\,\Omega^{+T}(y,\xi)\right]d\xi \\
&\quad -\sum_{j=1}^{N}\frac{2i}{s_{22}'^2(\bar{\zeta}_j)}\left[Z^-(x,\bar{\zeta}_j)\,\Theta^{-T}(y,\bar{\zeta}_j)+\dot{Z}^-(x,\bar{\zeta}_j)\,\Omega^{-T}(y,\bar{\zeta}_j)\right] \\
&\quad +\sum_{j=1}^{N}\frac{2i}{\hat{s}_{22}'^2(\zeta_j)}\left[Z^+(x,\zeta_j)\,\Theta^{+T}(y,\zeta_j)+\dot{Z}^+(x,\zeta_j)\,\Omega^{+T}(y,\zeta_j)\right]=\delta(x-y)I,
\end{aligned}
\tag{2.264}
$$

where

$$
\Theta^-(x,\bar{\zeta}_j)\equiv\dot{\Omega}^-(x,\bar{\zeta}_j)-\frac{s_{22}''(\bar{\zeta}_j)}{s_{22}'(\bar{\zeta}_j)}\,\Omega^-(x,\bar{\zeta}_j), \tag{2.265}
$$

$$
\Theta^+(x,\zeta_j)\equiv\dot{\Omega}^+(x,\zeta_j)-\frac{\hat{s}_{22}''(\zeta_j)}{\hat{s}_{22}'(\zeta_j)}\,\Omega^+(x,\zeta_j), \tag{2.266}
$$

and the dot above a variable represents its derivative with respect to ζ. Here $\{Z^-(x,\bar{\zeta}_j), \dot{Z}^-(x,\bar{\zeta}_j), Z^+(x,\zeta_j), \dot{Z}^+(x,\zeta_j)\}$ are discrete squared eigenfunctions, and $\{\Theta^-(x,\bar{\zeta}_j), \Omega^-(x,\bar{\zeta}_j), \Theta^+(x,\zeta_j), \Omega^+(x,\zeta_j)\}$ are discrete adjoint squared eigenfunctions. The inner products between these discrete and continuous squared eigenfunctions and their adjoints can be read off from this closure relation.

The above general closure relation for squared eigenfunctions can be rigorously proved in several different ways. One way is to repeat the calculations in the previous subsections, but with variations of the discrete scattering data taken into consideration. This way of proof was given by Newell (1980) for the Zakharov–Shabat system (2.2), and by Kaup and Yang (2009) for a third-order scattering operator. An easier way to prove this general closure is to find an operator whose eigenstates are these squared eigenfunctions (this operator is the recursion operator). Then the closure of eigenstates of this recursion operator (including discrete-eigenfunction contributions) can be established by the contour integration method as used in Sec. 2.5 (see Gerdjikov and Khristov (1980), Gerdjikov

(2005)). To execute this second way of proof for the Zakharov–Shabat system (2.2), we first notice that the recursion operator for this system is an integrodifferential operator which will be presented in Sec. 2.8. Even though this operator contains integral terms, these integral terms vanish as $x \to \pm\infty$ (for localized potentials). Thus one can still develop a scattering theory, define Jost solutions, and obtain their analytic properties in a way very similar to what we have done to the Zakharov–Shabat system. In this scattering theory, (Z^+, Z^-) plays the role of Φ and $(-\Omega^+, -\Omega^-)^T$ plays the role of Ψ^{-1}. Then repeating the closure proof of Zakharov–Shabat eigenstates in Sec. 2.5, one can obtain a closure relation for eigenstates of this recursion operator, where the continuous-integral part is precisely the left side of Eq. (2.258), and the discrete-eigenstate part is precisely the residues of integrands in that continuous integral. In other words, one gets the general closure relation (2.264) for squared eigenfunctions with discrete-spectrum contributions.

One may notice that in the closure relation (2.264), for each pair of the discrete eigenvalues $(\zeta_k, \bar{\zeta}_k)$, there are four discrete terms present. These four terms correspond to variations in the two complex parameters ζ_k and c_k (four real parameters) in the NLS solitons (2.122). When the solution is the single NLS soliton (2.128), then these four terms also correspond to variations in the amplitude, velocity, initial position, and initial phase of this single soliton.

2.7 Squared Eigenfunctions and the Linearization Operator

In this section, we explore the relation between squared eigenfunctions and the linearization operator of the NLS equation. Suppose a solution $u(x,t)$ of the NLS equation is perturbed to another solution of the NLS equation $u(x,t) + \delta u(x,t)$, where $\delta u \ll u$. Substituting the perturbed solution into the NLS equation (2.1) and neglecting higher-order terms in δu, then the linearized NLS equation for the infinitesimal perturbation $(\delta u, \delta u^*)^T$ is

$$\mathcal{L} \begin{pmatrix} \delta u(x,t) \\ \delta u^*(x,t) \end{pmatrix} = 0, \tag{2.267}$$

where

$$\mathcal{L} = \begin{pmatrix} i\partial_t + \partial_{xx} + 4|u|^2 & 2u^2 \\ -2u^{*2} & i\partial_t - \partial_{xx} - 4u^2 \end{pmatrix} \tag{2.268}$$

is the linearization operator. Notice that this operator contains time derivatives. In our treatment of squared eigenfunctions in the previous section, those squared eigenfunctions were defined in terms of $\Phi = J_- e^{-i\zeta\Lambda x}$ and its inverse, and the time was a dummy variable.

The Jost solutions Φ satisfy the Zakharov–Shabat equations (2.2), but they do not satisfy the temporal equation (2.3) in the Lax pair.

To better explore the relationship between squared eigenfunctions and the linearization operator $\mathcal{L}$, we introduce time-dependent Jost solutions

$$\Phi^{(t)} \equiv \Phi e^{-2i\zeta^2\Lambda t} = J_- e^{-i\zeta\Lambda x - 2i\zeta^2\Lambda t}. \tag{2.269}$$

These time-dependent Jost solutions satisfy both the scattering and temporal equations (2.2)–(2.3) of the Lax pair, because J_- satisfies the transformed Lax pair (2.6)–(2.7). Using these time-dependent Jost functions, we define time-dependent squared eigenfunctions analogously to (2.255), i.e.,

$$Z^{-(t)} \equiv \begin{pmatrix} \phi_{12}^{(t)}\widehat{\phi}_{12}^{(t)} \\ \phi_{22}^{(t)}\widehat{\phi}_{11}^{(t)} \end{pmatrix}, \quad Z^{+(t)} \equiv \begin{pmatrix} \phi_{11}^{(t)}\widehat{\phi}_{22}^{(t)} \\ \phi_{21}^{(t)}\widehat{\phi}_{21}^{(t)} \end{pmatrix}. \tag{2.270}$$

Here $\phi_{ij}^{(t)}$ are the matrix elements of $\Phi^{(t)}$, and $\widehat{\phi}_{ij}^{(t)}$ are the matrix elements of $(\Phi^{(t)})^{-1}$. It is easy to see that these time-dependent squared eigenfunctions are related to the original ones as

$$Z^{-(t)} = Z^- e^{4i\zeta^2 t}, \quad Z^{+(t)} = Z^+ e^{-4i\zeta^2 t}, \tag{2.271}$$

i.e., they differ by only a time-dependent factor. We also recall from Eqs. (2.104)–(2.105) that

$$\delta\rho(t,\zeta) = \delta\rho(0,\zeta)e^{4i\zeta^2 t}, \quad \delta\hat{\rho}(t,\zeta) = \delta\hat{\rho}(0,\zeta)e^{-4i\zeta^2 t}. \tag{2.272}$$

Now we assume the Zakharov–Shabat system has no discrete spectrum (for the convenience of discussion). Then when the time variable in the potential-variation formula (2.254) is restored and the above two relations utilized, we get the new expansion in terms of the time-dependent squared eigenfunctions:

$$\begin{pmatrix} \delta u(x,t) \\ \delta u^*(x,t) \end{pmatrix} = -\frac{1}{\pi}\int_{-\infty}^{\infty}\left[Z^{-(t)}(x,t,\xi)\,\delta\rho(0,\xi) + Z^{+(t)}(x,t,\xi)\,\delta\hat{\rho}(0,\xi)\right]d\xi. \tag{2.273}$$

Substituting this expansion into the linearized equation (2.267), we obtain

$$\int_{-\infty}^{\infty}\left[\mathcal{L}Z^{-(t)}(x,t,\xi)\,\delta\rho(0,\xi) + \mathcal{L}Z^{+(t)}(x,t,\xi)\,\delta\hat{\rho}(0,\xi)\right]d\xi = 0. \tag{2.274}$$

It is important to realize that initial variations of the scattering data $(\delta\rho,\delta\hat{\rho})(0,\xi)$ are arbitrary and linearly independent. Then in order for the above equation to hold, we must have

$$\mathcal{L}Z^{-(t)}(x,t,\xi) = \mathcal{L}Z^{+(t)}(x,t,\xi) = 0 \quad \text{for all } \xi \in \mathbb{R}. \tag{2.275}$$

In other words, the time-dependent squared eigenfunctions $\{Z^{-(t)}, Z^{+(t)}\}$ satisfy the linearized equation of the NLS equation. This is an important property of squared eigenfunctions, and it will play an important role in studies of the perturbed NLS equation (see Chapter 4).

Corresponding to these time-dependent squared eigenfunctions, we can also define time-dependent adjoint squared eigenfunctions analogously to (2.232) as

$$\Omega^{-(t)} \equiv \begin{pmatrix} \psi_{21}^{(t)}\widehat{\psi}_{21}^{(t)} \\ -\psi_{11}^{(t)}\widehat{\psi}_{22}^{(t)} \end{pmatrix}, \quad \Omega^{+(t)} \equiv \begin{pmatrix} -\psi_{22}^{(t)}\widehat{\psi}_{11}^{(t)} \\ \psi_{12}^{(t)}\widehat{\psi}_{12}^{(t)} \end{pmatrix}, \tag{2.276}$$

where $\psi_{ij}^{(t)}$ and $\widehat{\psi}_{ij}^{(t)}$ are matrix elements of the time-dependent Jost solutions

$$\Psi^{(t)} \equiv \Psi e^{-2i\zeta^2\Lambda t} = J_+ e^{-i\zeta\Lambda x - 2i\zeta^2\Lambda t} \tag{2.277}$$

and their inverse $(\Psi^{(t)})^{-1}$. These time-dependent adjoint squared eigenfunctions are related to the old ones only by a time-dependent factor:

$$\Omega^{-(t)} = \Omega^- e^{-4i\zeta^2 t}, \quad \Omega^{+(t)} = \Omega^+ e^{4i\zeta^2 t}. \tag{2.278}$$

Then by a general argument in Yang and Kaup (2009), we can show that these time-dependent adjoint squared eigenfunctions satisfy the adjoint linearized equation of the NLS equation, i.e.,

$$\mathcal{L}^A \Omega^{-(t)}(x,t,\xi) = \mathcal{L}^A \Omega^{+(t)}(x,t,\xi) = 0 \quad \text{for all } \xi \in \mathbb{R}, \tag{2.279}$$

where the adjoint linearization operator $\mathcal{L}^A$ is

$$\mathcal{L}^A = \begin{pmatrix} -i\partial_t + \partial_{xx} + 2|u|^2 & -u^{*2} \\ u^2 & -i\partial_t - \partial_{xx} - 2u^2 \end{pmatrix} \tag{2.280}$$

under the inner product definition (2.233). This is expected in view of the closure relation (2.258) as well as the relations (2.271) and (2.278), where one can see that $(\Omega^{-(t)}, \Omega^{+(t)})$ should be the adjoint functions of $(Z^{-(t)}, Z^{+(t)})$. Thus if $(Z^{-(t)}, Z^{+(t)})$ satisfy the linearized NLS equation, then $(\Omega^{-(t)}, \Omega^{+(t)})$ should satisfy the adjoint linearized NLS equation.

In relations (2.275) and (2.279), when ξ is taken as the zeros $(\zeta_j, \bar{\zeta}_j)$ of $(\hat{s}_{22}, s_{22})$ off the real axis $\mathbb{R}$, they are still satisfied. This means that time-dependent discrete squared eigenfunctions and their adjoints also satisfy the linearized NLS equation and its adjoint equation, respectively.

It is important to notice that the above proof of time-dependent squared eigenfunctions satisfying the linearized NLS equation apparently works also for any integrable equation whose scattering operator in the Lax pair is the Zakharov–Shabat system (2.2). All these integrable equations originating from the same Zakharov–Shabat scattering problem will be considered in the next section.

2.8 Recursion Operator and the AKNS Hierarchy

In the theory of integrable equations, an important task is to construct new equations which are solvable through the inverse scattering transform method. An important finding by Ablowitz, Kaup, Newell and Segur (1974) is that, associated with each scattering operator, one can actually construct an infinite hierarchy of integrable equations. For the Zakharov–Shabat system (2.2), they obtained this hierarchy which contains the NLS equation (2.1) as a particular case. This hierarchy is now called the Ablowitz–Kaup–Newell–Segur (AKNS) hierarchy, and it will be derived in this section. Our method will be a modification of the original one used by Ablowitz et al. (1974). Similar methods have been employed for the derivation of other integrable hierarchies associated with other scattering operators (see Fokas and Fuchssteiner (1981), Ma et al. (1996), Ma and Zhou (1999), Chen (2006) for examples).

To make the discussion more general, we write the Zakharov–Shabat system (2.2) in the following generalized form:

$$Y_x = MY, \quad M = \begin{pmatrix} -i\zeta & q \\ r & i\zeta \end{pmatrix}, \tag{2.281}$$

where (q,r) are functions of (x,t). Suppose the temporal equation for Y is

$$Y_t = NY, \quad N = \begin{pmatrix} A & B \\ C & D \end{pmatrix}; \tag{2.282}$$

then by cross-differentiation, we get the zero curvature equation,

$$M_t - N_x + [M,N] = 0, \tag{2.283}$$

where $[M,N] = MN - NM$ is the commutator. This equation is the compatibility condition for the linear spatial and temporal equations (2.281)–(2.282) (i.e., the Lax pair). The above zero-curvature equation gives the following four relations:

$$-A_x - rB + qC = 0, \tag{2.284}$$

$$q_t - B_x - 2i\zeta B - q(A-D) = 0, \tag{2.285}$$

$$r_t - C_x + 2i\zeta C + r(A-D) = 0, \tag{2.286}$$

$$-D_x + rB - qC = 0. \tag{2.287}$$

Here we have assumed that the spectral parameter ζ is time independent. From (2.284) and (2.287) we get

$$A = \partial^{-1}(qC - rB) + A_0, \tag{2.288}$$

$$D = \partial^{-1}(rB - qC) + D_0, \tag{2.289}$$

where ∂^{-1} is an antiderivative in x which can be taken as either $\partial^{-1} = \int_{-\infty}^{x} dy$ or $\partial^{-1} = -\int_{x}^{\infty} dy$, and (A_0, D_0) are x-independent. Inserting the above two equations into (2.285) and (2.286), we get

$$\begin{pmatrix} q \\ -r \end{pmatrix}_t = i(2\zeta - L_R)\begin{pmatrix} B \\ C \end{pmatrix} + (A_0 - D_0)\begin{pmatrix} q \\ r \end{pmatrix}, \tag{2.290}$$

where

$$L_R = i\begin{pmatrix} \partial - 2q\,\partial^{-1}r & 2q\,\partial^{-1}q \\ -2r\,\partial^{-1}r & -\partial + 2r\,\partial^{-1}q \end{pmatrix} \tag{2.291}$$

is called the recursion operator, and $\partial = \partial/\partial x$. To derive evolution equations, we expand

$$\begin{pmatrix} B \\ C \end{pmatrix} = \sum_{j=1}^{n} \begin{pmatrix} b_j \\ c_j \end{pmatrix} (2\zeta)^{n-j} \tag{2.292}$$

and take

$$D_0 = -A_0 = i\zeta(2\zeta)^{n-1}. \tag{2.293}$$

Inserting (2.292)–(2.293) into (2.290) and equating terms of the same power in 2ζ, we then get the following equations:

$$\begin{pmatrix} q \\ -r \end{pmatrix}_t = -iL_R\begin{pmatrix} b_n \\ c_n \end{pmatrix} \tag{2.294}$$

and

$$\begin{pmatrix} b_1 \\ c_1 \end{pmatrix} = \begin{pmatrix} q \\ r \end{pmatrix}, \quad \begin{pmatrix} b_j \\ c_j \end{pmatrix} = L_R\begin{pmatrix} b_{j-1} \\ c_{j-1} \end{pmatrix}, \quad j = 2, \ldots, n. \tag{2.295}$$

From these equations, we obtain the AKNS hierarchy

$$i\begin{pmatrix} q \\ -r \end{pmatrix}_t = L_R^n\begin{pmatrix} q \\ r \end{pmatrix}, \quad n = 1, 2, \ldots. \tag{2.296}$$

Taking $n = 2$ and $r = -q^*$, then (2.296) becomes the NLS equation. But the hierarchy (2.296) contains many more equations when one takes different integers n or imposes different relations between q and r. For instance, when $n = 3$ and $r = -1$, (2.296) gives the KdV equation

$$q_t + q_{xxx} + 6qq_x = 0. \tag{2.297}$$

When $n=3$ and $r=-q$, (2.296) gives the modified KdV equation

$$q_t+q_{xxx}+6q^2q_x=0. \tag{2.298}$$

And when $n=3$ and $r=-q^*$, (2.296) gives the complex modified KdV equation

$$q_t+q_{xxx}+6|q|^2q_x=0. \tag{2.299}$$

Consequently, the hierarchies of the NLS equation, the KdV equation, the modified KdV equation, and the complex modified KdV equation are all special cases of the AKNS hierarchy. As an example, by taking $r=-1$ and odd integers of n, i.e., $n=2m+1$, one can easily show by induction that the AKNS hierarchy (2.296) reduces to

$$\begin{pmatrix} q \\ 1 \end{pmatrix}_t = -\begin{pmatrix} K^m q_x \\ 0 \end{pmatrix}, \quad m=1,2,\ldots, \tag{2.300}$$

i.e.,

$$q_t+K^m q_x=0, \quad m=1,2,\ldots, \tag{2.301}$$

which is the KdV hierarchy. Here the recursion operator K is

$$K=\partial^2+4q+2q_x\partial^{-1}. \tag{2.302}$$

In the above examples of the AKNS hierarchy (2.296), some relations were always imposed on the potentials q and r. This does not have to be. In other words, one can also leave q and r independent. In that case, coupled integrable equations for (q,r) would be obtained. For instance, if we take $n=2$, then the AKNS hierarchy (2.296) becomes the following coupled Schrödinger equations:

$$iq_t+q_{xx}-2q^2r=0, \tag{2.303}$$

$$ir_t-r_{xx}+2r^2q=0. \tag{2.304}$$

The NLS equation (2.1) would result if one imposes $r=-q^*$. But if q and r are left independent, then the above equations can exhibit novel solution dynamics which are very different from that in the NLS equation (2.1). This novel solution dynamics will be presented in Sec. 2.10.

It should be noted that the AKNS hierarchy (2.296) can be further generalized. Indeed, for any polynomial function $\Omega(k)$, the equation

$$i\begin{pmatrix} q \\ -r \end{pmatrix}_t = \Omega(L_R)\begin{pmatrix} q \\ r \end{pmatrix} \tag{2.305}$$

is also integrable. Here $\omega = -\Omega(k)$ is the linear dispersion relation of the r equation for Fourier waves $e^{i(kx-\omega t)}$. To prove this, let us consider the example of $\Omega(k) = k^m + k^n$, where m and n are two integers (the proof for the general case would be similar). To show that Eq. (2.305) is still integrable for $\Omega(k) = k^m + k^n$, we only need to take

$$\begin{pmatrix} B \\ C \end{pmatrix} = \sum_{j=1}^{m} \begin{pmatrix} b_j \\ c_j \end{pmatrix} (2\zeta)^{m-j} + \sum_{j=1}^{n} \begin{pmatrix} \alpha_j \\ \beta_j \end{pmatrix} (2\zeta)^{n-j}, \tag{2.306}$$

$$D_0 = -A_0 = i\zeta \left[(2\zeta)^{m-1} + (2\zeta)^{n-1} \right], \tag{2.307}$$

and

$$\begin{pmatrix} b_j \\ c_j \end{pmatrix} = L_R^{j-1} \begin{pmatrix} q \\ r \end{pmatrix}, \quad \begin{pmatrix} \alpha_j \\ \beta_j \end{pmatrix} = L_R^{j-1} \begin{pmatrix} q \\ r \end{pmatrix}. \tag{2.308}$$

Inserting these equations into (2.290), we find that Eq. (2.290) is satisfied when

$$i \begin{pmatrix} q \\ -r \end{pmatrix}_t = \left(L_R^m + L_R^n \right) \begin{pmatrix} q \\ r \end{pmatrix}. \tag{2.309}$$

Thus this evolution equation is also integrable.

One remark we would like to make is that even though the recursion operator L_R is nonlocal in (q,r), nonetheless one can show that each equation in the AKNS hierarchy (2.296) or (2.305) is always local, i.e., the time derivatives (q_t, r_t) at any position x depend on the values (q,r) and their spatial derivatives at this x position only. In other words, the AKNS hierarchies (2.296) and (2.305) are all differential equations, not integrodifferential equations.

2.9 Squared Eigenfunctions and the Recursion Operator

Squared eigenfunctions of the AKNS hierarchy (2.296) are closely related to the recursion operator L_R of this hierarchy. Here squared eigenfunctions $\{Z^-, Z^+\}$ for this hierarchy (i.e., for the generalized Zakharov–Shabat system (2.281)) are defined similarly to Eq. (2.255), or equivalently (2.256), for the original Zakharov–Shabat system (2.2), and they are products of Jost solutions Φ whose boundary conditions are set at $x = -\infty$. In this connection, the antiderivative ∂^{-1} in L_R must be taken as $\partial^{-1} = \int_{-\infty}^{x} dy$ in accordance with these Jost solutions. The corresponding recursion operator L_R then is

$$L_R = i \begin{pmatrix} \partial - 2q \int_{-\infty}^{x} dy\, r & 2q \int_{-\infty}^{x} dy\, q \\ -2r \int_{-\infty}^{x} dy\, r & -\partial + 2r \int_{-\infty}^{x} dy\, q \end{pmatrix}. \tag{2.310}$$

Using the generalized Zakharov–Shabat equations (2.281), the definition (2.256) for squared eigenfunctions (Z^-, Z^+), and boundary conditions of Jost solutions Φ, one can verify that these squared eigenfunctions are also eigenfunctions of the recursion operator L_R with the following eigenrelations:

$$L_R Z^{\pm}(x,\zeta) = 2\zeta\, Z^{\pm}(x,\zeta), \quad \zeta \in \mathbb{R}. \tag{2.311}$$

Similarly, the adjoint squared eigenfunctions (Ω^-, Ω^+) as defined in (2.232), or equivalently in (2.236), are also eigenfunctions of the adjoint recursion operator L_R^A with the eigenrelations:

$$L_R^A \Omega^{\pm}(x,\zeta) = 2\zeta\, \Omega^{\pm}(x,\zeta), \quad \zeta \in \mathbb{R}, \tag{2.312}$$

where the adjoint recursion operator L_R^A is

$$L_R^A = i \begin{pmatrix} -\partial - 2r \int_x^{\infty} dy\, q & -2r \int_x^{\infty} dy\, r \\ 2q \int_x^{\infty} dy\, q & \partial + 2q \int_x^{\infty} dy\, r \end{pmatrix}. \tag{2.313}$$

In fact, in the original derivation of squared eigenfunctions (Ablowitz et al. (1974), Kaup (1976a, 1976b)), these eigenrelations were how the squared eigenfunctions were defined. When time dependence is introduced into these squared eigenfunctions and their adjoints through Eqs. (2.271) and (2.278), the time-dependent squared eigenfunctions $(Z^{-(t)}, Z^{+(t)})$ and their adjoints $(\Omega^{-(t)}, \Omega^{+(t)})$ satisfy the eigenrelations (2.311)–(2.312) of the recursion operators as well. Since these time-dependent squared eigenfunctions also satisfy the linearized equation of the AKNS hierarchy (see Sec. 2.7, especially the last paragraph), we see that

$$\mathcal{L} L_R\, Z^{\pm}(x,\zeta) = L_R \mathcal{L}\, Z^{\pm}(x,\zeta) = 0, \quad \zeta \in \mathbb{R}, \tag{2.314}$$

where $\mathcal{L}$ is the linearization operator of the AKNS hierarchy (2.296). This same relation holds for discrete squared eigenfunctions of $Z^{\pm}$ where ζ takes on zeros $(\zeta_j, \bar{\zeta}_j)$ of the scattering coefficients $(\hat{s}_{22}, s_{22})$. Since these squared eigenfunctions form a complete set, Eq. (2.314) indicates that

$$\mathcal{L} L_R = L_R \mathcal{L}, \tag{2.315}$$

i.e., the recursion operator L_R and the linearization operator $\mathcal{L}$ commute. Similarly the adjoints of these two operators commute as well, i.e.,

$$\mathcal{L}^A L_R^A = L_R^A \mathcal{L}^A. \tag{2.316}$$

These commutability relations for the AKNS hierarchy were proved by direct verifications in Yang (2002b). These relations for the KdV hierarchy have also been proved by the symmetry approach (Fokas (1980)) and direct verifications (Yang (2002b)).

2.10 Amplitude-Changing and Self-Collapsing Solitons in the AKNS Hierarchy

For many familiar equations in the AKNS hierarchy (2.296), such as the NLS equation (2.1) and the modified KdV equation (2.298), constraints were imposed on the potentials (q,r) (see Sec. 2.8). This potential reduction leads to involution properties of the scattering data. As a result, solitons in these equations exhibit very familiar behaviors. For instance, single-soliton solutions in these equations are solitary waves which move without change of shape (see Eq. (2.128)). However, if no potential reduction is imposed, solitons in the AKNS hierarchy can exhibit very different and unexpected behaviors. For example, single-soliton solutions now can change their shape or amplitude during propagation. In addition, solitons can blow up to infinity in finite time even if the initial condition is smooth. In this section, we demonstrate these unusual soliton dynamics.

We choose the coupled Schrödinger equations (2.303)–(2.304), i.e.,

$$iq_t + q_{xx} - 2q^2 r = 0, \tag{2.317}$$

$$ir_t - r_{xx} + 2r^2 q = 0 \tag{2.318}$$

in our demonstration. These equations come from the AKNS hierarchy (2.296) when $n = 2$ is taken. The general N-soliton solutions in this system can be readily obtained from the Riemann–Hilbert solutions in Theorem 2.1 of Sec. 2.1.2 (by setting $G = I$) and the potential reconstruction formula (2.48) as

$$q(x,t) = 2i\left(\sum_{j,k=1}^{N} v_j \left(M^{-1}\right)_{jk} \bar{v}_k\right)_{12} \tag{2.319}$$

and

$$r(x,t) = -2i\left(\sum_{j,k=1}^{N} v_j \left(M^{-1}\right)_{jk} \bar{v}_k\right)_{21}, \tag{2.320}$$

where

$$v_k(x,t) = e^{-i\zeta_k \Lambda x - 2i\zeta_k^2 \Lambda t}\, v_{k0}, \tag{2.321}$$

$$\bar{v}_k(x,t) = \bar{v}_{k0}\, e^{i\bar{\zeta}_k \Lambda x + 2i\bar{\zeta}_k^2 \Lambda t}, \tag{2.322}$$

M is an $N \times N$ matrix whose (j,k)th element is given by

$$M_{jk} = \frac{\bar{v}_j v_k}{\bar{\zeta}_j - \zeta_k}, \quad 1 \leq j,k \leq N, \tag{2.323}$$

$\Lambda = \text{diag}(1,-1)$, and $v_{k0}, \bar{v}_{k0}$ are constant vectors. Without loss of generality, we scale these two constant vectors as

$$v_{k0} = (c,1)^T, \quad \bar{v}_{k0} = (\bar{c},1), \tag{2.324}$$

where $c, \bar{c}$ are complex constants. These solution formulae are very similar to those of the N-soliton solutions in the NLS equation (see Sec. 2.2), except that the zeros $\zeta_k, \bar{\zeta}_k$ are now independent of each other, and so are the constant vectors $v_{k0}, \bar{v}_{k0}$ (i.e., $c, \bar{c}$), because the potentials q, r are now independent. Below we only examine single solitons in this system.

When taking $N = 1$ in the general soliton solutions (2.319)–(2.320), we get single-soliton solutions in Eqs. (2.317)–(2.318) as

$$q(x,t) = \quad 2i(\bar{\zeta}_1 - \zeta_1)\frac{c_1\, e^{\theta_1+\bar{\theta}_1}}{e^{\bar{\theta}_1-\theta_1} + c_1\bar{c}_1\, e^{\theta_1-\bar{\theta}_1}} \tag{2.325}$$

and

$$r(x,t) = -2i(\bar{\zeta}_1 - \zeta_1)\frac{\bar{c}_1\, e^{-\theta_1-\bar{\theta}_1}}{e^{\bar{\theta}_1-\theta_1} + c_1\bar{c}_1\, e^{\theta_1-\bar{\theta}_1}}, \tag{2.326}$$

where

$$\theta_1 = -i\zeta_1 x - 2i\zeta_1^2 t, \quad \bar{\theta}_1 = -i\bar{\zeta}_1 x - 2i\bar{\zeta}_1^2 t. \tag{2.327}$$

Denoting

$$\zeta_1 = \xi + i\eta, \quad \bar{\zeta}_1 = \bar{\xi} + i\bar{\eta}, \tag{2.328}$$

where $\xi, \eta, \bar{\xi}$, and $\bar{\eta}$ are the real and imaginary parts of ζ_1 and $\bar{\zeta}_1$, then since $\zeta_1 \in \mathbb{C}_+$ and $\bar{\zeta}_1 \in \mathbb{C}_-$, we have $\eta > 0$ and $\bar{\eta} < 0$. Now, we consider two special cases of these single-soliton solutions.

Amplitude-Changing Single Solitons

The first case we consider is where the discrete eigenvalues ζ_1 and $\bar{\zeta}_1$ have opposite imaginary parts but different real parts, i.e.,

$$\xi \neq \bar{\xi}, \quad \eta = -\bar{\eta}. \tag{2.329}$$

Also, for simplicity, we take $c_1 = \bar{c}_1 = 1$. Then the single-soliton solution (2.325)–(2.326) becomes

$$q(x,t) = i(\bar{\zeta}_1 - \zeta_1)\operatorname{sech}\big[\alpha\big(x + 2(\xi + \bar{\xi})t\big)\big]\, e^{4\eta(\xi-\bar{\xi})t + i\phi}, \tag{2.330}$$

$$r(x,t) = -i(\bar{\zeta}_1 - \zeta_1)\operatorname{sech}\big[\alpha\big(x + 2(\xi + \bar{\xi})t\big)\big]\, e^{-4\eta(\xi-\bar{\xi})t - i\phi}, \tag{2.331}$$

where

$$\alpha = 2\eta - i(\xi - \bar{\xi}), \quad \phi = -(\xi + \bar{\xi})x - 2(\xi^2 + \bar{\xi}^2 - 2\eta^2)t.$$

This soliton moves at velocity $-2(\xi + \bar{\xi})$. More importantly, at the center of this soliton, where $x + 2(\xi + \bar{\xi})t = 0$, its amplitude exponentially increases or decreases at the rate of $4\eta(\xi - \bar{\xi})$. To illustrate this amplitude-changing soliton, we take

$$\zeta_1 = i, \quad \bar{\zeta}_1 = 0.1 - i. \tag{2.332}$$

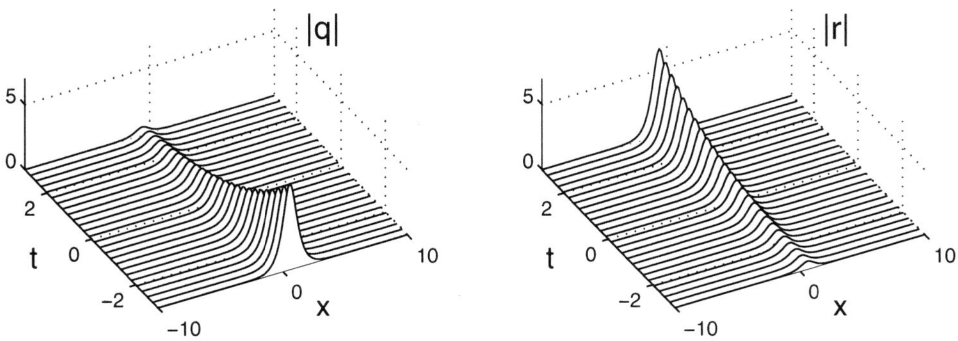

Figure 2.4. *An amplitude-changing single-soliton solution in the coupled Schrödinger equations* (2.317)–(2.318). *This soliton is obtained from the formulae* (2.330)–(2.331) *with parameters* (2.332).

The corresponding solution (2.330)–(2.331) is displayed in Fig. 2.4. It is seen that the q-component of the soliton is exponentially decaying, while the r-component of the soliton is exponentially growing. This behavior is very remarkable.

Self-Collapsing Single Solitons

The second case we consider is where the discrete eigenvalues ζ_1 and $\bar{\zeta}_1$ have equal real parts, but their imaginary parts are not opposite to each other, i.e.,

$$\xi = \bar{\xi}, \quad \eta \neq -\bar{\eta}. \tag{2.333}$$

As before, we also take $c_1 = \bar{c}_1 = 1$ for simplicity. Then the single-soliton solution (2.325)–(2.326) becomes

$$q(x,t) = (\eta - \bar{\eta})\,\text{sech}\Big[(\eta - \bar{\eta})(x + 4\xi t) + 2i(\eta^2 - \bar{\eta}^2)t\Big]\, e^{(\eta+\bar{\eta})(x+4\xi t)+i\phi}, \tag{2.334}$$

$$r(x,t) = -(\eta - \bar{\eta})\,\text{sech}\Big[(\eta - \bar{\eta})(x + 4\xi t) + 2i(\eta^2 - \bar{\eta}^2)t\Big]\, e^{-(\eta+\bar{\eta})(x+4\xi t)-i\phi}, \tag{2.335}$$

where

$$\phi = -2\xi x - 2(2\xi^2 - \eta^2 - \bar{\eta}^2)t.$$

This soliton moves at velocity -4ξ. But its more important feature is that this soliton blows up to infinity in finite time, even though the initial condition (at $t = 0$) is perfectly smooth. This is clear because the amplitudes of $|q|$ and $|r|$ at the soliton center $x + 4\xi t = 0$ are proportional to $1/\cos 2(\eta^2 - \bar{\eta}^2)t$, which blow up to infinity at finite time $t = \pi/4(\eta^2 - \bar{\eta}^2)$. After the collapse, the soliton miraculously recovers itself and becomes smooth and normal

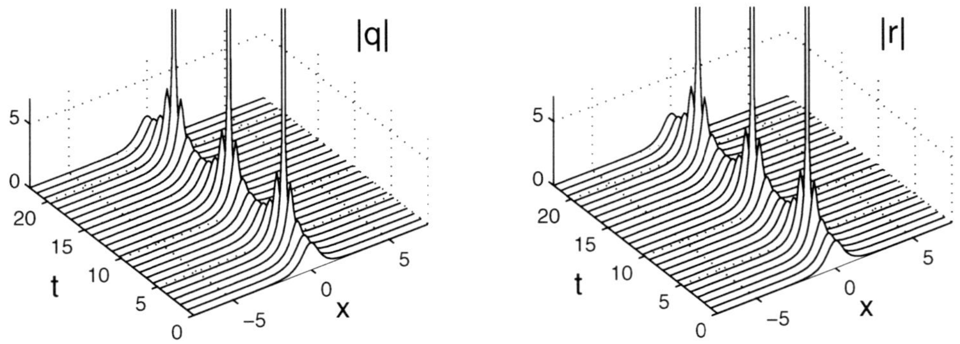

Figure 2.5. *A self-collapsing single-soliton solution in the coupled Schrödinger equations* (2.317)–(2.318). *This soliton is obtained from the formulae* (2.334)–(2.335) *with parameters* (2.336).

again. But then it will collapse again. This process repeats indefinitely. To illustrate, we take

$$\zeta_1 = i, \quad \bar{\zeta}_1 = -1.1i. \tag{2.336}$$

The corresponding solution (2.334)–(2.335) is displayed in Fig. 2.5. It is seen that the solution collapses in both q and r components, and this collapse is periodic in time. One can also see from the solution formulae (2.334)–(2.335) that near collapse, the soliton amplitude blows up to infinity as $(t-t_c)^{-1}$, where t_c is the time of collapse. We would like to mention that wave collapse is a familiar phenomenon in many nonintegrable equations (such as the 2D NLS equation; see Sec. 5.10). It is remarkable that this phenomenon appears in integrable equations such as (2.317)–(2.318) as well.

The singularity in the collapsing solitons (2.334)–(2.335) appears only at certain discrete times. We would like to add that Eqs. (2.317)–(2.318) also admit soliton solutions which are singular at all times. For instance, if we take

$$\bar{\zeta}_1 = \zeta_1^*, \quad c_1 = 1, \quad \bar{c}_1 = -1 \tag{2.337}$$

in the single-soliton (2.325)–(2.326), we would get the solution

$$q(x,t) = r^*(x,t) = -2\eta \operatorname{csch}[2\eta(x+4\xi t)]e^{-2i\xi x - 4i(\xi^2-\eta^2)t}, \tag{2.338}$$

that are singular at the soliton center $x+4\xi t = 0$ for all time t. Notice that under the potential reduction $q = r^*$, Eqs. (2.317)–(2.318) reduce to the self-defocusing NLS equation

$$iq_t + q_{xx} - 2|q|^2 q = 0. \tag{2.339}$$

Thus the singular soliton (2.338) above is a singular solution in this self-defocusing NLS equation. Such singular solitons, which are singular at all times, are well known in integrable equations. A familiar example is the singular soliton

$$u(x,t) = -\frac{1}{2}k^2 \operatorname{csch}^2 \frac{1}{2}k(x-k^2t) \tag{2.340}$$

in the KdV equation

$$u_t + u_{xxx} + 6uu_x = 0, \tag{2.341}$$

where k is a real constant, and $\operatorname{csch} x = 1/\sinh x$ is the hyperbolic cosecant function. Other examples of singular solitons in the KdV equation include the so-called positons and complexitons (Matveev (1992), Ma (2005)). The simplest KdV positon can be obtained by taking k to be purely imaginary in the above singular soliton (2.340), while the simplest KdV complexiton can be obtained by taking the two spectral eigenvalues in the KdV's two-soliton solution to be complex conjugate of each other. These singular solitons maintain their singularity for all time and thus are quite different from the self-collapsing soliton (2.334)–(2.335) presented above.

Chapter 3

Theories for Integrable Equations with Higher-Order Scattering Operators

The NLS equation and the AKNS hierarchy in the previous chapter are associated with the second-order Zakharov–Shabat scattering problem. In this chapter, we consider integrable equations which are associated with higher-order scattering problems. Our treatments will closely resemble those in the previous chapter (with only minor modifications). Thus these treatments are universal for a large class of integrable systems.

First, we consider a few examples. One is the Manakov system

$$iu_t + u_{xx} + (|u|^2 + |v|^2)u = 0, \tag{3.1}$$

$$iv_t + v_{xx} + (|u|^2 + |v|^2)v = 0, \tag{3.2}$$

which was shown to be integrable by Manakov (1973). The scattering operator of the Lax pair for this system is

$$Y_x = \begin{pmatrix} -i\zeta & 0 & u \\ 0 & -i\zeta & v \\ -u^* & -v^* & i\zeta \end{pmatrix} Y, \tag{3.3}$$

which is a third-order system. Another equation is the Sasa–Satsuma equation

$$u_t + u_{xxx} + 6|u|^2u_x + 3u(|u|^2)_x = 0, \tag{3.4}$$

which was shown to be integrable by Sasa and Satsuma (1991). Its scattering problem in the Lax pair is

$$Y_x = \begin{pmatrix} -i\zeta & 0 & u \\ 0 & -i\zeta & u^* \\ -u^* & -u & i\zeta \end{pmatrix} Y, \tag{3.5}$$

which is also a third-order system. The third equation is the matrix NLS equation

$$i\mathbf{U}_t + \mathbf{U}_{xx} + 2\mathbf{U}\mathbf{U}^\dagger\mathbf{U} = 0, \tag{3.6}$$

where $\mathbf{U}(x,t)$ is a $k\times k$ matrix function. This equation is also integrable (Tsuchida and Wadati (1998), Gerdjikov (2005)). Its scattering operator is

$$Y_x = \begin{pmatrix} -i\zeta\mathbf{I}_k & \mathbf{U} \\ -\mathbf{U}^\dagger & i\zeta\mathbf{I}_k \end{pmatrix} Y, \tag{3.7}$$

which is a $2k$th-order system. Here $\mathbf{I}_k$ is the kth-order identity matrix.

A common feature of these equations is that their scattering operators are higher-order systems, but these systems are degenerate in the sense that their diagonal elements are all repeated $i\zeta$ or $-i\zeta$. It turns out that all equations of this type belong to the same integrable hierarchy and can be treated together. This will be done in this chapter.

3.1 Integrable Hierarchy for a Higher-Order Scattering Operator

In this section, we derive the integrable hierarchy which will contain the Manakov equations, the Sasa–Satsuma equation, the matrix NLS equation, and many others as particular examples. The scattering operator associated with this hierarchy is the following higher-order degenerate system:

$$Y_x = \mathbf{M}Y, \quad \mathbf{M} = \begin{pmatrix} -i\zeta\mathbf{I}_k & \mathbf{q} \\ \mathbf{r}^T & i\zeta\mathbf{I}_m \end{pmatrix}, \tag{3.8}$$

where $\mathbf{q}$ and $\mathbf{r}$ are $k\times m$ matrix functions of (x,t), and $\mathbf{I}_k, \mathbf{I}_m$ are kth- and mth-order identity matrices. This derivation is an extension of that for the AKNS hierarchy in the previous chapter. The only difference is that we need to respect the order of multiplication now since the quantities involved are matrices and hence generally do not commute. Our calculation follows Wang et al. (2010).

Suppose the temporal equation of the Lax pair is

$$Y_t = \mathbf{N}Y, \quad \mathbf{N} = \begin{pmatrix} \mathbf{A} & \mathbf{B} \\ \mathbf{C}^T & \mathbf{D} \end{pmatrix}, \tag{3.9}$$

where the matrix $\mathbf{N}$ has the same block structure as the matrix $\mathbf{M}$. Then using the zero curvature equation

$$\mathbf{M}_t - \mathbf{N}_x + [\mathbf{M},\mathbf{N}] = 0 \tag{3.10}$$

and assuming the spectral parameter ζ to be time independent, we get

$$-\mathbf{A}_x - \mathbf{B}\mathbf{r}^T + \mathbf{q}\mathbf{C}^T = 0, \tag{3.11}$$

$$\mathbf{q}_t - \mathbf{B}_x - 2i\zeta\mathbf{B} - \mathbf{A}\mathbf{q} + \mathbf{q}\mathbf{D} = 0, \tag{3.12}$$

$$\mathbf{r}_t - \mathbf{C}_x + 2i\zeta\mathbf{C} + \mathbf{A}^T\mathbf{r} - \mathbf{r}\mathbf{D}^T = 0, \tag{3.13}$$

$$-\mathbf{D}_x + \mathbf{r}^T\mathbf{B} - \mathbf{C}^T\mathbf{q} = 0. \tag{3.14}$$

From (3.11) and (3.14) we have

$$\mathbf{A} = -\partial^{-1}(\mathbf{B}\mathbf{r}^T - \mathbf{q}\mathbf{C}^T) + \mathbf{A}_0, \tag{3.15}$$

$$\mathbf{D} = \partial^{-1}(\mathbf{r}^T\mathbf{B} - \mathbf{C}^T\mathbf{q}) + \mathbf{D}_0, \tag{3.16}$$

where ∂^{-1} is an antiderivative in x which can be taken as either $\partial^{-1} = \int_{-\infty}^{x} dy$ or $\partial^{-1} = -\int_{x}^{\infty} dy$, and $(\mathbf{A}_0, \mathbf{D}_0)$ are x-independent. Inserting the above two equations into (3.12) and (3.13), we get

$$\begin{pmatrix} \mathbf{q} \\ -\mathbf{r} \end{pmatrix}_t = i(2\zeta - \mathbf{L}_R)\begin{pmatrix} \mathbf{B} \\ \mathbf{C} \end{pmatrix} + \begin{pmatrix} \mathbf{A}_0\mathbf{q} - \mathbf{q}\mathbf{D}_0 \\ \mathbf{A}_0^T\mathbf{r} - \mathbf{r}\mathbf{D}_0^T \end{pmatrix}, \tag{3.17}$$

where the recursion operator $\mathbf{L}_R$ is

$$\mathbf{L}_R\begin{pmatrix} \mathbf{B} \\ \mathbf{C} \end{pmatrix} = i\begin{pmatrix} \mathbf{B}_x - \mathbf{q}\,\partial^{-1}(\mathbf{r}^T\mathbf{B} - \mathbf{C}^T\mathbf{q}) - [\partial^{-1}(\mathbf{B}\mathbf{r}^T - \mathbf{q}\mathbf{C}^T)]\mathbf{q} \\ -\mathbf{C}_x - \mathbf{r}\,\partial^{-1}(\mathbf{B}^T\mathbf{r} - \mathbf{q}^T\mathbf{C}) - [\partial^{-1}(\mathbf{r}\mathbf{B}^T - \mathbf{C}\mathbf{q}^T)]\mathbf{r} \end{pmatrix}. \tag{3.18}$$

It is noted that a special case of this recursion operator, for the vector NLS equations of the Manakov type, was given by Kodama and Mikhailov (2001).

Now we expand

$$\begin{pmatrix} \mathbf{B} \\ \mathbf{C} \end{pmatrix} = \sum_{j=1}^{n}\begin{pmatrix} \mathbf{b}_j \\ \mathbf{c}_j \end{pmatrix}(2\zeta)^{n-j} \tag{3.19}$$

and take

$$\mathbf{A}_0 = -\frac{1}{2}i(2\zeta)^n\mathbf{I}_k, \quad \mathbf{D}_0 = \frac{1}{2}i(2\zeta)^n\mathbf{I}_m.$$

Inserting these equations into (3.17) and equating terms of the same power in 2ζ, the same relations as (2.294)–(2.295) would be obtained. From those relations, the integrable hierarchy associated with the higher-order scattering problem (3.8) is then

$$i\begin{pmatrix} \mathbf{q} \\ -\mathbf{r} \end{pmatrix}_t = \mathbf{L}_R^n\begin{pmatrix} \mathbf{q} \\ \mathbf{r} \end{pmatrix}, \quad n = 1, 2, \ldots. \tag{3.20}$$

When $n = 2$, this hierarchy is (see Tsuchida and Wadati (1998), Gerdjikov (2005))

$$i\mathbf{q}_t + \mathbf{q}_{xx} - 2\mathbf{q}\mathbf{r}^T\mathbf{q} = 0, \tag{3.21}$$

$$-i\mathbf{r}_t + \mathbf{r}_{xx} - 2\mathbf{r}\mathbf{q}^T\mathbf{r} = 0. \tag{3.22}$$

When $n = 3$, this hierarchy is (see Tsuchida and Wadati (1998))

$$\mathbf{q}_t + \mathbf{q}_{xxx} - 3\mathbf{q}_x\mathbf{r}^T\mathbf{q} - 3\mathbf{q}\mathbf{r}^T\mathbf{q}_x = 0, \tag{3.23}$$

$$-\mathbf{r}_t - \mathbf{r}_{xxx} + 3\mathbf{r}_x\mathbf{q}^T\mathbf{r} + 3\mathbf{r}\mathbf{q}^T\mathbf{r}_x = 0. \tag{3.24}$$

3.2 Various Reductions of the Hierarchy

The hierarchy derived in the previous section admits various reductions. In this section, we will discuss reductions in the second and third hierarchy equations (3.21)–(3.22) and (3.23)–(3.24), and we show that these reductions give the Manakov equations, the Sasa–Satsuma equation, the matrix NLS equation, and many others as special cases.

First we consider reductions of the second hierarchy equation (3.21)–(3.22). A general reduction of this equation is that

$$\mathbf{r}^T = -\mathbf{B}_1\mathbf{q}^\dagger\mathbf{B}_2, \tag{3.25}$$

where $\mathbf{B}_1$ and $\mathbf{B}_2$ are constant Hermitian matrices, i.e., $\mathbf{B}_{1,2}^\dagger = \mathbf{B}_{1,2}$, and their orders are $m \times m$ and $k \times k$, respectively. The reduced equation is then

$$i\mathbf{q}_t + \mathbf{q}_{xx} + 2\mathbf{q}\mathbf{B}_1\mathbf{q}^\dagger\mathbf{B}_2\mathbf{q} = 0. \tag{3.26}$$

Below we consider a few special cases of this reduction.

1. If we take

$$\mathbf{q} = (u, v)^T, \quad \mathbf{B}_1 = 1, \quad \mathbf{B}_2 = \mathbf{I}_2, \tag{3.27}$$

then Eq. (3.26) becomes the Manakov equations (3.1)–(3.2).

2. If we take

$$\mathbf{q} = (u, v)^T, \quad \mathbf{B}_1 = 1, \quad \mathbf{B}_2 = \text{diag}(1, -1), \tag{3.28}$$

then (3.26) becomes a focusing-defocusing NLS system

$$iu_t + u_{xx} + (|u|^2 - |v|^2)u = 0, \tag{3.29}$$

$$iv_t + v_{xx} + (|u|^2 - |v|^2)v = 0, \tag{3.30}$$

where the u-component experiences self-focusing nonlinearity, while the v-component experiences self-defocusing nonlinearity.

3. If we take

$$\mathbf{q} = (u, v)^T, \quad \mathbf{B}_1 = 1, \quad \mathbf{B}_2 = \begin{pmatrix} 0 & 1 \\ 1 & 0 \end{pmatrix}, \tag{3.31}$$

then (3.26) gives

$$iu_t + u_{xx} + 2(|u|^2 v + u^2 v^*) = 0, \tag{3.32}$$

$$iv_t + v_{xx} + 2(|v|^2 u + v^2 u^*) = 0, \tag{3.33}$$

which is a specially coupled NLS-type system.

4. If we take

$$\mathbf{q} = \mathbf{U}, \quad \mathbf{B}_1 = \mathbf{B}_2 = \mathbf{I}_k, \tag{3.34}$$

where $\mathbf{U}$ is a $k \times k$ matrix function, then Eq. (3.26) becomes the matrix NLS equation (3.6).

Next we consider reductions of the third hierarchy equation (3.23)–(3.24). This equation also admits the general reduction (3.25), and the reduced equation is

$$\mathbf{q}_t + \mathbf{q}_{xxx} + 3\mathbf{q}_x \mathbf{B}_1 \mathbf{q}^\dagger \mathbf{B}_2 \mathbf{q} + 3\mathbf{q} \mathbf{B}_1 \mathbf{q}^\dagger \mathbf{B}_2 \mathbf{q}_x = 0. \tag{3.35}$$

A few special cases of this reduction are as below.

1. If we take (3.27), then this equation becomes

$$u_t + u_{xxx} + 6|u|^2 u_x + 3(uv)_x v^* = 0, \tag{3.36}$$

$$v_t + v_{xxx} + 6|v|^2 v_x + 3(uv)_x u^* = 0, \tag{3.37}$$

which is a coupled complex modified KdV-type system.

2. The above system admits a further reduction. Taking $v = u^*$, it then becomes the Sasa–Satsuma equation (3.4).

3. If we take

$$\mathbf{q} = \begin{pmatrix} u & v \\ v^* & u^* \end{pmatrix}, \quad \mathbf{B}_1 = \mathbf{B}_2 = \mathbf{I}_2, \tag{3.38}$$

then (3.35) becomes (Sakovich and Tsuchida (2000))

$$u_t + u_{xxx} + 6(|u|^2 + |v|^2) u_x + 6u(|v|^2)_x = 0, \tag{3.39}$$

$$v_t + v_{xxx} + 6(|u|^2 + |v|^2) v_x + 6v(|u|^2)_x = 0, \tag{3.40}$$

which is another coupled complex modified KdV system.

3.3 Riemann–Hilbert Problem for the Hierarchy

In this section, we present the inverse scattering transform method for the hierarchy (3.20). As before, the Jost solutions and their analytical properties in the scattering equation (3.8) of this hierarchy will play a crucial role. To simplify the notation, we will work with a 3×3 scattering equation (3.8). These results can be trivially extended to the higher-order scattering equation (3.8), and thus there is no loss of generality in our treatment.

The 3×3 scattering problem (3.8) can be written as

$$Y_x = -i\zeta\Lambda Y + QY, \tag{3.41}$$

where Y is a matrix function, $\Lambda = \mathrm{diag}(1,1,-1)$, and

$$Q = \begin{pmatrix} 0 & 0 & u \\ 0 & 0 & v \\ \hat{u} & \hat{v} & 0 \end{pmatrix} \tag{3.42}$$

is the potential matrix. In the notation of Eq. (3.8), $\mathbf{q} = (u,v)^T$ and $\mathbf{r}^T = (\hat{u},\hat{v})$. In our analysis, we always assume that the potential Q decays to zero sufficiently fast as $x\to\pm\infty$. The time evolution equation of the Lax pair for the hierarchy (3.20) can be seen from Sec. 3.1 as

$$Y_t = -\frac{1}{2}i(2\zeta)^n\Lambda Y + VY, \tag{3.43}$$

where V is a matrix function of Q and ζ and decays to zero as $x\to\pm\infty$. Introducing new variables

$$J = Y\exp\left\{i\zeta\Lambda x + \frac{1}{2}i(2\zeta)^n\Lambda t\right\}, \tag{3.44}$$

then the Lax pair (3.41) and (3.43) for J becomes

$$J_x = -i\zeta[\Lambda, J] + QJ \tag{3.45}$$

and

$$J_t = -\frac{1}{2}i(2\zeta)^n[\Lambda, J] + VJ, \tag{3.46}$$

where $[\Lambda, J] = \Lambda J - J\Lambda$. In this section, we only consider the scattering equation (3.45) and will treat time t as a dummy variable.

The matrix Jost solutions $J_\pm(x,\zeta)$ of Eq. (3.45) are defined by the large-x asymptotics

$$J_\pm(x,\zeta) \to \mathbf{I}_3, \qquad x\to\pm\infty. \tag{3.47}$$

Here $\mathbf{I}_3$ is the unit matrix of rank three. Also we introduce the notation

$$J_-E = \Phi = (\phi_1,\phi_2,\phi_3), \tag{3.48}$$

$$J_+E = \Psi = (\psi_1,\psi_2,\psi_3), \tag{3.49}$$

where $E = e^{-i\zeta\Lambda x}$. Since Φ and Ψ are both fundamental matrices of the scattering problem (3.41), they are related by an x-independent scattering matrix $S(\zeta) = [s_{ij}(\zeta)]$:

$$\Phi = \Psi S, \quad \zeta \in \mathbb{R}, \tag{3.50}$$

i.e.,

$$J_- = J_+ E S E^{-1}, \quad \zeta \in \mathbb{R}. \tag{3.51}$$

Since $\operatorname{tr} Q = 0$, using Abel's identity (2.12) for fundamental matrices Φ and Ψ of Eq. (3.41) and recalling the boundary conditions (3.47), we see that

$$\det J_\pm(x,\zeta) = \det S(\zeta) = 1 \tag{3.52}$$

for all x and $\zeta \in \mathbb{R}$.

Analytic properties of these Jost solutions and scattering elements can be determined analogously to what we did for the Zakharov–Shabat system in Sec. 2.1.1. Using boundary conditions (3.47), the scattering equation (3.45) for $J_\pm$ can be rewritten as Volterra-type integral equations of the form (2.20)–(2.21). Thus in view of the structure (3.42) of the potential matrix Q, we see that Eq. (2.20) for the first and second columns of J_- contains only the exponential factor $e^{i\zeta(x-y)}$ which decays when ζ is in the upper half plane $\mathbb{C}_+$, and Eq. (2.21) for the third column of J_+ contains only the exponential factor $e^{i\zeta(y-x)}$ which also falls off for $\zeta \in \mathbb{C}_+$. Thus these three columns can be analytically extended to $\zeta \in \mathbb{C}_+$. In other words, the Jost solutions

$$P^+ = (\phi_1, \phi_2, \psi_3) e^{i\zeta\Lambda x} = J_- H_1 + J_+ H_2 \tag{3.53}$$

are analytic in $\zeta \in \mathbb{C}_+$, where

$$H_1 = \operatorname{diag}(1,1,0), \qquad H_2 = \operatorname{diag}(0,0,1). \tag{3.54}$$

From the Volterra integral equations for P^+, we see that

$$P^+(x,\zeta) \to \mathbf{I}_3, \quad \zeta \in \mathbb{C}_+ \to \infty. \tag{3.55}$$

Similarly, the Jost solutions $(\psi_1, \psi_2, \phi_3) e^{i\zeta\Lambda x}$ are analytic in $\zeta \in \mathbb{C}_-$, and their large-ζ asymptotics is

$$(\psi_1, \psi_2, \phi_3) e^{i\zeta\Lambda x} \to \mathbf{I}_3, \quad \zeta \in \mathbb{C}_- \to \infty. \tag{3.56}$$

To obtain the analytic counterpart of P^+ in $\mathbb{C}_-$, we consider the adjoint scattering equation

$$K_x = -i\zeta[\Lambda, K] - KQ. \tag{3.57}$$

The inverse Jost matrices $J_{\pm}^{-1}$ satisfy this adjoint equation. Notice that

$$J_{-}^{-1} = E\Phi^{-1}, \quad J_{+}^{-1} = E\Psi^{-1}. \tag{3.58}$$

Let us express Φ^{-1} and Ψ^{-1} as a collection of rows,

$$\Phi^{-1} = \begin{pmatrix} \widehat{\phi}_1 \\ \widehat{\phi}_2 \\ \widehat{\phi}_3 \end{pmatrix}, \quad \Psi^{-1} = \begin{pmatrix} \widehat{\psi}_1 \\ \widehat{\psi}_2 \\ \widehat{\psi}_3 \end{pmatrix}. \tag{3.59}$$

then by techniques similar to those used above, we can show that the first and second rows of J_{-}^{-1} and the third row of J_{+}^{-1} are analytic in $\zeta \in \mathbb{C}_-$; i.e., the adjoint Jost solutions

$$P^- = e^{-i\zeta\Lambda x} \begin{pmatrix} \widehat{\phi}_1 \\ \widehat{\phi}_2 \\ \widehat{\psi}_3 \end{pmatrix} = H_1 J_{-}^{-1} + H_2 J_{+}^{-1} \tag{3.60}$$

are analytic in $\zeta \in \mathbb{C}_-$. In addition, their large-ζ asymptotics is

$$P^-(x,\zeta) \to \mathbf{I}_3, \quad \zeta \in \mathbb{C}_- \to \infty. \tag{3.61}$$

Similarly, the first and second rows of J_{+}^{-1} and the third row of J_{-}^{-1}, i.e., $e^{-i\zeta x}\widehat{\psi}_1$, $e^{-i\zeta x}\widehat{\psi}_2$, and $e^{i\zeta x}\widehat{\phi}_3$, are analytic in $\zeta \in \mathbb{C}_+$, and their large-ζ asymptotics is

$$e^{-i\zeta\Lambda x} \begin{pmatrix} \widehat{\psi}_1 \\ \widehat{\psi}_2 \\ \widehat{\phi}_3 \end{pmatrix} \to \mathbf{I}_3, \quad \zeta \in \mathbb{C}_+ \to \infty. \tag{3.62}$$

Using these analytical properties of Jost solutions and the relation $S = \Psi^{-1}\Phi$, analytic properties of the scattering matrix S and its inverse S^{-1} can be quickly obtained. All of these analytical properties of Jost solutions and the scattering matrix can be summarized as follows:

$$\Phi = (\phi_1^+, \phi_2^+, \phi_3^-), \quad \Psi = (\psi_1^-, \psi_2^-, \psi_3^+), \tag{3.63}$$

$$\Phi^{-1} = \begin{pmatrix} \widehat{\phi}_1^- \\ \widehat{\phi}_2^- \\ \widehat{\phi}_3^+ \end{pmatrix}, \quad \Psi^{-1} = \begin{pmatrix} \widehat{\psi}_1^+ \\ \widehat{\psi}_2^+ \\ \widehat{\psi}_3^- \end{pmatrix}, \tag{3.64}$$

$$S = \begin{pmatrix} s_{11}^{+} & s_{12}^{+} & s_{13} \\ s_{21}^{+} & s_{22}^{+} & s_{23} \\ s_{31} & s_{32} & s_{33}^{-} \end{pmatrix}, \quad S^{-1} = \begin{pmatrix} \hat{s}_{11}^{-} & \hat{s}_{12}^{-} & \hat{s}_{13} \\ \hat{s}_{21}^{-} & \hat{s}_{22}^{-} & \hat{s}_{23} \\ \hat{s}_{31} & \hat{s}_{32} & \hat{s}_{33}^{+} \end{pmatrix}. \tag{3.65}$$

Here the superscripts "±" indicate the half plane of analyticity for the underlying quantity.

Hence we have constructed two matrix functions P^+ and P^- which are analytic in $\mathbb{C}_+$ and $\mathbb{C}_-$, respectively. On the real line, using Eqs. (3.51), (3.53), and (3.60), we easily see that

$$P^-(\zeta)P^+(\zeta) = G(\zeta), \qquad \zeta \in \mathbb{R}, \tag{3.66}$$

where

$$G = E(H_1 + H_2 S)(H_1 + S^{-1}H_2)E^{-1} = E \begin{pmatrix} 1 & 0 & \hat{s}_{13} \\ 0 & 1 & \hat{s}_{23} \\ s_{31} & s_{32} & 1 \end{pmatrix} E^{-1}. \tag{3.67}$$

Equation (3.66) determines a matrix Riemann–Hilbert problem. The normalization condition for this Riemann–Hilbert problem can be seen from (3.55) and (3.61) as the canonical condition

$$P^{\pm}(x,\zeta) \to \mathbf{I}_3, \quad \zeta \in \mathbb{C}_{\pm} \to \infty. \tag{3.68}$$

If this problem can be solved, then the potential Q can be reconstructed from its solution $P^{\pm}$. Specifically, by expanding $P^{\pm}$ as

$$P^{\pm}(x,\zeta) = \mathbf{I}_3 + \zeta^{-1}P_1^{\pm}(x) + \mathcal{O}(\zeta^{-2}), \tag{3.69}$$

inserting it into Eq. (3.45), and comparing terms of the same order in ζ^{-1}, we find that the potential Q is given by

$$Q = i[\Lambda, P_1^{+}] = -i[\Lambda, P_1^{-}]. \tag{3.70}$$

The scattering matrices S, S^{-1} and the other Jost solutions $(\phi_3, \psi_1, \psi_2, \widehat{\phi}_3, \widehat{\psi}_1, \widehat{\psi}_2)$ can be obtained from the Riemann–Hilbert solutions $P^{\pm}$ as well. Indeed, from the relation $\Phi = \Psi S$ and the large-x asymptotics of the Jost functions (Φ, Ψ), we see that the first and second columns of S as well as the third row of S can be obtained from the large-x asymptotics of ϕ_1, ϕ_2, and $\widehat{\psi}_3$, while the first and second rows of S^{-1} as well as the third column of S^{-1} can be obtained from the large-x asymptotics of $\widehat{\phi}_1, \widehat{\phi}_2$, and ψ_3. The remaining elements in S and S^{-1} can then be determined from the relation $SS^{-1} = \mathbf{I}_3$. After scattering matrices S and S^{-1} have been obtained, the rest of the Jost solutions can be readily determined from the relation $\Phi = \Psi S$.

To solve the Riemann–Hilbert problem (3.66), in addition to the scattering coefficients $(s_{31}, s_{32}, \hat{s}_{13}, \hat{s}_{23})$ for $\zeta \in \mathbb{R}$, one also needs the locations of zeros for $\det P^{\pm}$ in $\mathbb{C}_{\pm}$ as well as the kernels of $P^{\pm}$ at these zeros. From the definitions of $P^{\pm}$ in (3.53) and (3.60) as well as the scattering relation (3.51), we see that

$$\det P^{+} = \hat{s}_{33}, \quad \det P^{-} = s_{33}. \tag{3.71}$$

Suppose $\hat{s}_{33}(\zeta)$ and $s_{33}(\zeta)$ have simple zeros at $\zeta_k \in \mathbb{C}_+$ and $\bar{\zeta}_k \in \mathbb{C}_-$ $(1 \le k \le N)$, respectively. In this case, each of the kernels of $P^+(\zeta_k)$ and $P^-(\bar{\zeta}_k)$ contains only a single column vector v_k or row vector $\bar{v}_k$:

$$P^{+}(\zeta_k)\, v_k = 0, \quad \bar{v}_k\, P^{-}(\bar{\zeta}_k) = 0, \quad 1 \le k \le N. \tag{3.72}$$

Then the minimal scattering data for solving the Riemann–Hilbert problem (3.66) is

$$\left\{ s_{31}(\zeta), s_{32}(\zeta), \hat{s}_{13}(\zeta), \hat{s}_{23}(\zeta), \zeta \in \mathbb{R}; \quad \zeta_k, \bar{\zeta}_k, v_k, \bar{v}_k, 1 \le k \le N \right\}. \tag{3.73}$$

Here the vectors $(v_k, \bar{v}_k)$ are x-dependent. Repeating the same calculations as in the end of Sec. 2.1.2, we find that the spatial dependence of these vectors is

$$v_k(x) = e^{-i\zeta_k \Lambda x}\, v_{k0}, \quad \bar{v}_k(x) = \bar{v}_{k0}\, e^{i\bar{\zeta}_k \Lambda x}, \tag{3.74}$$

where v_{k0} and $\bar{v}_{k0}$ are x-independent column and row vectors.

Solutions to this Riemann–Hilbert problem (3.66) have been given in Sec. 2.1.2, where a general matrix Riemann–Hilbert problem was solved. These solutions will be used to construct N-soliton solutions of the hierarchy (3.20) in Sec. 3.5.

If the potential matrix Q satisfies certain symmetry properties, then such properties will trickle down to the Jost solutions and scattering matrices. Let us consider the general reduction (3.25). In this case, the potential matrix satisfies the following symmetry property:

$$Q^{\dagger} = -\mathbf{B} Q \mathbf{B}^{-1}, \quad \mathbf{B} = \begin{pmatrix} \mathbf{B}_2 & \\ & \mathbf{B}_1^{-1} \end{pmatrix}. \tag{3.75}$$

Then, by taking the Hermitian to the scattering equation (3.45) and utilizing this symmetry in Q, one finds that matrices $J_{\pm}^{\dagger}(\zeta^*)\mathbf{B}$ also satisfy the adjoint equation (3.57). But $J_{\pm}^{-1}(\zeta)$ satisfy (3.57) as well, and hence $J_{\pm}^{\dagger}(\zeta^*)\mathbf{B}$ and $J_{\pm}^{-1}(\zeta)$ are linearly related. Then using the boundary conditions (3.47) of $J_{\pm}$, we see that $J_{\pm}^{\dagger}(\zeta^*)\mathbf{B} = \mathbf{B} J_{\pm}^{-1}(\zeta)$, and thus

$$J_{\pm}^{\dagger}(\zeta^*) = \mathbf{B} J_{\pm}^{-1}(\zeta) \mathbf{B}^{-1}. \tag{3.76}$$

From this, we get the involution relations for Φ and Ψ as

$$\Phi^{\dagger}(\zeta^*) = \mathbf{B}\Phi^{-1}(\zeta)\mathbf{B}^{-1}, \quad \Psi^{\dagger}(\zeta^*) = \mathbf{B}\Psi^{-1}(\zeta)\mathbf{B}^{-1}. \tag{3.77}$$

Then in view of the scattering relation $\Phi = \Psi S$, we see that S satisfies the involution property

$$S^{\dagger}(\zeta^{*}) = \mathbf{B}S^{-1}(\zeta)\mathbf{B}^{-1}. \tag{3.78}$$

From this, we get

$$s_{33}^{*}(\zeta^{*}) = \hat{s}_{33}(\zeta). \tag{3.79}$$

Thus zeros of the Riemann–Hilbert problem (3.66) come in conjugate pairs:

$$\bar{\zeta}_k = \zeta_k^{*}. \tag{3.80}$$

Similarly, analytic solutions $P^{\pm}$ satisfy the involution relation

$$(P^{+})^{\dagger}(\zeta^{*}) = \mathbf{B}P^{-}(\zeta)\mathbf{B}^{-1}. \tag{3.81}$$

As a result, vectors in the kernels of $P^{+}(\zeta_k)$ and $P^{-}(\bar{\zeta}_k)$ satisfy the involution relation

$$\bar{v}_k = v_k^{\dagger}\mathbf{B}. \tag{3.82}$$

These involution properties will play an important role in the calculation of solutions for each individual integrable equation in the hierarchy (3.20). It is noted by passing that for the 3×3 scattering problem (3.41) under consideration, $\mathbf{B}_1$ is a real constant, and $\mathbf{B}_2$ is a 2×2 constant Hermitian matrix.

3.4 Time Evolution of Scattering Data

Now we determine time evolution of the scattering data (3.73) for the hierarchy (3.20). First, we recall the scattering relation

$$J_{-}E = J_{+}ES, \quad \zeta \in \mathbb{R}. \tag{3.83}$$

Notice that $J_{\pm}$ satisfy the temporal equation (3.46). Hence $J_{-}E$, or $J_{+}ES$, satisfies this temporal equation as well since $E = e^{-i\zeta\Lambda x}$ is time independent and diagonal. Then by inserting $J_{+}ES$ into (3.46), taking the limit $x \to +\infty$, and recalling the boundary condition (3.47) for J_{+} as well as the fact that $V \to 0$ as $x \to \pm\infty$, we get

$$S_t = -\frac{1}{2}i(2\zeta)^n[\Lambda, S]. \tag{3.84}$$

Similarly,

$$(S^{-1})_t = -\frac{1}{2}i(2\zeta)^n[\Lambda, S^{-1}]. \tag{3.85}$$

From these two equations, we get

$$\frac{\partial \hat{s}_{33}}{\partial t} = \frac{\partial s_{33}}{\partial t} = 0, \tag{3.86}$$

$$\frac{\partial s_{31}}{\partial t} = i(2\zeta)^n s_{31}, \quad \frac{\partial s_{32}}{\partial t} = i(2\zeta)^n s_{32}, \tag{3.87}$$

$$\frac{\partial \hat{s}_{13}}{\partial t} = -i(2\zeta)^n \hat{s}_{13}, \quad \frac{\partial \hat{s}_{23}}{\partial t} = -i(2\zeta)^n \hat{s}_{23}. \tag{3.88}$$

Equation (3.86) shows that $\hat{s}_{33}$ and s_{33} are time independent, and thus their zeros ζ_k and $\bar{\zeta}_k$ are also time independent. The other equations give the time evolution for the scattering data $(s_{31}, s_{32}, \hat{s}_{13}, \hat{s}_{23})$.

The time dependence of the scattering data v_k and $\bar{v}_k$ can be determined in the same way as was done in Sec. 2.1.3. Repeating that calculation, we get

$$v_k(x,t) = \exp\left[-i\zeta_k \Lambda x - \frac{1}{2} i(2\zeta_k)^n \Lambda t\right] v_{k0} \tag{3.89}$$

and

$$\bar{v}_k(x,t) = \bar{v}_{k0} \exp\left[i\bar{\zeta}_k \Lambda x + \frac{1}{2} i(2\zeta_k)^n \Lambda t\right], \tag{3.90}$$

where $(v_{k0}, \bar{v}_{k0})$ are now constants. With this, the time evolution for the minimal scattering data (2.95) is all obtained.

3.5 N-Soliton Solutions

The N-soliton solutions in the hierarchy (3.20) can be obtained by solving the Riemann–Hilbert problem (3.66) with $G = I$ and the discrete scattering data $\{\zeta_k, \bar{\zeta}_k, v_{k0}, \bar{v}_{k0}, 1 \le k \le N\}$. The solution to this Riemann–Hilbert problem has been given in Theorem 2.1 of Sec. 2.1.2 as $P^+ = \Gamma$ and $P^- = \Gamma^{-1}$, where Γ is given in Eq. (2.72). Hence the function P_1^+ in the expansion (3.69) of P^+ is

$$P_1^+(x,t) = \sum_{j,k=1}^{N} v_j \left(M^{-1}\right)_{jk} \bar{v}_k, \tag{3.91}$$

where vectors $(v_j, \bar{v}_j)$ are given by (3.89)–(3.90), and the matrix M is given by (2.74), i.e.,

$$M_{jk} = \frac{\bar{v}_j v_k}{\bar{\zeta}_j - \zeta_k}, \quad 1 \le j,k \le N. \tag{3.92}$$

The N-soliton solutions are then given from (3.70) as

$$u(x,t) = 2i P_1^+[1,3], \quad v(x,t) = 2i P_1^+[2,3], \tag{3.93}$$

$$\hat{u}(x,t) = -2i P_1^+[3,1], \quad \hat{v}(x,t) = -2i P_1^+[3,2]. \tag{3.94}$$

Notice that the only information used to construct the N-soliton solutions in the hierarchy (3.20) is the discrete scattering data $\{\zeta_k, \bar{\zeta}_k, v_{k0}, \bar{v}_{k0}, 1 \leq k \leq N\}$ and the dispersion relation of the hierarchy (3.20) as reflected in the exponents of Eqs. (3.89)–(3.90). For a particular evolution equation in this hierarchy (3.20), the only caution the reader should heed is the symmetry reduction of this equation from the hierarchy. This symmetry reduction corresponds to symmetry properties of the potential matrix Q in Eq. (3.42), and it induces the corresponding involution properties in the discrete scattering data $\{\zeta_k, \bar{\zeta}_k, v_{k0}, \bar{v}_{k0}, 1 \leq k \leq N\}$. For instance, the symmetry reduction (3.35) leads to the symmetry (3.75) of the potential matrix Q and involution properties (3.80) and (3.82) of the discrete scattering data. For some evolution equations (such as the Sasa–Satsuma equation (3.4)), the potential matrix Q admits more than one symmetry; thus the discrete scattering data also admits more than one involution. These involution properties of the discrete scattering data must all be respected in order to obtain the correct N-soliton solutions. To illustrate, explicit N-soliton solutions for the Manakov equations (3.1)–(3.2), the focusing-defocusing NLS system (3.29)–(3.30), and the Sasa–Satsuma equation (3.4) will be given later in this chapter (see Secs. 3.10–3.12).

3.6 Infinite Number of Conservation Laws

The hierarchy (3.20) also admits an infinite number of conservation laws. These conservation laws can be derived analogously to what we did for the NLS equation in Sec. 2.3. The only main difference is that conservation laws here will be generated by a coupled Riccati system rather than a single Riccati equation.

Let us consider a solution $Y = (y_1, y_2, y_3)^T$ to the third-order scattering operator (3.41). Defining

$$\mu^{(1)} = y_1/y_3, \quad \mu^{(2)} = y_2/y_3, \tag{3.95}$$

then it is easy to find from (3.41) that

$$(\ln y_3)_x = i\zeta + \hat{u}\mu^{(1)} + \hat{v}\mu^{(2)}. \tag{3.96}$$

Using the temporal equation (3.9) of this hierarchy, an analogous equation for $(\ln y_3)_t$ can also be derived. Cross-differentiating these two equations with respect to t and x, respectively, we see that if we expand $\mu^{(1)}$ and $\mu^{(2)}$ into the power series,

$$\mu^{(j)}(x,t,\zeta) = -\sum_{n=1}^{\infty} \frac{\mu_n^{(j)}(x,t)}{(-2i\zeta)^n}, \quad j = 1,2, \tag{3.97}$$

then $\hat{u}\mu_n^{(1)} + \hat{v}\mu_n^{(2)}$ would be the density of a local conservation law, and an infinite number of conserved quantities are

$$I_n = \int_{-\infty}^{\infty} \left(\hat{u}\mu_n^{(1)} + \hat{v}\mu_n^{(2)}\right) dx, \quad n = 1,2,\ldots. \tag{3.98}$$

To determine the expansion coefficients $\mu_n^{(1,2)}$, we use the scattering equation (3.41). Simple calculations show that $\mu^{(1,2)}$ satisfy the following coupled Riccati system:

$$\mu_x^{(1)} = -\mu^{(1)}\left(\hat{u}\mu^{(1)} + \hat{v}\mu^{(2)}\right) - 2i\zeta\mu^{(1)} + u, \tag{3.99}$$

$$\mu_x^{(2)} = -\mu^{(2)}\left(\hat{u}\mu^{(1)} + \hat{v}\mu^{(2)}\right) - 2i\zeta\mu^{(2)} + v. \tag{3.100}$$

Then by inserting $\mu^{(1,2)}$'s expansions (3.97) into this Riccati system and equating terms of the same power in $(-2i\zeta)^{-1}$, we find that $\mu_n^{(1,2)}$ are given as

$$\mu_1^{(1)} = u, \quad \mu_1^{(2)} = v, \tag{3.101}$$

$$\mu_2^{(1)} = u_x, \quad \mu_2^{(2)} = v_x, \tag{3.102}$$

and

$$\mu_{n+1}^{(1)} = \mu_{n,x}^{(1)} - \sum_{k=1}^{n-1}\left(\hat{u}\mu_k^{(1)} + \hat{v}\mu_k^{(2)}\right)\mu_{n-k}^{(1)}, \quad n \geq 2, \tag{3.103}$$

$$\mu_{n+1}^{(2)} = \mu_{n,x}^{(2)} - \sum_{k=1}^{n-1}\left(\hat{u}\mu_k^{(1)} + \hat{v}\mu_k^{(2)}\right)\mu_{n-k}^{(2)}, \quad n \geq 2. \tag{3.104}$$

Then the infinite number of conserved quantities (3.98) are obtained. The first three conserved quantities are

$$I_1 = \int_{-\infty}^{\infty} (u\hat{u} + v\hat{v})\,dx, \tag{3.105}$$

$$I_2 = \int_{-\infty}^{\infty} (u_x\hat{u} + v_x\hat{v})\,dx, \tag{3.106}$$

$$I_3 = \int_{-\infty}^{\infty} \left\{u_{xx}\hat{u} + v_{xx}\hat{v} - (u\hat{u} + v\hat{v})^2\right\} dx, \tag{3.107}$$

which are the mass (or power), momentum, and energy of the hierarchy (3.20). Higher conserved quantities can be similarly calculated. If one also wishes to obtain the flux functions of local conservation laws, then one can first use the temporal equation (3.9) of the Lax pair to derive the $(\ln y_3)_t$ equation, then insert the expansions (3.97). The coefficients of $(\ln y_3)_t$ at various orders of ζ^{-n} would then be the fluxes of local conservation laws.

For a particular evolution equation in the hierarchy (3.20), if the potential Q admits symmetry reductions, then these symmetries should also be inserted into the above general conservation laws. For instance, the Manakov system (3.1)–(3.2) admits the symmetry reductions $\hat{u} = -u^*, \hat{v} = -v^*$. Utilizing these symmetries, the first three conserved quantities of the Manakov system are then

$$I_1 = \int_{-\infty}^{\infty} (|u|^2 + |v|^2)\,dx, \tag{3.108}$$

$$I_2 = \int_{-\infty}^{\infty} (u^* u_x + v^* v_x) dx, \tag{3.109}$$

$$I_3 = \int_{-\infty}^{\infty} \left\{ u^* u_{xx} + v^* v_{xx} + (|u|^2 + |v|^2)^2 \right\} dx. \tag{3.110}$$

One may notice that these conserved quantities are not all the conserved quantities in the Manakov system. For instance, the powers of each component

$$I_{1u} = \int_{-\infty}^{\infty} |u|^2 dx, \quad I_{1v} = \int_{-\infty}^{\infty} |v|^2 dx \tag{3.111}$$

are individually also the conserved quantities of the Manakov system, but these conserved quantities are not in the infinite number of conserved quantities (3.98) given above. How all the conservation laws of the hierarchy (3.20) can be derived is still not clear.

3.7 Closure of Eigenstates in the Higher-Order Scattering Operator

In this section, we establish the closure of eigenstates in the third-order scattering operator (3.41). This proof is a simple extension of the one for the Zakharov–Shabat system in Sec. 2.5 and will be only sketched below. The proof of closure for the more general scattering operator (3.8) could be similarly given.

First, we define functions

$$R^+(x,y,\zeta) = \chi^+(x,\zeta)\,[\theta(y-x)H_1 - \theta(x-y)H_2](\chi^+)^{-1}(y,\zeta), \tag{3.112}$$

$$R^-(x,y,\zeta) = \chi^-(x,\zeta)\,[\theta(x-y)H_1 - \theta(y-x)H_2](\chi^-)^{-1}(y,\zeta), \tag{3.113}$$

where

$$\chi^+ = (\phi_1, \phi_2, \psi_3), \quad \chi^- = (\psi_1, \psi_2, \phi_3), \tag{3.114}$$

and $\theta(x)$ is the standard step function given in Eq. (2.187) of the previous chapter. Functions $R^\pm$ are meromorphic for $\zeta \in \mathbb{C}_\pm$, respectively, and are bounded as $\zeta \to \infty$. From similar relations as (2.188)–(2.189) but with $H_{1,2}$ replaced by (3.54), we see that

$$\det \chi^+(x,\zeta) = e^{-i\zeta x}\hat{s}_{33}(\zeta), \quad \det \chi^-(x,\zeta) = e^{-i\zeta x} s_{33}(\zeta). \tag{3.115}$$

Thus $R^\pm$ has pole singularities at the zeros of $\hat{s}_{33}$ and s_{33}, respectively. Then we define two complex contours, γ_+ and γ_-, with γ_+ starting from $\zeta = -\infty + i0^+$, passing over all zeros of $\hat{s}_{33}(\zeta)$ in $\mathbb{C}_+$, and ending at $\zeta = \infty + i0^+$, and with γ_- starting from $\zeta = -\infty + i0^-$, passing under all zeros of $s_{33}(\zeta)$ in $\mathbb{C}_-$, and ending at $\zeta = \infty + i0^-$. Using the large-ζ asymptotics

(3.55)–(3.56) and (3.61)–(3.62) of Jost solutions and then bringing the contours to the real axis, we get

$$\int_{\gamma_+} R^+(x,y,\zeta)d\zeta + \int_{\gamma_-} R^-(x,y,\zeta)d\zeta = 2\pi\delta(x-y)\Lambda. \tag{3.116}$$

Using the residue theorem, the left side of the above equation is found to be

$$\begin{aligned}\int_{\gamma_+} R^+(x,y,\zeta)d\zeta + \int_{\gamma_-} R^-(x,y,\zeta)d\zeta &= \int_{-\infty}^{\infty} \left[R^+(x,y,\xi)+R^-(x,y,\xi)\right]d\xi \\ &+2\pi i\sum_j \left\{\text{Res}\left[R^-(x,y,\zeta),\bar{\zeta}_j\right] - \text{Res}\left[R^+(x,y,\zeta),\zeta_j\right]\right\},\end{aligned} \tag{3.117}$$

where $\zeta_j \in \mathbb{C}_+$ and $\bar{\zeta}_j \in \mathbb{C}_-$ are the zeros of $\hat{s}_{33}(\zeta)$ and $s_{33}(\zeta)$. Following similar calculations as in Sec. 2.5, we also have

$$\begin{aligned}\int_{-\infty}^{\infty} \left[R^+(x,y,\xi)+R^-(x,y,\xi)\right]d\xi = \int_{-\infty}^{\infty} \Big[\chi^+(x,\xi)\,H_1(\chi^+)^{-1}(y,\xi) \\ -\chi^-(x,\xi)\,H_2(\chi^-)^{-1}(y,\xi)\Big]d\xi,\end{aligned} \tag{3.118}$$

$$\text{Res}\left[R^+(x,y,\zeta),\zeta_j\right] = \text{Res}\left[\chi^+(x,\zeta)\,H_1(\chi^+)^{-1}(y,\zeta),\zeta_j\right], \tag{3.119}$$

and

$$\text{Res}\left[R^-(x,y,\zeta),\bar{\zeta}_j\right] = -\text{Res}\left[\chi^-(x,\zeta)\,H_2(\chi^-)^{-1}(y,\zeta),\bar{\zeta}_j\right]. \tag{3.120}$$

These expressions can be further simplified. Notice that

$$(\chi^+)^{-1} = (\phi_1,\phi_2,\psi_3)^{-1} = (\Phi H_1 + \Psi H_2)^{-1} \tag{3.121}$$

and

$$\begin{pmatrix} \widehat{\psi}_1 \\ \widehat{\psi}_2^+ \\ \widehat{\phi}_3^+ \end{pmatrix} = H_1\Psi^{-1} + H_2\Phi^{-1}. \tag{3.122}$$

Using the relation $\Phi = \Psi S$, we get

$$\left(H_1\Psi^{-1} + H_2\Phi^{-1}\right)(\Phi H_1 + \Psi H_2) = H_1 S H_1 + H_2 S^{-1} S_2, \tag{3.123}$$

and thus

$$(\chi^+)^{-1} = \begin{pmatrix} s_{11} & s_{12} & \\ s_{21} & s_{22} & \\ & & \hat{s}_{33} \end{pmatrix}^{-1} \begin{pmatrix} \widehat{\psi}_1 \\ \widehat{\psi}_2 \\ \widehat{\phi}_3 \end{pmatrix}. \tag{3.124}$$

Similarly,

$$(\chi^{-})^{-1} = \begin{pmatrix} \hat{s}_{11} & \hat{s}_{12} & \\ \hat{s}_{21} & \hat{s}_{22} & \\ & & s_{33} \end{pmatrix}^{-1} \begin{pmatrix} \widehat{\phi}_1 \\ \widehat{\phi}_2 \\ \widehat{\psi}_3 \end{pmatrix}. \tag{3.125}$$

Substituting the above two equations into (3.116)–(3.120), the final closure relation is then obtained. Assuming the zeros $(\zeta_j, \bar{\zeta}_j)$ are all simple, then the closure relation reads

$$\begin{aligned} &\frac{1}{2\pi}\int_{-\infty}^{\infty}\left\{(\phi_1(x,\xi),\phi_2(x,\xi))\mathbf{s}^{-1}(\xi)\begin{pmatrix}\widehat{\psi}_1(y,\xi)\\ \widehat{\psi}_2(y,\xi)\end{pmatrix}-\frac{1}{s_{33}(\xi)}\phi_3(x,\xi)\widehat{\psi}_3(y,\xi)\right\}d\xi \\ &\quad -i\sum_j\left\{(\phi_1(x,\zeta_j),\phi_2(x,\zeta_j))\,\mathbf{s}_j^{-}\begin{pmatrix}\widehat{\psi}_1(y,\zeta_j)\\ \widehat{\psi}_2(y,\zeta_j)\end{pmatrix}+\frac{1}{s_{33}'(\bar{\zeta}_j)}\phi_3(x,\bar{\zeta}_j)\widehat{\psi}_3(y,\bar{\zeta}_j)\right\} \\ &= \delta(x-y)\Lambda. \end{aligned} \tag{3.126}$$

Here

$$\mathbf{s}(\zeta) \equiv \begin{pmatrix} s_{11}(\zeta) & s_{12}(\zeta) \\ s_{21}(\zeta) & s_{22}(\zeta) \end{pmatrix}, \quad \mathbf{s}_j^{-} \equiv \lim_{\zeta\to\zeta_j}(\zeta-\zeta_j)\mathbf{s}^{-1}(\zeta). \tag{3.127}$$

Note that $\det(\mathbf{s}) = \hat{s}_{33}$, thus $\mathbf{s}^{-1}(\zeta)$ has pole singularity at the zeros ζ_j of $\hat{s}_{33}$, and $\mathbf{s}_j^{-}$ in the above equation is the residue of $\mathbf{s}^{-1}(\zeta)$ at ζ_j. The closure relation (3.126) shows that for the third-order scattering operator (3.41), its discrete and continuous eigenstates also form a complete set. For multiple zeros of $(\zeta_j, \bar{\zeta}_j)$, the closure relation is similar, except that the residue terms in (3.126) need to be calculated differently.

3.8 Squared Eigenfunctions of the Higher-Order Scattering Operator

In this section, we derive the squared eigenfunctions for the third-order scattering operator (3.41) and its integrable hierarchy. Most of the derivations are analogous to those for the Zakharov–Shabat system (2.2) in Sec. 2.6, so our derivation will be brief in general. Some new features do arise though. Such features will be elaborated in detail. We will first assume the potential functions $(u, v, \hat{u}, \hat{v})$ in (3.42) to be independent, and hence derive the generic squared eigenfunctions for the third-order scattering operator (3.41). Afterwards, we will discuss the case when $(u, v, \hat{u}, \hat{v})$ are dependent on each other. We will describe how to obtain the corresponding squared eigenfunctions from the generic ones through symmetry reduction, and we explain how squared eigenfunctions can become *sums* of products of Jost functions due to the potential reduction.

3.8.1 Squared Eigenfunctions for Generic Potentials

We first consider the generic case when the potential functions $(u, v, \hat{u}, \hat{v})$ are independent from each other. We also assume as before that s_{33} and $\hat{s}_{33}$ have no zeros (the general case of s_{33} and $\hat{s}_{33}$ having zeros will be added later). Following our procedure described in Sec. 2.6, we first calculate variations of the scattering data in terms of variation of the potential. Repeating the same calculation as in Sec. 2.6, we also get the variation relations (2.221)–(2.224), i.e.,

$$\delta s_{ij}(\xi) = \int_{-\infty}^{\infty} \widehat{\psi}_i(x,\xi)\,\delta Q(x)\,\phi_j(x,\xi)\,dx \tag{3.128}$$

and

$$\delta \hat{s}_{ij}(\xi) = -\int_{-\infty}^{\infty} \widehat{\phi}_i(x,\xi)\,\delta Q(x)\,\psi_j(x,\xi)\,dx, \tag{3.129}$$

except that δQ is now

$$\delta Q = \begin{pmatrix} 0 & 0 & \delta u \\ 0 & 0 & \delta v \\ \delta \hat{u} & \delta \hat{v} & 0 \end{pmatrix}. \tag{3.130}$$

Defining new scattering coefficients

$$\rho_1 \equiv \frac{s_{31}}{s_{33}}, \quad \rho_2 \equiv \frac{s_{32}}{s_{33}}, \quad \hat{\rho}_1 \equiv \frac{\hat{s}_{13}}{\hat{s}_{33}}, \quad \hat{\rho}_2 \equiv \frac{\hat{s}_{23}}{\hat{s}_{33}}, \tag{3.131}$$

we can then find from the variation relations (3.128)–(3.129) that

$$\delta \rho_j(\xi) = \frac{1}{s_{33}^2(\xi)} \int_{-\infty}^{\infty} \widehat{\psi}_3(x,\xi)\,\delta Q(x)\,\varphi_j(x,\xi)\,dx \tag{3.132}$$

and

$$\delta \hat{\rho}_j(\xi) = -\frac{1}{\hat{s}_{33}^2(\xi)} \int_{-\infty}^{\infty} \widehat{\varphi}_j(x,\xi)\,\delta Q(x)\,\psi_3(x,\xi)\,dx, \tag{3.133}$$

where

$$\varphi_j = s_{33}\phi_j - s_{3j}\phi_3, \quad \widehat{\varphi}_j = \hat{s}_{33}\widehat{\phi}_j - \hat{s}_{j3}\widehat{\phi}_3, \quad j = 1,2. \tag{3.134}$$

Now we simplify these expressions of φ_j and $\widehat{\varphi}_j$ and show that they are analytic in the same half planes of $\widehat{\psi}_3$ and ψ_3, respectively. For this purpose, we first rewrite φ_j and $\widehat{\varphi}_j$ in terms of functions:

$$\chi^- = (\psi_1, \psi_2, \phi_3), \quad \hat{\chi}^+ = \begin{pmatrix} \widehat{\psi}_1 \\ \widehat{\psi}_2 \\ \widehat{\phi}_3 \end{pmatrix}. \tag{3.135}$$

Notice that

$$\chi^- = \Psi H_1 + \Phi H_2 = \Phi(S^{-1} H_1 + H_2) \tag{3.136}$$

and

$$\hat{\chi}^+ = H_1 \Psi^{-1} + H_2 \Phi^{-1} = (H_1 S + H_2)\Phi^{-1}; \tag{3.137}$$

thus

$$\Phi = \chi^- B, \quad \Phi^{-1} = \widehat{B}\hat{\chi}^+, \tag{3.138}$$

where

$$B = (S^{-1} H_1 + H_2)^{-1} = (H_1 + S H_2)^{-1} S \tag{3.139}$$

and

$$\widehat{B} = (H_1 S + H_2)^{-1} = S^{-1}(H_1 + H_2 S^{-1})^{-1}. \tag{3.140}$$

From the first formula of B in (3.139), we readily obtain the first two rows of B as the inverse of the 2×2 block in the upper left corner of S^{-1}, followed by a zero column. From the second formula of B in (3.139), we readily obtain the third row of B as the third row of S divided by s_{33}. The matrix $\widehat{B}$ can be similarly determined from the two formulae in (3.140). Thus matrices B and $\widehat{B}$ are found to be

$$B = \frac{1}{s_{33}} \begin{pmatrix} \hat{s}_{22} & -\hat{s}_{12} & 0 \\ -\hat{s}_{21} & \hat{s}_{11} & 0 \\ s_{31} & s_{32} & s_{33} \end{pmatrix}, \quad \widehat{B} = \frac{1}{\hat{s}_{33}} \begin{pmatrix} s_{22} & -s_{12} & \hat{s}_{13} \\ -s_{21} & s_{11} & \hat{s}_{23} \\ 0 & 0 & \hat{s}_{33} \end{pmatrix}. \tag{3.141}$$

Inserting Eqs. (3.138) and (3.141) into (3.134), the expressions for φ_j and $\widehat{\varphi}_j$ then reduce to

$$(\varphi_1, \varphi_2) = (\psi_1, \psi_2) \begin{pmatrix} \hat{s}_{22} & -\hat{s}_{12} \\ -\hat{s}_{21} & \hat{s}_{11} \end{pmatrix} \tag{3.142}$$

and

$$\begin{pmatrix} \widehat{\varphi}_1 \\ \widehat{\varphi}_2 \end{pmatrix} = \begin{pmatrix} s_{22} & -s_{12} \\ -s_{21} & s_{11} \end{pmatrix} \begin{pmatrix} \widehat{\psi}_1 \\ \widehat{\psi}_2 \end{pmatrix}. \tag{3.143}$$

Recalling the analytic properties of Jost solutions and scattering matrices summarized in (3.63)–(3.65), we see that (φ_1, φ_2) are indeed analytic in $\mathbb{C}_-$ as $\widehat{\psi}_3$, and $(\widehat{\varphi}_1, \widehat{\varphi}_2)$ are indeed analytic in $\mathbb{C}_+$ as ψ_3; thus products of Jost solutions and adjoint Jost solutions in the variation formulae (3.132)–(3.133) for $\delta\rho_j$ and $\delta\hat{\rho}_j$ are analytic, as we desired. Furthermore, these products are only between Jost solutions Ψ and their inverse Ψ^{-1} whose boundary conditions are both set at $x = +\infty$.

The variation relations (3.132)–(3.133) can be rewritten into the following more convenient form:

$$\delta\rho_j(\xi) = \frac{1}{s_{33}^2(\xi)} \left\langle \Omega_j^-(x,\xi),\, (\delta u,\, \delta v,\, -\delta\hat{u},\, -\delta\hat{v})^T(x) \right\rangle, \tag{3.144}$$

$$\delta\hat{\rho}_j(\xi) = \frac{1}{\hat{s}_{33}^2(\xi)} \left\langle \Omega_j^+(x,\xi),\, (\delta u,\, \delta v,\, -\delta\hat{u},\, -\delta\hat{v})^T(x) \right\rangle, \tag{3.145}$$

where $j = 1,2$,

$$\Omega_j^- = \begin{pmatrix} \widehat{\psi}_{31}\varphi_{3j} \\ \widehat{\psi}_{32}\varphi_{3j} \\ -\widehat{\psi}_{33}\varphi_{1j} \\ -\widehat{\psi}_{33}\varphi_{2j} \end{pmatrix}, \quad \Omega_j^+ = \begin{pmatrix} -\widehat{\varphi}_{j1}\psi_{33} \\ -\widehat{\varphi}_{j2}\psi_{33} \\ \widehat{\varphi}_{j3}\psi_{13} \\ \widehat{\varphi}_{j3}\psi_{23} \end{pmatrix}, \tag{3.146}$$

φ_{ij}, $\widehat{\varphi}_{jk}$ are elements in vectors φ_j, $\widehat{\varphi}_j$ as

$$\varphi_j = (\varphi_{1j}, \varphi_{2j}, \varphi_{3j})^T, \quad \widehat{\varphi}_j = (\widehat{\varphi}_{j1}, \widehat{\varphi}_{j2}, \widehat{\varphi}_{j3}), \tag{3.147}$$

and the inner product is defined as

$$\langle \mathbf{f}, \mathbf{g} \rangle = \int_{-\infty}^{\infty} \mathbf{f}^T(x)\, \mathbf{g}(x)\, dx. \tag{3.148}$$

These functions $\{\Omega_1^\pm, \Omega_2^\pm\}$ are the adjoint squared eigenfunctions. Notice that in the variation relations (3.144)–(3.145), the potential variation takes the form of $(\delta u,\, \delta v,\, -\delta\hat{u},\, -\delta\hat{v})^T$ rather than $(\delta u,\, \delta v,\, \delta\hat{u},\, \delta\hat{v})^T$. The same goes to the expansion of the potential variation (3.156) below. This form of the potential variation corresponds to the form of the integrable hierarchy (3.20) and is necessary so that the corresponding squared eigenfunctions and their adjoints are also eigenfunctions of the recursion operator $\mathbf{L}_R$ and its adjoint operator $\mathbf{L}_R^A$ (see the next section). For the Zakharov–Shabat system (2.2), $\hat{u} = -u^*$. Thus the above form of the potential variation is consistent with that in variation relations (2.230)–(2.231) and (2.254) for the Zakharov–Shabat system.

Next, we calculate variation of the potential in terms of variations of the scattering data. Defining new Jost solutions

$$F^+ = P^+ \operatorname{diag}\left(1, 1, \frac{1}{\hat{s}_{33}}\right), \qquad F^- = (P^-)^{-1} \operatorname{diag}(1, 1, s_{33}), \tag{3.149}$$

we find that the Riemann–Hilbert problem (3.66) becomes

$$F^{+}(\zeta)=F^{-}(\zeta)\widetilde{G}(\zeta), \qquad \zeta\in\mathbb{R}, \tag{3.150}$$

where

$$\widetilde{G}=E\begin{pmatrix}1 & 0 & \hat{\rho}_1\\ 0 & 1 & \hat{\rho}_2\\ \rho_1 & \rho_2 & 1+\rho_1\hat{\rho}_1+\rho_2\hat{\rho}_2\end{pmatrix}E^{-1}. \tag{3.151}$$

Taking variation to this Riemann–Hilbert problem and repeating the same calculations as in Sec. 2.6, we get (Yang and Kaup (2009))

$$\delta u(x)=-\frac{1}{\pi}\int_{-\infty}^{\infty}\Pi_{13}(x,\xi)\,d\xi, \qquad \delta v(x)=-\frac{1}{\pi}\int_{-\infty}^{\infty}\Pi_{23}(x,\xi)\,d\xi, \tag{3.152}$$

$$\delta\hat{u}(x)=\frac{1}{\pi}\int_{-\infty}^{\infty}\Pi_{31}(x,\xi)\,d\xi, \qquad \delta\hat{v}(x)=\frac{1}{\pi}\int_{-\infty}^{\infty}\Pi_{32}(x,\xi)\,d\xi, \tag{3.153}$$

where

$$\Pi=\Phi\begin{pmatrix}0 & 0 & \delta\hat{\rho}_1\\ 0 & 0 & \delta\hat{\rho}_2\\ \delta\rho_1 & \delta\rho_2 & 0\end{pmatrix}\Phi^{-1}, \tag{3.154}$$

i.e.,

$$\Pi=\phi_3\widehat{\phi}_1\,\delta\rho_1+\phi_3\widehat{\phi}_2\,\delta\rho_2+\phi_1\widehat{\phi}_3\,\delta\hat{\rho}_1+\phi_2\widehat{\phi}_3\,\delta\hat{\rho}_2. \tag{3.155}$$

Thus variation of the potential via variations of the scattering data is

$$(\delta u,\delta v,-\delta\hat{u},-\delta\hat{v})^{T}(x)=-\frac{1}{\pi}\int_{-\infty}^{\infty}\sum_{j=1}^{2}\left[Z_j^{-}(x,\xi)\delta\rho_j(\xi)+Z_j^{+}(x,\xi)\delta\hat{\rho}_j(\xi)\right]d\xi, \tag{3.156}$$

where

$$Z_j^{-}=\begin{pmatrix}\phi_{13}\widehat{\phi}_{j3}\\ \phi_{23}\widehat{\phi}_{j3}\\ \phi_{33}\widehat{\phi}_{j1}\\ \phi_{33}\widehat{\phi}_{j2}\end{pmatrix}, \quad Z_j^{+}=\begin{pmatrix}\phi_{1j}\widehat{\phi}_{33}\\ \phi_{2j}\widehat{\phi}_{33}\\ \phi_{3j}\widehat{\phi}_{31}\\ \phi_{3j}\widehat{\phi}_{32}\end{pmatrix}, \quad j=1,2, \tag{3.157}$$

are the squared eigenfunctions, and ϕ_{ij}, $\widehat{\phi}_{ij}$ are the (i,j)th elements of matrices Φ, Φ^{-1}.

When the potential expansion (3.156) is inserted into the variation relations (3.144)–(3.145), we get the inner products between squared eigenfunctions and their adjoints as

$$\left\langle \Omega_j^-(x,\xi), Z_j^-(x,\xi')\right\rangle = -\pi s_{33}^2(\xi)\delta(\xi-\xi'), \quad j=1,2, \tag{3.158}$$

$$\left\langle \Omega_j^+(x,\xi), Z_j^+(x,\xi')\right\rangle = -\pi \hat{s}_{33}^2(\xi)\delta(\xi-\xi'), \quad j=1,2, \tag{3.159}$$

and other inner products are zero. Inserting the variation relations (3.144)–(3.145) into the potential expansion (3.156) and exchanging orders of integration, we get the closure relation

$$\begin{aligned}&-\frac{1}{\pi}\int_{-\infty}^{\infty}\sum_{k=1}^{2}\left[\frac{1}{s_{33}^2(\xi)}Z_k^-(x,\xi)\Omega_k^{-T}(y,\xi)+\frac{1}{\hat{s}_{33}^2(\xi)}Z_k^+(x,\xi)\Omega_k^{+T}(y,\xi)\right]d\xi\\ &=\delta(x-y)\mathbf{I}_4,\end{aligned} \tag{3.160}$$

where $\mathbf{I}_4$ is the unit matrix of rank 4.

In the general case where s_{33} and $\hat{s}_{33}$ have zeros, the above closure relation needs to include the discrete-spectrum contribution. As before, this discrete-spectrum contribution is nothing but the residues of integrand functions in the above closure relation. Notice from the above expressions of squared eigenfunctions that $Z_1^-\Omega_1^{-T}$ and $Z_2^-\Omega_2^{-T}$ are analytic in the lower half plane, while $Z_1^+\Omega_1^{+T}$ and $Z_2^+\Omega_2^{+T}$ are analytic in the upper half plane. Also notice that $1/s_{33}^2$ and $1/\hat{s}_{33}^2$ are meromorphic in the lower and upper half planes, respectively. Thus the residues of integrand functions in (3.160) can be easily obtained. These residue terms are similar to those in Eq. (2.264) of Sec. 2.6. Specifically, these terms are

$$\begin{aligned}&-\sum_{j=1}^{N}\frac{2i}{s_{33}'^2(\bar{\zeta}_j)}\sum_{k=1}^{2}\left[Z_k^-(x,\bar{\zeta}_j)\Theta_k^{-T}(y,\bar{\zeta}_j)+\dot{Z}_k^-(x,\bar{\zeta}_j)\Omega_k^{-T}(y,\bar{\zeta}_j)\right]\\ &+\sum_{j=1}^{N}\frac{2i}{\hat{s}_{33}'^2(\zeta_j)}\sum_{k=1}^{2}\left[Z_k^+(x,\zeta_j)\Theta_k^{+T}(y,\zeta_j)+\dot{Z}_k^+(x,\zeta_j)\Omega_k^{+T}(y,\zeta_j)\right],\end{aligned} \tag{3.161}$$

where

$$\Theta_k^-(x,\bar{\zeta}_j)=\dot{\Omega}_k^-(x,\bar{\zeta}_j)-\frac{s_{33}''(\bar{\zeta}_j)}{s_{33}'(\bar{\zeta}_j)}\Omega_k^-(x,\bar{\zeta}_j), \tag{3.162}$$

$$\Theta_k^+(x,\zeta_j)=\dot{\Omega}_k^+(x,\zeta_j)-\frac{\hat{s}_{33}''(\zeta_j)}{\hat{s}_{33}'(\zeta_j)}\Omega_k^+(x,\zeta_j), \tag{3.163}$$

and the dot above a variable represents its derivative with respect to ζ. However, for each pair of zeros $(\zeta_j,\bar{\zeta}_j)$, Eq. (3.161) contains eight terms, but we only expect six. The reason for our expecting only six terms can be heuristically understood as follows. Let us consider the reduction (3.27) which gives the Manakov equations. In this case, due to the involution properties of zeros $(\zeta_j,\bar{\zeta}_j)$ and vectors $\{\bar{v}_{j0}, v_{j0}\}$, the discrete spectral data in Eqs. (3.73)–(3.74) for each pair of zeros contains only six real parameters, i.e., ζ_j and two complex parameters in v_{j0} (since the third complex parameter in v_{j0} can be normalized to unity). This means that a Manakov single soliton contains only six real parameters. As a result, the linearized Manakov equations around their single soliton only have six discrete (localized) solutions which are induced by variation of the soliton to its six free parameters. Since discrete squared eigenfunctions also satisfy the linearized Manakov equations (see Secs. 2.7 and 3.9), their number then can only be six, not eight. This means that the eight terms for each pair of zeros $(\zeta_j, \bar{\zeta}_j)$ in (3.161) are not linearly independent. This is indeed the case, and it is an important fact of the third-order scattering problem (3.41).

Below, we will show that $\Omega_1^-(x,\bar{\zeta}_j)$ and $\Omega_2^-(x,\bar{\zeta}_j)$ are proportional to each other; while $\Omega_1^+(x,\zeta_j)$ and $\Omega_2^+(x,\zeta_j)$ are proportional to each other; thus there are only six linearly independent terms in Eq. (3.161), as one would expect. Our proof will follow Kaup and Yang (2009). Even though this proof is possible for a general localized potential, below we assume the potential to have compact support for simplicity. In this case, all Jost solutions and the scattering coefficients exist in the entire complex plane. Defining

$$\chi^+ = (\phi_1, \phi_2, \psi_3), \tag{3.164}$$

then

$$\chi^+ = \Phi H_1 + \Psi H_2 = \Phi(H_1 + S^{-1} H_2). \tag{3.165}$$

Utilizing this equation and (3.138), we get

$$\chi^+ = \chi^- A, \tag{3.166}$$

where χ^- is defined in Eq. (3.135), and the matrix A is

$$A = B(H_1 + S^{-1} H_2) = (H_1 + S H_2)^{-1}(S H_1 + H_2). \tag{3.167}$$

From the first formula of A above as well as the form (3.141) of matrix B, we can readily obtain the first and second columns of A. From the second formula of A in (3.167), we can readily obtain the third column of A. Putting these results together, we find the matrix A to be

$$A = \frac{1}{s_{33}} \begin{pmatrix} \hat{s}_{22} & -\hat{s}_{12} & -s_{13} \\ -\hat{s}_{21} & \hat{s}_{11} & -s_{23} \\ s_{31} & s_{32} & 1 \end{pmatrix}. \tag{3.168}$$

Recall that $\bar{\zeta}_j$ is a zero of $s_{33}(\zeta)$. Thus from Eqs. (3.135), (3.142), (3.166), and (3.168), we find that at $\zeta = \bar{\zeta}_j$,

$$\varphi_k(x,\bar{\zeta}_j) = -s_{3k}(\bar{\zeta}_j)\,\phi_3(x,\bar{\zeta}_j), \quad k = 1,2, \tag{3.169}$$

i.e., φ_1 and φ_2 are proportional to each other. Consequently, in view of the definitions (3.146) of Ω_1^- and Ω_2^-, we get

$$\Omega_k^-(x,\bar{\zeta}_j) = -s_{3k}(\bar{\zeta}_j)\,\Omega_0^-(x,\bar{\zeta}_j), \quad k = 1,2, \tag{3.170}$$

where

$$\Omega_0^- \equiv \left(\widehat{\psi}_{31}\phi_{33},\ \widehat{\psi}_{32}\phi_{33},\ -\widehat{\psi}_{33}\phi_{13},\ -\widehat{\psi}_{33}\phi_{23}\right)^T. \tag{3.171}$$

Hence $\Omega_1^-(x,\bar{\zeta}_j)$ and $\Omega_2^-(x,\bar{\zeta}_j)$ are proportional to each other. Carrying out similar calculations to $\Omega_1^+(x,\zeta_j)$ and $\Omega_2^+(x,\zeta_j)$, we find that

$$\widehat{\varphi}_k(x,\zeta_j) = -\hat{s}_{k3}(\zeta_j)\widehat{\phi}_3(x,\zeta_j), \quad k=1,2, \tag{3.172}$$

and

$$\Omega_k^+(x,\zeta_j) = -\hat{s}_{k3}(\zeta_j)\,\Omega_0^+(x,\zeta_j), \quad k=1,2, \tag{3.173}$$

where

$$\Omega_0^+ \equiv \left(-\widehat{\phi}_{31}\psi_{33},\ -\widehat{\phi}_{32}\psi_{33},\ \widehat{\phi}_{33}\psi_{13},\ \widehat{\phi}_{33}\psi_{23}\right)^T. \tag{3.174}$$

Thus $\Omega_1^+(x,\zeta_j)$ and $\Omega_2^+(x,\zeta_j)$ are proportional to each other. Inserting the relations (3.170) and (3.173) into the residues (3.161), and adding them to the continuous contributions in Eq. (3.160), the final closure relation of squared eigenfunctions in the presence of discrete-spectrum contributions is then

$$\begin{aligned}
&-\frac{1}{\pi}\int_{-\infty}^{\infty}\sum_{k=1}^{2}\left[\frac{1}{s_{33}^2(\xi)}Z_k^-(x,\xi)\Omega_k^{-T}(y,\xi)+\frac{1}{\hat{s}_{33}^2(\xi)}Z_k^+(x,\xi)\Omega_k^{+T}(y,\xi)\right]d\xi \\
&-\sum_{j=1}^{N}\frac{2i}{s_{33}'^2(\bar{\zeta}_j)}\left[Z_d^-(x,\bar{\zeta}_j)\Omega_0^{-T}(y,\bar{\zeta}_j)+\sum_{k=1}^{2}Z_k^-(x,\bar{\zeta}_j)\Theta_k^{-T}(y,\bar{\zeta}_j)\right] \\
&+\sum_{j=1}^{N}\frac{2i}{\hat{s}_{33}'^2(\zeta_j)}\left[Z_d^+(x,\zeta_j)\Omega_0^{+T}(y,\zeta_j)+\sum_{k=1}^{2}Z_k^+(x,\zeta_j)\Theta_k^{+T}(y,\zeta_j)\right]=\delta(x-y)\mathbf{I}_4,
\end{aligned} \tag{3.175}$$

where

$$Z_d^- \equiv -\sum_{k=1}^{2}s_{3k}\dot{Z}_k^-, \quad Z_d^+ \equiv -\sum_{k=1}^{2}\hat{s}_{k3}\dot{Z}_k^+. \tag{3.176}$$

In the above closure relation, there are now only six discrete terms for each pair of $(\zeta_j,\bar{\zeta}_j)$, which is the correct number of terms to be expected.

3.8.2 Squared Eigenfunctions under Potential Reductions

Next, we consider the case when the potential functions $(u,v,\hat{u},\hat{v})$ are related to each other. Here we consider two representative cases. One is that

$$\hat{u} = -u^*, \quad \hat{v} = -v^*, \tag{3.177}$$

as in the Manakov system (3.1)–(3.2). This case corresponds to the general reduction (3.25) (i.e., (3.75)), with $\mathbf{B}_1 = 1$ and $\mathbf{B}_2 = \mathbf{I}_2$. In this case, there are two independent complex variables (u, v) in the potential matrix Q. The other case is that

$$v = u^*, \quad \hat{u} = -u^*, \quad \hat{v} = -u, \tag{3.178}$$

as in the Sasa–Satsuma equation (3.4). This case corresponds to two simultaneous reductions: one is (3.25) with $\mathbf{B}_1 = 1$ and $\mathbf{B}_2 = \mathbf{I}_2$, and the other one is $v = u^*$. In this case, there is only one independent complex variable u in the potential matrix Q.

In the first case (3.177), since complex conjugates (u^*, v^*) should be treated as being different from (u, v), squared eigenfunctions for this case are exactly the same as the generic ones obtained above. The only difference is that due to the symmetry reduction (3.177), Jost solutions and scattering elements satisfy the involution properties (see (3.77)–(3.80))

$$\Phi^\dagger(\zeta^*) = \Phi^{-1}(\zeta), \quad \Psi^\dagger(\zeta^*) = \Psi^{-1}(\zeta), \tag{3.179}$$

and

$$S^\dagger(\zeta^*) = S^{-1}(\zeta), \quad s_{33}^*(\zeta^*) = \hat{s}_{33}(\zeta), \quad \bar{\zeta}_k = \zeta_k^*. \tag{3.180}$$

These involution properties should be kept in mind when building the squared eigenfunctions and their adjoints from the generic expressions given above.

In the second case (3.178), the situation is quite different. In this case, due to the further reduction of the potential, the squared eigenfunctions are no longer single products of Jost solutions and their adjoints (as in the generic case of the previous subsection). Instead, they become sums of products of Jost solutions and their adjoints. In addition, the squared eigenfunctions are no longer vectors of length four as in the generic case, but rather reduce to vectors of length two. These are distinctive features which we will elucidate below.

Before we derive squared eigenfunctions under the potential reduction (3.178), we need to detail the involution properties of Jost solutions and scattering elements under this reduction. This reduction (3.178) corresponds to two simultaneous symmetry reductions,

$$Q^\dagger = -Q \tag{3.181}$$

and

$$Q^* = \sigma Q \sigma^{-1}, \qquad \text{where} \quad \sigma = \begin{pmatrix} 0 & 1 & 0 \\ 1 & 0 & 0 \\ 0 & 0 & 1 \end{pmatrix}. \tag{3.182}$$

Due to the first reduction (3.181), Jost solutions and scattering elements satisfy the involution properties (3.179)–(3.180) above. Due to the second reduction (3.182), we see that

$\sigma^{-1}\Phi^*(-\zeta^*)\sigma$ and $\sigma^{-1}\Psi^*(-\zeta^*)\sigma$ also satisfy the scattering equation (3.41). Then in view of the boundary conditions of Jost solutions Φ and Ψ at $x=\pm\infty$, we see that these matrix solutions must be equal to $\Phi(\zeta)$ and $\Psi(\zeta)$, respectively. Thus, Φ and Ψ satisfy the additional involution properties

$$\Phi^*(-\zeta^*)=\sigma\Phi(\zeta)\sigma^{-1}, \quad \Psi^*(-\zeta^*)=\sigma\Psi(\zeta)\sigma^{-1}. \tag{3.183}$$

Then the corresponding involution relation for the scattering matrix S can be obtained from $\Phi=\Psi S$ as

$$S^*(-\zeta^*)=\sigma S(\zeta)\sigma^{-1}. \tag{3.184}$$

From these involution relations, we see that

$$\rho_2(\zeta)=\hat{\rho}_1(-\zeta), \quad \hat{\rho}_2(\zeta)=\rho_1(-\zeta), \quad \zeta\in\mathbb{R}, \tag{3.185}$$

and

$$\sigma\phi_1(\zeta)=\widehat{\phi}_2^T(-\zeta), \quad \sigma\phi_2(\zeta)=\widehat{\phi}_1^T(-\zeta), \quad \sigma\phi_3(\zeta)=\widehat{\phi}_3^T(-\zeta). \tag{3.186}$$

Now we derive the squared eigenfunctions under the potential reduction (3.178). First, we re-examine the formulae (3.144)–(3.145) for variations of the scattering data due to variation of the potential. Under the reduction (3.178),

$$-\delta\hat{u}=\delta v=\delta u^*, \quad -\delta\hat{v}=\delta u. \tag{3.187}$$

Thus formulae (3.144)–(3.145) reduce to

$$\delta\rho_1(\xi)=\frac{1}{s_{33}^2(\xi)}\Big\langle \Xi^-(x,\xi),\,\big(\delta u,\,\delta u^*\big)^T(x)\Big\rangle, \tag{3.188}$$

$$\delta\hat{\rho}_1(\xi)=\frac{1}{\hat{s}_{33}^2(\xi)}\Big\langle \Xi^+(x,\xi),\,\big(\delta u,\,\delta u^*\big)^T(x)\Big\rangle, \tag{3.189}$$

where

$$\Xi^-=\begin{pmatrix}\widehat{\psi}_{31}\varphi_{31}-\widehat{\psi}_{33}\varphi_{21}\\ \widehat{\psi}_{32}\varphi_{31}-\widehat{\psi}_{33}\varphi_{11}\end{pmatrix}, \quad \Xi^+=\begin{pmatrix}\widehat{\varphi}_{13}\psi_{23}-\widehat{\varphi}_{11}\psi_{33}\\ \widehat{\varphi}_{13}\psi_{13}-\widehat{\varphi}_{12}\psi_{33}\end{pmatrix}. \tag{3.190}$$

Here φ_{ij} and $\widehat{\varphi}_{ij}$ are as defined in (3.142) and (3.147). The formulae for $\delta\rho_2$ and $\delta\hat{\rho}_2$ are redundant due to the above involution relations and thus can be dropped. These functions Ξ^- and Ξ^+ are the adjoint squared eigenfunctions under the symmetry reduction (3.178). These adjoint squared eigenfunctions become *sums* of products of Jost solutions. The reason is clearly due to the potential reduction (3.187), which leads to the combination of terms in the generic formulae (3.144)–(3.145).

Next, we re-examine the formula (3.156) for variation of the potential due to variations in the scattering data. In view of the involution relations (3.185), we see that the terms on the right-hand side of (3.156) can also be combined so that (3.156) becomes

$$(\delta u, \delta v, -\delta\hat{u}, -\delta\hat{v})^T(x) = -\frac{1}{\pi}\int_{-\infty}^{\infty}\Big\{\big[Z_1^-(x,\xi)+Z_2^+(x,-\xi)\big]\delta\rho_1(\xi) + \big[Z_1^+(x,\xi)+Z_2^-(x,-\xi)\big]\delta\hat{\rho}_1(\xi)\Big\}d\xi. \quad (3.191)$$

Then using the involution properties (3.186) and (3.187), the above expansion further reduces to

$$\left(\delta u, \delta u^*\right)^T(x) = -\frac{1}{\pi}\int_{-\infty}^{\infty}\left[W^-(x,\xi)\delta\rho_1(\xi)+W^+(x,\xi)\delta\hat{\rho}_1(\xi)\right]d\xi, \quad (3.192)$$

where

$$W^- = \begin{pmatrix} \phi_{13}\widehat{\phi}_{13}+\phi_{33}\widehat{\phi}_{12} \\ \phi_{23}\widehat{\phi}_{13}+\phi_{33}\widehat{\phi}_{11} \end{pmatrix}, \quad W^+ = \begin{pmatrix} \phi_{11}\widehat{\phi}_{33}+\phi_{31}\widehat{\phi}_{32} \\ \phi_{21}\widehat{\phi}_{33}+\phi_{31}\widehat{\phi}_{31} \end{pmatrix} \quad (3.193)$$

are the squared eigenfunctions under the symmetry reduction (3.178). The last two rows in Eq. (3.191) are redundant due to the involution relations and can be dropped. These squared eigenfunctions are also *sums* of products of Jost functions. The reason here is due to the involution relations (3.185) between scattering coefficients, which lead to the combination of terms in the generic potential expansion (3.156). Of course, the involution relations (3.185) are induced by the potential reduction (3.178) in the first place. The inner products and closure relations of these squared eigenfunctions can be trivially obtained by inserting the potential expansion (3.192) and scattering-data variation formulae (3.188)–(3.189) into each other as we did before for the generic case, and so they will not be presented.

3.9 Squared Eigenfunctions, the Linearization Operator, and the Recursion Operator

As we have seen for the NLS equation in the previous chapter, squared eigenfunctions are intimately related to the linearization operator and the recursion operator of the integrable equation. Specifically, squared eigenfunctions are the eigenfunctions of the recursion operator, and time-dependent squared eigenfunctions satisfy the linearized integrable equation. These relations hold for the hierarchy (3.20) associated with the higher-order scattering equation (3.8) too. To demonstrate, we will show these relations for the hierarchy associated with the third-order scattering equation (3.41) in this section. In our discussions, we will first consider the generic case where the potential functions $(u, v, \hat{u}, \hat{v})$ are independent of each other. The nongeneric case with potential reductions will be considered at the end of the section.

When potentials $(u, v, \hat{u}, \hat{v})$ are independent of each other, squared eigenfunctions $\{Z_1^{\pm}, Z_2^{\pm}\}$ are given in Eq. (3.157), and they are products of Jost solutions Φ and their adjoints Φ^{-1} whose boundary conditions are all set at $x = -\infty$. The recursion operator $\mathbf{L}_R$ is given in Eq. (3.18) with

$$\mathbf{q} = \begin{pmatrix} u \\ v \end{pmatrix}, \quad \mathbf{r} = \begin{pmatrix} \hat{u} \\ \hat{v} \end{pmatrix}. \tag{3.194}$$

In order for these squared eigenfunctions to be related to this recursion operator, the antiderivative ∂^{-1} in the definition (3.18) of this operator must be taken as $\partial^{-1} = \int_{-\infty}^{x} dy$. In that case, these squared eigenfunctions are also eigenfunctions of this recursion operator $\mathbf{L}_R$. To show this, we first notice that Jost solutions Φ satisfy the scattering equation (3.41), i.e.,

$$\Phi_x = -i\zeta\Lambda\Phi + Q\Phi, \tag{3.195}$$

while adjoint Jost solutions Φ^{-1} satisfy the adjoint scattering equation of (3.41), i.e.,

$$-(\Phi^{-1})_x = -i\zeta\Phi^{-1}\Lambda + \Phi^{-1}Q. \tag{3.196}$$

Then using the definitions (3.157) of squared eigenfunctions and the above two equations for Jost solutions, together with these Jost solutions' boundary conditions of $\Phi(x) \to e^{-i\zeta\Lambda x}$ as $x \to -\infty$, we can easily verify that these squared eigenfunctions are indeed eigenfunctions of the recursion operator with the eigenrelations

$$\mathbf{L}_R Z_j^{\pm} = 2\zeta Z_j^{\pm}, \quad j = 1, 2. \tag{3.197}$$

Similarly one can verify that adjoint squared eigenfunctions $\Omega_j^{\pm}$ are eigenfunctions of the adjoint recursion operator $\mathbf{L}_R^A$ as well.

Next, we consider the relation between squared eigenfunctions and the linearization operator of the hierarchy (3.20). This hierarchy is

$$i\begin{pmatrix} \mathbf{q} \\ -\mathbf{r} \end{pmatrix}_t = \mathbf{L}_R^n \begin{pmatrix} \mathbf{q} \\ \mathbf{r} \end{pmatrix}, \quad n = 1, 2, \ldots, \tag{3.198}$$

where variables $(\mathbf{q}, \mathbf{r})$ are given in Eq. (3.194) above. Perturbing the solutions $(\mathbf{q}, \mathbf{r})$ to $(\mathbf{q} + \delta\mathbf{q}, \mathbf{r} + \delta\mathbf{r})$ with $(\delta\mathbf{q}, \delta\mathbf{r}) \ll 1$, and linearizing this hierarchy, we get the linearized equation which can be written as

$$\mathcal{L}\begin{pmatrix} \delta\mathbf{q}(x,t) \\ -\delta\mathbf{r}(x,t) \end{pmatrix} = 0, \tag{3.199}$$

where $\mathcal{L}$ is the linearization operator. It is important to emphasize that this linearized equation is formulated for the perturbations $(\delta\mathbf{q}, -\delta\mathbf{r})$ rather than $(\delta\mathbf{q}, \delta\mathbf{r})$. This is necessary so that the form of perturbations $(\delta\mathbf{q}, -\delta\mathbf{r})$ is consistent with the form of potential variations $(\delta u, \delta v, -\delta\hat{u}, -\delta\hat{v})^T$ in the expansion (3.156) of the previous section. It is also necessary so that the form of the linearized equation (3.199) is consistent with the form of the hierarchy (3.198). These consistencies must be followed in order to establish the relations between the linearization operator, the recursion operator, and squared eigenfunctions below.

Now we follow the general procedure of Sec. 2.7 to make the connection between squared eigenfunctions and the linearization operator $\mathcal{L}$ above. For this purpose, we substitute the potential-variation expansion (3.156) into the above linearized equation (3.199) and get

$$\int_{-\infty}^{\infty}\sum_{j=1}^{2}\left[\mathcal{L}Z_j^-(x,t,\xi)\,\delta\rho_j(\xi,t)+\mathcal{L}Z_j^+(x,t,\xi)\,\delta\hat{\rho}_j(\xi,t)\right]d\xi=0. \tag{3.200}$$

Here the time variable has been restored in the squared eigenfunctions and the scattering-data variations since the potentials $(\mathbf{q},\mathbf{r})$ are now varying with time (see (3.199)). For simplicity, we have also assumed that the potentials do not contain discrete eigenvalues. Recalling the time evolution equations (3.86)–(3.88) of the scattering data in this hierarchy (3.20), we see that for $j=1,2$,

$$\delta\rho_j(\xi,t)=\delta\rho_j(\xi,0)e^{i(2\xi)^n t},\quad \delta\hat{\rho}_j(\xi,t)=\delta\hat{\rho}_j(\xi,0)e^{-i(2\xi)^n t}. \tag{3.201}$$

Then defining time-dependent squared eigenfunctions

$$Z_j^{-(t)}=Z_j^- e^{i(2\zeta)^n t},\quad Z_j^{+(t)}=Z_j^+ e^{-i(2\zeta)^n t},\quad j=1,2, \tag{3.202}$$

Eq. (3.200) then becomes

$$\int_{-\infty}^{\infty}\sum_{j=1}^{2}\left[\mathcal{L}Z_j^{-(t)}(x,t,\xi)\,\delta\rho_j(\xi,0)+\mathcal{L}Z_j^{+(t)}(x,t,\xi)\,\delta\hat{\rho}_j(\xi,0)\right]d\xi=0. \tag{3.203}$$

Because the initial scattering-data variations can be arbitrary, in order for the above equation to hold, we must have

$$\mathcal{L}Z_j^{-(t)}(x,t,\xi)=\mathcal{L}Z_j^{+(t)}(x,t,\xi)=0,\quad j=1,2, \tag{3.204}$$

for all $x,t,\xi\in\mathbb{R}$. In other words, the time-dependent squared eigenfunctions $Z_j^{\pm(t)}$ satisfy the linearized equation of the hierarchy (3.20). Note that these time-dependent squared eigenfunctions are nothing but the original squared eigenfunctions (3.157) with Jost solutions Φ replaced by the time-dependent Jost solutions

$$\Phi^{(t)}\equiv\Phi\exp\left\{-\frac{1}{2}i(2\zeta)^n\Lambda t\right\}=J_-\exp\left\{-i\zeta\Lambda x-\frac{1}{2}i(2\zeta)^n\Lambda t\right\}. \tag{3.205}$$

These time-dependent Jost solutions satisfy both of the Lax pair (3.41) and (3.43). Similarly, by introducing time-dependent adjoint squared eigenfunctions

$$\Omega_j^{-(t)} = \Omega_j^{-} e^{-i(2\zeta)^n t}, \quad \Omega_j^{+(t)} = \Omega_j^{+} e^{i(2\zeta)^n t}, \quad j = 1,2, \tag{3.206}$$

then these time-dependent adjoint squared eigenfunctions satisfy the adjoint linearized equation, i.e.,

$$\mathcal{L}^A \Omega_j^{-(t)}(x,t,\xi) = \mathcal{L}^A \Omega_j^{+(t)}(x,t,\xi) = 0, \quad j = 1,2, \tag{3.207}$$

for all $x,t,\xi \in \mathbb{R}$, where $\mathcal{L}^A$ is the adjoint operator of $\mathcal{L}$ under the inner product (3.148).

Now we know that these time-dependent squared eigenfunctions satisfy the linearized equation (3.199). In addition, they are also eigenfunctions of the recursion operator $\mathbf{L}_R$ in view of Eqs. (3.197) and (3.202), and thus we have

$$(\mathcal{L}\mathbf{L}_R - \mathbf{L}_R\mathcal{L})\, Z_j^{\pm(t)} = 0, \quad j = 1,2. \tag{3.208}$$

Then since these squared eigenfunctions form a complete set, we see that

$$\mathcal{L}\mathbf{L}_R = \mathbf{L}_R\mathcal{L}; \tag{3.209}$$

i.e., the recursion operator and the linearization operator of the hierarchy (3.20) commute. Similarly, their adjoint operators also commute.

Next, we discuss the case where the potentials admit reductions. If the reduction is only (3.25), i.e.,

$$\mathbf{r}^T = -\mathbf{B}_1 \mathbf{q}^{\dagger} \mathbf{B}_2, \tag{3.210}$$

where $\mathbf{B}_1$ and $\mathbf{B}_2$ are constant Hermitian matrices, then since $\mathbf{q}$ and its complex conjugate are treated as independent variables, all the above relations still hold without change under this reduction. An example is the Manakov equations (3.1)–(3.2) where all the results above still apply. But if there are further reductions in the potentials beyond (3.210), then certain modifications will need to be made in order for the above relations to hold. Let us take the Sasa–Satsuma equation (3.4) as an example. This equation has two simultaneous potential reductions. One is $\mathbf{r}^T = -\mathbf{q}^{\dagger}$, which is of the type (3.210), and the other one is $v = u^*$ (see Sec. 3.2). For this equation, the squared eigenfunctions become vectors of length two (rather than vectors of length four), as was shown in the previous section. Its linearization operator, for the variables $(\delta u, \delta u^*)^T$, is also a 2×2 matrix operator rather than the generic 4×4 matrix operator. In accordance with these length reductions, the recursion operator for the Sasa–Satsuma equation should also be reduced from the generic 4×4 operator $\mathbf{L}_R$ in Eq. (3.18) to a 2×2 operator (this 2×2 recursion operator was given by Sergyeyev and Demskoi (2007)). After these reductions, analogous relations as obtained earlier in this section would then carry over to the Sasa–Satsuma equation as well.

3.10 Solutions in the Manakov System

In the remaining three sections, we consider three special integrable equations in the hierarchy (3.20) of the third-order scattering operator (3.41) and illustrate their soliton and multisoliton dynamics. These equations are chosen partially due to their physical relevance (or potential physical relevance). But more importantly, they are chosen because their soliton and multisoliton solutions exhibit interesting features which are quite different from those in the NLS equation of the previous chapter.

In this section, we consider the Manakov system (3.1)–(3.2), i.e.,

$$iu_t + u_{xx} + (|u|^2 + |v|^2)u = 0, \tag{3.211}$$

$$iv_t + v_{xx} + (|u|^2 + |v|^2)v = 0, \tag{3.212}$$

which arises frequently in nonlinear optics and nonlinear water waves (see Sec. 1.3). The N-soliton solutions in this system can be readily derived from the general formulae (3.91)–(3.93) after proper involution properties are included. From Secs. 3.2 and 3.3, we see that this system possesses the following involution properties:

$$\bar{\zeta}_k = \zeta_k^*, \quad \bar{v}_k = v_k^{\dagger}. \tag{3.213}$$

Also $n = 2$ in the temporal equation (3.43) of the Lax pair. Inserting these relations into the general formulae (3.91)–(3.93), the N-soliton solution in the Manakov system (3.211)–(3.212) would be obtained. Without loss of generality, we let

$$v_{k0} = (\alpha_k, \beta_k, 1)^T. \tag{3.214}$$

In addition, we introduce the notation

$$\theta_k = -i\zeta_k x - 2i\zeta_k^2 t. \tag{3.215}$$

Then these N-soliton solutions can be written out more explicitly as

$$\begin{pmatrix} u(x,t) \\ v(x,t) \end{pmatrix} = 2i \sum_{j,k=1}^{N} \begin{pmatrix} \alpha_j \\ \beta_j \end{pmatrix} e^{\theta_j - \theta_k^*} \left(M^{-1}\right)_{jk}, \tag{3.216}$$

where M is an $N \times N$ matrix whose elements are given by

$$M_{jk} = \frac{1}{\zeta_j^* - \zeta_k}\left[e^{-\left(\theta_j^* + \theta_k\right)} + (\alpha_j^*\alpha_k + \beta_j^*\beta_k)e^{\theta_j^* + \theta_k}\right]. \tag{3.217}$$

Next, we examine single-soliton and two-soliton solutions in the Manakov system in more detail.

Single-Soliton Solutions

To get single-soliton solutions, we set $N = 1$ in the above formulae. Letting

$$\zeta_1 = \xi + i\eta, \quad \sqrt{|\alpha_1|^2 + |\beta_1|^2} = e^{-2\eta x_0}, \tag{3.218}$$

and introducing the unit polarization vector

$$\mathbf{c} = \frac{1}{\sqrt{|\alpha_1|^2 + |\beta_1|^2}} \begin{pmatrix} \alpha_1 \\ \beta_1 \end{pmatrix}, \tag{3.219}$$

where $\mathbf{c}^\dagger \mathbf{c} = 1$, then the single-soliton solution in the Manakov system can be rewritten as

$$(u,v)^T(x,t) = \mathbf{c} \cdot 2\eta \operatorname{sech}[2\eta(x + 4\xi t - x_0)] \exp\left\{-2i\xi x - 4i\left(\xi^2 - \eta^2\right)t\right\}. \tag{3.220}$$

Without the polarization vector $\mathbf{c}$, this Manakov soliton would just be the NLS soliton (2.128), which has a sech profile and moves at velocity -4ξ. The role of this polarization vector is to control the relative power distribution between the two components in this soliton. However, the total power of this Manakov soliton, defined as the sum of its two components' individual powers, is

$$P = \int_{-\infty}^{\infty} (|u|^2 + |v|^2)dx = 4\eta, \tag{3.221}$$

which depends only on the discrete eigenvalue ζ_1, but not on the polarization vector $\mathbf{c}$. This is an important feature of the Manakov system, and it has direct implications for soliton collisions below.

Collisions of Manakov Solitons

When $N = 2$ and $\xi_1 \neq \xi_2$, where $\xi_k = \mathrm{Re}(\zeta_k)$, solutions (3.216) describe the collision of two Manakov solitons. In this collision process, while the discrete eigenvalue ζ_k of each soliton does not change, its polarization vector $\mathbf{c}_k$ does. Thus after collision, while the power of each Manakov soliton remains the same as before, its polarization can rotate, causing a redistribution of power among its two components. This polarization rotation after collision is a distinctive feature of Manakov solitons which has no counterpart in the NLS equation. To illustrate, we take

$$\zeta_1 = 0.1 + 0.7i, \quad \zeta_2 = -0.1 + 0.7i, \quad \alpha_1 = \alpha_2 = 1, \quad \beta_1 = 0.25, \quad \beta_2 = 0, \tag{3.222}$$

and the corresponding solution (3.216) is plotted in Fig. 3.1. As we can see, the left soliton before collision has only a u-component and no v-component. But after collision when

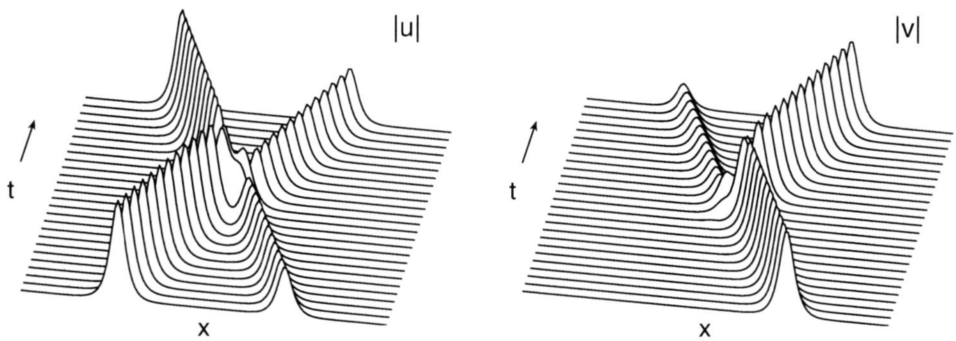

Figure 3.1. *Polarization rotations in the collision of two Manakov solitons. The solution parameters are given in Eq.* (3.222).

this soliton has re-emerged on the right side, its v-component appears. In other words, its polarization has rotated due to the collision. The polarization of the other soliton has rotated as well since the relative power distribution between its two components has also changed after collision (see Fig. 3.1). This phenomenon of polarization rotation provides another scenario of soliton collisions which is different from the elastic collision of NLS solitons. However, the power of each Manakov soliton remains the same before and after collision. In other words, the power of each Manakov soliton still passes through completely.

This polarization rotation of Manakov solitons can be explicitly analyzed. Repeating the same asymptotic analysis as for NLS-soliton collisions in Sec. 2.2, we can find that when $\xi_1 \neq \xi_2$, the two-Manakov-soliton solution from (3.216) decomposes into two single Manakov solitons as $t \to \pm\infty$. Let us assume that $\xi_1 > \xi_2$, i.e., at $t = -\infty$, soliton-1 is on the right side of soliton-2 and moves slower. Then polarizations of the two single Manakov solitons before and after collision are related by the following formulae (Manakov (1973), Tsuchida (2004)):

$$\mathbf{c}_1^+ = \frac{1}{\chi}\frac{\zeta_2^* - \zeta_1}{\zeta_2 - \zeta_1}\left[\mathbf{c}_1^- - \frac{\zeta_2 - \zeta_2^*}{\zeta_2 - \zeta_1}(\mathbf{c}_2^{-\dagger}\mathbf{c}_1^-)\mathbf{c}_2^-\right], \tag{3.223}$$

$$\mathbf{c}_2^+ = \frac{1}{\chi}\frac{\zeta_1 - \zeta_2^*}{\zeta_1^* - \zeta_2^*}\left[\mathbf{c}_2^- + \frac{\zeta_1 - \zeta_1^*}{\zeta_1^* - \zeta_2^*}(\mathbf{c}_1^{-\dagger}\mathbf{c}_2^-)\mathbf{c}_1^-\right], \tag{3.224}$$

where $\mathbf{c}_k^-$ and $\mathbf{c}_k^+$ ($k = 1, 2$) are the unit polarization vectors for the two solitons before and after collision, and

$$\chi = \left|\frac{\zeta_1 - \zeta_2^*}{\zeta_1 - \zeta_2}\right|\left[1 - \frac{(\zeta_1 - \zeta_1^*)(\zeta_2 - \zeta_2^*)}{|\zeta_1 - \zeta_2|^2}|\mathbf{c}_1^{-\dagger}\mathbf{c}_2^-|^2\right]^{1/2}. \tag{3.225}$$

These formulae show that polarizations of Manakov solitons always rotate after collision, except two special cases where the original polarization vectors $\mathbf{c}_1^-$ and $\mathbf{c}_2^-$ are parallel (i.e., $\mathbf{c}_2^- = \alpha\,\mathbf{c}_1^-$, where α is a complex constant), or orthogonal to each other (i.e., $\mathbf{c}_1^{-\dagger}\mathbf{c}_2^- = 0$). In the first case, the Manakov system becomes equivalent to the NLS equation. Positions and phases of the two Manakov solitons shift after collision as well. The amounts of position shifts are $-\ln\chi/2\eta_1$ and $\ln\chi/2\eta_2$, respectively, while the amounts of phase shifts can be obtained from changes in the unit polarization vectors through the formulae (3.223)–(3.224).

3.11 Solutions in a Coupled Focusing-Defocusing NLS System

In this section, we consider the coupled focusing-defocusing nonlinear Schrödinger (FDNLS) equations (3.29)–(3.30), i.e.,

$$iu_t + u_{xx} + (|u|^2 - |v|^2)u = 0, \tag{3.226}$$

$$iv_t + v_{xx} + (|u|^2 - |v|^2)v = 0. \tag{3.227}$$

This system resembles the Manakov system (3.211)–(3.212), but with two sign differences. These sign differences are physically significant because, unlike the Manakov system where both components exhibit self-focusing nonlinearities, this FDNLS system exhibits self-focusing nonlinearity in its u-component but self-defocusing nonlinearity in its v-component. As a consequence, solution dynamics in this FDNLS system exhibits significant differences from that in the Manakov system.

The N-soliton solutions in this FDNLS system can be derived from the general formulae (3.91)–(3.93) after proper involution properties are incorporated. For this system, while zeros $(\zeta_k, \bar{\zeta}_k)$ are still related by $\bar{\zeta}_k = \zeta_k^*$ (see Eq. (3.80)), the involution property for eigenvectors $(v_k, \bar{v}_k)$ now becomes

$$\bar{v}_k = v_k^\dagger \mathbf{B}, \quad \mathbf{B} = \mathrm{diag}(1, -1, 1) \tag{3.228}$$

in view of Eqs. (3.28), (3.75), and (3.82). Then letting

$$v_{k0} = (\alpha_k, \beta_k, 1)^T, \quad \theta_k = -i\zeta_k x - 2i\zeta_k^2 t \tag{3.229}$$

as before, the general N-soliton solution in this FDNLS system can be written out explicitly as

$$\begin{pmatrix} u(x,t) \\ v(x,t) \end{pmatrix} = 2i \sum_{j,k=1}^{N} \begin{pmatrix} \alpha_j \\ \beta_j \end{pmatrix} e^{\theta_j - \theta_k^*} \left(M^{-1}\right)_{jk}, \tag{3.230}$$

where M is an $N \times N$ matrix given by

$$M_{jk} = \frac{1}{\zeta_j^* - \zeta_k}\left[e^{-\left(\theta_j^*+\theta_k\right)} + (\alpha_j^*\alpha_k - \beta_j^*\beta_k)e^{\theta_j^*+\theta_k}\right]. \tag{3.231}$$

Notice that these soliton formulae differ from those of Manakov equations only by a sign in the expression of M_{jk}.

Single-Soliton Solutions

When $N = 1$, we get single-soliton solutions from the above formulae as

$$\begin{pmatrix} u \\ v \end{pmatrix} = \begin{pmatrix} \alpha_1 \\ \beta_1 \end{pmatrix} 2i(\zeta_1^* - \zeta_1)\frac{e^{\theta_1-\theta_1^*}}{\left(|\alpha_1|^2 - |\beta_1|^2\right)e^{\theta_1^*+\theta_1} + e^{-(\theta_1^*+\theta_1)}}. \tag{3.232}$$

Notice that this soliton is nonsingular only when $|\alpha_1| > |\beta_1|$, i.e., the u-component must be stronger than the v-component. Under this restriction, introducing the notation

$$\zeta_1 = \xi + i\eta, \quad \sqrt{|\alpha_1|^2 - |\beta_1|^2} = e^{-2\eta x_0}, \tag{3.233}$$

and the polarization vector

$$\mathbf{c} = \frac{1}{\sqrt{|\alpha_1|^2 - |\beta_1|^2}}\begin{pmatrix} \alpha_1 \\ \beta_1 \end{pmatrix}, \tag{3.234}$$

then this single-soliton solution can be rewritten as

$$(u,v)^T(x,t) = \mathbf{c}\cdot 2\eta\,\mathrm{sech}\,[2\eta(x+4\xi t - x_0)]\exp\left\{-2i\xi x - 4i\left(\xi^2-\eta^2\right)t\right\}, \tag{3.235}$$

which is a familiar sech-shaped solitary wave similar to the Manakov case. This polarization vector $\mathbf{c}$ also controls the relative power distribution between the two components in this soliton, but it is not a unit vector anymore (i.e., $\mathbf{c}^\dagger\mathbf{c} \neq 1$). As before, we also define the power of this FDNLS soliton as the sum of its two components' individual powers. Then

$$P = \int_{-\infty}^{\infty}(|u|^2+|v|^2)dx = 4\eta\frac{|\alpha_1|^2+|\beta_1|^2}{|\alpha_1|^2-|\beta_1|^2}. \tag{3.236}$$

It is important to notice that this power depends not only on the soliton's discrete eigenvalue ζ_1, but also on its polarization vector $\mathbf{c}$. This power dependence on the polarization vector is a new feature of this FDNLS system, and it leads to new scenarios of soliton collisions different from those in the Manakov system. This will be discussed next.

Collisions of FDNLS Solitons

When two FDNLS solitons collide with each other, the discrete eigenvalues ζ_k associated with each soliton will not change after collision, but polarization vectors $\mathbf{c}_k$ of the two solitons will change as in the Manakov case. The new feature here is that, since each soliton's power depends on its polarization, this power will then change after collision. In other words, the two FDNLS solitons will transfer power from one to the other due to the collision (Kanna et al. (2006)). This contrasts the Manakov case where such power transfer does not take place. In addition to this power transfer between the two solitons, relative distributions of each soliton's power among its two components will change as well. This feature is analogous to the polarization rotation in collisions of Manakov solitons.

To illustrate this power-transfer collision in the FDNLS system, we take parameter values

$$\zeta_1 = 0.1 + 0.7i, \quad \zeta_2 = -0.1 + 0.7i, \quad \alpha_1 = 1, \quad \alpha_2 = 0.8, \quad \beta_1 = 0.25, \quad \beta_2 = 0.1. \tag{3.237}$$

The corresponding solution (3.230) is illustrated in Fig. 3.2. This solution describes the collision between two FDNLS solitons. As we can see from this figure, before the collision, the right soliton (soliton-1) has much higher power than the left one (soliton-2). But when soliton-1 emerges out of collision (now on the left side), its power drops significantly. On the other hand, soliton-2, which originally has little power, acquires a lot of power from soliton-1 and hence emerges out of collision as a much stronger soliton. Thus power transfer has taken place from soliton-1 to soliton-2 during collision. This collision visually looks like the two solitons are reflected off by each other—a phenomenon which can be called soliton reflection. This is a new scenario of soliton collisions which has no counterpart in the NLS and Manakov systems.

3.12 Solutions in the Sasa–Satsuma Equation

In this section, we consider solutions in the Sasa–Satsuma equation (3.4), i.e.,

$$u_t + u_{xxx} + 6|u|^2 u_x + 3u(|u|^2)_x = 0. \tag{3.238}$$

This equation is interesting since its single-soliton solutions have more complicated profiles and can be double humped (Sasa and Satsuma (1991)). This contrasts the NLS equation, the Manakov equations, the FDNLS equations, and many others where single solitons are all single humped. This double-hump structure of single solitons is induced by the double symmetries (3.181)–(3.182) of the potential matrix in this equation, which in turn impose further constraints on the discrete eigenvalues so that they must appear as quadruples rather than pairs in the complex plane.

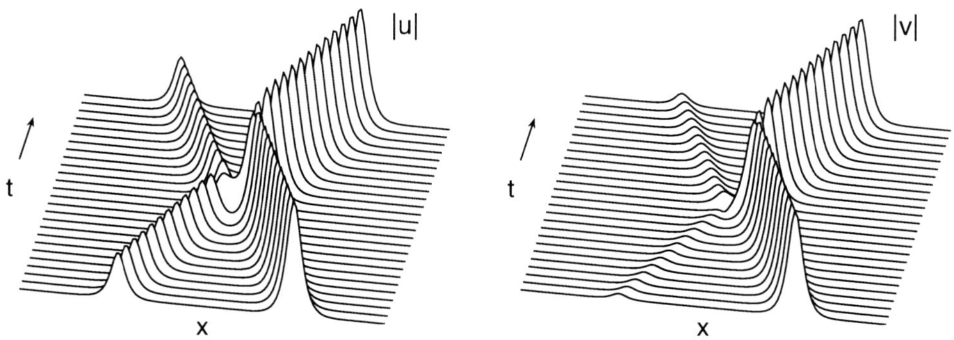

Figure 3.2. *Soliton reflection in the FDNLS system* (3.226)–(3.227). *The solution parameters are given in Eq.* (3.237).

To construct N-soliton solutions in the Sasa–Satsuma equation, we first detail its involution properties on eigenvalues and eigenvectors. Since the potential matrix in this equation satisfies the double symmetries (3.181)–(3.182), involution properties (3.179)–(3.180) and (3.183)–(3.184) on Jost solutions and the scattering matrix then are both satisfied. From the involution relation (3.184), we get

$$s_{33}^*(-\zeta^*) = s_{33}(\zeta). \tag{3.239}$$

Thus if $\bar{\zeta}_k$ is a zero of $s_{33}(\zeta)$, so is $-\bar{\zeta}_k^*$. Combining this relation with the involution relation $\bar{\zeta}_k = \zeta_k^*$ in (3.180), we see that discrete eigenvalues in the Sasa–Satsuma equation always appear in quadruples $(\zeta_k, \zeta_k^*, -\zeta_k, -\zeta_k^*)$. This differs from the previous systems where discrete eigenvalues appear in pairs (ζ_k, ζ_k^*) under the single potential symmetry (3.75).

Eigenvectors in the kernels of Riemann–Hilbert solutions $P^\pm$ at the zeros $(\zeta_k, \zeta_k^*, -\zeta_k, -\zeta_k^*)$ satisfy additional constraints too. From the double involution properties (3.179) and (3.183) of Jost solutions as well as the definitions (3.53) and (3.60) of $P^\pm$, we see that $P^\pm$ satisfy double involutions relations

$$(P^+)^\dagger(\zeta^*) = P^-(\zeta), \quad (P^\pm)^*(-\zeta^*) = \sigma P^\pm(\zeta)\sigma^{-1}. \tag{3.240}$$

Thus if eigenvectors in the kernels of $P^\pm$ at the zeros $(\zeta_k, \zeta_k^*, -\zeta_k, -\zeta_k^*)$ are such that

$$P^+(\zeta_k)v_k^{(1)} = 0, \quad P^+(-\zeta_k^*)v_k^{(2)} = 0, \tag{3.241}$$

and

$$\bar{v}_k^{(1)} P^-(\zeta_k^*) = 0, \quad \bar{v}_k^{(2)} P^-(-\zeta_k) = 0, \tag{3.242}$$

then using the above involutions relations (3.240), we get involution properties of eigenvectors as

$$\bar{v}_k^{(1)} = v_k^{(1)\dagger}, \quad \bar{v}_k^{(2)} = v_k^{(2)\dagger}, \quad v_k^{(2)} = \sigma v_k^{(1)*}, \quad \bar{v}_k^{(2)} = \bar{v}_k^{(1)*}\sigma. \tag{3.243}$$

Incorporating the above involution properties of eigenvalues and eigenvectors into the general formulae (3.91)–(3.93), multisoliton solutions in the Sasa–Satsuma equation will be obtained. Assuming that there are N quadruples of eigenvalues $\{(\zeta_k, \zeta_k^*, -\zeta_k, -\zeta_k^*), 1 \leq k \leq N\}$, and letting

$$v_{k0}^{(1)} = (\alpha_k, \beta_k, 1)^T, \quad \theta_k = -i\zeta_k x - 4i\zeta_k^3 t, \tag{3.244}$$

then the N-soliton solutions of the Sasa–Satsuma equation are given by

$$u(x,t) = 2i \sum_{j,k=1}^{2N} \alpha_j \, e^{\theta_j - \theta_k^*} \left(M^{-1}\right)_{jk}, \tag{3.245}$$

where M is a $2N \times 2N$ matrix with

$$M_{jk} = \frac{1}{\zeta_j^* - \zeta_k} \left[(\alpha_j^* \alpha_k + \beta_j^* \beta_k) e^{\theta_j^* + \theta_k} + e^{-\left(\theta_j^* + \theta_k\right)} \right] \tag{3.246}$$

and

$$\zeta_{N+k} = -\zeta_k^*, \quad \theta_{N+k} = \theta_k^*, \quad \alpha_{N+k} = \beta_k^*, \quad \beta_{N+k} = \alpha_k^*, \quad 1 \leq k \leq N. \tag{3.247}$$

Next, we set $N = 1$ in the above formula and examine its single-soliton solutions. First, we consider the special cases where $\alpha_1 \beta_1 = 0$. Solutions for the two cases of $\alpha_1 = 0$ and $\beta_1 = 0$ are equivalent, and thus we take $\beta_1 = 0$ below. After simple algebra, the single-soliton solution from formula (3.245) becomes

$$u(x,t) = 2i\alpha_1(\zeta_1^* - \zeta_1) \frac{|\alpha_1|^2 e^{2\theta_1} + \gamma\, e^{-2\theta_1^*}}{|\alpha_1|^4 e^{2(\theta_1 + \theta_1^*)} + 2|\alpha_1|^2 + |\gamma|^2 e^{-2(\theta_1 + \theta_1^*)}}, \tag{3.248}$$

where $\gamma = (\zeta_1 + \zeta_1^*)/2\zeta_1^*$. Let

$$\zeta_1 = \xi + i\eta, \quad \alpha_1 = e^{-2\eta x_0 + i\sigma_0}; \tag{3.249}$$

then the above solution can be rewritten as

$$u(x,t) = \psi(x - vt - x_0)\, e^{-2i\xi x - i\lambda t + i\sigma_0}, \tag{3.250}$$

where

$$\psi(x) = 4\eta \frac{e^{2\eta x} + \gamma\, e^{-2\eta x}}{e^{4\eta x} + 2 + |\gamma|^2 e^{-4\eta x}} \tag{3.251}$$

and

$$v = 4(\eta^2 - 3\xi^2), \quad \lambda = 8\xi(\xi^2 - 3\eta^2). \tag{3.252}$$

This solution is a moving solitary wave. Its intensity profile is

$$|\psi(x)|^2 = 16\eta^2 \frac{e^{4\eta x}+2|\gamma|^2+|\gamma|^2 e^{-4\eta x}}{\left(e^{4\eta x}+2+|\gamma|^2 e^{-4\eta x}\right)^2}, \tag{3.253}$$

which depends only on η and $|\gamma|$. The parameter η controls the width of this soliton, while $|\gamma|$ controls its shape. An interesting property of this soliton is that its shape can be single or double humped depending on the parameter $|\gamma|$. Indeed, it can be easily checked that this soliton is single humped when $|\gamma| > 0.5$, but becomes double humped when $|\gamma| < 0.5$. To illustrate, we take $\eta = 1$ and $|\gamma| = 0.1$. The corresponding shape function $|\psi(x)|$ is plotted in Fig. 3.3(a). It is seen that this soliton has two intensity peaks. This is quite unusual in integrable systems where single-soliton solutions are often single humped.

The solitary waves (3.250) presented above are only part of the single-soliton solutions (3.245) (with $N = 1$). When $\alpha_1\beta_1 \neq 0$, these single-soliton solutions are no longer solitary waves. Instead, they become spatially localized and temporally periodic bound states (Mihalache et al. (1993a)). To illustrate, we take

$$\zeta_1 = 0.5+i, \quad \alpha_1 = \beta_1 = 1. \tag{3.254}$$

The corresponding single-soliton solution (3.245) is displayed in Fig. 3.3(b). This solution is a moving breather. The reason for this breather is that this single soliton has four discrete eigenvalues $(\zeta_1, \zeta_1^*, -\zeta_1, -\zeta_1^*)$, which double those in single solitons of many other integrable systems (such as the NLS equation and the Manakov system). Thus this breather may be viewed as the counterpart of a two-soliton breather state of the NLS equation in

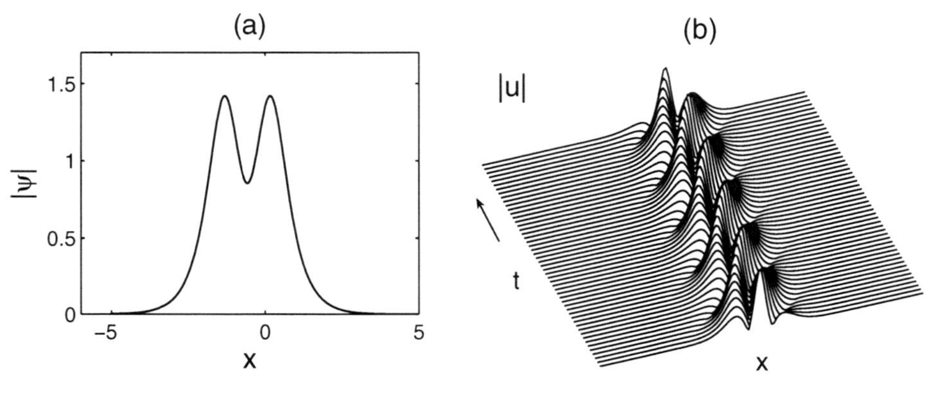

Figure 3.3. *Single-soliton solutions in the Sasa–Satsuma equation* (3.238) : (a) *shape function of a double-hump solitary wave* (3.250) *with* $|\gamma| = 0.1$ *and* $\eta = 1$; (b) *a bound state with* $\zeta_1 = 0.5+i$ *and* $\alpha_1 = \beta_1 = 1$.

Fig. 2.1(b). However, these single-soliton breathers in the Sasa–Satsuma equation are robust and will persist if the initial conditions are perturbed. This contrasts the NLS breathers which generally will break up under perturbations (because the two constituent solitons in those breathers will generically acquire different velocities under perturbations). Note that this Sasa–Satsuma breather solution in Fig. 3.3(b) resembles the complexiton solutions in a coupled KdV system; see Hu et al. (2006).

Chapter 4

Soliton Perturbation Theories and Applications

Integrable equations as considered in the previous two chapters can describe certain physical wave systems at the lowest order of approximation. For instance, pulse transmission in optical fibers is governed by the NLS equation at the leading order. Integrable equations can support soliton solutions which travel without change of shape. When perturbations such as damping, higher-order dispersion, and higher-order nonlinearity are brought into consideration, a physical system is then modeled by a perturbed integrable equation (see Eq. (1.50) for optical pulses in fibers for instance). In a perturbed system, solitons may not propagate stationarily anymore. Their shapes may be distorted over time. In addition, energy radiation can be excited which can affect the soliton's evolution in nontrivial ways. In order to describe soliton evolution under perturbations, a soliton perturbation theory is required.

Perturbation theories for single solitons have been developed for many integrable equations. Examples include the KdV equation (Karpman and Maslov (1977), Kaup and Newell (1978a), Kodama (1985), Herman (1990), Grimshaw and Mitsudera (1993), Yan and Tang (1996)), the NLS equation (Kaup (1976a), Keener and Mclaughlin (1977), Kaup and Newell (1978a), Karpman and Maslov (1978), Kaup (1990), Hasegawa and Kodama (1995)), the sine-Gordon equation (Fogel et al. (1977)), the Benjamin–Ono equation (Kaup et al. (1999)), the derivative NLS equation (and the related modified NLS equation) (Shchesnovich and Doktorov (1999), Chen and Yang (2002)), the Manakov equations (Lakoba and Kaup (1997), Shchesnovich and Doktorov (1997)), the massive Thirring model (Kaup and Lakoba (1996)), the fifth-order KdV equation (Yang (2001a)), the complex modified KdV equation (Yang (2003), Hoseini and Marchant (2009)), and many others (see also the review article by Kivshar and Malomed (1989) and the references therein). In addition to these single-soliton perturbation theories, perturbation theories for soliton collisions have also been developed for the KdV equation (Kodama (1987)), the NLS equation (Kano (1989)), the Benjamin–Ono equation (Matsuno (1995)), the Manakov equations (Yang (1999)), and the complex modified KdV equation (Hoseini and Marchant (2006)). All these perturbation theories can be separated into several categories. One category is the perturbation theory based on the inverse scattering transform (Kaup (1976a), Karpman and Maslov (1977, 1978), Kaup and Newell (1978a)), which has a recent development based on the Riemann–Hilbert problem (Shchesnovich and Doktorov (1997, 1999)). In this method, one calculates the shift of the scattering data of the soliton due to the perturbation, and then uses inverse scattering to reconstruct the perturbed solution. Another category is the direct soliton perturbation theory (Fogel et al. (1977), Keener and Mclaughlin (1977), Kaup (1990), Herman (1990), Grimshaw and Mitsudera (1993), Matsuno (1995), Yan and Tang (1996), Kaup and Lakoba (1996), Lakoba and Kaup (1997), Kaup et al. (1999), Yang (1999, 2001a, 2003), Chen and Yang (2002), Hoseini and Marchant (2009)). In this method, one solves the linearized wave equation around a soliton by expanding its solution into a set of complete eigenfunctions of the linearization operator. Suppression of secular growth in the linearized solution gives the evolution equations of soliton parameters, and the radiation is given by the linearized solution through integrals of these complete eigenfunctions. This method does not rely explicitly on the inverse scattering transform and is often easier to apply. But its connection to the integrable theory is still visible since these eigenfunctions of the linearized equation are simply the squared eigenfunctions of the underlying scattering operator (see the previous two chapters). Both of these two methods can give not only the evolution of soliton parameters, but also the perturbation-induced radiation. The third category is the normal form method (Kodama (1985, 1987), Kano (1989), Hoseini and Marchant (2006)). In this method, one transforms the perturbed equation into a normal form, and then uses the conserved quantities of the normal form to derive evolution equations of soliton parameters. Among these different categories of perturbation theories, the direct soliton perturbation theory is notable for its simplicity and versatility. This perturbation theory and its applications will be described in this chapter.

4.1 Direct Soliton Perturbation Theory for the NLS Equation

We consider the perturbed NLS equation

$$iu_t + u_{xx} + 2|u|^2 u = \varepsilon F(u), \tag{4.1}$$

where F is the perturbation term, and $\varepsilon \ll 1$. When $\varepsilon = 0$, this equation has a soliton solution (2.128) which can be rewritten as

$$u(x,t) = e^{i\phi}\Phi(\theta), \tag{4.2}$$

where

$$\Phi(\theta) = r\ \operatorname{sech} r\theta, \tag{4.3}$$

$$\phi = \frac{1}{2}v\theta + \sigma, \quad \theta = x - \nu, \tag{4.4}$$

and

$$\nu = vt + x_0, \quad \sigma = \left(r^2 + \frac{1}{4}v^2\right)t + \sigma_0. \tag{4.5}$$

This soliton has four free parameters: amplitude r, velocity v, initial position x_0, and initial phase σ_0. We now consider how this soliton evolves under small perturbations (i.e., $0 \neq \varepsilon \ll 1$). In the presence of perturbations, the four parameters of the soliton will slowly change with time, the shape of the soliton will be modified, and energy radiation will be emitted. These effects will be calculated below by a direct soliton perturbation theory.

The essence of the direct soliton perturbation theory is a multiscale perturbation analysis. Here the solution contains two time scales. The four soliton parameters evolve on the slow time scale $T = \varepsilon t$, while the other aspects of the solution (such as energy radiation) evolve on the fast time scale t. According to a standard multiscale perturbation analysis, we expand the solution $u(x,t)$ into the following perturbation series:

$$u(x,t) = e^{i\phi}(\Phi + \varepsilon\Phi_1 + \varepsilon^2\Phi_2 + \cdots). \tag{4.6}$$

Here Φ, ϕ, and θ are as given above, but ν and σ are now defined by

$$\nu = \int_0^t v\,d\tau + x_0, \quad \sigma = \int_0^t \left(r^2 + \frac{1}{4}v^2\right)d\tau + \sigma_0, \tag{4.7}$$

and

$$r = r(T), \quad v = v(T), \quad x_0 = x_0(T), \quad \sigma_0 = \sigma_0(T) \tag{4.8}$$

are slow variables. Substituting the above perturbation series into Eq. (4.1), we find that the $O(1)$ terms all cancel out because the soliton (4.2) is a solution of the NLS equation. At $O(\varepsilon)$, the function $A_1 \equiv (\Phi_1, \Phi_1^*)^T$ satisfies the following linear inhomogeneous equation:

$$(i\partial_t + L)A_1 = W, \tag{4.9}$$

where $W = (w, -w^*)^T$,

$$w = F_0 - i\Phi_{,r} r_T + i x_{0T}\Phi_\theta - \left(\frac{1}{2}v x_{0T} - \frac{1}{2}v_T\theta - \sigma_{0T}\right)\Phi, \tag{4.10}$$

$$F_0 = e^{-i\phi}F(\Phi e^{i\phi}), \tag{4.11}$$

the operator L is

$$L = \begin{pmatrix} \partial_{\theta\theta} - r^2 + 4\Phi^2 & 2\Phi^2 \\ -2\Phi^2 & -\partial_{\theta\theta} + r^2 - 4\Phi^2 \end{pmatrix}, \tag{4.12}$$

and the partial time derivative ∂_t in (4.9) is with respect to the fast time t. The operator L is closely related to the linearization operator $\mathcal{L}$ in (2.268) for the NLS equation, and it will also be called the linearization operator below. The initial condition for the first-order equation (4.9) is

$$A_1|_{t=0} = 0 \tag{4.13}$$

since we are considering the evolution of an initial soliton under perturbations. Equation (4.9) can be solved by the eigenfunction expansion method, where one expands the inhomogeneous term W as well as the solution A_1 into the complete set of operator L's eigenfunctions. This method is analogous to the Fourier transform method for solving linear PDEs with constant coefficients. In this method, both eigenfunctions and adjoint eigenfunctions of L are needed. Below, we first derive L's eigenfunctions and adjoint eigenfunctions, then we solve Eq. (4.9) and finish the soliton perturbation theory.

4.1.1 Eigenfunctions and Adjoint Eigenfunctions of the Linearization Operator

In this subsection, we show that eigenfunctions and adjoint eigenfunctions of the linearization operator L are simply squared eigenfunctions and adjoint squared eigenfunctions of the NLS equation (2.1) for the soliton potential (4.2). Explicit expressions of these eigenfunctions will also be derived.

In order to obtain L's eigenfunctions, we notice that L is related to the linearization operator $\mathcal{L}$ of the NLS equation around the soliton solution $u = \Phi(\theta)e^{ir^2t}$ (see (2.268) with x replaced by θ):

$$\mathcal{L} = \begin{pmatrix} i\partial_t + \partial_{\theta\theta} + 4\Phi^2 & 2\Phi^2 e^{2ir^2t} \\ -2\Phi^2 e^{-2ir^2t} & i\partial_t - \partial_{\theta\theta} - 4\Phi^2 \end{pmatrix}. \tag{4.14}$$

We have shown in Sec. 2.7 that time-dependent squared eigenfunctions satisfy the linearized NLS equation. Specifically,

$$\mathcal{L}Z^{\pm(t)}(\theta,t,\xi) = 0, \qquad \xi \in \mathbb{R}, \tag{4.15}$$

where $Z^{\pm(t)}$ are given by (2.256) and (2.271). For the soliton potential $u = \Phi(\theta)e^{ir^2t}$, the corresponding analytical functions $P^{\pm}$ are simply Γ and Γ^{-1} given by (2.72), (2.73), and (2.109), with $N = 1$, $\zeta_1 = ir/2$, $v_{10} = (1,1)^T$, and with x replaced by θ. In addition, the scattering matrix S is diagonal. From this information, we can obtain all Jost solutions as well as the scattering matrix. In particular, we get $J_- E = (\phi_1, \phi_2)$ as

$$\phi_1(\theta,t,\zeta) = \frac{e^{-i\zeta\theta}}{2i\zeta - r}\begin{pmatrix} 2i\zeta + r\tanh r\theta \\ r\,\mathrm{sech} r\theta\, e^{-ir^2t} \end{pmatrix}, \qquad \zeta \in \mathbb{R}, \tag{4.16}$$

$$\phi_2(\theta,t,\zeta)=\frac{e^{i\zeta\theta}}{2i\zeta+r}\begin{pmatrix} r\,\mathrm{sech}\,r\theta\, e^{ir^2t} \\ 2i\zeta-r\tanh r\theta \end{pmatrix}, \quad \zeta\in\mathbb{R}. \tag{4.17}$$

Let $\zeta=rk/2\in\mathbb{R}$, then $Z^{\pm(t)}$ can be obtained from Eqs. (2.256) and (2.271) as

$$Z^{-(t)}(\theta,t,k)=\frac{e^{ir^2k^2t}}{(ik+1)^2}\,\mathrm{diag}(e^{2ir^2t},1)\,\mathcal{Z}_1(\theta,k), \tag{4.18}$$

$$Z^{+(t)}(\theta,t,k)=\frac{e^{-ir^2k^2t}}{(ik-1)^2}\,\mathrm{diag}(1,e^{-2ir^2t})\,\mathcal{Z}_2(\theta,k), \tag{4.19}$$

where

$$\mathcal{Z}_1(\theta,k)=\begin{pmatrix} -\mathrm{sech}^2 r\theta \\ (ik-\tanh r\theta)^2 \end{pmatrix} e^{irk\theta}, \tag{4.20}$$

$$\mathcal{Z}_2(\theta,k)=\begin{pmatrix} (ik+\tanh r\theta)^2 \\ -\mathrm{sech}^2 r\theta \end{pmatrix} e^{-irk\theta}. \tag{4.21}$$

Substituting (4.18)–(4.19) into (4.15), we get

$$L\mathcal{Z}_1(\theta,k)=\lambda(k)\mathcal{Z}_1(\theta,k), \quad L\mathcal{Z}_2(\theta,k)=-\lambda(k)\mathcal{Z}_2(\theta,k), \tag{4.22}$$

where

$$\lambda(k)=r^2(k^2+1). \tag{4.23}$$

Thus $\mathcal{Z}_1(\theta,k)$ and $\mathcal{Z}_2(\theta,k)$ are L's continuous eigenfunctions. Note from (4.18)–(4.19) that these continuous eigenfunctions of L are simply squared eigenfunctions $Z^{-(t)}$ and $Z^{+(t)}$ evaluated at time $t=0$, multiplied by a constant which does not affect eigenrelations. Likewise, L's discrete eigenfunctions are also proportional to the discrete squared eigenfunctions $Z^-(\zeta_1^*)$, $\dot{Z}^-(\zeta_1^*)$, $Z^+(\zeta_1)$, and $\dot{Z}^+(\zeta_1)$ (as seen in the closure relation (2.264)) with $\zeta_1=ir/2$ and time set to $t=0$. These discrete eigenfunctions can be more easily derived directly from the continuous eigenrelations (4.22). Evaluating (4.22) and their derivative equations (with respect to k) at $k=\mp i$, we obtain

$$L\mathcal{Z}_1(\theta,-i)=L\mathcal{Z}_2(\theta,i)=0, \tag{4.24}$$

$$L\dot{\mathcal{Z}}_1(\theta,-i)=-2ir^2\mathcal{Z}_1(\theta,-i), \tag{4.25}$$

$$L\dot{\mathcal{Z}}_2(\theta,i)=-2ir^2\mathcal{Z}_2(\theta,i). \tag{4.26}$$

Here the dot above $\mathcal{Z}_j$ represents its derivative with respect to k. Thus $\mathcal{Z}_1(\theta,-i)$ and $\mathcal{Z}_2(\theta,i)$ are L's discrete eigenfunctions with zero eigenvalue, while $\dot{\mathcal{Z}}_1(\theta,-i)$ and $\dot{\mathcal{Z}}_2(\theta,i)$

are L's generalized discrete eigenfunctions at the zero eigenvalue. Since the set of squared eigenfunctions is complete for arbitrary time t, while L's eigenfunctions are proportional to these squared eigenfunctions with time set to $t = 0$, L's eigenfunctions hence form a complete set as well.

Regarding L's discrete eigenfunctions, it is more convenient to use the following equivalent but simpler functions:

$$\mathcal{Z}_{D,1}(\theta) = \Phi_\theta \begin{pmatrix} 1 \\ 1 \end{pmatrix}, \qquad \mathcal{Z}_{D,2}(\theta) = \Phi \begin{pmatrix} 1 \\ -1 \end{pmatrix}, \tag{4.27}$$

$$\mathcal{Z}_{G,1}(\theta) = \frac{1}{2}\theta\Phi \begin{pmatrix} 1 \\ -1 \end{pmatrix}, \qquad \mathcal{Z}_{G,2}(\theta) = \Phi_r \begin{pmatrix} 1 \\ 1 \end{pmatrix}, \tag{4.28}$$

which are linearly related to the above discrete eigenfunctions $\mathcal{Z}_1(\theta,-i)$, $\mathcal{Z}_2(\theta,i)$, $\dot{\mathcal{Z}}_1(\theta,-i)$, and $\dot{\mathcal{Z}}_2(\theta,i)$; see Yang (2000). Here $\Phi(\theta)$ is given in (4.3). It is easy to verify that

$$L\mathcal{Z}_{D,1}(\theta) = L\mathcal{Z}_{D,2}(\theta) = 0, \tag{4.29}$$

$$L\mathcal{Z}_{G,1}(\theta) = \mathcal{Z}_{D,1}(\theta), \quad L\mathcal{Z}_{G,2}(\theta) = 2r\mathcal{Z}_{D,2}(\theta). \tag{4.30}$$

Thus $\mathcal{Z}_{D,1}(\theta)$ and $\mathcal{Z}_{D,2}(\theta)$ are L's discrete eigenfunctions with zero eigenvalue, while $\mathcal{Z}_{G,1}(\theta)$ and $\mathcal{Z}_{G,2}(\theta)$ are L's generalized discrete eigenfunctions at the zero eigenvalue. These four new discrete eigenfunctions have clear physical meanings. They are the so-called Goldstein modes which are generated by the four free parameters in the NLS soliton (4.2): initial phase, initial position, amplitude, and velocity. Indeed, we know that the unperturbed NLS equation (4.1) (with x replaced by θ) has the following soliton solution:

$$u(\theta,t) = \Phi(\theta - vt + \theta_0)\exp\left\{\frac{1}{2}iv\theta + i(r^2 - \frac{1}{4}v^2)t + i\sigma_0\right\}, \tag{4.31}$$

where θ_0, σ_0, v, r are the four free parameters. Due to these free parameters, the partial derivatives $\partial u/\partial\theta_0$, $\partial u/\partial\sigma_0$, $\partial u/\partial v$, and $\partial u/\partial r$ satisfy the linearized NLS equation. Substituting these four functions into the linearized NLS equation and setting $v = \theta_0 = \sigma_0 = t = 0$, we will get the eigenrelations (4.29)–(4.30).

To solve the first-order equation (4.9), we also need L's adjoint eigenfunctions, which are eigenfunctions of the adjoint operator L^A. To define this adjoint operator, the inner product is taken as

$$\langle \mathbf{f}, \mathbf{g} \rangle = \int_{-\infty}^{\infty} \mathbf{f}^T(x)\,\mathbf{g}(x)\,dx, \tag{4.32}$$

which is the same as that used in the previous two chapters (see (2.233) and (3.148)). Under this inner product, L^A is equal to the transpose of L. Analogously to eigenfunctions of L, adjoint eigenfunctions $\Upsilon_{1,2}(\theta,k)$ are related to time-dependent adjoint squared eigenfunctions $\Omega^{\pm(t)}(\theta,t,\zeta)$ as

$$\Omega^{-(t)}(\theta,t,\zeta) = -\frac{e^{-ir^2k^2t}}{(ik+1)^2}\mathrm{diag}(e^{-2ir^2t},1)\Upsilon_1(\theta,k), \tag{4.33}$$

$$\Omega^{+(t)}(\theta,t,\zeta) = -\frac{e^{ir^2k^2t}}{(ik-1)^2}\mathrm{diag}(1,e^{2ir^2t})\Upsilon_2(\theta,k), \tag{4.34}$$

where $\zeta = rk/2 \in \mathbb{R}$,

$$\Upsilon_1(\theta,k) = \begin{pmatrix} \mathrm{sech}^2 r\theta \\ (ik+\tanh r\theta)^2 \end{pmatrix} e^{-irk\theta}, \tag{4.35}$$

$$\Upsilon_2(\theta,k) = \begin{pmatrix} (ik-\tanh r\theta)^2 \\ \mathrm{sech}^2 r\theta \end{pmatrix} e^{irk\theta}, \tag{4.36}$$

with the eigenrelations

$$L^A\Upsilon_1(\theta,k) = \lambda(k)\Upsilon_1(\theta,k), \quad L^A\Upsilon_2(\theta,k) = -\lambda(k)\Upsilon_2(\theta,k). \tag{4.37}$$

Here $\lambda(k)$ is given in (4.23). Discrete adjoint eigenfunctions can be obtained from discrete adjoint squared eigenfunctions in the closure relation (2.264) with time set to zero; or they can be obtained from continuous adjoint eigenrelations (4.37) above by setting $k = \mp i$. Their more convenient forms can be obtained by noticing that operator $\sigma_3 L$ is self-adjoint under the inner product (4.32), i.e., $(\sigma_3 L)^A = \sigma_3 L$, where $\sigma_3 = \mathrm{diag}(1,-1)$. Thus $L^A = \sigma_3 L \sigma_3^{-1}$. As a result, adjoint discrete eigenfunctions are simply

$$\Upsilon_{D,1}(\theta) = \sigma_3 Z_{D,1}(\theta), \qquad \Upsilon_{D,2}(\theta) = \sigma_3 Z_{D,2}(\theta), \tag{4.38}$$

$$\Upsilon_{G,1}(\theta) = \sigma_3 Z_{G,1}(\theta), \qquad \Upsilon_{G,2}(\theta) = \sigma_3 Z_{G,2}(\theta), \tag{4.39}$$

and their eigenrelations are

$$L^A\Upsilon_{D,1}(\theta) = L^A\Upsilon_{D,2}(\theta) = 0, \tag{4.40}$$

$$L^A\Upsilon_{G,1}(\theta) = \Upsilon_{D,1}(\theta), \quad L^A\Upsilon_{G,2}(\theta) = 2r\Upsilon_{D,2}(\theta). \tag{4.41}$$

Inner products between L's continuous eigenfunctions and adjoint continuous eigenfunctions can be obtained from the inner products of continuous squared eigenfunctions

(2.261)–(2.263). Here for the soliton potential, $s_{11}(\zeta) = s_{22}^{-1}(\zeta) = (\zeta - ir/2)/(\zeta + ir/2)$. Then by setting $\xi = rk/2$, Eqs. (2.261)–(2.263) give the following inner products:

$$\langle \mathcal{Z}_1(\theta,k), \Upsilon_1(\theta,k')\rangle = \langle \mathcal{Z}_2(\theta,k), \Upsilon_2(\theta,k')\rangle = \frac{2\pi}{r}(k^2+1)^2\delta(k-k'), \tag{4.42}$$

and the other two inner products are zero. Here $\delta(\cdot)$ is the Dirac delta function. These inner product relations can also be obtained directly. For instance, to derive the first inner product in (4.42), notice from the eigenrelations (4.22) and (4.37) that

$$\begin{aligned} &\left[\lambda(k) - \lambda(k')\right] \langle \mathcal{Z}_1(\theta,k), \Upsilon_1(\theta,k')\rangle \\ &= \langle L\mathcal{Z}_1(\theta,k), \Upsilon_1(\theta,k')\rangle - \langle \mathcal{Z}_1(\theta,k), L^A \Upsilon_1(\theta,k')\rangle \\ &= \left[\mathcal{Z}_{1\theta}^T(\theta,k)\sigma_3\Upsilon_1(\theta,k') - \mathcal{Z}_1^T(\theta,k)\sigma_3\Upsilon_{1\theta}(\theta,k')\right]_{\theta=-\infty}^{\theta=\infty}. \end{aligned} \tag{4.43}$$

Here the second step is obtained through integration by parts. Inserting the large-θ asymptotics of $\mathcal{Z}_1$ and Υ_1 into the right-hand side of the above equation and simplifying, we find that

$$\begin{aligned} \langle \mathcal{Z}_1(\theta,k), \Upsilon_1(\theta,k')\rangle = (k^2+1)^2 \left. \frac{e^{ir(k-k')\theta}}{ir(k-k')} \right|_{\theta=\infty} \\ - (k'^2+1)^2 \left. \frac{e^{ir(k-k')\theta}}{ir(k-k')} \right|_{\theta=-\infty}. \end{aligned} \tag{4.44}$$

Using the well-known result

$$\lim_{\theta\to\pm\infty} P\left(\frac{e^{ik\theta}}{ik}\right) = \pm\pi\,\delta(k), \tag{4.45}$$

where $P(\cdot)$ represents the principal value, we get

$$\langle \mathcal{Z}_1(\theta,k), \Upsilon_1(\theta,k')\rangle = 2\pi(k^2+1)^2\delta[r(k-k')], \tag{4.46}$$

which is the same as the first inner product in Eq. (4.42). The other inner product in (4.42) can be obtained similarly. The above derivation of inner products between continuous eigenfunctions and adjoint eigenfunctions is a standard method which applies to any linear operator.

Regarding inner products between L's discrete eigenfunctions and adjoint discrete eigenfunctions, they can be calculated directly. It is easy to verify that the only nonzero inner products are

$$\langle \mathcal{Z}_{D,1}(\theta), \Upsilon_{G,1}(\theta)\rangle = \langle \mathcal{Z}_{G,1}(\theta), \Upsilon_{D,1}(\theta)\rangle = -r, \tag{4.47}$$

$$\langle \mathcal{Z}_{D,2}(\theta), \Upsilon_{G,2}(\theta)\rangle = \langle \mathcal{Z}_{G,2}(\theta), \Upsilon_{D,2}(\theta)\rangle = 2. \tag{4.48}$$

Other inner products between L's discrete eigenfunctions and adjoint continuous eigenfunctions are all zero.

4.1.2 Solution for the Perturbed Soliton

After the eigenfunctions and adjoint eigenfunctions of the linearization operator L have been obtained, we can now solve the first-order equation (4.9) and derive the solution for the soliton under perturbations.

To solve Eq. (4.9), we first expand the forcing term W into the complete set of L's eigenfunctions,

$$W = c_1 Z_{D,1}(\theta) + c_2 Z_{D,2}(\theta) + c_3 Z_{G,1}(\theta) + c_4 Z_{G,2}(\theta) + \int_{-\infty}^{\infty} [\alpha_1(k) Z_1(\theta,k) + \alpha_2(k) Z_2(\theta,k)]\, dk. \tag{4.49}$$

Utilizing the orthogonality relations (4.42), (4.47), and (4.48), we find that

$$c_1 = -\frac{1}{r}\langle W, \Upsilon_{G,1}\rangle, \qquad c_2 = \frac{1}{2}\langle W, \Upsilon_{G,2}\rangle, \tag{4.50}$$

$$c_3 = -\frac{1}{r}\langle W, \Upsilon_{D,1}\rangle, \qquad c_4 = \frac{1}{2}\langle W, \Upsilon_{D,2}\rangle, \tag{4.51}$$

$$\alpha_j(k) = \frac{r}{2\pi(k^2+1)^2}\langle \mathcal{F}, \Upsilon_j\rangle, \quad j = 1,2, \tag{4.52}$$

where

$$\mathcal{F} \equiv (F_0, -F_0^*)^T. \tag{4.53}$$

Next, we expand the solution A_1 into the complete set of L's eigenfunctions as well,

$$A_1 = h_1(t) Z_{D,1}(\theta) + h_2(t) Z_{D,2}(\theta) + h_3(t) Z_{G,1}(\theta) + h_4(t) Z_{G,2}(\theta) + \int_{-\infty}^{\infty} [g_1(k,t) Z_1(\theta,k) + g_2(k,t) Z_2(\theta,k)]\, dk. \tag{4.54}$$

Inserting the two expansions (4.49) and (4.54) into the first-order equation (4.9) and equating the coefficients of the same eigenfunction of L, we obtain the following relations:

$$i h_{1t} + h_3 = c_1, \quad i h_{2t} + 2r h_4 = c_2, \tag{4.55}$$

$$i h_{3t} = c_3, \quad i h_{4t} = c_4, \tag{4.56}$$

$$i g_{1t} + \lambda(k) g_1 = \alpha_1(k), \quad i g_{2t} - \lambda(k) g_2 = \alpha_2(k). \tag{4.57}$$

As in (4.9), the partial time derivatives in these equations are with respect to the fast time t. Due to the initial condition $A_1|_{t=0} = 0$, initial conditions for h_j and g_j in the above equations are all zero. Notice that the coefficients c_j $(1 \le j \le 4)$ depend on the slow time T only. Thus solutions to Eq. (4.56) are $h_3 = -ic_3 t$ and $h_4 = -ic_4 t$, which grow linearly with time. Such linearly growing terms are called secular terms in the literature. These terms would make the perturbation series (4.6) invalid over the long time scale $t = O(\varepsilon^{-1})$ and thus must be suppressed. Suppression of such secular terms in h_3 and h_4 requires $c_3 = c_4 = 0$. Similarly, suppression of secular terms in h_1 and h_2 requires $c_1 = c_2 = 0$. Thus, we get

$$c_j = 0, \quad 1 \le j \le 4, \tag{4.58}$$

in which case the solutions for h_j are

$$h_j(t) = 0, \quad 1 \le j \le 4. \tag{4.59}$$

Inserting the expression (4.10) for W into the c_j formulae (4.50)–(4.51) and making simple calculations, requirements (4.58) lead to the following dynamical equations for the soliton parameters:

$$\frac{dr}{dT} = \int_{-\infty}^{\infty} \mathrm{Im}(F_0)\cdot r\,\mathrm{sech}\,r\theta\,d\theta, \tag{4.60}$$

$$\frac{dv}{dT} = -2\int_{-\infty}^{\infty} \mathrm{Re}(F_0)\cdot r\,\mathrm{sech}\,r\theta\,\tanh r\theta\,d\theta, \tag{4.61}$$

$$\frac{dx_0}{dT} = \int_{-\infty}^{\infty} \mathrm{Im}(F_0)\cdot \theta\,\mathrm{sech}\,r\theta\,d\theta, \tag{4.62}$$

$$\frac{d\sigma_0}{dT} = \frac{1}{2}v\frac{dx_0}{dT} - \int_{-\infty}^{\infty} \mathrm{Re}(F_0)\cdot \mathrm{sech}\,r\theta\,(1 - r\theta\tanh r\theta)\,d\theta. \tag{4.63}$$

Evolution equations for the position ν and phase σ then can be found from these equations and (4.7) as

$$\frac{d\nu}{dt} = v + \varepsilon\int_{-\infty}^{\infty} \mathrm{Im}(F_0)\cdot \theta\,\mathrm{sech}\,r\theta\,d\theta, \tag{4.64}$$

$$\begin{aligned}\frac{d\sigma}{dt} &= r^2 + \frac{1}{4}v^2 + \varepsilon\left\{\frac{1}{2}v\int_{-\infty}^{\infty} \mathrm{Im}(F_0)\cdot\theta\,\mathrm{sech}\,r\theta\,d\theta\right.\\ &\left.\quad - \int_{-\infty}^{\infty} \mathrm{Re}(F_0)\cdot\mathrm{sech}\,r\theta\,(1 - r\theta\tanh r\theta)\,d\theta\right\}.\end{aligned} \tag{4.65}$$

Solving Eqs. (4.57) with zero initial conditions, we get

$$g_1(k,t) = \frac{\alpha_1(k)(1 - e^{i\lambda(k)t})}{\lambda(k)}, \quad g_2(k,t) = -\frac{\alpha_2(k)(1 - e^{-i\lambda(k)t})}{\lambda(k)}, \tag{4.66}$$

which are bounded for all times. Substituting solutions (4.59) and (4.66) into (4.54) and (4.6), we finally obtain the perturbed soliton solution (up to $O(\varepsilon)$) as

$$\begin{aligned}u(x,t) = e^{i\phi}&\left\{r\,\mathrm{sech}\,r\theta - \varepsilon\,\mathrm{sech}^2 r\theta\int_{-\infty}^{\infty}\frac{1 - e^{ir^2(k^2+1)t}}{2\pi r(k^2+1)^3}e^{irk\theta}\langle\mathcal{F},\Upsilon_1\rangle\,dk\right.\\ &\left.-\varepsilon\int_{-\infty}^{\infty}\frac{1 - e^{-ir^2(k^2+1)t}}{2\pi r(k^2+1)^3}(ik + \tanh r\theta)^2 e^{-irk\theta}\langle\mathcal{F},\Upsilon_2\rangle\,dk\right\},\end{aligned} \tag{4.67}$$

where ϕ and θ are given in (4.4), the perturbation function $\mathcal{F}$ is given in (4.53), and evolution equations for r, v, ν, and σ are given by Eqs. (4.60)–(4.61) and (4.64)–(4.65). One can see from this expression that when $t \to \infty$, the radiation terms, which involve $e^{\pm ir^2(k^2+1)t}$ in the integrals of Eq. (4.67), disperse and decay at the rate of $t^{-1/2}$ (see also Whitham (1974)). Since group velocities of these radiation modes in the NLS equation are proportional to the wavenumber k, after a short time, the high-wavenumber modes quickly escape to the far field; hence the remaining radiation modes near the soliton have wavenumbers k approximately zero. For these $k \approx 0$ modes, their oscillation frequency is approximately r^2 (see (4.67)). Thus these radiation modes cause the shape of the perturbed soliton to oscillate at the frequency r^2. After a long time, when all the energy radiation has dispersed and the shape oscillation of the perturbed soliton stopped, the solution will asymptotically approach a stationary state,

$$u(x,t) \longrightarrow e^{i\phi}\left\{ r\,\mathrm{sech} r\theta - \varepsilon\,\mathrm{sech}^2 r\theta \int_{-\infty}^{\infty} \frac{\langle \mathcal{F}, \Upsilon_1 \rangle}{2\pi r(k^2+1)^3} e^{irk\theta} dk \right.$$
$$\left. - \varepsilon \int_{-\infty}^{\infty} \frac{\langle \mathcal{F}, \Upsilon_2 \rangle}{2\pi r(k^2+1)^3} (ik + \tanh r\theta)^2 e^{-irk\theta} dk \right\}, \qquad (4.68)$$

which is a deformed soliton induced by the perturbations. By now, a complete description for the evolution of an NLS soliton under perturbations has been obtained.

It is noted that in the above soliton perturbation theory, most of the effort was spent on calculating squared eigenfunctions and the first-order radiation field of the perturbed soliton. If one is interested only in deriving the dynamical equations (4.60)–(4.65) for the soliton's parameters, then he can do so quite easily by utilizing the soliton's Goldstein modes (4.27)–(4.28) and their adjoint modes (4.38)–(4.39), and requiring the forcing term W to be orthogonal to those adjoint Goldstein modes. These requirements lead to the same equations (4.58) and hence the same dynamical equations (4.60)–(4.65); thus expressions of squared eigenfunctions are not needed at all. Indeed, formulae for squared eigenfunctions are necessary only when energy radiation needs to be calculated. The fundamental reason for this is that dynamical equations for the soliton's parameters, at time scale $T = \varepsilon t$, are decoupled from the first-order radiation field of the perturbed soliton, and thus can be calculated separately. This phenomenon occurs for all perturbed integrable equations. Hence dynamical equations for the soliton's parameters, at time scale $T = \varepsilon t$, can always be derived readily without the knowledge of squared eigenfunctions. In fact, even for perturbed *nonintegrable* equations, derivation of evolution equations for the solitary wave's parameters (at time scale $T = \varepsilon t$) is an easy matter by utilizing the solitary wave's Goldstein modes (Kodama and Ablowitz (1981), Yang and Kaup (2000)). In some situations, however, interesting dynamics of the perturbed soliton occurs at the longer time scale $T_2 = \varepsilon^2 t$. At this time scale, evolution of the soliton's parameters will be coupled with the first-order radiation field, and hence explicit calculations of squared eigenfunctions and the first-order radiation field then become necessary. Such a soliton perturbation theory will be presented in Sec. 4.4.

4.1.3 Evolution of a Perturbed Soliton in the NLS Equation

In the integrable NLS equation

$$iu_t + u_{xx} + 2|u|^2 u = 0, \tag{4.69}$$

when the initial condition is a perturbed single soliton, the solution will relax oscillatorily into an NLS soliton, and a certain amount of radiation will be emitted into the far field at the same time. In this section, we give a complete and accurate description of this solution dynamics by applying the soliton perturbation theory of the previous section. Note that oscillatory relaxation of an initial condition into a final stationary state is a ubiquitous phenomenon which occurs not only in the NLS equation (4.69), but also in many other integrable and nonintegrable wave equations. Thus our characterization of this relaxation process below is of universal value.

For the purpose of demonstration, we consider the following initial condition:

$$u(x,0) = (1+\epsilon)\operatorname{sech} x, \quad \epsilon \ll 1, \tag{4.70}$$

which is a slightly amplified NLS soliton. We want to determine the final state of the solution, and also characterize how this initial condition relaxes to the final state. This problem has been treated before by other techniques (see Anderson (1983), Kath and Smyth (1995), Kuznetsov et al. (1995)), but our treatment below gives simpler and more accurate results.

First we turn this problem into a soliton perturbation problem. For this purpose, we define a scaled variable

$$U(x,t) = u(x,t)/(1+\epsilon), \tag{4.71}$$

which satisfies the perturbed NLS equation

$$iU_t + U_{xx} + 2|U|^2 U = -4\epsilon|U|^2 U. \tag{4.72}$$

Here the $O(\epsilon^2)$ term has been neglected. The initial condition for U is

$$U(x,0) = \operatorname{sech} x, \tag{4.73}$$

which is soliton solution of the unperturbed NLS equation (4.72) with initial amplitude $r = 1$ and initial position $x_0 = 0$. Thus this problem has become a soliton perturbation problem analyzed in the previous section. Here $F_0 = -4r^2\operatorname{sech}^3\theta$. Substituting this F_0 into dynamical equations (4.60)–(4.61) for r and v, we find that

$$dr/dT = dv/dT = 0, \tag{4.74}$$

thus $r=1$, and $v=0$ for all slow time T. Therefore, we will replace r by one and v by zero in all calculations below. Similarly, we find that $dv/dt=0$, and hence the position of this soliton does not change, and $\theta=x$. Its phase equation (4.65) can be calculated through simple algebra to be

$$\frac{d\sigma}{dt}=r^2+4\epsilon r^2=1+4\epsilon. \tag{4.75}$$

Thus

$$\phi(t)=\sigma(t)=(1+4\epsilon)t. \tag{4.76}$$

To calculate w from (4.10), notice from dynamical equations (4.62)–(4.63) that $dx_0/dT=0$, and $d\sigma_0/dT=4r^2$; thus

$$w=F_0+\sigma_{0T}\Phi=4\operatorname{sech}\theta(1-\operatorname{sech}^2\theta). \tag{4.77}$$

Then utilizing the formula

$$\int_{-\infty}^{\infty}e^{ikx}\operatorname{sech}x\,dx=\pi\operatorname{sech}\frac{1}{2}k\pi \tag{4.78}$$

and performing integration by parts, it is easy to find that

$$\langle\mathcal{F},\Upsilon_1\rangle=-\langle\mathcal{F},\Upsilon_2\rangle=-\pi(k^2+1)^2\operatorname{sech}\frac{1}{2}k\pi. \tag{4.79}$$

Inserting these expressions into (4.67) and recalling the scaling (4.71), the NLS solution for the initial condition (4.70) will be obtained.

First, we consider the final state of this solution as $t\to\infty$. Inserting (4.79) into (4.68), we get, up to $O(\epsilon^2)$,

$$U(x,t)\to e^{i(1+4\epsilon)t}\left\{\operatorname{sech}x+\epsilon\int_{-\infty}^{\infty}\frac{\operatorname{sech}\frac{1}{2}k\pi}{2(k^2+1)}\left[\operatorname{sech}^2x-(ik-\tanh x)^2\right]e^{ikx}dk\right\}. \tag{4.80}$$

It is easy to verify that the integral in the above expression is equal to $\operatorname{sech}x(1-2x\tanh x)$. Thus the asymptotic solution $u(x,t)$ of the NLS equation (4.69) for the initial condition (4.70) as $t\to\infty$ is

$$\begin{aligned}u(x,t)&\to e^{i(1+4\epsilon)t}(1+\epsilon)\left[\operatorname{sech}x+\epsilon\operatorname{sech}x(1-2x\tanh x)+O(\epsilon^2)\right]\\&=e^{i(1+4\epsilon)t}\left[(1+2\epsilon)\operatorname{sech}(1+2\epsilon)x+O(\epsilon^2)\right].\end{aligned} \tag{4.81}$$

Note that if one only wants to determine this final state, then a simpler way is that after the phase function $\phi(t)=(1+4\epsilon)t$ has been obtained from the dynamical equations (4.60)–(4.65), one substitutes the stationary soliton solution $u(x,t)=f(x)e^{i(1+4\epsilon)t}$ into the NLS equation (4.69) and directly gets $f(x) = \sqrt{1+4\epsilon}\,\text{sech}\sqrt{1+4\epsilon}\,x$, which agrees with (4.81) up to $O(\epsilon)$. The above asymptotic solution (4.81) can also be verified by the exact result of Satsuma and Yajima (1974) (see Sec. 2.4.1), as the initial condition (4.70) is of Satsuma–Yajima type (2.159). According to their exact result, the initial condition (4.70) contains a single discrete eigenvalue $\zeta_1 = i(\frac{1}{2}+\epsilon)$. Thus as $t \to \infty$, the solution will evolve into an NLS soliton,

$$u(x,t) \to e^{i(1+2\epsilon)^2 t}(1+2\epsilon)\text{sech}(1+2\epsilon)x, \tag{4.82}$$

which agrees with the perturbation-theory solution (4.81) up to $O(\epsilon)$. Their difference, which is of $O(\epsilon^2)$, is expected as our perturbation-theory solution was derived with accuracy $O(\epsilon^2)$.

Next, we examine the relaxation process of the initial solution (4.70) to its final state (4.81). This relaxation process involves interaction between the NLS soliton and energy radiation and cannot be captured by the Satsuma–Yajima analysis (see Sec. 2.4.1). On this problem, the soliton perturbation treatment has a distinctive advantage. To proceed, we insert (4.79) into the solution formula (4.67). Utilizing the above steady-state calculation result (see (4.80)–(4.81)), we find that the full perturbed solution in the relaxation process is

$$\begin{aligned} u(x,t) = e^{i\phi(t)}\Bigg\{ &(1+2\epsilon)\text{sech}(1+2\epsilon)x - \epsilon\int_{-\infty}^{\infty}\frac{\text{sech}\frac{1}{2}k\pi}{2(k^2+1)} \\ &\times\left[\text{sech}^2 x\, e^{i(k^2+1)t} - (ik-\tanh x)^2 e^{-i(k^2+1)t}\right]e^{ikx}dk + O(\epsilon^2)\Bigg\}. \end{aligned} \tag{4.83}$$

For $x = O(1)$ and $t \gg 1$, the main contribution to the integral in the above equation comes from the $k \approx 0$ region. Thus the wavenumber k in this integrand, except inside functions $e^{\pm i(k^2+1)t}$, can be approximated by $k=0$. The reduced integral then can be calculated exactly, and one finds that the large-time asymptotics of this transient solution at $x = O(1)$ is

$$\begin{aligned} u(x,t) = e^{i\phi(t)}\Bigg\{ &(1+2\epsilon)\text{sech}(1+2\epsilon)x \\ &-\epsilon\sqrt{\frac{\pi}{4t}}\left[\text{sech}^2 x\, e^{i(t+\frac{\pi}{4})} - \tanh^2 x\, e^{-i(t+\frac{\pi}{4})}\right] + O(\epsilon^2)\Bigg\}. \end{aligned} \tag{4.84}$$

This formula shows that the shape of the perturbed soliton oscillates at frequency one, and this oscillation decays with time at the rate of $t^{-1/2}$. It also shows that the perturbed soliton contains a long flat radiation shelf of height $\epsilon\sqrt{\pi/4t}$ at the foot of the soliton. This radiation shelf has wavenumber $k=0$ and zero group velocity; hence it does not leak energy from the soliton to the far field. This analytical solution compares well with the numerical solution when ϵ is small. For larger values of ϵ, while the decay rates of shape oscillations in the analytical and numerical solutions still agree well with each other, frequencies of those oscillations start to deviate from each other. In order to improve the accuracy of the analytical formula (4.84), we notice that according to the discussions in the end of the previous subsection, the frequency of shape oscillations is r^2, where r is the amplitude of the soliton. In the present case, the perturbed soliton eventually relaxes to a new soliton with amplitude $r = 1+2\epsilon$ (see (4.81)–(4.82) and (4.84)). Thus we modify the oscillation frequency in (4.84) from 1 to $(1+2\epsilon)^2$ and get a new formula for the perturbed solution at $x = O(1)$ and $t \gg 1$, up to $O(\epsilon)$, as

$$u(x,t) = e^{i\phi(t)}\left\{(1+2\epsilon)\operatorname{sech}(1+2\epsilon)x \right.$$
$$\left. -\epsilon\sqrt{\frac{\pi}{4t}}\left[\operatorname{sech}^2 x\, e^{i\left[(1+2\epsilon)^2 t+\frac{\pi}{4}\right]} - \tanh^2 x\, e^{-i\left[(1+2\epsilon)^2 t+\frac{\pi}{4}\right]}\right]\right\}. \tag{4.85}$$

This modified solution formula can also be obtained in the framework of the multiscale soliton perturbation theory described earlier in this section by proceeding to the $O(\epsilon^2)$ equation for Φ_2 and suppressing its linear growth in time. The intensity peak of this perturbed solution is located at $x = 0$; hence the time evolution for the amplitude $|u(0,t)|$ of the perturbed solution (4.85), up to $O(\epsilon)$, is

$$|u(0,t)| = 1+2\epsilon - \epsilon\sqrt{\frac{\pi}{4t}}\cos\left[(1+2\epsilon)^2 t + \frac{\pi}{4}\right]. \tag{4.86}$$

These analytical solution formulae for the relaxation process, which were derived for $\epsilon \ll 1$ and $t \gg 1$, prove to be very accurate even when ϵ is not so small and t not so large. To demonstrate, we take $\epsilon = 0.2$ in the initial condition (4.70). In this case, the numerically computed amplitude function $|u(0,t)|$ and the analytical formula (4.86) are both plotted in Fig. 4.1(a). We see that they are almost indistinguishable from each other after only one shape oscillation! Solution profiles $|u(x,t)|$ at time $t = 15$ from numerical computations and the analytical formula (4.85) are also plotted together in Fig. 4.1(b). We see that the analytical and numerical profiles are also almost indistinguishable from each other. In this solution profile, the long radiation shelf is clearly visible at the foot of the soliton.

4.2 Higher-Order Effects on Optical Solitons

We have known that when a light pulse travels in an optical fiber, its electric field is given by (1.22), where the envelope function $A(z,t)$ satisfies the NLS equation (1.23) at the lowest order of approximation. If the light pulse is very short, some higher-order linear and nonlinear effects may become significant and need to be considered. These higher-order effects include the third-order dispersion, self-steepening (also called nonlinear dispersion),

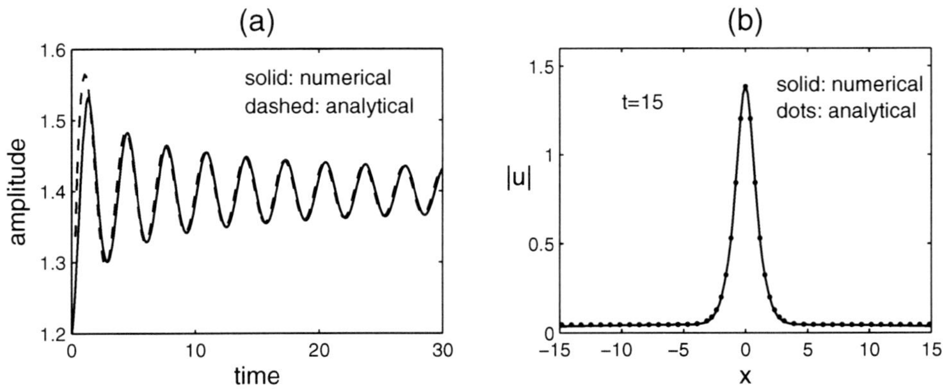

Figure 4.1. *Evolution of a perturbed soliton* (4.70) *in the NLS equation* (4.69). *Here* $\epsilon = 0.2$. (a) *Time evolution of the solution's amplitude* $|u(0,t)|$; (b) *solution profile* $|u(x;t)|$ *at time* $t = 15$. *Solid lines are numerical results, and dashed line and dots are the analytical formulae* (4.85)–(4.86). *In* (b), *the formation of a radiation shelf at the foot of the soliton can be observed.*

and the Raman effect. The governing equation with these higher-order corrections is given in (1.50) (see also Hasegawa and Kodama (1995) and Agrawal (2007)). If we define the following nondimensional variables:

$$\tau = T/T_0, \quad Z = z/L_D, \quad u = \sqrt{\gamma L_D}U, \quad L_D = T_0^2/|\beta_2|, \tag{4.87}$$

where T_0 is the pulse width and L_D the dispersion length, then assuming the dispersion is anomalous ($\beta_2 < 0$), the nondimensional equation containing these higher-order effects is then

$$iu_Z + \frac{1}{2}u_{\tau\tau} + |u|^2 u = i\delta_3 u_{\tau\tau\tau} - is(|u|^2 u)_\tau + \tau_R u(|u|^2)_\tau. \tag{4.88}$$

Here δ_3, s, and τ_R are the third-order dispersion, self-steepening, and Raman coefficients, and their expressions are

$$\delta_3 = \frac{\beta'''(\omega_0)}{6|\beta_2|T_0}, \quad s = \frac{1}{\omega_0 T_0}, \quad \tau_R = \frac{T_R}{T_0}. \tag{4.89}$$

These coefficients are all inversely proportional to the pulse width T_0. If $T_0 \gg 1$ ps (picosecond), then all these coefficients are very small, thus these higher-order effects are all very weak. But for femtosecond pulses, these coefficients may become appreciable. For instance, in a standard silica fiber with typical values $\beta_2 = -20\,\text{ps}^2/\text{km}$ and $\beta''' = 0.1\,\text{ps}^3/\text{km}$ at wavelength $\lambda = 1.55\,\mu m$, for a pulse with $T_0 \sim 30$ fs (femtoseconds) and assuming $T_R = 3$ fs, then $\delta_3 \sim 0.03$, $s \sim 0.03$, $\tau_R \sim 0.1$. It is seen that these higher-order effects, especially the

Raman effect, become nonnegligible. Below we examine each of these three higher-order effects on optical solitons separately.

4.2.1 Raman Effect

The Raman effect is the dominant higher-order effect on optical solitons. When monochromatic light travels in an optical fiber, it emits light with a certain lower frequency due to the inelastic scattering of the fiber atoms. This is the Raman effect discovered by C.V. Raman in 1928. When a soliton is very narrow, its spectrum becomes very broad. Hence the higher-frequency components of the soliton will continuously transfer energy to its lower-frequency components through the Raman effect, and this leads to the soliton frequency downshift. This phenomenon was first observed by Mitschke and Mollenauer (1986), and theoretically explained by Gordon (1986). Below we use the soliton perturbation theory to quantitatively explain this phenomenon.

The model equation (4.88) including only the Raman effect is

$$iu_Z + \frac{1}{2}u_{\tau\tau} + |u|^2 u = \tau_R u(|u|^2)_\tau. \tag{4.90}$$

Here we treat τ_R as small, and thus the Raman term is considered as a perturbation. Notice that the coefficients in the above unperturbed NLS equation are a little different from those in (4.1), and this difference amounts to a rescaling of the distance Z by a factor of 2. Solitons in this NLS equation are

$$u(Z,\tau) = r(Z)\,\mathrm{sech}\, r(Z)[\tau - \nu(Z)]\, e^{-i\delta(Z)[\tau-\nu(Z)]+i\sigma(Z)}, \tag{4.91}$$

where

$$\nu = -\int_0^Z \delta\, ds + \tau_0, \quad \sigma = \frac{1}{2}\int_0^Z (r^2+\delta^2)\, ds + \sigma_0. \tag{4.92}$$

Here r is the soliton amplitude, and δ is a frequency parameter ($-\delta$ is the counterpart of the soliton's velocity in the previous section). In the absence of the Raman perturbation, soliton parameters r, δ, τ_0, and σ_0 are constants. In the presence of this perturbation, some of these parameters will change over distance. To illustrate, we take the initial condition as a sech soliton with unit amplitude, i.e.,

$$u(0,\tau) = \mathrm{sech}\, \tau, \tag{4.93}$$

and we take the Raman coefficient as $\tau_R = 0.1$. By numerically simulating Eq. (4.90), the evolution of this soliton under Raman effect is obtained and displayed in Fig. 4.2. The most distinctive Raman effect in this evolution is that the soliton's velocity (i.e., $-\delta$) steadily increases; i.e., the soliton's frequency δ steadily decreases during propagation. This is the phenomenon of soliton frequency downshift reported by Mitschke and Mollenauer (1986).

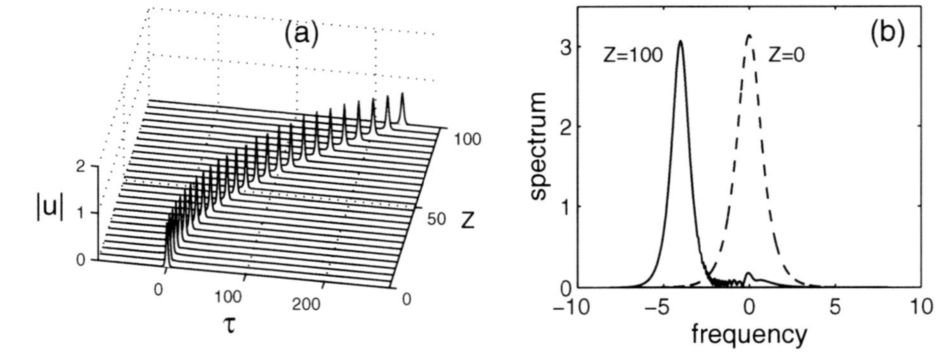

Figure 4.2. *Evolution of an NLS soliton under the Raman effect. The initial condition is the NLS soliton with unit amplitude* (4.93)*, and* $\tau_R = 0.1$.

Another Raman effect is that the amplitude of the soliton also decreases, but this decay is very slow during propagation.

The Raman effect on NLS solitons can be quantitatively analyzed by the soliton perturbation theory of the previous section. Using this theory and recalling the Z-variable rescaling, we get evolution equations for soliton parameters as

$$\frac{dr}{dZ} = \frac{d\tau_0}{dZ} = \frac{d\sigma_0}{dZ} = 0, \tag{4.94}$$

$$\frac{d\delta}{dZ} = -\frac{8}{15}\tau_R r^4. \tag{4.95}$$

These results show that under the Raman effect, the soliton amplitude does not change (to the leading order of the soliton perturbation theory), but its frequency decreases linearly with distance at the rate of $8\tau_R r^4/15$. For the initial condition (4.93) where $r = 1$, which means that we consider a physical initial pulse $\mathrm{sech}(T/T_0)$, by recalling variable normalizations (4.87), the rate of frequency downshift in physical units is then

$$\frac{d\omega}{dz} = -\frac{8T_R|\beta_2|}{15T_0^4}. \tag{4.96}$$

This rate is inversely proportional to T_0^4. Thus for very short pulses, the frequency downshift can be very significant. For instance, consider a normal optical fiber with $\beta_2 = -20\,\mathrm{ps}^2/\mathrm{km}$ and $T_R = 3$ fs. Then for a pulse with $T_0 = 0.5$ ps, its rate of frequency downshift is about 0.5 THz/km. Considering that the spectral width of a $\mathrm{sech}(T/T_0)$ pulse with $T_0 = 0.5$ ps is under 0.5 THz, this Raman frequency downshift is very noticeable. In the experiment by Mitschke and Mollenauer (1986), the frequency of a $T_0 \sim 0.3$ ps pulse downshifted by 8 THz under 0.4 km, which is very large.

The above analytical results tell us that under the Raman effect, the frequency of the soliton is downshifting, but its amplitude does not change. These results were obtained within the first-order soliton perturbation theory, which is valid when $\tau_R \ll 1$ and $Z = O(1/\tau_R)$. For longer distances Z or not-small values of τ_R, will these analytical results still hold? To address this question, we notice that Eq. (4.90) for arbitrary values of τ_R admits exact "frequency-shifting" solitons which are the counterpart of the self-bending photorefractive screening solitons reported by Facao and Parker (2003). These "frequency-shifting" solitons have the form

$$u(Z,\tau) = e^{i\phi(Z,\eta)}F(\eta), \tag{4.97}$$

where

$$\eta = \tau + \delta Z - \frac{1}{2}aZ^2, \tag{4.98}$$

$$\phi = -(\delta - aZ)\eta + \frac{1}{6}a^2Z^3 - \frac{1}{2}a\delta Z^2 + \frac{1}{2}\delta^2 Z + bZ, \tag{4.99}$$

and the real function $F(\eta)$ satisfies the following ordinary differential equation (ODE):

$$\frac{1}{2}\frac{d^2F}{d\eta^2} - (a\eta + b)F + F^3 = \tau_R F(F^2)_\eta. \tag{4.100}$$

Here δ is a frequency parameter, a is the frequency-downshifting rate which determines the amplitude of the solution F (see also (4.95)), and δ, a, b are all constants. The constant b can be absorbed by a shift in η, and so we set $b = 0$ without loss of generality. When $\tau_R \neq 0$, Eq. (4.100) admits solutions F which are vanishing at $\eta = \pm\infty$ with $a \neq 0$. In the tails of $F(\eta)$ where $F \ll 1$, nonlinear terms in (4.100) can be neglected, and hence F satisfies a linear Airy equation. As $\eta \to +\infty$, F decays very rapidly in proportion to the first Airy function, while as $\eta \to -\infty$, F decays very slowly as a linear combination of the first and second Airy functions. Specifically, the asymptotic behaviors of the function F at $|\eta| \gg 1$ are

$$F(\eta) \to \begin{cases} \alpha \, \mathrm{Ai}[(2a)^{1/3}\eta], & \eta \to +\infty, \\ \beta \left\{\cos\theta \, \mathrm{Ai}[(2a)^{1/3}\eta] + \sin\theta \, \mathrm{Bi}[(2a)^{1/3}\eta]\right\}, & \eta \to -\infty, \end{cases} \tag{4.101}$$

where $\mathrm{Ai}(x)$ and $\mathrm{Bi}(x)$ are the first and second Airy functions, α is the coefficient of the rapidly decaying tail at $\eta \gg 1$, and (β, θ) are the amplitude and phase parameters of the slowly decaying tail at $\eta \ll -1$. For each τ_R, a, and α value, one can integrate Eq. (4.100) from $\eta = +\infty$ to $-\infty$ and obtain a localized solution $F(\eta)$ with the above asymptotics (4.101). Two such solutions with $\tau_R = 0.1, a = 0.05$ and two different values of α are displayed in Fig. 4.3. The solution in Fig. 4.3(a) has a dominant sech-like hump and a weak Airy tail, while the solution in Fig. 4.3(b) has a strong Airy tail. Even though the size β of this

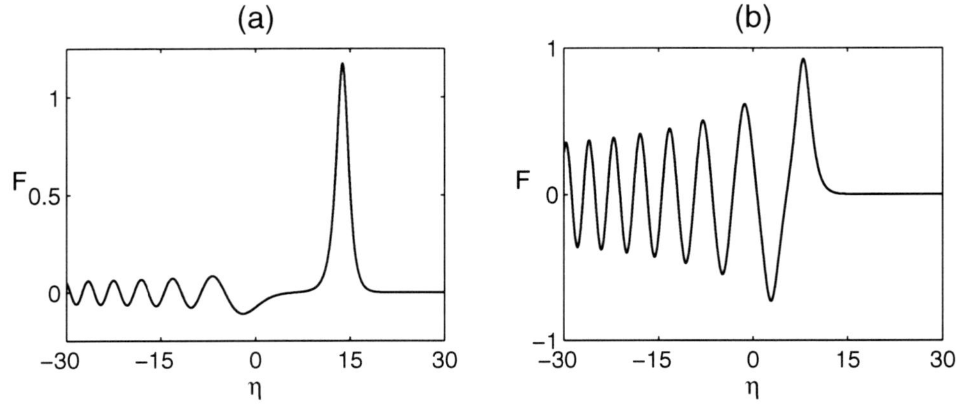

Figure 4.3. *Two exact frequency-shifting solitons in Eq.* (4.100) *with* $\tau_R = 0.1$, $a = 0.05$, *and* $b = 0$*:* (a) $\alpha = 666{,}000$*;* (b) $\alpha = 1{,}000$.

Airy tail at the left side depends on the parameter α at the right side, this tail never vanishes (i.e., $\beta \neq 0$) for any $\alpha \neq 0$. Thus, since the Airy functions in (4.101) decay in proportion to $|\eta|^{-1/4}$ when $\eta \ll -1$, the power of this solution, $\int_{-\infty}^{\infty} |F|^2 d\eta$, is then infinite. This indicates that the "frequency-shifting" solitons (4.97) always have slowly decaying Airy tails on its left side and have infinite power. Because of this, when one starts with an NLS soliton which has finite power, this solution, in an attempt to approach the above "frequency-shifting" solitons under the Raman effect, will not only downshift its frequency, but also emit Airy tails at its leading (left) edge. This energy loss eventually will lead to the decay of the soliton. Indeed, this decay is visible in the numerical solution of Eq. (4.90) in Fig. 4.2, where the amplitude of the soliton drops from the initial value 1 to the final value 0.68 (at $Z = 100$). This decay of the soliton is not captured by the above first-order perturbation theory, because this decay occurs over distance scales longer than $O(1/\tau_R)$. How one can analytically calculate the decay rate of solitons under the Raman effect is still unclear.

It is noted that in the absence of the self-phase modulation and Raman effects, Eq. (4.90) becomes the linear Schrödinger equation

$$iu_Z + \frac{1}{2}u_{\tau\tau} = 0, \tag{4.102}$$

whose localized but frequency-shifting solution is

$$u(Z,\tau) = \alpha\, e^{i\phi(Z,\eta)} \mathrm{Ai}[(2a)^{1/3}\eta], \tag{4.103}$$

where functions η and ϕ are given in Eqs. (4.98)–(4.99), and b has been set to zero. This Airy-function solution of the linear Schrödinger equation (4.102) is the fundamental reason for the existence of frequency-shifting solitons in the NLS equation under the Raman effect (4.90). Another important fact to be noted is that in the absence of the Raman effect, the NLS equation (i.e., (4.90) with $\tau_R = 0$) still admits frequency-shifting solitons of the form (4.97). These solitons are the nonlinear counterpart of the Airy solution (4.103) in the linear Schrödinger equation (4.102), and their shapes are analogous to that in Fig. 4.3(b).

4.2.2 Self-Steepening Effect

Now we consider the NLS equation under the self-steepening effect:

$$iu_Z + \frac{1}{2}u_{\tau\tau} + |u|^2 u = -is\,(|u|^2 u)_\tau. \tag{4.104}$$

In the absence of the second-order dispersion term $u_{\tau\tau}$, the self-steepening term on the right-hand side of Eq. (4.104) causes an optical pulse to become asymmetric and steepen up at the trailing edge (i.e., the right side of the pulse); see Agrawal (2007). The second-order dispersion reduces this self-steepening considerably. However, this self-steepening term still has its consequences. To illustrate its effect, we take the initial condition as the sech soliton with unit amplitude (4.93), and take the self-steepening coefficient as $s = 0.2$. Evolution of this soliton under the self-steepening effect is numerically determined and displayed in Fig. 4.4. It is seen that the main effect of the self-steepening term is a position shift of the soliton.

The effect of self-steepening on an NLS soliton can be analyzed by the perturbation theory of the previous section as well. Using that theory, we find that

$$\frac{dr}{dZ} = \frac{d\delta}{dZ} = \frac{d\sigma_0}{dZ} = 0, \quad \frac{d\tau_0}{dZ} = sr^2. \tag{4.105}$$

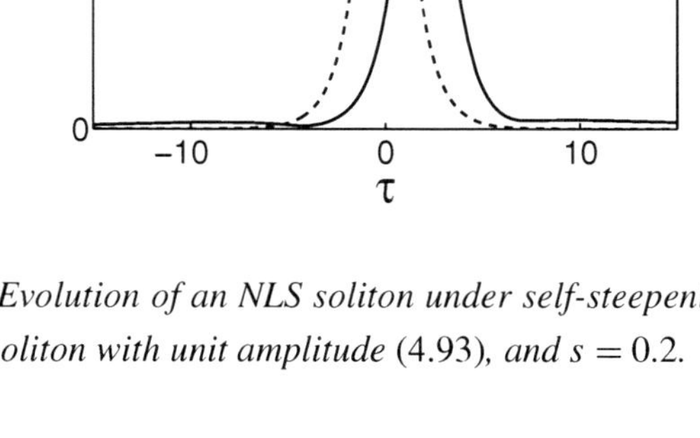

Figure 4.4. *Evolution of an NLS soliton under self-steepening effect. The initial condition is the NLS soliton with unit amplitude* (4.93), *and* $s = 0.2$.

These results confirm analytically that the main effect of self-steepening on the soliton is indeed a position shift in the amount of $\tau_0 = sr^2 Z$. This position shift linearly increases with distance.

It is noted that the NLS equation with the self-steepening term, i.e., Eq. (4.104), is actually integrable for arbitrary values of the steepening coefficient s (Wadati, et al. (1979)). This equation is sometimes called the modified NLS equation in the literature. Through a gauge transformation (Mihalache et al. (1993b)), this equation can also be transformed to the derivative NLS equation which is integrable as well (Kaup and Newell (1978b)). Because of this, an NLS soliton under the self-steepening effect will eventually evolve into a single soliton of the modified NLS equation.

4.2.3 Third-Order Dispersion Effect

In this subsection, we discuss the effect of third-order dispersion on optical solitons. The model equation for this study is

$$iu_Z + \frac{1}{2}u_{\tau\tau} + |u|^2 u = i\delta_3 u_{\tau\tau\tau}, \tag{4.106}$$

where δ_3 is the third-order dispersion coefficient. To illustrate this effect, we first simulate this equation with $\delta_3 = 0.1$ and the initial condition as an NLS soliton with unit amplitude (4.93). The amplitude and spectrum of the solution at $Z = 15$ are displayed in Fig. 4.5. It is seen that the third-order dispersion causes a position shift to the soliton. More significantly, a high-frequency continuous-wave (cw) tail with frequency peak approximately at $1/2\delta_3$ appears on the trailing edge (right side) of the soliton (this cw tail corresponds to the frequency peak on the right side of the spectrum in Fig. 4.5(b)). Emission of this cw tail drains energy from the soliton and causes it to decay.

The position shift of the soliton due to third-order dispersion can be analyzed by the soliton perturbation theory of the previous section. Following that theory, we find that

$$\frac{dr}{dZ} = \frac{d\delta}{dZ} = 0 \tag{4.107}$$

and

$$\frac{d\tau_0}{dZ} = \delta_3(r^2 + 3\delta^2), \quad \frac{d\sigma_0}{dZ} = 2\delta_3\delta(r^2 - \delta^2). \tag{4.108}$$

When the soliton frequency is initially $\delta = 0$, then the only third-order dispersion effect on the soliton is a position shift in the amount of $\delta_3 r^2 Z$, which grows linearly with distance.

Regarding the high-frequency cw tail emitted at the trailing edge of the soliton, it originates from a resonance between the soliton and the continuous spectrum of the system. Notice that for a linear Fourier component $e^{i(\beta Z - \omega\tau)}$ in Eq. (4.106), the dispersion relation

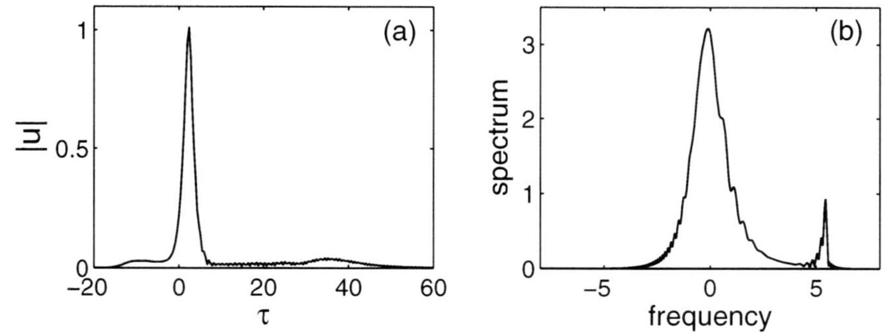

Figure 4.5. *Amplitude and spectrum of an NLS soliton under third-order dispersion effect at $Z = 15$. The initial condition is the NLS soliton with unit amplitude* (4.93), *and* $\delta_3 = 0.1$.

is $\beta = -\omega^2/2 + \delta_3\omega^3$. For any wavenumber β, this dispersion relation admits a real frequency value ω. Hence all wavenumbers (including the soliton's wavenumber) lie inside the continuous spectrum of Eq. (4.106). As a result, this soliton is in resonance with the continuous spectrum, and thus it feeds energy into the resonant Fourier frequency. Another way to understand this resonance is the following. Let us consider a general frequency mode $f(Z)e^{-i\omega\tau}$. Inserting this mode into Eq. (4.106), we find that the amplitude $f(Z)$ of this mode satisfies an inhomogeneous equation. The homogeneous part of this equation is $if'(Z) + (-\omega^2/2 + \delta_3\omega^3)f(Z)$, and the inhomogeneous (i.e., forcing) term comes from the nonlinear interaction between various frequency modes. When $\omega = \omega_c = 1/2\delta_3$, the second term in the homogeneous part vanishes. Then this inhomogeneous equation shows that $f(Z)$ will linearly grow, and hence resonance occurs. When this happens, energy is fed into the resonant frequency mode ω_c, and thus a cw tail with this frequency appears. This cw tail cannot be captured by the soliton perturbation theory of the previous section. The reason is that the amplitude of this cw tail turns out to be exponentially small in δ_3, while all corrections in the perturbation theory of the previous section were only algebraically small in δ_3. This cw tail will be thoroughly analyzed by the exponential asymptotics technique in Sec. 5.5.

4.3 Weak Interactions of NLS Solitons

In the NLS equation

$$iu_t + u_{xx} + 2|u|^2u = 0, \tag{4.109}$$

if the initial condition is two well-separated single solitons, these two solitons will interact with each other due to tail interference. This interaction is slow (at least initially) since the tail overlap between the two solitons is weak. However, over a long time, the accumulative effect of this weak interaction becomes significant. Under this interaction, the two solitons can attract each other and collide, or repel each other and separate, depending on initial conditions of the two solitons. This issue of weak interactions arises in a number of physical situations, most notably in soliton-based fiber communications, where interference between neighboring soliton pulses is a detrimental effect and needs to be suppressed. To study this weak interaction, one method is to calculate discrete eigenvalues of the Zakharov–Shabat system for the initial condition of two well-separated single solitons (Satsuma and Yajima (1974)). Based on these discrete eigenvalues, one can predict the long-time (final) states of the solution. But this method cannot describe the intermediate interaction process. Furthermore, this method cannot be generalized to study weak interactions of solitary waves in nonintegrable equations. Below, we study this weak interaction by another method proposed by Karpman and Solovev (1981) (see also Gorshkov and Ostrovsky (1981)), which can overcome these two restrictions.

In weak interactions, the initial condition is two well-separated single solitons,

$$u(x,0) = u_1(x,0) + u_2(x,0). \tag{4.110}$$

Here

$$u_k(x,0) = r_{k0}\,\mathrm{sech}\, r_{k0}(x - x_{k0})\, e^{\frac{1}{2}i v_{k0}(x-x_{k0})+i\sigma_{k0}}, \quad k = 1,2, \tag{4.111}$$

are two NLS solitons with initial amplitude r_{k0}, velocity v_{k0}, position x_{k0}, and phase σ_{k0}. We will determine how these two solitons evolve under their weak tail interactions. Before the analysis, we first show a few interaction scenarios by numerical simulations. We select $v_{10} = v_{20} = 0$, $x_{20} - x_{10} = 6$, and the following four sets of initial parameters:

$$\text{(a)}\ r_{10} = r_{20} = 1, \quad \sigma_{20} - \sigma_{10} = 0; \tag{4.112}$$

$$\text{(b)}\ r_{10} = r_{20} = 1, \quad \sigma_{20} - \sigma_{10} = \pi; \tag{4.113}$$

$$\text{(c)}\ r_{10} = r_{20} = 1, \quad \sigma_{20} - \sigma_{10} = \frac{1}{4}\pi; \tag{4.114}$$

$$\text{(d)}\ r_{10} = 0.95, \quad r_{20} = 1.05, \quad \sigma_{20} - \sigma_{10} = 0. \tag{4.115}$$

The interaction results for these four initial conditions are displayed in Fig. 4.6. We see that in the first case where the two solitons initially have the same amplitude, velocity, and phase, they attract each other and collide. After the collision, they separate, but then return and collide again. This process repeats indefinitely. In the second case, where the two solitons initially have the same amplitude and velocity but with π-phase difference (out of phase), they repel each other and move away. In the third case, where the two solitons initially have the same amplitude and velocity and have a $\pi/4$-phase difference, they first attract each other, but then repel each other and move away. In the fourth case, where the

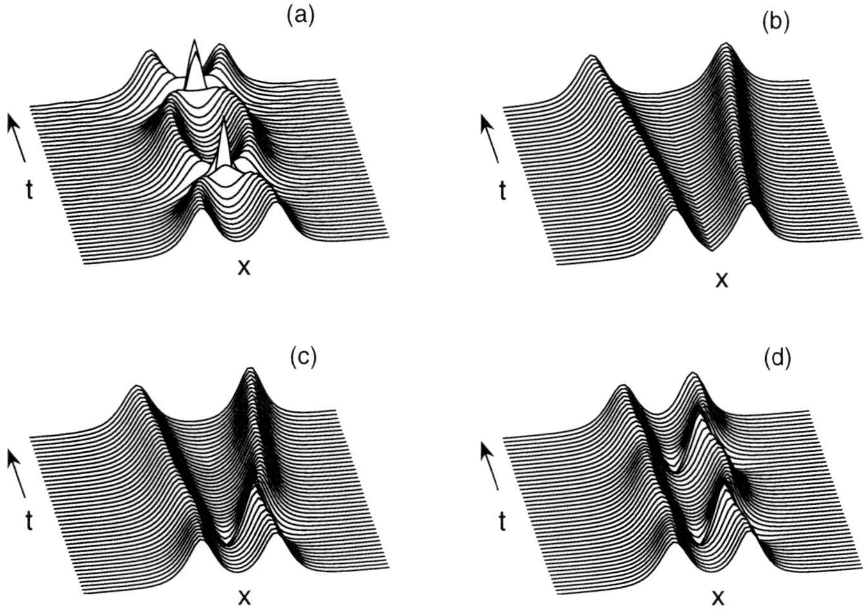

Figure 4.6. *Four scenarios of weak interactions between two NLS solitons for the four initial conditions* (4.112)–(4.115).

two solitons initially have the same velocity but different amplitude, they oscillate around their initial positions and form an oscillating bound state. These numerical results reveal the rich dynamics of weak interactions between two NLS solitons.

Below we present the analytical theory for weak interactions of NLS solitons and explain the above numerical results. The basic idea is to treat tail interference between solitons as small perturbations to each soliton, and then use the soliton perturbation theory to derive dynamical equations for each soliton's parameters. When the two solitons are well separated, the solution $u(x,t)$ can be approximated by a linear superposition of these two solitons,

$$u(x,t) \approx u_1(x,t) + u_2(x,t), \tag{4.116}$$

where

$$u_k(x,t) = r_k \operatorname{sech} r_k\theta_k \, e^{i\phi_k}, \quad k = 1,2, \tag{4.117}$$

$$\phi_k = \frac{1}{2}v_k\theta_k + \sigma_k, \quad \theta_k = x - \xi_k, \tag{4.118}$$

$$\xi_k = \int_0^t v_k \, d\tau + x_{k0}, \quad \sigma_k = \int_0^t \left(r_k^2 + \frac{1}{4}v_k^2\right) d\tau + \sigma_{k0}. \tag{4.119}$$

Here ξ_k is the location of the kth soliton. The parameters r_k, v_k, x_{k0}, and σ_{k0} in these solitons are slowly varying since the interaction is weak. Below we refer to the initially left soliton as the first soliton with index $k = 1$, and the initially right soliton as the second soliton with index $k = 2$. Thus initially $\xi_2 > \xi_1$. In the approximate solution (4.116), energy radiation is neglected. This is justified as energy radiation emitted during weak interactions is indeed very weak. In our analysis, we also assume that the two solitons are well separated and have approximately the same velocities and amplitudes. Mathematically, these assumptions are expressed as

$$|\Delta r \Delta \xi| \ll 1 \ll r\Delta\xi, \quad |\Delta v \Delta \xi| \ll 1, \tag{4.120}$$

where the notation is

$$r = \frac{1}{2}(r_1 + r_2), \quad v = \frac{1}{2}(v_1 + v_2),$$

and

$$\Delta r = r_2 - r_1, \quad \Delta v = v_2 - v_1, \quad \Delta\xi = \xi_2 - \xi_1, \quad \Delta\phi = \phi_2 - \phi_1.$$

The former part of this assumption requires that $r\Delta\xi$ be large, i.e., the two solitons are well separated. Hence $\varepsilon \equiv e^{-r\Delta\xi} \ll 1$. In addition, it requires that Δr be very small, i.e., the two solitons have approximately the same amplitude. The latter part of the assumption (4.120) requires that Δv be very small, i.e., the two solitons have approximately the same velocity. Now we substitute the superposition (4.116) into the NLS equation (4.109). The nonlinear term $|u|^2 u$ produces six nonlinear terms in u_1 and u_2. In the vicinity of the first soliton (where $r_1(x - \xi_1) = O(1)$), we can see that the three nonlinear terms, $u_2^2 u_1^*$, $|u_2|^2 u_1$, and $|u_2|^2 u_2$, are $O(\varepsilon^2)$ or $O(\varepsilon^3)$, which are higher order than the other three nonlinear terms which are either $O(1)$ or $O(\varepsilon)$. In addition, the term $iu_{2t} + u_{2xx}$ is roughly proportional to $|u_2|^2 u_2$, which is also $O(\varepsilon^3)$. Neglecting these higher-order terms, the equation in the vicinity of the first soliton becomes

$$iu_{1t} + u_{1xx} + 2|u_1|^2 u_1 = -2u_1^2 u_2^* - 4|u_1|^2 u_2. \tag{4.121}$$

The right-hand side of this equation is $O(\varepsilon)$; thus this equation is a perturbed NLS equation for the first soliton. The equation for the second soliton u_2 can be similarly derived, and it is (4.121) with indices 1 and 2 exchanged.

Now we apply the soliton perturbation theory to Eq. (4.121) and derive the dynamical equations for the first soliton's parameters. Inserting the u_k expression (4.117) into the right-hand side of (4.121), we find that the perturbation term εF_0 in the soliton perturbation theory is

$$\varepsilon F_0 = -2r_1^2 r_2 \operatorname{sech}^2 r_1\theta_1 \operatorname{sech} r_2\theta_2 (e^{-i\Delta\phi} + 2e^{i\Delta\phi}). \tag{4.122}$$

When this expression is substituted into the dynamical equation (4.60) for r_1, we get

$$\frac{dr_1}{dt} = -2r_1^3 r_2 \int_{-\infty}^{\infty} \operatorname{sech}^3 r_1(x-\xi_1)\operatorname{sech} r_2(x-\xi_2)\sin\Delta\phi\, dx. \tag{4.123}$$

The integral in the above equation can be simplified asymptotically under our assumption (4.120). First, we rewrite

$$r_1(x-\xi_1) = r(x-\xi_1) - \frac{1}{2}\Delta r(x-\xi_1). \tag{4.124}$$

Since the dominant contribution to the above integral comes from the $x-\xi_1 = O(\Delta\xi)$ region, then under our assumption $|\Delta r\Delta\xi| \ll 1$ in (4.120), $r_1(x-\xi_1)$ in (4.123) can be approximated by $r(x-\xi_1)$. Likewise, $r_2(x-\xi_2)$ in (4.123) can be approximated by $r(x-\xi_2)$. Second, we know that

$$\begin{aligned}\Delta\phi &= \frac{1}{2}v_2(x-\xi_2) - \frac{1}{2}v_1(x-\xi_1) + \Delta\sigma \\ &= \frac{1}{2}\Delta v\left[x - \frac{1}{2}(\xi_1+\xi_2)\right] - \frac{1}{2}v\Delta\xi + \Delta\sigma.\end{aligned} \tag{4.125}$$

Since $|\Delta v\Delta\xi| \ll 1$ due to our assumption, and the dominant contribution to the integral of (4.123) comes from the region where $x - \frac{1}{2}(\xi_1+\xi_2) = O(\Delta\xi)$, then

$$\Delta\phi \approx -\frac{1}{2}v\Delta\xi + \Delta\sigma, \tag{4.126}$$

which is independent of x; thus the $\sin\Delta\phi$ term can be taken out of the integral in (4.123). Third, the remaining integral $\int_{-\infty}^{\infty}\operatorname{sech}^3 r(x-\xi_1)\operatorname{sech} r(x-\xi_2)dx$ can be evaluated asymptotically under our assumption that $r\Delta\xi \gg 1$. In this case, the main contribution to the integral comes from the region of $r(x-\xi_1) = O(1)$ where $\operatorname{sech} r(x-\xi_2) \approx 2e^{r(x-\xi_2)}$. The resulting integral then can be easily found to be $4e^{-r\Delta\xi}/r$. This result can also be obtained rigorously by reducing the above integral through variable transformations as

$$\int_{-\infty}^{\infty}\operatorname{sech}^3 r(x-\xi_1)\operatorname{sech} r(x-\xi_2)dx = \frac{8e^{-r\Delta\xi}}{r}\int_0^{\infty}\frac{y}{(y+1)^3(e^{-2r\Delta\xi}y+1)}dy; \tag{4.127}$$

hence its exact expression can be readily obtained. In the limit of $r\Delta\xi \gg 1$, the value of this integral asymptotically becomes $4e^{-r\Delta\xi}/r$, which agrees with the above simpler derivation.

Under the above asymptotic approximations, the r_1 equation (4.123) then simplifies to

$$\frac{dr_1}{dt} = -8r^3 e^{-r\Delta\xi}\sin\Delta\phi. \tag{4.128}$$

Dynamical equations for the other parameters of the first soliton as well as parameters of the second soliton can be similarly obtained. The results are

$$\frac{dr_k}{dt} = (-1)^k 8r^3 e^{-r\Delta\xi} \sin\Delta\phi, \tag{4.129}$$

$$\frac{dv_k}{dt} = (-1)^{k+1} 16r^3 e^{-r\Delta\xi} \cos\Delta\phi, \tag{4.130}$$

$$\frac{d\xi_k}{dt} = v_k - 4re^{-r\Delta\xi} \sin\Delta\phi, \tag{4.131}$$

$$\frac{d\sigma_k}{dt} = r_k^2 + \frac{1}{4}v_k^2 + 2re^{-r\Delta\xi} \sin\Delta\phi + 12r^2 e^{-r\Delta\xi} \cos\Delta\phi. \tag{4.132}$$

Here, $k = 1,2$. From these equations, we find that dynamical equations for r, v, $\Delta\xi$, Δr, Δv, and $\Delta\phi$ are even simpler:

$$r_t = v_t = 0, \tag{4.133}$$

$$\Delta r_t = 16r^3 e^{-r\Delta\xi} \sin\Delta\phi, \tag{4.134}$$

$$\Delta v_t = -32r^3 e^{-r\Delta\xi} \cos\Delta\phi, \tag{4.135}$$

$$\Delta\xi_t = \Delta v, \tag{4.136}$$

$$\Delta\phi_t = 2r\Delta r. \tag{4.137}$$

Here the approximation (4.126) has been used in the calculation of $\Delta\phi_t$. From these dynamical equations, we see that during weak interactions of two NLS solitons, the average amplitude and velocity do not change with time. This reflects the conservation of total power and momentum in the NLS equation. But individual amplitudes and velocities may change, which means that power and momentum can redistribute between the two solitons during and after weak interactions. If we define a new complex variable

$$Y = -r\Delta\xi + i\Delta\phi, \tag{4.138}$$

then the four dynamical equations (4.134)–(4.137) can be compressed into a single equation

$$\ddot{Y} = 32r^4 e^Y, \tag{4.139}$$

where $\ddot{Y}$ represents the second time derivative of Y. This equation is clearly integrable. Multiplying $\dot{Y}$ by the above equation and integrating once, we find that

$$\dot{Y}^2 - 64r^4 e^Y = 4C_0^2, \tag{4.140}$$

where

$$C_0 = \frac{1}{2}\sqrt{\dot{Y}_0^2 - 64r^4 e^{Y_0}} \tag{4.141}$$

is a constant, and $(Y_0, \dot{Y}_0)$ are the initial conditions of $(Y, \dot{Y})$. Combining (4.139) and (4.140), we get

$$\ddot{Y} = -\frac{1}{2}(4C_0^2 - \dot{Y}^2), \tag{4.142}$$

whose solution for $\dot{Y}$ is

$$\dot{Y}(t) = -2C_0 \tanh C_0(t + B_0), \tag{4.143}$$

where

$$B_0 = -\frac{1}{C_0}\operatorname{arctanh}\frac{\dot{Y}_0}{2C_0}. \tag{4.144}$$

Here, arctanh(z) is the inverse hyperbolic cotangent function. Integrating (4.143) one more time, we get the solution $Y(t)$ as

$$Y(t) = Y_0 - 2\ln\left|\frac{\cosh C_0(t + B_0)}{\cosh C_0 B_0}\right|. \tag{4.145}$$

Below we give more explicit solutions corresponding to the four interaction scenarios shown in Fig. 4.6.

(1) Let $r_{10} = r_{20} = 1$, $\Delta v_0 = \Delta\phi_0 = 0$; i.e., the two solitons initially have the same amplitude, velocity, and phase. This case corresponds to Fig. 4.6(a). In this case, $\dot{Y}_0 = B_0 = 0$, and $C_0 = 4ie^{-\Delta\xi_0/2}$ which is purely imaginary. Hence the solution (4.145) shows that $\Delta\phi(t) = 0$ for all time; i.e., the two solitons always remain in phase. The soliton separation $\Delta\xi$ is

$$\Delta\xi(t) = \Delta\xi_0 + 2\ln|\cos 4e^{-\Delta\xi_0/2}t|, \tag{4.146}$$

which is a periodic function with period $t_p = \pi e^{\Delta\xi_0/2}/4$. At the beginning, $\Delta\xi$ decreases with time, which means that the two solitons attract each other and move together. At time

$$t_c = \frac{\arccos(e^{-\Delta\xi_0/2})}{4e^{-\Delta\xi_0/2}} \cong \frac{1}{2}t_p, \tag{4.147}$$

$\Delta\xi = 0$; i.e., the two solitons collide. Shortly after the collision, $\Delta\xi$ starts to increase again, which means that the two solitons move away from each other. At time t_p, the two solitons return to their initial state and start to move toward each other again. This process repeats indefinitely. When compared to the numerical solution in Fig. 4.6(a), we find that the analytical formula (4.146) gives a very accurate description of the interaction dynamics for all times, except when t is in a small neighborhood of $t_c + nt_p$ ($n = 1, 2, \ldots$), where our assumption $r\Delta\xi \gg 1$ for weak interactions breaks down. This can be seen more clearly in Fig. 4.7(a), where the soliton separations $\Delta\xi$ from the numerical solution and the analytical formula (4.146) are compared. Normally when $t \sim t_c$ where the two solitons

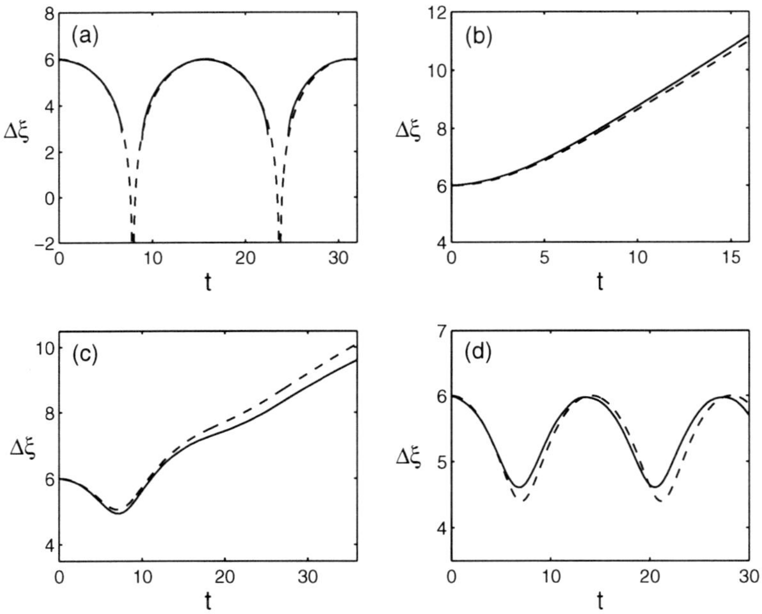

Figure 4.7. *Comparison of soliton separations $\Delta\xi(t)$ between analytical formulae and numerical results for the four weak interactions shown in Fig. 4.6. Solid lines: numerical values; dashed lines: analytical values.*

collide (and hence the interaction is not weak), the analytical prediction beyond the collision time t_c should not be expected to be reliable. But in the present case, our analytical formula still gives an accurate prediction for almost all times, which is incidental and surprising.

(2) Let $r_{10} = r_{20} = 1$, $\Delta v_0 = 0$, and $\Delta\phi_0 = \pi$, i.e.; the two solitons initially have the same amplitude and velocity but a π-phase difference. This case corresponds to Fig. 4.6(b). In this case, $\dot{Y}_0 = B_0 = 0$, and $C_0 = 4e^{-\Delta\xi_0/2}$ which is real. Thus $\Delta\phi(t) = \pi$ for all time, and the soliton separation $\Delta\xi$ is

$$\Delta\xi(t) = \Delta\xi_0 + 2\ln|\cosh 4e^{-\Delta\xi_0/2}t|. \tag{4.148}$$

This function increases for all time, which means that the two solitons repel each other and move away to infinity. This agrees with the numerical result shown in Fig. 4.6(b). Quantitative comparison between the analytical and numerical results on the soliton separation $\Delta\xi$ is given in Fig. 4.7(b). This figure shows that the analytical formula (4.148) gives a very good prediction of $\Delta\xi$ for all times.

(3) Let $r_{10} = r_{20} = 1$, $\Delta v_0 = 0$, and $\Delta\phi_0$ take an arbitrary value. This case corresponds to Fig. 4.6(c). In this case, $\dot{Y}_0 = B_0 = 0$ and $C_0 = 4ie^{(-\Delta\xi_0 + i\Delta\phi_0)/2}$. Hence the solution

(4.145) gives

$$-\Delta\xi + i\Delta\phi = -\Delta\xi_0 + i\Delta\phi_0 - 2\ln\left|\cosh 4ie^{(-\Delta\xi_0+i\Delta\phi_0)/2}t\right|. \tag{4.149}$$

From this formula, we see that when $\Delta\phi_0$ is not equal to 0 or π, then $\Delta\phi$ will not remain a constant anymore. In addition, if $\Delta\phi_0 \neq 0$, then $\Delta\xi \to \infty$ as $t \to \infty$, and thus the two solitons will escape from each other and move to infinity at large times. But the transient interaction process depends on the value of the initial-phase difference $\Delta\phi_0$. For the initial parameters chosen in Fig. 4.6(c), a comparison of $\Delta\xi$ between numerical and analytical results is shown in Fig. 4.7(c). In this case, the agreement is a bit worse at large times mainly because the two solitons first move toward each other and thus make our assumption $r\Delta\xi$ less true.

(4) Let $r_1 = \frac{1}{2}(1-\Delta r_0)$, $r_2 = \frac{1}{2}(1+\Delta r_0)$, and $\Delta v_0 = \Delta\phi_0 = 0$; i.e., the two solitons initially have the same velocity and phase but different amplitude. This case corresponds to Fig. 4.6(d). In this case, one can show from (4.145) that the formula for $\Delta\xi$ is

$$\Delta\xi(t) = \Delta\xi_0 + \ln\frac{\Delta r_0^2 + 8e^{-\Delta\xi_0}(1+\cos\gamma t)}{\Delta r_0^2 + 16e^{-\Delta\xi_0}}, \tag{4.150}$$

where $\gamma = 2\sqrt{\Delta r_0^2 + 16e^{-\Delta\xi_0}}$. This formula indicates that the two solitons form an oscillating bound state, which agrees with Fig. 4.6(d). Physically, the formation of this bound state can be understood intuitively as follows. We have known that when two solitons have the same phase, they attract each other; while if they have π-phase difference, they repel each other. In the present case, since $\Delta r \neq 0$, the phase difference $\Delta\phi$ between the two solitons linearly increases with time in view of the relation (4.137). When $\Delta\phi$ is close to 0 mod 2π, the two solitons attract each other. But at a later time, $\Delta\phi$ becomes close to π mod 2π, when they start to repel each other. This process repeats continuously, and the solitonic attraction and repulsion balance each other on average, and thus an oscillating bound state is formed. A quantitative comparison of $\Delta\xi$ between numerical and analytical results for Fig. 4.6(d) is shown in Fig. 4.7(d). The agreement is reasonable in light of the fact that $\Delta\xi$ is not very large in certain time intervals.

Lastly, the above analysis for the weak interaction of two NLS solitons can be generalized to the weak interaction of an arbitrary number of NLS solitons. By suitably defining variables associated with soliton parameters, it has been shown by Gerdjikov et al. (1996) that the weak interaction of N solitons is governed by a complex Toda chain with N nodes. The complex Toda chain is exactly integrable (Moser (1975), Toda (1989)), and thus the weak interaction of N solitons can be well described. The weak interaction theory has also been developed for other integrable equations such as the modified NLS equation (Gerdjikov et al. (2001)) and the Manakov equations (Yang (2002c)). Even though these integrable equations are different from the NLS equation, dynamical equations for the weak interaction of their solitons remain the same as those in the NLS equation. Specifically, the weak interaction of two solitons in those equations is still governed by the dynamical equation (4.139), while the weak interaction of N solitons in those equations is still governed by the complex Toda chain. In the case of Manakov equations, it was shown analytically that weak interactions of Manakov solitons are greatly suppressed when the polarizations of Manakov solitons are orthogonal to each other (if their polarizations are parallel, their weak interactions become the same as those of NLS solitons). This analytical result explains an earlier experimental observation in fiber communications (Evangelides et al. (1992)), where it was found that by launching adjacent solitons along orthogonal polarizations of the fiber, the transmission speed doubled due to reduced interference between adjacent soliton bits. The weak interaction theory has also been developed for nonintegrable equations such as the coupled NLS equations and the generalized NLS equations (Yang (2001b), Zhu and Yang (2007)). A fascinating phenomenon on weak interactions of solitary waves in nonintegrable equations is the appearance of fractal scatterings, where the interaction outcome depends on initial conditions in a sensitive fractal manner (Dmitriev and Shigenari (2002), Zhu and Yang (2007)). This phenomenon and its theoretical analysis will be covered in the next chapter.

4.4 Soliton Perturbation Theory for the Complex Modified KdV Equation

In the soliton perturbation theory for the NLS equation developed earlier in this chapter, energy radiation in the perturbed solution (4.67) decays at the normal rate of $t^{-1/2}$ (see (4.84)–(4.86)). In such cases, this radiation is often unimportant, and one can derive evolution equations of soliton parameters (such as (4.60)–(4.63)) by ignoring this radiation. However, for certain perturbed integrable equations, this energy radiation in the perturbed solution either does not decay, or decays much slower than the $t^{-1/2}$ rate. In such cases, the radiation contains a cw shelf which propagates to far distances with its amplitude undiminished. Generation of this nondiminishing radiation shelf is a distinctive phenomenon which was first observed for the KdV soliton under damping perturbations (Karpman and Maslov (1977), Kaup and Newell (1978a), Kodama (1985), Grimshaw and Mitsudera (1993), Yan and Tang (1996)). It also appears in other perturbed integrable equations whenever there is a resonance between the soliton and the continuous spectrum (Yang (2001a), Yang (2003), Hoseini and Marchant (2009)). This radiation shelf constitutes a dominant mechanism of energy loss from the soliton, and thus it will dictate the soliton's long-time evolution. The soliton perturbation theory for such systems exhibits a number of new features which are absent in the NLS perturbation theory. In this section, we develop such a perturbation theory for the complex modified KdV equation. Our presentation follows Yang (2003), but more details are given here.

The perturbed complex modified KdV equation we consider is

$$u_t + u_{xxx} + 6|u|^2 u_x = i\epsilon\alpha |u|^2 u - \epsilon\gamma(|u|^2)_x u, \tag{4.151}$$

where $\epsilon \ll 1$, and α, γ are real parameters of $O(1)$. This equation is relevant for ultrashort pulse transmission near zero second-order dispersion point, where the third-order dispersion becomes important (see Eq. (1.50)). Small coefficients of the self-phase modulation and nonlinear dispersion terms in this equation are chosen for the convenience of the analysis. When $\epsilon = 0$, this equation is the complex modified KdV equation which is completely integrable by the inverse scattering transform method (see Sec. 2.8). Its single-soliton solution is

$$u(x,t) = \Phi(\theta) e^{-i\omega t + i\sigma_0}, \tag{4.152}$$

where $\theta = x - vt - x_0$,

$$\Phi(\theta) = r\,\text{sech} r\theta\, e^{ik\theta}, \tag{4.153}$$

and

$$v = r^2 - 3k^2, \quad \omega = 2k(r^2 + k^2). \tag{4.154}$$

Here amplitude r, wavenumber k, initial position x_0, and initial phase σ_0 are four free real parameters. When $0 \neq |\epsilon| \ll 1$, this soliton will deform due to perturbations on the right-hand side of Eq. (4.151), and its four parameters r, k, x_0, and σ_0 will undergo slow time evolution. Below, we use the soliton perturbation theory to derive this evolution.

First, we write the perturbed solution in the form

$$u(x,t) = \widehat{\Phi}(\theta,t) \exp\left\{-i\int_0^t \omega\, d\tau + i\sigma_0\right\}, \tag{4.155}$$

where

$$\theta = x - \int_0^t v\, d\tau - x_0. \tag{4.156}$$

Substituting this form into (4.151), we find the equation for $\widehat{\Phi}(\theta,t)$ as

$$\widehat{\Phi}_t - i\omega\widehat{\Phi} - v\widehat{\Phi}_\theta + \widehat{\Phi}_{\theta\theta\theta} + 6|\widehat{\Phi}|^2\widehat{\Phi}_\theta = \epsilon F(\widehat{\Phi}) - i\sigma_{0t}\widehat{\Phi} + x_{0t}\widehat{\Phi}_\theta, \tag{4.157}$$

where

$$F(\widehat{\Phi}) = \left[i\alpha|\widehat{\Phi}|^2 - \gamma(|\widehat{\Phi}|^2)_\theta\right]\widehat{\Phi}. \tag{4.158}$$

Next, we expand the solution $\widehat{\Phi}$ into a two-time-scale perturbation series

$$\widehat{\Phi}(\theta,t) = \Phi(\theta) + \epsilon\Phi_1(\theta,t) + \epsilon^2\Phi_2(\theta,t) + \cdots, \tag{4.159}$$

where $\Phi(\theta)$ is given in (4.153), and the parameters r, k, x_0, and σ_0 in $\Phi(\theta)$ are functions of the slow time $T = \epsilon t$. When this series is substituted into Eq. (4.157), at $O(1)$, the equation is satisfied automatically. At $O(\epsilon)$, we obtain the linear inhomogeneous equation for $A_1 \equiv (\Phi_1, \Phi_1^*)^T$ as

$$(\partial_t + L)A_1 = W, \tag{4.160}$$

where $W = (w, w^*)^T$,

$$w = F(\Phi) - r_T \Phi_r - k_T \Phi_k + x_{0T}\Phi_\theta - i\sigma_{0T}\Phi, \tag{4.161}$$

the partial time derivative ∂_t on the left side of (4.160) is with respect to the fast time t, and

$$L = \begin{pmatrix} \partial_{\theta\theta\theta} + (6|\Phi|^2 - v)\partial_\theta + 6\Phi^*\Phi_\theta - i\omega & 6\Phi\Phi_\theta \\ 6\Phi^*\Phi_\theta^* & \partial_{\theta\theta\theta} + (6|\Phi|^2 - v)\partial_\theta + 6\Phi\Phi_\theta^* + i\omega \end{pmatrix} \tag{4.162}$$

is the linearization operator of the unperturbed complex modified KdV equation (4.157). Equation (4.160) is to be solved with the zero initial condition

$$A_1|_{t=0} = 0. \tag{4.163}$$

To solve Eq. (4.160), we first determine the complete sets of eigenfunctions and adjoint eigenfunctions for the linearization operator L. Note that the complex modified KdV equation belongs to the AKNS hierarchy (see Sec. 2.8), thus eigenfunctions and adjoint eigenfunctions of L are squared Zakharov–Shabat eigenstates just like the NLS equation. Consequently, these eigenfunctions can be derived in the same way as we had done for the NLS soliton in Sec. 4.1.1. The only difference is that the soliton (4.153) here has an extra $e^{ik\theta}$ factor. For this potential, analytical functions $P^\pm$ of the Zakharov–Shabat system are Γ and Γ^{-1} given by (2.72), (2.73), and (2.99), with $N = 1$, $\zeta_1 = (ir - k)/2$, and $v_{10} = (1, 1)^T$. From these Jost solutions, we can obtain the squared eigenfunctions for the complex modified KdV equation as we had done before for the NLS equation. The continuous squared eigenfunctions are

$$\mathcal{Z}_1(\theta, \xi) = \begin{pmatrix} -r^2 \text{sech}^2 r\theta\, e^{2ik\theta} \\ (2i\xi + ik - r\tanh r\theta)^2 \end{pmatrix} e^{2i\xi\theta}, \tag{4.164}$$

$$\mathcal{Z}_2(\theta, \xi) = \begin{pmatrix} (2i\xi + ik + r\tanh r\theta)^2 \\ -r^2 \text{sech}^2 r\theta\, e^{-2ik\theta} \end{pmatrix} e^{-2i\xi\theta}, \tag{4.165}$$

where $\xi \in \mathbb{R}$, and the eigenrelations are

$$L\mathcal{Z}_1(\theta, \xi) = -i\lambda(\xi)\mathcal{Z}_1(\theta, \xi), \quad L\mathcal{Z}_2(\theta, \xi) = i\lambda(\xi)\mathcal{Z}_2(\theta, \xi), \tag{4.166}$$

with

$$\lambda(\xi) = (2\xi)^3 + 2\xi v - \omega = 2(\xi - k)[(2\xi + k)^2 + r^2]. \tag{4.167}$$

Notice that $\lambda(k) = 0$; that is., the kernel of L contains continuous eigenfunctions $Z_{1,2}(\theta, k)$ which are oscillatory at large distances. This fact will be important in later analysis.

Discrete eigenmodes of operator L can be conveniently derived by taking variations of the soliton (4.153) with respect to its four free parameters r, k, x_0, and σ_0, similar to what we had done in the NLS perturbation theory. Equivalently, by inserting (4.153) into the complex modified KdV equation, we see that $\Phi(\theta)$ satisfies the equation

$$-i\omega\Phi - v\Phi_\theta + \Phi_{\theta\theta\theta} + 6|\Phi|^2\Phi_\theta = 0. \tag{4.168}$$

Using this equation and its derivatives with respect to θ, k, and r, we find that the L's two discrete eigenfunctions and two discrete generalized eigenfunctions are

$$Z_{D,1}(\theta) = \begin{pmatrix} \Phi_\theta \\ \Phi_\theta^* \end{pmatrix}, \qquad Z_{D,2}(\theta) = \begin{pmatrix} \Phi \\ -\Phi^* \end{pmatrix}, \tag{4.169}$$

$$Z_{G,1}(\theta) = \begin{pmatrix} \Phi_k \\ \Phi_k^* \end{pmatrix}, \qquad Z_{G,2}(\theta) = \begin{pmatrix} \Phi_r \\ \Phi_r^* \end{pmatrix}, \tag{4.170}$$

and the eigenrelations are

$$LZ_{D,1}(\theta) = LZ_{D,2}(\theta) = 0, \tag{4.171}$$

$$LZ_{G,1}(\theta) = 2i(r^2 + 3k^2)Z_{D,2}(\theta) - 6kZ_{D,1}(\theta), \tag{4.172}$$

$$LZ_{G,2}(\theta) = 4ikrZ_{D,2}(\theta) + 2rZ_{D,1}(\theta). \tag{4.173}$$

Next we consider eigenfunctions of the adjoint linearization operator L^A. As before, the inner product is defined as

$$\langle \mathbf{f}, \mathbf{g} \rangle = \int_{-\infty}^{\infty} \mathbf{f}^T(x)\,\mathbf{g}(x)\,dx. \tag{4.174}$$

We have known that the eigenfunctions of L^A are the adjoint squared eigenstates of the Zakharov–Shabat system (see Eq. (2.236)). But in the present case, these adjoint eigenfunctions can be obtained more easily by noticing that L^A is related to L as

$$L^A = -\sigma_2 L \sigma_2^{-1}, \quad \sigma_2 = \begin{pmatrix} 0 & -1 \\ 1 & 0 \end{pmatrix}. \tag{4.175}$$

Thus, continuous eigenfunctions of L^A are simply

$$\Upsilon_1(\theta,\xi) = \sigma_2 \mathcal{Z}_2(\theta,\xi), \quad \Upsilon_2(\theta,\xi) = \sigma_2 \mathcal{Z}_1(\theta,\xi), \quad \xi \in \mathbb{R}, \tag{4.176}$$

and discrete eigenfunctions of L^A are

$$\Upsilon_{D,1}(\theta) = \sigma_2 \mathcal{Z}_{D,1}(\theta), \qquad \Upsilon_{D,2}(\theta) = \sigma_2 \mathcal{Z}_{D,2}(\theta), \tag{4.177}$$

$$\Upsilon_{G,1}(\theta) = \sigma_2 \mathcal{Z}_{G,1}(\theta), \qquad \Upsilon_{G,2}(\theta) = \sigma_2 \mathcal{Z}_{G,2}(\theta). \tag{4.178}$$

Inner products between continuous eigenfunctions of L and L^A can be obtained from the inner product relations (2.261)–(2.262) when eigenfunction scalings are taken into account (see also Sec. 4.1.1), and inner products between discrete eigenfunctions can be calculated directly. The results of these inner products are

$$\langle \mathcal{Z}_1(\theta,\xi), \Upsilon_1(\theta,\xi') \rangle = -\langle \mathcal{Z}_2(\theta,\xi), \Upsilon_2(\theta,\xi') \rangle = \pi[(2\xi+k)^2 + r^2]^2 \delta(\xi - \xi'), \tag{4.179}$$

$$\langle \mathcal{Z}_{D,1}(\theta), \Upsilon_{G,1}(\theta) \rangle = -\langle \mathcal{Z}_{G,1}(\theta), \Upsilon_{D,1}(\theta) \rangle = -2ir, \tag{4.180}$$

$$\langle \mathcal{Z}_{D,1}(\theta), \Upsilon_{G,2}(\theta) \rangle = -\langle \mathcal{Z}_{G,2}(\theta), \Upsilon_{D,1}(\theta) \rangle = -2ik, \tag{4.181}$$

$$\langle \mathcal{Z}_{D,2}(\theta), \Upsilon_{G,2}(\theta) \rangle = -\langle \mathcal{Z}_{G,2}(\theta), \Upsilon_{D,2}(\theta) \rangle = -2, \tag{4.182}$$

and all other inner products are zero.

Now we solve Eq. (4.160). The method is to expand the solution A_1 as well as the inhomogeneous term W into the complete set of eigenfunctions of the linearization operator L, similar to what we had done in the NLS perturbation theory. In order to suppress secular growth in A_1, the coefficients of L's discrete eigenfunctions in W must be zero. In view of the above inner product relations, this means that W must be orthogonal to the four adjoint discrete eigenfunctions, i.e.,

$$\langle W, \Upsilon_{D,1} \rangle = \langle W, \Upsilon_{D,2} \rangle = \langle W, \Upsilon_{G,1} \rangle = \langle W, \Upsilon_{G,2} \rangle = 0. \tag{4.183}$$

Inserting the expression (4.161) of W into these orthogonality conditions, we obtain the following dynamical equations for the soliton's slowly varying parameters:

$$k_T = r_T = 0, \tag{4.184}$$

$$x_{0T} = \frac{2}{3}\gamma r^2, \quad \sigma_{0T} = r^2\left(\alpha + \frac{2}{3}\gamma k\right). \tag{4.185}$$

These equations show that due to perturbations, the position x_0 and phase σ_0 of the soliton acquire a shift which increases linearly with the slow time $T = \epsilon t$, but the amplitude r and wavenumber k of the soliton do not change at the slow time scale $T = O(1)$.

In this soliton perturbation theory, the more important part is the first-order solution A_1. Using the above eigenfunction expansion method, we can readily obtain the formula for A_1 as

$$A_1(\theta,t) = i\int_{-\infty}^{\infty}\left\{\frac{\alpha_1(\xi)\left[1-e^{i\lambda(\xi)t}\right]}{\lambda(\xi)}\mathcal{Z}_1(\theta,\xi) - \frac{\alpha_2(\xi)\left[1-e^{-i\lambda(\xi)t}\right]}{\lambda(\xi)}\mathcal{Z}_2(\theta,\xi)\right\}d\xi, \tag{4.186}$$

where

$$\alpha_j(\xi) = (-1)^{j-1}\frac{\langle \mathcal{F}(\theta), \Upsilon_j(\theta,\xi)\rangle}{\pi[(2\xi+k)^2+r^2]^2}, \quad j=1,2, \tag{4.187}$$

and

$$\mathcal{F} = \left(F(\Phi), F^*(\Phi)\right)^T. \tag{4.188}$$

Notice that even though the denominator $\lambda(\xi)$ in (4.186) vanishes at $\xi = k$ in view of its factorization (4.167), the integrand of (4.186) is nonsingular at $\xi = k$, since functions $1-e^{\pm i\lambda(\xi)t}$ in the numerators vanish at $\xi = k$ as well. However, this vanishing of the denominator $\lambda(\xi)$ at a real ξ value in the A_1-solution (4.186) is a novel and important feature which was not present in the NLS perturbation theory. As we will show below, this feature signals a resonance phenomenon, where a cw shelf is excited which travels to far distances with its height undiminished.

Next we examine this first-order solution at large distances and times. When $|\theta| \to \infty$, since the first component of $\mathcal{Z}_1(\theta,\xi)$ is exponentially small in $|\theta|$, formula (4.186) then shows that the first-order solution Φ_1 becomes

$$\Phi_1(\theta,t) \to -i\int_{-\infty}^{\infty}\frac{\alpha_2(\xi)\left[1-e^{-i\lambda(\xi)t}\right]}{\lambda(\xi)}[2i\xi+ik+r\,\mathrm{sgn}(\theta)]^2 e^{-2i\xi\theta}d\xi. \tag{4.189}$$

In this Φ_1 formula, since the integrand is nonsingular at $\xi = k$, this integral is then equal to its principal value at $\xi = k$. Thus the integral in (4.189) will be treated as its principal value below. This treatment is useful because when the two terms in $1-e^{-i\lambda(\xi)t}$ are split apart (as we will do below), the integrand for each term in (4.189) will be singular at $\xi = k$; thus we need to use the principal value to give those improper integrals a definite meaning. To calculate the integral in (4.189), we recall that for any continuous and bounded function $f(z)$, due to the well-known result (4.45), the following formula holds:

$$P\int_{-\infty}^{\infty} f(z)\frac{e^{iz\theta}}{z}dz \to i\pi f(0)\mathrm{sgn}(\theta), \quad |\theta| \to \infty, \tag{4.190}$$

where P represents the principal value of an improper integral. This formula can also be heuristically understood as follows. According to the Riemann–Lebesgue lemma, for any function $y(x)$, if the integral $\int_a^b |y(x)|dx$ exists and is finite, then

$$\int_a^b y(x)e^{i\theta x}dx \to 0, \quad |\theta| \to \infty. \tag{4.191}$$

The reason for this is that as $\theta \to \pm\infty$, the integrand in the above equation oscillates rapidly; thus contributions from adjacent oscillations cancel out, which causes the integral to approach zero. Due to this lemma, when $|\theta| \to \infty$, contributions to the integral in Eq. (4.190) on intervals away from $z = 0$ approach zero. Hence the main contribution to this integral comes from the $z \approx 0$ region. Consequently, $f(z)$ inside this integral can be approximated by $f(0)$ and then taken out. The remaining principal-value integral can be evaluated exactly as $i\pi\,\mathrm{sgn}(\theta)$, and hence formula (4.190) is proved. Using this formula and the approximation of $\lambda(\xi)$ near its zero $\xi = k$,

$$\lambda(\xi) \approx \lambda(k) + \lambda'(k)(\xi - k) = 2(9k^2 + r^2)(\xi - k), \quad \text{when } \xi \approx k, \tag{4.192}$$

and splitting the term $1 - e^{-i\lambda(\xi)t}$ in (4.189) and treating the resulting two integrals as their principal values at $\xi = k$, we find that when $|\theta| \gg 1$ and $|\theta + (9k^2 + r^2)t| \gg 1$,

$$\Phi_1(\theta, t) \to -\frac{\pi\alpha_2(k)[3ik + r\,\mathrm{sgn}(\theta)]^2}{2(9k^2 + r^2)} \left\{\mathrm{sgn}(\theta) - \mathrm{sgn}[\theta + (9k^2 + r^2)t]\right\} e^{-2ik\theta}. \tag{4.193}$$

In other words,

$$\Phi_1(\theta, t) \to \begin{cases} 0, & \theta \ll -(9k^2 + r^2)t \ \text{ and } \ \theta \gg 1, \\ h\,e^{-2ik\theta}, & -(9k^2 + r^2)t \ll \theta \ll -1, \end{cases} \tag{4.194}$$

where

$$h = \frac{\pi\alpha_2(k)(3ik - r)^2}{9k^2 + r^2}. \tag{4.195}$$

The value of $\alpha_2(k)$ can be explicitly calculated from the formula (4.187) and the Fourier transform formula of the sech function (4.78). When the result is substituted into (4.195), the final formula for h is

$$h = -\frac{1}{4} i\pi(\alpha + 2\gamma k) \frac{r - 3ik}{r + 3ik} \operatorname{sech} \frac{3\pi k}{2r}. \tag{4.196}$$

The behavior (4.194) of the first-order solution Φ_1 is very important. It shows that the perturbation to the complex modified KdV equation excites a cw shelf of height ϵh and wavenumber $-2k$, which travels at the speed of $-(9k^2 + r^2)$ along the negative θ direction. The key feature of this cw shelf is that its height does not decrease at the fast time scale $t = O(1)$. This is very different from typical dispersive energy radiation and its zero-wavenumber radiation shelf as in the perturbed NLS soliton, which decay at the fast time scale t (at the rate of $t^{-1/2}$; see Secs. 4.1.2–4.1.3, and Eqs. (4.67), (4.83)–(4.86) in particular).

Why is the cw shelf excited in the perturbed complex modified KdV soliton? The reason can be understood from the linear dispersion relation of this equation. In the moving frame θ, when we seek infinitesimal Fourier modes $e^{i(p\theta-\omega t)}$ in this equation, we get the linear dispersion relation

$$\omega = -p^3 - vp. \tag{4.197}$$

For any value of the frequency ω, this dispersion equation admits a real wavenumber p, indicating that a Fourier oscillation mode is admissible. Hence the continuous spectrum of this equation is the entire ω-axis. For ordinary solitons, its frequency is expected to lie *outside* the continuous spectrum of the underlying wave system. However, for the soliton (4.152) in the complex modified KdV equation, its frequency ω lies in the equation's continuous spectrum and is thus an embedded soliton (Yang et al. (1999)) (also see Sec. 5.6). For this frequency ω (and velocity v) as in (4.154), the real solution p of the dispersion equation (4.197) is $p = -2k$; thus this soliton is in resonance with the Fourier mode $e^{-2ik\theta - i\omega t}$. As a result, when this soliton is perturbed, this Fourier mode will be resonantly excited. This situation is similar to the generation of cw tails in the NLS soliton under third-order dispersion perturbation (see Sec. 4.2.3). Mathematically, this resonance is manifested by the vanishing of the denominator $\lambda(\xi)$ at $\xi = k$ in the radiation integral (4.186). This is equivalent to the above dispersion equation (4.197) having a real root $p = -2k$ when one recognizes that the equation $\lambda(\xi) = 0$ is the same as the above dispersion relation (4.197) under the substitution $p = -2\xi$. Due to this vanishing denominator in the radiation integral, a cw with wavenumber $-2k$ arises in the radiation solution as Eq. (4.194) shows. This cw travels to $\theta = -\infty$ at its group velocity $\omega'(p)|_{p=-2k} = -(9k^2 + r^2)$, which agrees with the asymptotic behavior (4.194) of the radiation shelf derived above.

The behavior (4.194) of the first-order solution Φ_1 also confirms the Sommerfeld radiation condition, which says that the energy which is radiated from the perturbed soliton (source) must scatter to infinity, and no energy may be radiated from infinity into the field. In the present case, the dominant wavenumber in the first-order radiation is $-2k$, whose group velocity $\omega'(-2k) = -(9k^2 + r^2)$, which is negative. Thus, according to the Sommerfeld condition, this radiation shelf can only propagate from the perturbed soliton to $\theta = -\infty$, but *not* from $\theta = +\infty$ to the perturbed soliton. This is exactly what Eq. (4.194) shows.

It is noted that in Yang (2003), the height of this cw shelf was derived by considering the steady-state equation (4.160) of the first-order solution A_1 (by dropping the time derivative term in (4.160)). That was motivated by the fact that as $t \to \infty$, the solution A_1 of Eq. (4.160) approaches a steady state where Φ_1 vanishes as $\theta \to +\infty$ and tends to $he^{-2ik\theta}$ as $\theta \to -\infty$, see Eq. (4.194). Thus by taking the inner product of the steady-state equation $LA_1 = W$ with the adjoint continuous eigenfunction $\Upsilon_2(\theta, k)$, performing integration by parts to the left integral, and utilizing the above large-θ asymptotics of $\Phi_1(\theta)$, one gets

$$-(9k^2 + r^2)(3ik + r)^2 h = \langle \mathcal{F}(\theta), \Upsilon_2(\theta, k) \rangle. \tag{4.198}$$

This relation gives the same formula for h as (4.195) above. This derivation of the shelf's height is simpler. But our longer derivation above sheds more light on the transient dynamics of the radiation shelf. In addition, it proves that this cw shelf appears only on the left side of the soliton, and its front indeed travels at its group velocity $-(9k^2+r^2)$.

This cw radiation constitutes a main energy loss to the soliton. As a result, the amplitude r and wavenumber k of the soliton will have to change over time. Since the soliton's power is proportional to its amplitude r, and the rate of the power loss through this cw radiation is equal to $\epsilon^2 h^2$ multiplying the magnitude of the group velocity $9k^2+r^2$ of this continuous wave, we anticipate that r and k will change on the time scale of $\epsilon^2 t$. This long-time dynamics will be derived below. One may be tempted to predict that, due to this energy loss, the soliton will eventually decay to zero. This is not entirely correct, however. The reason is that the height h of this cw shelf vanishes when $k=-\alpha/2\gamma$; see Eq. (4.196). Thus as the soliton evolves due to cw radiation, if its wavenumber k approaches $-\alpha/2\gamma$, then this cw radiation will vanish, and hence the soliton will tend to a steady state. On the other hand, if k does not approach $-\alpha/2\gamma$, then this cw radiation will persist, and hence the soliton will eventually decay. Which of these two evolution scenarios prevails depends on how the wavenumber k evolves. This evolution of k is coupled with the evolution of r; thus dynamical equations for both k and r need to be derived. To obtain these dynamical equations, one can introduce the longer time scale $T_2=\epsilon^2 t$ and pursue the above soliton perturbation theory to the second order (this method was used by Yang (2001a) in his treatment of a similar problem of KdV type). But an easier and physically more intuitive way is to use conservation laws (Yang (2001a), Pelinovsky and Yang (2002, 2005), Tan et al. (2002), Yang (2003)). This will be employed below.

The perturbed complex modified KdV equation (4.151) admits the following two conserved quantities:

$$I_1=\int_{-\infty}^{\infty}|u|^2dx, \tag{4.199}$$

$$I_2=\int_{-\infty}^{\infty}\left\{\frac{1}{2}i\epsilon\alpha(u_xu^*-uu_x^*)-\epsilon\gamma|u_x|^2+\epsilon\gamma\left(1+\frac{1}{3}\epsilon\gamma\right)|u|^4\right\}dx, \tag{4.200}$$

which hold for arbitrary values of ϵ. The first conservation law is the power conservation, and the second conservation law has no clear physical meaning. When Eq. (4.155) and the perturbation series (4.159) are substituted into these conserved quantities and terms up to order ϵ^2 retained, the first conservation law becomes

$$\frac{d}{dt}\int_{-\infty}^{\infty}\left\{|\Phi|^2+\epsilon(\Phi\Phi_1^*+\Phi^*\Phi_1)+\epsilon^2(\Phi\Phi_2^*+\Phi^*\Phi_2)+\epsilon^2|\Phi_1|^2\right\}d\theta=0. \tag{4.201}$$

To proceed further, the following three facts are noted: (i) $|\Phi|^2$ changes slowly (on the time scale $\epsilon^2 t$); (ii) the Φ_1 field is driven by the inhomogeneous term in Eq. (4.160) and quickly becomes stationary in the soliton region (where $\Phi = O(1)$); the Φ_2 field is similar; (iii) the Φ_1 field develops a cw shelf of height h on the left side of the soliton; this shelf travels to $\theta = -\infty$ at its group velocity $-(r^2+9k^2)$. The first two facts tell us that the time derivatives of the integrals of the second and third terms in Eq. (4.201) are of $O(\epsilon^3)$ or higher and thus can be neglected. The third fact shows that the time derivative of the integral of the last term in (4.201) is $\epsilon^2(r^2+9k^2)|h|^2$. When these results and the Φ formula (4.153) are utilized, the first conservation law (4.201) gives the dynamical equation for the soliton amplitude r. Similar calculations for the second conservation law produces the dynamical equation for the soliton's wavenumber k. When the h formula (4.196) is substituted, the dynamical equations for r and k finally take the form

$$\frac{dr}{dt} = -\frac{1}{32}\epsilon^2\pi^2(\alpha+2\gamma k)^2(r^2+9k^2)\text{sech}^2\frac{3\pi k}{2r}, \tag{4.202}$$

$$\frac{dk}{dt} = \frac{1}{32r}\epsilon^2\pi^2(\alpha+2\gamma k)[3\alpha k-\gamma(r^2+3k^2)](r^2+9k^2)\text{sech}^2\frac{3\pi k}{2r}, \tag{4.203}$$

and they govern the soliton's slow evolution in the weakly perturbed complex modified KdV equation (4.151).

Dynamical equations (4.202)–(4.203) admit a continuous family of fixed points:

$$k = -\frac{\alpha}{2\gamma}, \quad r \text{ is arbitrary.} \tag{4.204}$$

These fixed points correspond to a continuous family of solitons (4.152) in the perturbed complex modified KdV equation (4.151). These solitons' amplitudes r are arbitrary, while their speeds v depend on r and the above k value through the relation (4.154). Linear and nonlinear stabilities of these fixed points in the dynamical equations (4.202)–(4.203) can be readily determined, and one can find that these fixed points are stable both linearly and nonlinearly. The simplest way to demonstrate their stabilities is by drawing a phase portrait in the (r,k) plane. Such a phase portrait with $\alpha = -1$ and $\gamma = 1$ is shown in Fig. 4.8. At these parameter values, the fixed points are the $k = 1/2$ line (dashed). From this phase portrait, one can see that these fixed points are nonlinearly stable, and they attract a large region of initial conditions (except the lower left corner of the phase space). This means that initial solitons (4.152) with (r,k) in that large basin of initial conditions will approach a steady-state soliton with $k = 1/2$. In this process, the cw radiation which is emitted from the initial soliton steadily lessens and eventually disappears. This phenomenon is quite remarkable. On the other hand, Fig. 4.8 also shows that when the (r,k) values of the initial soliton (4.152) lie in the lower left corner of the phase space, this soliton does decay to zero since r approaches zero. Thus the final state of the soliton (4.152) in Eq. (4.151) depends on the initial (r,k) values of this soliton.

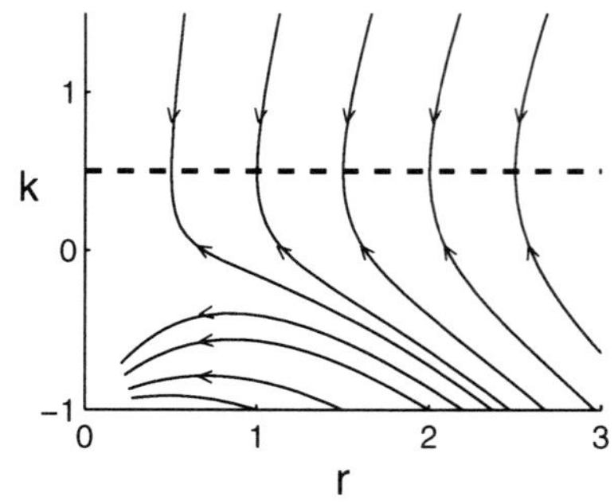

Figure 4.8. *Phase portrait of the dynamical system* (4.202)–(4.203) *for* $\alpha = -1$ *and* $\gamma = 1$. *(After Yang* (2003).*)*

It is noted that Eq. (4.151) admits a family of exact solitary wave solutions for arbitrary values of ϵ (Gromov and Tyutin (1998)):

$$u(x,t) = r\,\mathrm{sech}(rb\theta)e^{ik_*\theta - i\omega t}, \tag{4.205}$$

where

$$\theta = x - vt, \quad k_* = -\frac{\alpha}{2\gamma}, \quad b = \sqrt{1 + \frac{\epsilon\gamma}{3}}, \tag{4.206}$$

and

$$v = b^2 r^2 - 3k_*^2, \quad \omega = 2k_*(b^2 r^2 + k_*^2). \tag{4.207}$$

Here the soliton amplitude r is arbitrary. When $\epsilon \ll 1$, this solution family reduces to the soliton (4.152) with parameters (4.204) to the leading order of ϵ. The existence of this soliton family in Eq. (4.151) is remarkable since the frequency ω of these solitons resides in the continuous spectrum of this wave equation, and thus is in resonance with the continuous spectrum. Ordinarily, solitons which are in resonance with the continuous spectrum will excite cw shelf radiation and thus cannot be truly localized (see Sec. 5.5). However, the solitons (4.205), or their counterparts (4.152) with parameters (4.204), are truly localized, because the heights of their cw radiation shelves happen to be zero in view of Eq. (4.196). These solitons which reside inside the continuous spectrum are quite unusual and were termed embedded solitons by Yang et al. (1999) (see Sec. 5.6 for more details). What is more remarkable about these embedded solitons (4.205) is that they are linearly and nonlinearly stable, as the above soliton perturbation theory shows (numerical evidence of this stability and its analytical proof for arbitrary values of ϵ in (4.151) can be found in Gromov et al. (1999) and Pelinovsky and Yang (2005)). This contrasts embedded solitons in some other physical systems, where they were found to be semistable (Yang et al. (1999), Yang and Akylas (2003)).

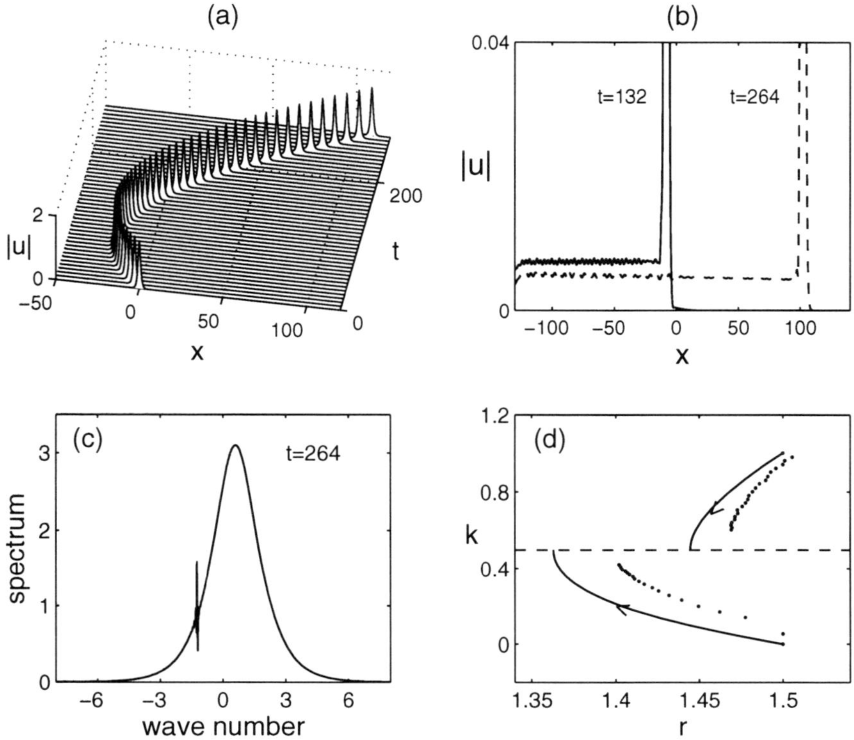

Figure 4.9. *Numerical evolutions of initial solitons* (4.152) *in the perturbed complex modified KdV equation* (4.151) *with perturbations of* $\epsilon\alpha = -0.1$ *and* $\epsilon\gamma = 0.1$. (a) *Time evolution of the solution with* $r_0 = 1.5$, $k_0 = 1$; (b) *profiles of the solution in* (a) *at two time values of* $t = 132$ *and* 264; (c) *spectrum of the solution in* (a) *at* $t = 264$; (d) *trajectories of* (r,k) *(dotted lines) for two initial solitons with* $(r_0,k_0) = (1.5,1)$ *and* $(1.5,0)$, *respectively; solid lines are trajectories from theoretical equations* (4.202)–(4.203). *(After Yang* (2003).*)*

Lastly, we confirm the above analytical results by direct numerical simulations. A number of numerical simulations of Eq. (4.151), starting with the soliton solution (4.152) at $t = 0$, have been performed for various small $\epsilon\alpha$ and $\epsilon\gamma$ values. Good agreement between the numerics and analysis has been observed. In addition, as ϵ approaches zero, the difference between the leading-order perturbation results (i.e., Eqs. (4.202)–(4.203)) and numerical results approaches zero, meaning that our perturbation theory is asymptotically accurate. For illustration purpose, we take $\epsilon\alpha = -0.1$ and $\epsilon\gamma = 0.1$, and choose two sets of initial amplitude and wavenumber values, $(r_0,k_0) = (1.5,1)$ and $(1.5, 0)$, which are on the opposite sides of the line of fixed points $k = -\alpha/2\gamma = 1/2$ in Fig. 4.8. The simulation results are displayed in Fig. 4.9. For $(r_0,k_0) = (1.5,1)$ which is above this fixed-point line, the pulse evolution is shown in Fig. 4.9(a). At these initial (r,k) values, the velocity v from formula (4.154) is negative, and thus the soliton initially moves toward the left. But due to perturbations in Eq. (4.151), the speed of the pulse slowly increases and eventually becomes positive and constant, and thus the pulse turns around and eventually moves toward the right at a constant speed. In Fig. 4.9(b), the field profiles at two times $t = 132$ and 264 are plotted. As predicted, a cw radiation shelf develops on the left of the soliton. Note that the height of this cw shelf slowly decreases with time, meaning that the soliton is stabilizing. In Fig. 4.9(c), the spectrum of the solution at $t = 264$ is plotted. At this time, the soliton's central wavenumber (i.e., the peak of the spectrum) is at $k \approx 0.6$. Notice that this spectrum contains a spike at the wavenumber which is approximately -1.2. This spike is caused by the cw radiation shelf, and its wavenumber -1.2 agrees with the theoretical value of $-2k$ in the above analysis. A direct comparison between the numerical result and the leading-order perturbation theory in the phase plane is shown in Fig. 4.9(d). We see that the numerical wavenumber k of the soliton indeed approaches the theoretical value $-\alpha/2\gamma = 1/2$. In addition, trajectories of (r,k) from the numerical result and from the theoretical equations (4.202)–(4.203) are in good agreement. When $(r_0, k_0) = (1.5, 0)$ below the fixed-point line, the soliton numerically also approaches a steady state with k on this fixed-point line. For this case, the numerical trajectory of (r,k) in the phase plane is also shown in Fig. 4.9(d), and it agrees with the theoretical trajectory as well.

Chapter 5

Theories for Nonintegrable Equations

Most nonlinear wave phenomena in science and engineering are governed by nonintegrable equations, and these equations are not near integrable systems either. In such cases, the inverse scattering theory and the soliton perturbation theory as presented in the previous three chapters are not applicable. Theories for nonintegrable equations started to appear forty years ago. By now, great advances have been made in this area, and many new phenomena unique to nonintegrable equations have been reported and theoretically explained. For instance, it has been found that solitary waves in nonintegrable equations can be unstable; collisions and weak interactions of solitary waves in nonintegrable equations can exhibit fractal-type dependence on initial conditions; and so on. These behaviors differ significantly from those in integrable systems. In this chapter, we present various theories for nonintegrable equations.

5.1 Solitary Waves in Nonintegrable Equations

Nonintegrable equations often admit solitary waves which are localized in space and propagate stationarily with time. These solitary waves are the counterpart of single solitons in integrable equations. These solitary waves, when exist, will play an important role in the solution dynamics. In this section, we use a generalized NLS equation to introduce these solitary waves.

The generalized NLS equation we consider is

$$iU_t + U_{xx} + F(|U|^2)U = 0. \tag{5.1}$$

Here $F(\cdot)$ is a real function, and we assume $F(0) = 0$. This equation describes nonlinear light propagation in a non-Kerr medium (see Chapter 1). It is nonintegrable in general (except the special case of $F(|U|^2) = |U|^2$, where it reduces to the NLS equation). This equation has three conserved quantities: the power P, momentum M, and Hamiltonian energy H, whose expressions are

$$P \equiv \int_{-\infty}^{\infty} |U|^2 dx, \tag{5.2}$$

$$M \equiv i \int_{-\infty}^{\infty} (U_x^* U - U_x U^*) dx, \tag{5.3}$$

$$H \equiv \int_{-\infty}^{\infty} \left[|U_x|^2 - G(|U|^2) \right] dx. \tag{5.4}$$

Here $G(y)$ is the antiderivative of $F(y)$ with $G(0) = 0$, and the superscript "*" represents complex conjugation.

The generalized NLS equation (5.1) admits the following stationary solitary waves:

$$U(x,t) = u(x) e^{i\mu t}, \tag{5.5}$$

where $u(x)$ is a localized real function (i.e., $u \to 0$ as $|x| \to \infty$), and $\mu > 0$ is a propagation constant. Substituting this solution into (5.1), we find that $u(x)$ satisfies the following ODE:

$$u_{xx} - \mu u + F(u^2) u = 0. \tag{5.6}$$

Notice that Eq. (5.1) admits Galilean invariance, that is, if $U(x,t)$ is a solution, so is $U(x - vt, t) e^{\frac{1}{2} i v x - \frac{1}{4} i v^2 t}$, where v is an arbitrary velocity parameter. From this Galilean invariance and the above stationary solitary waves, we can obtain the moving solitary waves

$$U_{moving}(x,t) = u(x - vt) e^{\frac{1}{2} i v x - \frac{1}{4} i v^2 t + i\mu t}. \tag{5.7}$$

For certain special forms of the nonlinearity $F(|U|^2)$, the function $u(x)$ admits explicit expressions. For instance, when the nonlinearity is of the following dual-power type,

$$F(|U|^2) = \alpha |U|^\sigma + \gamma |U|^{2\sigma}, \tag{5.8}$$

where α, γ, and σ are constants, then multiplying Eq. (5.6) by u_x and integrating once, we get

$$\left(\frac{du}{dx} \right)^2 = \mu u^2 - \frac{2\alpha}{\sigma + 2} u^{\sigma + 2} - \frac{\gamma}{\sigma + 1} u^{2(\sigma + 1)}. \tag{5.9}$$

Here the integration constant has been eliminated from the localization condition of $u|_{x=\infty} = u'|_{x=\infty} = 0$. Equation (5.9) can be solved by the variable transformation $y = u^{-\sigma}$. After simple algebra, we get the explicit expression of $u(x)$ as

$$u(x) = \left[\frac{A}{B + \cosh(Dx)} \right]^{1/\sigma}, \tag{5.10}$$

where

$$A = \frac{(2+\sigma)B\mu}{\alpha}, \quad D = \sigma\sqrt{\mu}, \quad B = \text{sgn}(\alpha)\left[1 + \frac{(2+\sigma)^2\gamma}{(1+\sigma)\alpha^2}\mu\right]^{-1/2}, \tag{5.11}$$

and the propagation constant μ is a free parameter (Kivshar and Agrawal (2003)). This solitary wave exists and is nonsingular (for at least a certain range of the propagation constant μ) as long as α and γ are not negative simultaneously. In that case, this wave is positive everywhere and has a single hump.

When $\mu \to 0^+$ and $\alpha < 0$, solution (5.10) degenerates into an algebraic solitary wave,

$$u_{al}(x) = \left[\frac{2(2+\sigma)(1+\sigma)/|\alpha|}{\sigma^2(1+\sigma)x^2 + (2+\sigma)^2\gamma/\alpha^2}\right]^{1/\sigma}, \tag{5.12}$$

which decays algebraically at large distances (Micallef et al. (1996)).

It is noted that in the physics literature, solitary waves are often just called solitons. In the remaining part of this book, we occasionally adopt this terminology as well.

5.2 Linearization Spectrum of Solitary Waves

When we consider the behavior of a solitary wave under small perturbations, it will lead us to the linearization spectrum of the solitary wave. This spectrum contains important information on the stability and other aspects of the solitary wave, and it will be discussed in this section.

Let us again use the generalized NLS equation (5.1) as an example to discuss this issue. When the solitary wave (5.5) is perturbed, i.e.,

$$U(x,t) = \left[u(x) + \widetilde{U}(x,t)\right]e^{i\mu t}, \quad \widetilde{U} \ll 1, \tag{5.13}$$

substituting this perturbed solution into (5.1) and linearizing (i.e., neglecting terms of $O(\widetilde{U}^2)$ and higher), we obtain a linearized equation for the perturbation $\widetilde{U}$:

$$i\widetilde{U}_t + \widetilde{U}_{xx} - \mu\widetilde{U} + \left[F(u^2) + u^2F'(u^2)\right]\widetilde{U} + u^2F'(u^2)\widetilde{U}^* = 0. \tag{5.14}$$

Solutions to this linear equation can be decomposed into a summation of various normal modes of the form

$$\widetilde{U}_{mode}(x,t) = f(x)e^{\lambda t} + g^*(x)e^{\lambda^* t}, \tag{5.15}$$

where λ is the eigenvalue of this normal mode. Inserting (5.15) into the linearized equation (5.14), we find that these normal modes are determined by the following linear eigenvalue problem:

$$f_{xx} + \left[F(u^2) + u^2F'(u^2) - \mu\right]f + u^2F'(u^2)g = -i\lambda f, \tag{5.16}$$

$$-g_{xx} - \left[F(u^2) + u^2F'(u^2) - \mu\right]g - u^2F'(u^2)f = -i\lambda g. \tag{5.17}$$

For convenience, we introduce the variable transformation:

$$f(x) = v(x) + w(x), \quad g(x) = v(x) - w(x). \tag{5.18}$$

Under this transformation, the above eigenvalue problem reduces to a simpler form,

$$LY = -i\lambda Y, \tag{5.19}$$

where

$$L = \begin{bmatrix} 0 & L_0 \\ L_1 & 0 \end{bmatrix}, \quad Y = \begin{pmatrix} v \\ w \end{pmatrix}, \tag{5.20}$$

and

$$L_0 = \partial_{xx} - \mu + F(u^2), \tag{5.21}$$

$$L_1 = \partial_{xx} - \mu + F(u^2) + 2u^2 F'(u^2). \tag{5.22}$$

Equation (5.19) is the linear-stability eigenvalue problem for the solitary wave (5.5). Its λ-spectrum is called the linear-stability spectrum for this solitary wave. It is easy to see that eigenvalues of Eq. (5.19) have the following symmetry properties: if λ is an eigenvalue, then so are $\lambda^*, -\lambda$, and $-\lambda^*$. This means that these eigenvalues always appear in pairs or quadruples.

The continuous spectrum of Eq. (5.19) can be readily obtained from the large-distance limit of $|x| \to \infty$. Under this limit, L_0 and L_1 approach $\partial_{xx} - \mu$, and thus L becomes a differential operator with constant coefficients. In order for λ to be in the continuous spectrum, the corresponding eigenfunction Y at large distances must be a Fourier mode, i.e., $Y = (c_1, c_2)^T e^{ikx}$, where k is a real wavenumber, and (c_1, c_2) are constants. Inserting this Fourier mode into the eigenvalue equation (5.19) and recalling the above asymptotics of operators L_0 and L_1, we find that the continuous eigenvalues are $\lambda = \pm i(k^2 + \mu)$. Then taking k to be from $-\infty$ to $+\infty$, we find the continuous spectrum of Eq. (5.19) to be the two line segments $\pm i[\mu, \infty)$ on the imaginary axis.

Regarding the discrete eigenvalues, first of all, $\lambda = 0$ is always a discrete eigenvalue. This zero eigenvalue has two discrete eigenfunctions and two generalized discrete eigenfunctions,

$$Y_{1d} = \begin{bmatrix} 0 \\ u \end{bmatrix}, \quad Y_{2d} = \begin{bmatrix} u_x \\ 0 \end{bmatrix} \tag{5.23}$$

and

$$Y_{1g} = \begin{pmatrix} u_\mu \\ 0 \end{pmatrix}, \quad Y_{2g} = \begin{pmatrix} 0 \\ xu/2 \end{pmatrix}, \tag{5.24}$$

where the eigenrelations are

$$LY_{1d} = LY_{2d} = 0 \tag{5.25}$$

and

$$LY_{1g} = Y_{1d}, \quad LY_{2g} = Y_{2d}. \tag{5.26}$$

These eigenfunctions and generalized eigenfunctions are induced by the translation invariance, phase invariance, and the free propagation constant μ and velocity v in the solitary waves (5.5) and (5.7). For instance, since a translated wave $u(x-x_0)e^{i\mu t}$ also satisfies Eq. (5.1), then by taking $x_0 \ll 1$ and linearizing this translated wave around the original position $x_0 = 0$, we find that $\widetilde{U}(x,t) = u'(x)$ would satisfy the linearized equation (5.14), which then gives the eigenvalue relation $LY_{1d} = 0$ in Eq. (5.25). The other eigenrelations can be obtained similarly.

In addition to this zero eigenvalue, the spectrum of Eq. (5.19) may contain other discrete eigenvalues as well. To illustrate, let us consider three examples of the generalized NLS equation (5.1) with dual-power nonlinearity (5.8):

(1) $F(|U|^2) = |U|^2,$ (5.27)

(2) $F(|U|^2) = |U|^2 + 3|U|^4,$ (5.28)

(3) $F(|U|^2) = -1.5|U|^{1.5} + |U|^3.$ (5.29)

Taking $\mu = 1$ in all three cases, the spectra of solitary waves for these three nonlinearities are displayed in Fig. 5.1. In the first case where Eq. (5.1) is the integrable NLS equation, the spectrum does not contain any nonzero discrete eigenvalues, see Fig. 5.1(a). This behavior holds for all integrable equations. In the other two nonintegrable cases, the situations are different. In the second case, (5.28), the spectrum contains a pair of discrete eigenvalues on the imaginary axis. These discrete imaginary eigenvalues are called internal modes in the

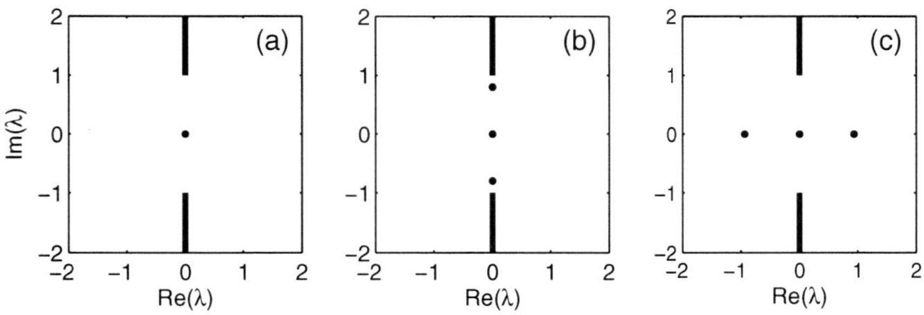

Figure 5.1. *Linear-stability spectra of the solitary wave* (5.5) *(with* $\mu = 1$*) in the generalized NLS equation* (5.1) *under three different nonlinearities* (5.27)–(5.29)*, respectively.*

literature. In the third case, (5.29), the spectrum contains a pair of real eigenvalues. The appearance of nonzero discrete eigenvalues in the linearization spectrum of solitary waves is a signature of nonintegrable equations, and these eigenvalues play an important role in the dynamics of solitary waves under perturbations. Specifically, in the third case, since the spectrum contains a real positive eigenvalue, the corresponding eigenmode (5.15) in the perturbed solution (5.13) would exponentially grow with time; hence the solitary wave (5.5) is linearly unstable. More generally, if the spectrum contains any eigenvalues with positive real parts, then such eigenvalues are unstable, and the solitary wave is linearly unstable. In the second case, since the spectrum contains a pair of internal modes, the corresponding eigenmode (5.15) in the perturbed solution (5.13) would oscillate with time. This oscillation leads to shape oscillations in the perturbed solitary wave (Pelinovsky et al. (1998); see also Etrich et al. (1996), Yang (1997, 2002a), Kivshar et al. (1998), and Barashenkov et al. (1998) for internal-mode-induced shape oscillations in other physical systems). These shape oscillations decay slowly with time due to resonant energy radiation, which is caused by nonlinearity-induced higher harmonic frequencies that enter the continuous spectrum of the system and hence resonantly excite radiation modes. The perturbed solitary wave eventually relaxes to a nearby solitary wave and emits some radiation to the far field (Yang (1997), Pelinovsky et al. (1998), Pelinovsky and Yang (2000)).

What are the origins of these internal modes and unstable eigenvalues? For the internal modes in Fig. 5.1(b), the nonlinearity is cubic quintic with $F(|U|^2) = |U|^2 + \gamma |U|^4$ and $\gamma = 3$. When the quintic nonlinearity coefficient γ approaches zero, where Eq. (5.1) reduces to the standard NLS equation, one finds that these internal modes approach the edges of the continuous spectrum $\pm i\mu$ (Pelinovsky et al. (1998)). This indicates that the internal modes in Fig. 5.1(b) bifurcate out from edges of the continuous spectrum when Eq. (5.1) changes from integrable to nonintegrable. The origin of unstable eigenvalue in Fig. 5.1(c) is quite different, and it is intimately linked to the slope of the power curve of the solitary waves. These unstable eigenvalues will be further examined in the next section.

5.3 Vakhitov–Kolokolov Stability Criterion and Its Generalization

Linear stability of solitary waves is clearly an important issue. In Fig. 5.1(c), we observed that the spectrum of that solitary wave contains a real positive eigenvalue, and hence that solitary wave is linearly unstable. To better understand this instability, we examine how this unstable eigenvalue moves as the propagation constant μ of the solitary wave varies. At the same time, we monitor the power of the solitary wave. Here the power P is defined as

$$P(\mu) = \int_{-\infty}^{\infty} |u(x;\mu)|^2 dx. \tag{5.30}$$

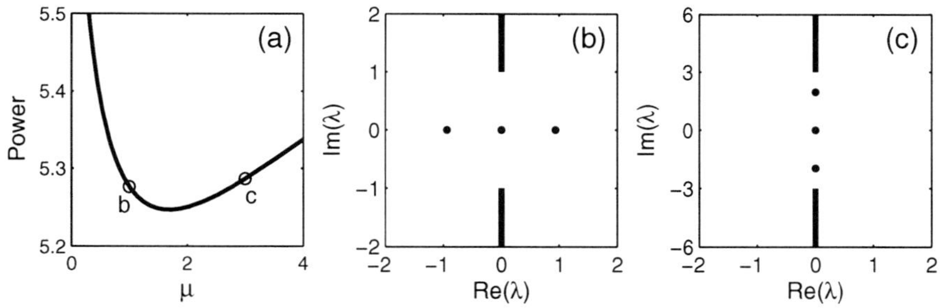

Figure 5.2. (a) *Power curve of solitary waves in the generalized NLS equation* (5.1) *with nonlinearity* (5.29)*;* (b, c) *linear-stability spectra of the solitary waves at* $\mu = 1$ *and* 3 *(marked by circles in* (a)*), respectively.*

Using the solitary-wave formula (5.10) for the underlying dual-power nonlinearity (5.29), this power function can be readily obtained, and it is plotted in Fig. 5.2(a). It is seen that this function has a minimum at $\mu_c \approx 1.7$. When $\mu < \mu_c$, $P'(\mu) < 0$, and when $\mu > \mu_c$, $P'(\mu) > 0$. At two μ values 1 and 3 which lie on the two sides of this power minimum, spectra of the solitary waves are displayed in Fig. 5.2(b, c), respectively. These spectra show that the solitary wave at $\mu = 1$ is linearly unstable, while the one at $\mu = 3$ is linearly stable. When μ changes continuously from 1 to 3, the pair of real eigenvalues moves to the origin, collides there, and then bifurcates off along the imaginary axis. The bifurcation point where the eigenvalue changes from unstable to stable turns out to be precisely the power minimum point. These behaviors suggest that linear stability of these solitary waves is intimately related to the sign of the slope of the power curve. This is indeed the case. This remarkable connection between linear stability and the slope of the power curve was first discovered by Vakhitov and Kolokolov (1973) and is now called the Vakhitov–Kolokolov stability criterion. Similar or related results were also derived by Weinstein (1986), Jones (1988), and Grillakis (1988) for the generalized NLS equations (see Eq. (5.31) below), and by Bona et al. (1987) and Pego and Weinstein (1992) for the KdV-type equations, using different analytical methods.

5.3.1 Vakhitov–Kolokolov Stability Criterion

To introduce the Vakhitov–Kolokolov stability criterion, let us consider the multidimensional NLS equation with a general form of nonlinearity:

$$iU_t + \nabla^2 U + F(|U|^2)U = 0, \tag{5.31}$$

where $\nabla^2 = \partial^2/\partial x_1^2 + \partial^2/\partial x_2^2 + \cdots + \partial^2/\partial x_N^2$ is the Laplacian in the N-dimensional space $\mathbf{x} = (x_1, x_2, \ldots, x_N)$, and $F(\cdot)$ is a real function with $F(0) = 0$. This equation is a multidimensional extension of the 1D generalized NLS equation (5.1) considered in the previous two sections. For a large class of nonlinearities $F(|U|^2)$, Eq. (5.31) admits a ground-state solitary wave

$$U(\mathbf{x},t) = u(\mathbf{x})e^{i\mu t}, \tag{5.32}$$

where $u(\mathbf{x})$ is positive everywhere and has a single hump. The function $u(\mathbf{x})$ satisfies the equation

$$\nabla^2 u - \mu u + F(u^2)u = 0. \tag{5.33}$$

To study the linear stability of this solitary wave, we use the same linearization procedure as developed in the previous section. That procedure can be reduced to the following more concise form. By utilizing the normal mode perturbation (5.15) and the variable transformation (5.18), we can perturb this solitary wave directly by normal modes as

$$U(\mathbf{x},t) = \left\{u(\mathbf{x}) + [v(\mathbf{x}) + w(\mathbf{x})]e^{\lambda t} + [v^*(\mathbf{x}) - w^*(\mathbf{x})]e^{\lambda^* t}\right\} e^{i\mu t}, \tag{5.34}$$

where $v, w \ll 1$ are normal-mode perturbations, and λ is the mode's eigenvalue. Inserting this perturbed solution into (5.31) and linearizing, we obtain the following linear eigenvalue problem:

$$L_0 w = -i\lambda v, \quad L_1 v = -i\lambda w, \tag{5.35}$$

where

$$L_0 = \nabla^2 - \mu + F(u^2), \tag{5.36}$$

$$L_1 = \nabla^2 - \mu + F(u^2) + 2u^2 F'(u^2) \tag{5.37}$$

are multidimensional counterparts of the 1D operators in (5.21)–(5.22). Both L_0 and L_1 are linear Schrödinger operators and are Hermitian, and thus all their eigenvalues are real. In addition, from Eq. (5.33) and its derivative with respect to x_n, we get

$$L_0 u = 0, \quad L_1 u_{x_n} = 0, \quad 1 \le n \le N. \tag{5.38}$$

Thus, zero is an eigenvalue of both operators. Since the ground state $u(\mathbf{x})$ is all positive and $L_0 u = 0$, according to properties of linear Schrödinger operators, zero is the largest eigenvalue of L_0, and the kernel of L_0 contains only one eigenfunction $u(\mathbf{x})$ (see Struwe (2000), specifically the paragraph just below Theorem B.4 on page 246). Regarding operator L_1, since $u(\mathbf{x})$ is the ground state, L_1 has only one positive eigenvalue (Vakhitov and Kolokolov (1973); Weinstein (1986), Proposition 4.2), and the corresponding eigenfunction is positive everywhere (Struwe (2000)). These facts will be important for the following analysis.

Now we present the Vakhitov–Kolokolov stability criterion for the ground state (5.32) in the multidimensional generalized NLS equation (5.31).

Theorem 5.1. *The ground state* (5.32) *in Eq.* (5.31) *is linearly unstable if and only if* $P'(\mu) < 0$. *When unstable, there is a single unstable eigenvalue which is purely real.*

Proof. First of all, by taking the $|\mathbf{x}| \gg 1$ limit in the eigenvalue problem (5.35), we easily see that continuous eigenvalues of λ all lie on the imaginary axis (see the previous section). Thus we need to consider only the discrete eigenvalues. For this purpose, we first eliminate the eigenfunction w in the linear eigenvalue problem (5.35) and reformulate it as

$$L_0 L_1 v = -\lambda^2 v. \tag{5.39}$$

We also normalize each discrete eigenfunction $v(\mathbf{x})$ by

$$\langle v, v\rangle = 1, \tag{5.40}$$

where the inner product is the standard one in the square-integrable functional space:

$$\langle f, g\rangle \equiv \int_{-\infty}^{\infty} f^*(\mathbf{x})\, g(\mathbf{x}) d\mathbf{x}. \tag{5.41}$$

Next, we take the inner product of Eq. (5.39) with $u(\mathbf{x})$. Recalling that L_0 is Hermitian and $L_0 u = 0$, we find that for any nonzero eigenvalue λ, the eigenfunction $v(\mathbf{x})$ is orthogonal to $u(\mathbf{x})$, i.e.,

$$\langle v, u\rangle = 0. \tag{5.42}$$

For stability analysis, we only need to consider nonzero eigenvalues; thus the above equation means that we need to consider only eigenfunctions $v(\mathbf{x})$ in the space

$$S \equiv \{v(\mathbf{x}) : \langle v, u\rangle = 0\}. \tag{5.43}$$

Since the kernel of L_0 contains only the function $u(\mathbf{x})$, L_0 is hence invertible in the space S. Taking the inverse of L_0 in Eq. (5.39), we then get

$$L_1 v = -\lambda^2 L_0^{-1} v. \tag{5.44}$$

Now we take the inner product of this equation with eigenfunction v, which gives

$$\lambda^2 = -\frac{\langle v, L_1 v\rangle}{\langle v, L_0^{-1} v\rangle}. \tag{5.45}$$

Since zero is L_0's largest eigenvalue, L_0 and L_0^{-1} are then negative definite in the space S, and hence $\langle v, L_0^{-1} v\rangle < 0$. The numerator $\langle v, L_1 v\rangle$ is clearly real as L_1 is Hermitian. Thus λ^2 is always real. This means that all eigenvalues λ of the linearization operator are either purely real or purely imaginary. In addition, since λ^2 is real, the eigenfunction v in Eq. (5.39), or equivalently (5.44), can be made purely real as well. Hence we will restrict the space S to contain only real functions which are orthogonal to u.

Linear stability of the solitary wave (5.32) is determined by the sign of $(\lambda^2)_{\max}$. Specifically, the solitary wave is linearly unstable if and only if $(\lambda^2)_{\max} > 0$. From the eigenvalue relation (5.44), we can easily see that for any two distinct eigenvalues λ_m and λ_n, their eigenfunctions v_m and v_n satisfy the orthogonality relation $\langle v_m, L_0^{-1} v_n\rangle = 0$. Hence by expanding a general function $v \in S$ into the complete set of eigenfunctions of the eigenvalue problem (5.44), we find from (5.45) that

$$\left(\lambda^2\right)_{\max} = -\min_{v\in S} \frac{\langle v, L_1 v\rangle}{\langle v, L_0^{-1} v\rangle}. \tag{5.46}$$

Notice that v in (5.45) is an eigenfunction of the linear-stability eigenvalue problem (5.44), but v in the minimization problem (5.46) is an arbitrary function in the space S. Since $\langle v, L_0^{-1} v\rangle < 0$, this wave is then linearly unstable if and only if

$$\min_{v\in S} \langle v, L_1 v\rangle > 0. \tag{5.47}$$

To determine the minimum of $\langle v, L_1 v\rangle$ subject to the two constraints (5.40) and (5.42), we use the method of Lagrange multipliers. Defining the Lagrange functional

$$\mathcal{L}(v,\kappa,\eta) = \langle v, L_1 v\rangle - \kappa\left[\langle v, v\rangle - 1\right] - 2\eta\langle v, u\rangle, \tag{5.48}$$

where κ and η are Lagrange multipliers, the problem then becomes seeking the stationary points of this functional $\mathcal{L}$. At the stationary point, the variation or partial derivative of $\mathcal{L}$ with respect to v, κ, and η will vanish. Since $v(\mathbf{x})$ is real, by taking the variation of this functional with respect to v and setting it to zero, one obtains

$$L_1 v - \kappa v = \eta u. \tag{5.49}$$

By taking the partial derivatives of $\mathcal{L}$ with respect to κ and η and setting them to zero, the constraints (5.40) and (5.42) will be recovered. When taking the inner product of Eq. (5.49) with u and recalling the constraint (5.42), we get

$$\langle v, L_1 v\rangle = \kappa\langle v, v\rangle. \tag{5.50}$$

Thus the instability condition (5.47) will be met if and only if the Lagrange functional $\mathcal{L}$ admits stationary points with $\kappa > 0$.

Next, we determine the stationary points of the Lagrange functional $\mathcal{L}$; i.e., we solve Eqs. (5.40), (5.42), and (5.49). If $\eta = 0$, then it is seen from Eqs. (5.42) and (5.49) that κ coincides with one of L_1's eigenvalues, and the eigenfunction of this eigenvalue must be orthogonal to u. But L_1 has only one positive eigenvalue, and its eigenfunction is positive and thus cannot be orthogonal to u (which is positive as well). Consequently, κ cannot be this positive eigenvalue of L_1, and hence no instability can occur in this case.

Below, we consider the other case where $\eta \neq 0$. To solve Eqs. (5.40), (5.42), and (5.49), we will expand $v(\mathbf{x})$ into the complete set of eigenfunctions of the linear Schrödinger operator L_1. For this operator, its continuous spectrum is the interval $\Lambda \in (-\infty, -\mu]$. For each Λ in this interval, the corresponding continuous eigenfunction is denoted as $\varphi(\mathbf{x};\Lambda)$ (actually, at each continuous eigenvalue Λ, there is more than one continuous eigenfunction; these continuous eigenfunctions can all be listed properly, and the following analysis will still hold). Operator L_1 also has a number of discrete eigenvalues $-\mu < \cdots < \Lambda_2 < \Lambda_1$, and their corresponding eigenfunctions are denoted as $\varphi_n(\mathbf{x})$, $n = 1, 2 \ldots$. Thus

$$L_1 \varphi_n(\mathbf{x}) = \Lambda_n \varphi_n(\mathbf{x}), \quad L_1 \varphi(\mathbf{x};\Lambda) = \Lambda \varphi(\mathbf{x};\Lambda). \tag{5.51}$$

Since $u(\mathbf{x})$ is the ground state and $L_1 u_{x_j} = 0$, from earlier discussions, L_1 has only one positive eigenvalue $\Lambda_1 > 0$. In addition, $\Lambda_2 = 0$ and $\varphi_2 = u_{x_j}$.

For the Schrödinger operator L_1, these continuous and discrete eigenfunctions form a complete set. Normalizing these eigenfunctions properly, their inner products are

$$\langle \varphi_m(\mathbf{x}),\ \varphi_n(\mathbf{x}) \rangle = \delta_{mn}, \tag{5.52}$$

$$\langle \varphi(\mathbf{x};\Lambda),\ \varphi(\mathbf{x};\Lambda') \rangle = \delta(\Lambda - \Lambda'), \tag{5.53}$$

where $\delta(\cdot)$ is the Dirac function, and δ_{mn} is either 1 or 0 depending on whether indices m and n are equal or not.

Now we determine the stationary points of the Lagrange functional $\mathcal{L}$. To do so, we expand u and v into this complete set of L_1's eigenfunctions:

$$u(\mathbf{x}) = \sum_{n \geq 1} c_n \varphi_n(\mathbf{x}) + \int_{-\infty}^{-\mu} c(\Lambda) \varphi(\mathbf{x};\Lambda) d\Lambda, \tag{5.54}$$

$$v(\mathbf{x}) = \sum_{n \geq 1} D_n \varphi_n(\mathbf{x}) + \int_{-\infty}^{-\mu} D(\Lambda) \varphi(\mathbf{x};\Lambda) d\Lambda. \tag{5.55}$$

Here the expansion coefficients can be obtained from the inner product relations (5.52)–(5.53) as

$$c_n = \langle \varphi_n, u \rangle, \quad c(\Lambda) = \langle \varphi(\mathbf{x};\Lambda), u(\mathbf{x}) \rangle, \tag{5.56}$$

and

$$D_n = \langle \varphi_n, v \rangle, \quad D(\Lambda) = \langle \varphi(\mathbf{x};\Lambda), v(\mathbf{x}) \rangle. \tag{5.57}$$

Inserting expansions (5.54)–(5.55) into the stationary-point equation (5.49), we get the expansion coefficients in $v(\mathbf{x})$ as

$$D_n = \frac{\eta c_n}{\Lambda_n - \kappa}, \quad D(\Lambda) = \frac{\eta c(\Lambda)}{\Lambda - \kappa}. \tag{5.58}$$

Then, using expansions (5.54)–(5.55), inner products (5.52)–(5.53), as well as Eq. (5.58), the orthogonality condition $\langle v,u\rangle = 0$ in (5.42) becomes

$$Q(\kappa) \equiv \sum_{n\geq 1} \frac{|c_n|^2}{\Lambda_n - \kappa} + \int_{-\infty}^{-\mu} \frac{|c(\Lambda)|^2}{\Lambda - \kappa} d\Lambda = 0. \tag{5.59}$$

In other words, κ is a root of the function $Q(\kappa)$.

Now we examine the locations of these roots. First we notice that $c_1 = \langle \varphi_1, u\rangle \neq 0$, as both φ_1 and u are positive. In addition, $c_2 = \langle \varphi_2, u\rangle = \langle u_{x_j}, u\rangle = 0$. Thus the largest root $\kappa_{\max}$ of $Q(\kappa)$ is located between $\Lambda_3 < 0$ and $\Lambda_1 > 0$. In the interval $\Lambda_3 < \Lambda < \Lambda_1$, $Q(\kappa)$ increases monotonically from $-\infty$ to $+\infty$. Therefore the sign of $\kappa_{\max}$ is determined by the sign of $Q(0)$. If $Q(0) > 0$, then $\kappa_{\max} < 0$, and if $Q(0) < 0$, then $\kappa_{\max} > 0$. From (5.54) and (5.59), it follows that

$$Q(0) = \sum_{n\geq 1} \frac{|c_n|^2}{\Lambda_n} + \int_{-\infty}^{-\mu} \frac{|c(\Lambda)|^2}{\Lambda} d\Lambda = \langle u, L_1^{-1} u\rangle. \tag{5.60}$$

To calculate $L_1^{-1}u$, we notice that by differentiating the solitary wave equation (5.33) with respect to μ, we obtain

$$L_1 u_\mu = u, \tag{5.61}$$

and thus $L_1^{-1}u = u_\mu$. Inserting this result into (5.60), we then get

$$Q(0) = \langle u, u_\mu\rangle = \frac{1}{2}P'(\mu). \tag{5.62}$$

Consequently, when $P'(\mu) < 0$, the ground state $u(\mathbf{x})$ is linearly unstable, and when $P'(\mu) > 0$, it is linearly stable. When it is unstable, the unstable eigenvalue λ is purely real (see earlier part of the proof). This concludes the proof of Theorem 5.1. □

When a ground state $u(\mathbf{x})$ is linearly unstable according to the Vakhitov–Kolokolov criterion, this instability exhibits characteristic features which are signatures of this instability. To illustrate these features, let us take the solitary wave in Fig. 5.2(b) as an example. This wave is linearly unstable since $P'(\mu) < 0$. To show its instability development, we perturb this solitary wave as

$$U(x,0) = (1+\epsilon)u(x), \tag{5.63}$$

where ϵ is a perturbation parameter. Taking $\epsilon = 0.02$ and -0.02, evolutions of the perturbed solitary waves are displayed in Fig. 5.3(a, b), respectively. In the former case, where the perturbed state has a higher power than the solitary wave, the pulse initially shrinks its width and steepens up. Then it retracts back somewhat and steepens up again. Hence the perturbed wave evolves into an oscillatory state. In the latter case where the perturbed state

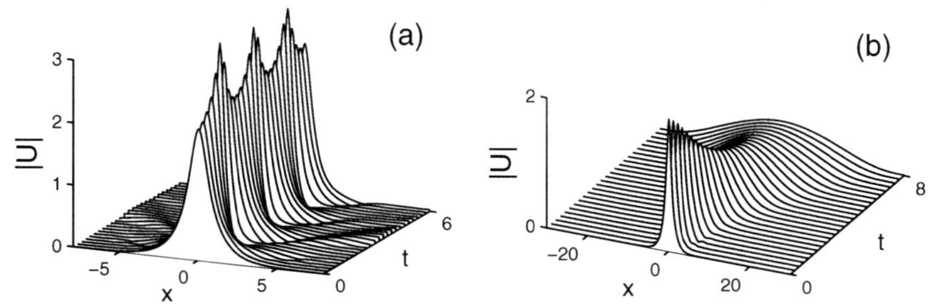

Figure 5.3. *Development of Vakhitov–Kolokolov instability for the solitary wave in Fig.* 5.2(b) *under perturbations* (5.63)*:* (a) $\epsilon = 0.02$*;* (b) $\epsilon = -0.02$.

has a lower power, the pulse steadily broadens and decays to zero. These two scenarios of instability development for different initial perturbations are signatures of Vakhitov–Kolokolov instability.

From Fig. 5.3, we see that the Vakhitov–Kolokolov instability can be viewed as a width instability. Indeed, during initial development of this instability, the width of the pulse either shrinks or expands depending on the initial perturbation. This width dynamics in the instability development has been examined by Pelinovsky et al. (1996). Through asymptotic analysis near the power minimum point, a dynamical equation for the propagation constant of the solitary wave was derived. Using this dynamical equation, all major features of Vakhitov–Kolokolov instability as seen in Fig. 5.3 can be explained.

The Vakhitov–Kolokolov stability criterion in Theorem 5.1 was obtained only for the ground state of the generalized NLS equation (5.31). It is well known that this generalized NLS equation, in two and higher spatial dimensions, also admits radially symmetric higher-order solitary waves (5.32) where $u(\mathbf{x})$ is still real valued but crosses zeros along the radial direction. For these higher-order (excited) states, L_0 is not negative definite in the space S, and thus the Vakhitov–Kolokolov analysis does not apply. This problem was studied by Jones (1988) and Grillakis (1988). Using different methods, Jones (1988) showed that if $n_+(L_1) - n_+(L_0) > 1$, where $n_+(L)$ represents the number of positive eigenvalues in an operator L, then the higher-order wave would possess a real positive eigenvalue and is thus linearly unstable. Grillakis (1988) showed that if $P'(\mu) < 0$, then the higher-order wave would possess a real positive eigenvalue and is thus linearly unstable.

5.3.2 Generalization of Vakhitov–Kolokolov Criterion

In the proof of the Vakhitov–Kolokolov stability criterion in the previous subsection, a key step is that operator L_0 is negative definite in space S. From properties of linear Schrödinger operators and the relation $L_0 u = 0$, we know that for any positive solitary wave $u(\mathbf{x})$ (whether ground state or not), zero is still the largest eigenvalue of L_0 (Struwe (2000)), and thus L_0 is still negative definite in S. On this basis, we can generalize Vakhitov–Kolokolov criterion to any positive solitary wave in a large class of Schrödinger-type nonlinear equations with potentials. This will be done in this subsection.

Let us consider the multidimensional NLS equation with a general form of nonlinearity and external potential,

$$iU_t + \nabla^2 U + F(|U|^2, \mathbf{x})U = 0, \tag{5.64}$$

where $F(\cdot,\cdot)$ is a real-valued function. For a large class of nonlinearities and potentials, this equation admits solitary waves

$$U(\mathbf{x},t) = e^{i\mu t} u(\mathbf{x}), \tag{5.65}$$

where $u(\mathbf{x})$ is a real function satisfying the equation

$$\nabla^2 u - \mu u + F(u^2, \mathbf{x})u = 0. \tag{5.66}$$

To study the linear stability of these solitary waves, we follow the linearization procedure of the previous subsection. The solitary wave (5.65) is perturbed by normal modes as

$$U(\mathbf{x},t) = \left\{ u(\mathbf{x}) + [v(\mathbf{x}) + w(\mathbf{x})]e^{\lambda t} + [v^*(\mathbf{x}) - w^*(\mathbf{x})]e^{\lambda^* t} \right\} e^{i\mu t}, \tag{5.67}$$

where $v, w \ll 1$, and λ is the normal mode's eigenvalue. The resulting eigenvalue problem for the normal mode is

$$L_0 w = -i\lambda v, \quad L_1 v = -i\lambda w, \tag{5.68}$$

where

$$L_0 = \nabla^2 - \mu + F(u^2, \mathbf{x}), \tag{5.69}$$

$$L_1 = \nabla^2 - \mu + F(u^2, \mathbf{x}) + 2u^2 F_{u^2}(u^2, \mathbf{x}). \tag{5.70}$$

Here F_{u^2} is the partial derivative of $F(u^2, \mathbf{x})$ with respect to u^2. These operators L_0, L_1 are counterparts of similar operators (5.36)–(5.37) in the previous subsection. Let us denote $n_+(L_1)$ as the number of positive discrete eigenvalues in L_1. Then if $u(\mathbf{x}) > 0$ everywhere (not necessarily a ground state), then the following generalized Vakhitov–Kolokolov stability criterion holds.

Theorem 5.2. *For a positive solitary wave $u(\mathbf{x})$ in Eq.* (5.64),

1. *if $n_+(L_1) = 0$, this solitary wave is linearly stable;*

2. *if $n_+(L_1) = 1$, and zero is not a discrete eigenvalue of L_1, then this solitary wave is linearly unstable if and only if $P'(\mu) < 0$;*

3. *if $n_+(L_1) = 1$, and zero is a discrete eigenvalue of L_1, then when the eigenfunction of this zero eigenvalue is orthogonal to $u(\mathbf{x})$, the solitary wave is linearly unstable if and only if $P'(\mu) < 0$; but when this eigenfunction is not orthogonal to $u(\mathbf{x})$, the solitary wave is linearly unstable;*

4. *if $n_+(L_1) \geq 2$, this solitary wave is linearly unstable.*

In addition, when the solitary wave is unstable, all unstable eigenvalues are purely real.

Proof. Here we adopt the same notation as in the previous subsection. Then as we mentioned at the beginning of this subsection, for a positive solitary wave $u(\mathbf{x})$, L_0 is still negative definite in S. Because of this, all analysis in the previous subsection (up to Eq. (5.59)) still holds. In particular, all eigenvalues are purely real or purely imaginary, and unstable eigenvalues are purely real. In addition, this solitary wave is linearly unstable if and only if the function $Q(\kappa)$ as defined in Eq. (5.59) has positive roots.

The difference from the previous subsection is on the structure of discrete eigenvalues in operator L_1. In the previous subsection, Eq. (5.31) was translation invariant, and the solitary wave $u(\mathbf{x})$ was a ground state. Hence L_1 had only one positive eigenvalue $\Lambda_1 > 0$ and the next discrete eigenvalue was $\Lambda_2 = 0$. In addition, $c_1 = \langle \varphi_1, u \rangle \neq 0$ and $c_2 = \langle \varphi_2, u \rangle = 0$. In the present case, while $c_1 \neq 0$ still holds, many other aspects become different. For instance, due to presence of the potential in Eq. (5.64), zero may not be an eigenvalue of L_1. In addition, for a general positive solitary wave $u(\mathbf{x})$, L_1 may not have positive eigenvalues at all, or may have more than one positive eigenvalue. These new situations require us to examine the roots of $Q(\kappa)$ a little more carefully. Below, we consider the various cases listed in Theorem 5.2.

1. If $n_+(L_1) = 0$, i.e., L_1 does not have any positive eigenvalues, then for any $\kappa > 0$, $Q(\kappa) > 0$. Thus, $Q(\kappa)$ has no positive roots, hence the solitary wave is linearly stable.

2. If $n_+(L_1) = 1$, and zero is not a discrete eigenvalue of L_1, then the largest root $\kappa_{\max}$ of $Q(\kappa)$ lies between $\Lambda_2 < 0$ and $\Lambda_1 > 0$. As in the previous subsection, the sign of $\kappa_{\max}$ depends on the sign of $Q(0)$. In the present case, relation (5.61) still holds. Then, by repeating the analysis of the previous subsection, we can show that the solitary wave is linearly unstable if and only if $P'(\mu) < 0$.

3. If $n_+(L_1) = 1$, and $\Lambda_2 = 0$ is a discrete eigenvalue of L_1, then when $c_2 = \langle \varphi_2, u \rangle = 0$, this situation is the same as in the previous subsection, and hence the solitary wave is linearly unstable if and only if $P'(\mu) < 0$. But when $c_2 \neq 0$, $Q(\kappa)$ monotonically increases from $-\infty$ to $+\infty$ in the interval $(0, \Lambda_1)$. Thus $\kappa_{\max} > 0$, and the solitary wave is linearly unstable.

4. If $n_+(L_1) \geq 2$, then L_1 has at least two positive eigenvalues, i.e., $0 < \Lambda_2 < \Lambda_1$. When $c_2 \neq 0$, $Q(\kappa)$ monotonically increases from $-\infty$ to $+\infty$ in the positive interval (Λ_2, Λ_1), hence $\kappa_{\max} > 0$, and the solitary wave is linearly unstable. If $c_2 = 0$, then one can easily verify that $\kappa_{\max} = \Lambda_2$, $\eta = 0$, and $v = \varphi_2$ is a stationary point of the Lagrange functional $\mathcal{L}$. Since $\kappa_{\max} = \Lambda_2 > 0$, the solitary wave is then linearly unstable as well. This concludes the proof of Theorem 5.2. □

One may notice that a certain part of this theorem resembles the original Vakhitov–Kolokolov criterion, where $P'(\mu) < 0$ implies instability and $P'(\mu) > 0$ implies stability. Especially, for the ground states of Eq. (5.31) in the previous subsection, $n_+(L_1) = 1$ and $c_2 = 0$, hence this theorem reproduces the original Vakhitov–Kolokolov criterion as a special case. However, significant differences exist between the original criterion and the extended one above. In particular, this extended criterion reveals that for the more general class of equations (5.64), $P'(\mu) < 0$ does not necessarily imply instability, and $P'(\mu) > 0$ does not necessarily imply stability. Here the number of positive eigenvalues in L_1 also plays a critical role.

Below we illustrate this generalized Vakhitov–Kolokolov criterion in Theorem 5.2 by two examples.

Example 1. Consider the 1D NLS equation with a harmonic potential and self-defocusing nonlinearity,

$$iU_t + U_{xx} - x^2 U - |U|^2 U = 0, \tag{5.71}$$

which arises frequently in Bose–Einstein condensates trapped in a harmonic potential under repulsive nonlinearities. This equation admits a family of ground-state solitary waves where $u(x)$ is positive and single humped. The power curve of this solution family is shown in Fig. 5.4(a), and the solution profile at $\mu = -2$ is displayed in Fig. 5.4(b). Notice that for this solution family, $P'(\mu) < 0$ on the entire power curve. However, this solution family is still linearly stable because $n_+(L_1) = 0$. For example, for the solution shown in Fig. 5.4(b), the spectrum of L_1 is plotted in Fig. 5.4(c). It is seen that all eigenvalues of L_1 are negative. A general analytical proof of this fact can be found in Yang and Lakoba (2008). Thus, according to the generalized stability criterion in Theorem 5.2, these ground states are linearly stable.

Example 2. Consider the 1D NLS equation with a double-well potential and self-focusing nonlinearity,

$$iU_t + U_{xx} - V(x)U + |U|^2 U = 0, \tag{5.72}$$

where the potential

$$V(x) = 3\left[e^{-(x-1.5)^2} + e^{-(x+1.5)^2}\right] \tag{5.73}$$

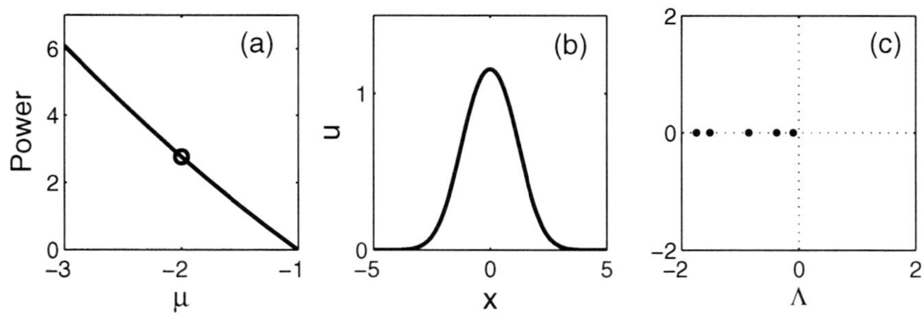

Figure 5.4. *Ground states in the* 1*D NLS equation with a harmonic potential* (5.71) *and their* L_1 *spectrum:* (a) *the power curve;* (b) *solution profile at* $\mu = -2$ *(marked by a circle in* (a)*);* (c) *spectrum of* L_1 *for the soliton in* (b).

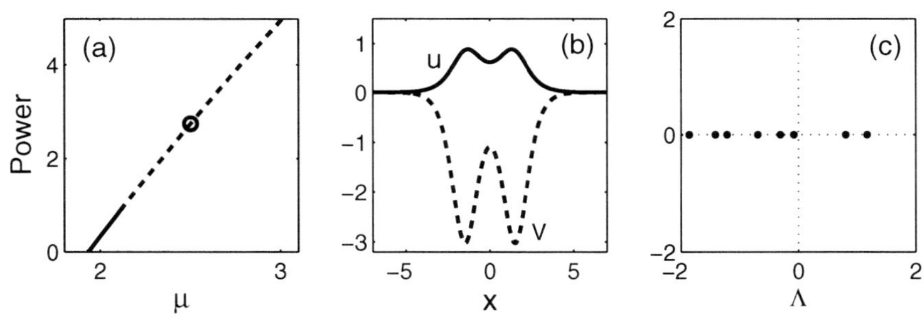

Figure 5.5. *Double-hump solitary waves in the* 1*D NLS equation with a double-hump potential* (5.72)–(5.73) *and their linear-stability behaviors:* (a) *the power curve; the solid part is stable, and the dashed part unstable;* (b) *solution profile* $u(x)$ *at* $\mu = 2.5$ *(marked by a circle in* (a)*); the dashed curve is the potential* $V(x)$*;* (c) *spectrum of* L_1 *for the soliton* $u(x)$ *in* (b).

has a double-well profile, see Fig. 5.5(b). This equation admits a family of double-hump positive solitary waves whose power curve is shown in Fig. 5.5(a). Notice that $P'(\mu) > 0$ everywhere on the power curve. However, only these solitary waves at low amplitudes ($\mu < \mu_c \approx 2.11$) are linearly stable, and those at higher amplitudes ($\mu > \mu_c$) are linearly unstable. The reason is that when $\mu < \mu_c$, $n_+(L_1) = 1$, but when $\mu > \mu_c$, $n_+(L_1) \geq 2$. For instance, when $\mu = 2.5$, the corresponding solitary wave is plotted in Fig. 5.5(b), and its L_1 spectrum is plotted in Fig. 5.5(c). It is seen that this L_1 spectrum contains two positive eigenvalues; thus according to the generalized Vakhitov–Kolokolov criterion in Theorem 5.2, this wave is linearly unstable.

The above two examples tell us that, when studying the linear stability of positive solitary waves in Eq. (5.64), both the spectral condition $n_+(L_1)$ and the slope condition $P'(\mu)$ must be used in order to determine the linear stability of the underlying solitary wave.

In addition to Eq. (5.64), the Vakhitov–Kolokolov criterion can be generalized to other types of equations as well. For instance, this criterion has been generalized to the vector NLS equations (without potential terms) by Pelinovsky and Kivshar (2000).

The Vakhitov–Kolokolov stability criterion and its generalization are important analytical results for soliton stabilities in NLS-type equations. Their main limitation is that the solitary wave must be positive (i.e., does not cross zeros). As such, they do not apply to zero-crossing real solitary waves (such as higher-order states) and complex-valued solitary waves (such as vortex solitons). For such solitons, linear stability eigenvalues may not be purely real or purely imaginary. Rather, they can be complex-valued (with λ^2 not real). Even if their eigenvalues are purely real or purely imaginary, they cannot be treated by the Vakhitov–Kolokolov analysis, and other analytical methods are necessary. For real-valued higher-order states in the potential-free generalized NLS equation (5.31), Jones (1988) and Grillakis (1988) have shown that if $n_+(L_1) - n_+(L_0) > 1$ or $P'(\mu) < 0$, these waves would be linearly unstable due to the existence of a positive eigenvalue (see the end of the previous subsection). But those results do not hold for the generalized NLS equations with potentials (5.64). For zero-crossing solitary waves in these equations (5.64), their linear stability will be examined in the next section and the next chapter.

5.4 Stability Switching at a Power Extremum

In this section, we consider linear stability of solitary waves which cross zeros. In this case, we will show that the power curve also plays an important role in the stability result. Specifically, if the power curve has a local extremum, then a positive (unstable) eigenvalue bifurcates out on one side of this power extremum, creating width instability. If no other instabilities exist in the system, then a change of stability occurs at the power extremum. Our analysis follows Shi and Yang (2008).

Let us use the multidimensional generalized NLS equation with potentials,

$$iU_t + \nabla^2 U + F(|U|^2, \mathbf{x})U = 0, \tag{5.74}$$

as an example to present the theory. Here $F(\cdot,\cdot)$ is a real-valued function. This equation admits solitary waves of the form

$$U(\mathbf{x},t) = u(\mathbf{x})e^{i\mu t}, \tag{5.75}$$

where $u(\mathbf{x})$ is a real function which may cross zeros, and μ is the propagation constant. For linear stability analysis of these solitary waves in this section, it is convenient to perturb these waves by normal modes as

$$U(\mathbf{x},t) = \left\{u(\mathbf{x}) + [v(\mathbf{x}) + i\lambda^{-1}w(\mathbf{x})]e^{\lambda t} + \left[v^*(\mathbf{x}) - [i\lambda^{-1}w(\mathbf{x})]^*\right]e^{\lambda^* t}\right\}e^{i\mu t}, \tag{5.76}$$

where $v, w \ll 1$ are normal-mode perturbations, and λ is the mode's eigenvalue. This normal mode differs from that in Eq. (5.67) by a scaling in the w-component. This scaling proves to be appropriate for the instability near a power extremum. Corresponding to these normal-mode perturbations, the linear stability eigenvalue problem is

$$L_0 w = -\lambda^2 v, \quad L_1 v = w, \tag{5.77}$$

where operators L_0 and L_1 are as given in Eqs. (5.69)–(5.70). For these operators, we have the relations

$$L_0 u = 0, \quad L_1 u_\mu = u. \tag{5.78}$$

Here, the second relation is obtained by differentiating the first relation with respect to μ. Since the eigenvalue problem (5.77) is linear, its eigenfunctions (v, w) will be treated below as $O(u)$. For these eigenvalue equations, by taking their inner products with u and u_μ, respectively, and utilizing the relations in (5.78), we find that for nonzero eigenvalues λ, eigenfunctions (v, w) satisfy the following orthogonality constraints:

$$\langle v, u\rangle = \langle w, u_\mu\rangle = 0. \tag{5.79}$$

At a power extremum $\mu = \mu_0$, $P'(\mu_0) = 0$, and hence

$$\langle u_0, u_{0\mu}\rangle = 0, \tag{5.80}$$

where $u_0 = u|_{\mu=\mu_0}$ and $u_{0\mu} = u_\mu\big|_{\mu=\mu_0}$. At this power extremum, the eigenvalue problem (5.77) has a zero-eigenvalue mode,

$$v_0 = u_{0\mu}, \quad w_0 = u_0, \quad \lambda_0 = 0. \tag{5.81}$$

Indeed, one can first see from relations (5.78) that $(v, w) = (u_\mu, u)$ and $\lambda = 0$ satisfy the eigenvalue equations (5.77) for all solitary waves with arbitrary μ. The reason this zero eigenmode only appears at a power extremum $\mu = \mu_0$ is that, away from the power extremum, this eigenvalue will move away from zero; hence its eigenfunction must satisfy the orthogonality constraints (5.79). Then using the continuity argument, this eigenfunction at the zero eigenvalue must satisfy these orthogonality constraints as well, which means that $\langle u, u_\mu\rangle = 0$. This is possible only at a power extremum in view of Eq. (5.80).

Near this power extremum, i.e., $|\mu - \mu_0| \ll 1$, we expand the eigenvalue and eigenfunctions into a power series in $\mu - \mu_0$:

$$v(\mathbf{x}, \mu) = v_0(\mathbf{x}) + (\mu - \mu_0)v_1(\mathbf{x}) + \cdots, \tag{5.82}$$

$$w(\mathbf{x}, \mu) = w_0(\mathbf{x}) + (\mu - \mu_0)w_1(\mathbf{x}) + \cdots, \tag{5.83}$$

$$\lambda^2(\mu) = (\mu - \mu_0)\eta_1 + (\mu - \mu_0)^2\eta_2 + \cdots. \tag{5.84}$$

In addition, we expand the solitary wave $u(\mathbf{x},\mu)$, $u_\mu(\mathbf{x},\mu)$ and operator L_0 into a power series in $\mu-\mu_0$ as well:

$$u(\mathbf{x},\mu) = u_0(\mathbf{x}) + (\mu-\mu_0)u_{0\mu}(\mathbf{x}) + \cdots, \tag{5.85}$$

$$u_\mu(\mathbf{x},\mu) = u_{0\mu}(\mathbf{x}) + (\mu-\mu_0)u_{0\mu\mu}(\mathbf{x}) + \cdots, \tag{5.86}$$

$$L_0 = L_{00} + (\mu-\mu_0)L_{01} + \cdots. \tag{5.87}$$

Here L_{00} is the operator L_0 at $\mu=\mu_0$ and $u=u_0$. Substituting expansions of u and L_0 into the relation $L_0 u = 0$, at $O(\mu-\mu_0)$ we get

$$L_{01}u_0 = -L_{00}u_{0\mu}. \tag{5.88}$$

Substituting eigenmode expansions (5.82)–(5.84) and the zeroth-order eigenmode (5.81) into the eigenvalue relation $L_0 w = -\lambda^2 v$ in (5.77), at $O(\mu-\mu_0)$ we get

$$L_{00}w_1 = -\eta_1 u_{0\mu} - L_{01}u_0. \tag{5.89}$$

Using the relation (5.88), we find that the solution w_1 to the above equation is

$$w_1 = u_{0\mu} - \eta_1 L_{00}^{-1}u_{0\mu}. \tag{5.90}$$

Here L_{00} is invertible in the space orthogonal to u_0, and hence $L_{00}^{-1}u_{0\mu}$ exists due to (5.80). Now we insert expansions (5.83) and (5.86) of w and u_μ and the above w_1 formula into the constraint $\langle w, u_\mu\rangle = 0$ in (5.79). At $O(\mu-\mu_0)$ we get

$$\langle u_0, u_{0\mu\mu}\rangle + \langle u_{0\mu}, u_{0\mu}\rangle = \eta_1\langle u_{0\mu}, L_{00}^{-1}u_{0\mu}\rangle. \tag{5.91}$$

Notice that

$$P''(\mu) = \langle u,u\rangle_{\mu\mu} = 2\left[\langle u, u_{\mu\mu}\rangle + \langle u_\mu, u_\mu\rangle\right]. \tag{5.92}$$

Evaluating this relation at the power extremum $\mu=\mu_0$ and inserting it into Eq. (5.91), we then obtain the formula for the coefficient η_1. Inserting this η_1 formula into the eigenvalue expansion (5.84), we obtain the leading-order formula for λ^2 as

$$\lambda^2 = \eta_1(\mu-\mu_0), \qquad \eta_1 = \frac{P_0''}{2\langle u_{0\mu}, L_{00}^{-1}u_{0\mu}\rangle}. \tag{5.93}$$

Here $P_0'' = P''(\mu_0)$. For any sign of η_1 in the above formula, λ^2 is always positive on one side of the power extremum μ_0 and negative on the other side. Thus this eigenvalue λ is always purely real (unstable) on one side of the power extremum and purely imaginary (stable) on the other side. In other words, this eigenvalue changes from stable to unstable at a power extremum. If no other instabilities exist for these solitary waves, then a switching of stability occurs at a power extremum. If $\langle u_{0\mu}, L_{00}^{-1} u_{0\mu} \rangle < 0$, as is the case when $u(\mathbf{x}) > 0$ everywhere (see the previous section), this eigenvalue would be unstable at the left of μ_0 when $P_0'' > 0$ (a power minimum) and at the right of μ_0 when $P_0'' < 0$ (a power maximum). In other words, this eigenvalue would be unstable on the side of μ_0 where $P'(\mu) < 0$. This is consistent with the Vakhitov–Kolokolov stability criterion in the previous section. However, if $\langle u_{0\mu}, L_{00}^{-1} u_{0\mu} \rangle > 0$, the situation would be just the opposite; i.e., this eigenvalue would be unstable on the side of μ_0 where $P'(\mu) > 0$. This is the opposite of the Vakhitov–Kolokolov slope condition for instability. Examples for both cases will be seen in Sec. 6.4 of the next chapter.

The instability which appears near a power extremum is a width instability. In other words, when this instability develops, the solitary wave either contracts or expands in width, depending on the initial perturbations. This is analogous to the Vakhitov–Kolokolov instability illustrated in Fig. 5.3. The reason for this width instability can be clearly seen from the expressions of unstable eigenfunctions in Eqs. (5.81)–(5.83). When the leading-order terms of these unstable eigenfunctions are scaled back to small perturbations and substituted into the perturbed solution (5.76), since $\lambda > 0$ for the unstable eigenvalue, we see that the perturbed solution is approximately a solitary wave with its propagation constant μ shifted. The amount of shift is proportional to $e^{\lambda t}$, which exponentially grows with time, and the direction of shift can be either upward or downward depending on the initial perturbations. Since the propagation constant controls the width of the solitary wave (see Eq. (5.66)), it is then clear that this instability is a width instability, where the perturbed solitary wave either shrinks or expands during the instability development.

It is noted that in some equations arising in 2D shallow water waves, it was argued that a necessary condition for a change of linear stability is that the power be at a local extremum (Akylas and Cho (2008), Akers and Milewski (2008)). For general equations such as NLS-type equations with potentials (5.64), that statement is invalid, because a change of stability can also occur when the power is not at a local extremum (an example is in Fig. 5.5 for Eq. (5.72) in the previous section). Our result in this section, on the other hand, shows that a power extremum is a *sufficient* condition for bifurcation of an unstable eigenvalue. Thus, in the absence of other instabilities in the system, a power extremum then is a *sufficient* condition for stability switching.

5.5 Nonlocal Waves and the Exponential Asymptotics Technique

In Sec. 4.2.3, we observed that when an NLS soliton is perturbed by a third-order dispersion term, the soliton emits a cw tail. This cw tail is generated because the third-order derivative perturbation qualitatively changes the dispersion relation. As a consequence, the soliton is in resonance with the system's continuous spectrum, which in turn creates the cw tail. This emission of cw tails turns out to be a common phenomenon, and arises generically when a solitary wave is in resonance with the continuous spectrum of the underlying wave system. For instance, this phenomenon occurs in the KdV equation perturbed by a fifth-order derivative term as well (Pomeau et al. (1988)). Other physical examples of this phenomenon include gravity-capillary water waves (with the Bond number between 0 and 1/3) and internal water waves (with more than one long-wave mode) (Vanden-Broeck (1991), Beale (1991), Sun (1991), Vanden-Broeck and Turner (1992), Akylas and Grimshaw (1992), Yang and Akylas (1996)). A review of this subject can also be found in Boyd (1998). When this resonance is between a soliton and a short wave in the continuous spectrum, the amplitude of the emitted short-wave tail turns out to be exponentially small with the resonant wavenumber, and thus it cannot be captured by a regular perturbation expansion. Instead, an exponential asymptotics technique is needed to treat this problem. This method was first used by Segur and Kruskal (1987) in their treatment of small-amplitude breather solutions in the ϕ^4 model. It was subsequently further developed and applied to the KdV equation perturbed by a fifth-order dispersion term (Pomeau et al. (1988), Grimshaw and Joshi (1995)), and the NLS equation perturbed by a third-order dispersion term (Wai et al. (1990), Grimshaw (1995), Calvo and Akylas (1997)). Calculations of exponentially small cw tails have also been performed for internal water waves (Akylas and Grimshaw (1992)) and gravity-capillary water waves (Yang and Akylas (1996)). In this section, we use the exponential asymptotics technique to calculate cw tails of NLS solitons under a third-order dispersion perturbation.

The NLS equation under a third-order dispersion perturbation can be written as

$$iU_t + \frac{1}{2}U_{xx} + |U|^2 U = i\epsilon U_{xxx}, \tag{5.94}$$

where $0 < \epsilon \ll 1$ is a small positive third-order dispersion coefficient (if $\epsilon < 0$, it can be made positive by changing x to $-x$). We first look for steady-state solutions in this equation in the form

$$U(x,t) = u(\theta)e^{i\mu t/2}, \quad \theta = x - \epsilon v t, \tag{5.95}$$

where μ is the propagation constant, and v is a velocity parameter. Here we allow this solution to move at a small velocity ϵv because, under the third-order dispersion perturbation, an initially stationary NLS soliton would move at $O(\epsilon)$ velocity (see Sec. 4.2.3). Through a simple variable rescaling, we can normalize the propagation constant μ to be unity, thus we will set $\mu = 1$ below. Then inserting (5.95) into (5.94), one obtains the stationary wave equation

$$\frac{1}{2}u_{\theta\theta} - \frac{1}{2}u + |u|^2 u = i\epsilon(u_{\theta\theta\theta} + v u_\theta). \tag{5.96}$$

For any fixed value of v, this equation admits bounded solutions which satisfy the symmetry condition

$$u^*(\theta) = u(-\theta), \tag{5.97}$$

i.e., $\mathrm{Re}(u)$ is symmetric, and $\mathrm{Im}(u)$ is antisymmetric. We will call these solutions symmetric solutions and determine them next.

Symmetric Nonlocal Waves

To characterize these symmetric solutions, we examine their tail behaviors at $|\theta| \gg 1$. When $\epsilon \ll 1$, these tails are very small; thus the nonlinear term in Eq. (5.96) can be neglected, and the solution to the remaining linear equation is a combination of admissible modes $e^{ik\theta}$, where k is a root of the cubic equation

$$\frac{1}{2}(k^2+1)+\epsilon(k^3 - vk) = 0. \tag{5.98}$$

This cubic equation has one real root k_0 and a complex conjugate pair $k_\pm$ where $k_- = k_+^*$ (here we let $\mathrm{Im}(k_+) > 0$). For $\epsilon \ll 1$, these roots have the perturbation expansions

$$k_0 = -\frac{1}{2\epsilon} - 2(v+1)\epsilon + \cdots, \tag{5.99}$$

$$k_\pm = \pm i + (v+1)\epsilon + \cdots. \tag{5.100}$$

For bounded solutions in Eq. (5.96), the tails at $\theta \to \pm\infty$ cannot contain the growing linear modes $e^{ik_\mp\theta}$. Then, in view of the symmetry condition (5.97), these tails must be of the form

$$u(\theta) \to R\, e^{ik_0\theta \pm i\sigma} + \alpha_\pm\, e^{ik_\pm\theta}, \quad \theta \to \pm\infty, \tag{5.101}$$

where R and σ are the amplitude and phase of the cw oscillations, and $\alpha_\pm$ are complex coefficients of the decaying modes $e^{ik_\pm x}$ with $\alpha_- = \alpha_+^*$. Since these tails contain cw oscillations, these bounded solutions are no longer localized in space, and hence they will be called nonlocal waves.

This tail analysis above not only characterizes the bounded solutions in Eq. (5.96), but also sets the stage for numerically computing these solutions by the shooting method. From the symmetry condition (5.97), we see that

$$\text{at } \theta = 0: \quad \mathrm{Im}(u) = \mathrm{Re}(u_\theta) = \mathrm{Im}(u_{\theta\theta}) = 0. \tag{5.102}$$

Thus for each given phase σ, we numerically integrate Eq. (5.96) from $\theta = -\infty$ to $\theta = 0$, starting from the initial condition (5.101). Our goal is to determine the three real constants R, $\mathrm{Re}(\alpha_+)$, and $\mathrm{Im}(\alpha_+)$ in these tails so that the three real conditions in (5.102) are satisfied. This shooting method is very effective and can obtain the oscillation amplitude R very quickly and accurately. The numerical results will be shown later in this section. This shooting procedure also reveals that the oscillation phase σ in these solutions is a free parameter, and hence Eq. (5.96) admits a continuous family of symmetric nonlocal waves which are parameterized by the phase σ (Grimshaw (1995)).

It is noted that in all previous studies on this problem (Wai et al. (1990), Grimshaw (1995), Calvo and Akylas (1997)), the velocity v in Eq. (5.96) was required to be equal to -1, and it was implied that nonlocal waves in (5.96) would not exist when $v \neq -1$. This is not true, however. Nonlocal waves in Eq. (5.96) actually can exist at any value of the velocity v. To reflect this fact, previous analysis will need to be modified so that it can apply to arbitrary values of v.

Below we analytically calculate amplitude R of cw tails in these nonlocal waves. If the solution to Eq. (5.96) is expanded into a regular perturbation series

$$u(\theta;\epsilon) = u_0(\theta) + \epsilon u_1(\theta) + \epsilon^2 u_2(\theta) + \cdots, \tag{5.103}$$

then one would find that

$$u_0(\theta) = \operatorname{sech}\theta, \tag{5.104}$$

$$u_1(\theta) = -3i \operatorname{sech}\theta \tanh\theta + i(v+1)\theta \operatorname{sech}\theta, \tag{5.105}$$

and so on. All the $u_n(\theta)$ terms are localized, and thus (5.103) represents a localized solution to Eq. (5.96), and it does not contain any cw tails. This means that these cw tails must be smaller than any power of ϵ; i.e., they are exponentially small in ϵ.

To calculate these exponentially small cw tails, we will work in the Fourier space (as in Calvo and Akylas (1997)). We will see that this approach essentially converts an exponential asymptotics problem (in the physical domain) to a multiscale expansion problem (in the wavenumber domain). In this approach, we will take the Fourier transform of $u(\theta)$. Since $u(\theta)$ is nonlocal, we first need to clarify how its Fourier transform is defined, and what feature its Fourier transform will exhibit. Recall from (5.101) that

$$u(\theta) \to R\, e^{ik_0\theta + i\sigma\,\operatorname{sgn}(\theta)}, \quad |\theta| \to \infty. \tag{5.106}$$

Let us split $u(\theta)$ into local and nonlocal parts as

$$u(\theta) = \psi(\theta) + u_L(\theta), \tag{5.107}$$

where

$$\psi(\theta) = R\, e^{ik_0\theta + i\sigma\,\operatorname{sgn}(\theta)} = [R\cos\sigma + iR\sin\sigma\,\operatorname{sgn}(\theta)]\, e^{ik_0\theta} \tag{5.108}$$

is the nonlocal part of $u(\theta)$, and $u_L(\theta)$ is its localized part. Since $u(\theta)$ is nonlocal, to define its Fourier transform, we first multiply it by a decaying function $e^{-a|\theta|}$ with $a > 0$, then take the Fourier transform of the product function, and then take the limit of $a \to 0^+$. In other words, we define the Fourier transform of the nonlocal function $u(\theta)$ as

$$\hat{u}(k) = \frac{1}{2\pi} \lim_{a\to 0^+} \int_{-\infty}^{\infty} u(\theta) e^{-a|\theta|} e^{-ik\theta}\, d\theta, \qquad k \in \mathbb{R}. \tag{5.109}$$

Here the coefficient $1/2\pi$ is included, since under this definition, the Fourier transform of a product $f(\theta)g(\theta)$ is simply the convolution $(\hat{f} * \hat{g})(k)$, which is convenient in later calculations. Under this definition, the Fourier transform of $\psi(\theta)$ can be easily found to be

$$\hat{\psi}(k) = R\cos\sigma\, \delta(k-k_0) + \frac{R\sin\sigma}{\pi} \frac{1}{k-k_0}, \tag{5.110}$$

where $\delta(k-k_0)$ is the Dirac delta function. The Fourier transform of the localized function $u_L(\theta)$ is bounded. Thus the Fourier transform $\hat{u}(k)$ of the nonlocal function $u(\theta)$ contains a simple-pole singularity at $k = k_0$, whose residue is $R\sin\sigma/\pi$. In view of this, if we can determine the residue of the pole singularity in $\hat{u}(k)$, then $R\sin\sigma$ would be obtained, and hence the tail amplitude R would be derived (with phase σ as a free parameter). This is how we will proceed.

First, we derive the governing equation for the Fourier transform $\hat{u}(k)$. For this purpose, we denote $w = u^*$ and recast Eq. (5.96) as a coupled system in (u, w):

$$\frac{1}{2}u_{\theta\theta} - \frac{1}{2}u + u^2 w = i\epsilon(u_{\theta\theta\theta} + v u_\theta), \tag{5.111}$$

$$\frac{1}{2}w_{\theta\theta} - \frac{1}{2}w + w^2 u = -i\epsilon(w_{\theta\theta\theta} + v w_\theta). \tag{5.112}$$

Then we take the Fourier transform of these equations, with the Fourier transform as defined in (5.109). The transformed system for $(\hat{u}, \hat{w})$ is

$$\frac{1}{2}(k^2+1)\hat{u} + \epsilon k(k^2 - v)\hat{u} = \widehat{u^2 w}, \tag{5.113}$$

$$\frac{1}{2}(k^2+1)\hat{w} - \epsilon k(k^2 - v)\hat{w} = \widehat{w^2 u}. \tag{5.114}$$

Corresponding to the symmetry condition (5.97) for (u, w), functions $(\hat{u}, \hat{w})$ satisfy the following symmetry condition:

$$\hat{w}(k) = \hat{u}(-k). \tag{5.115}$$

Next, we will calculate $\hat{u}(k)$ and show that it has a simple-pole singularity at $k \approx -1/2\epsilon$, which is the leading-order wavenumber of cw tails, see (5.99). To do so, we notice that by taking the Fourier transform to the straightforward expansion (5.103), one gets

$$\hat{u}(k) = \frac{1}{2}\text{sech}\frac{\pi k}{2}\left[1 - 3\epsilon k + \frac{1}{2}\epsilon(v+1)\pi \tanh\frac{\pi k}{2} + O(\epsilon^2)\right], \tag{5.116}$$

$$\hat{w}(k) = \frac{1}{2}\text{sech}\frac{\pi k}{2}\left[1 + 3\epsilon k - \frac{1}{2}\epsilon(v+1)\pi \tanh\frac{\pi k}{2} + O(\epsilon^2)\right]. \tag{5.117}$$

These expansions become nonuniform when $\kappa \equiv \epsilon k = O(1)$; thus they are replaced by the uniformly valid expressions

$$\hat{u}(k) = \frac{1}{2}\text{sech}\frac{\pi k}{2} F(\kappa, k), \tag{5.118}$$

$$\hat{w}(k) = \frac{1}{2}\text{sech}\frac{\pi k}{2} G(\kappa, k). \tag{5.119}$$

In previous studies (Wai et al. (1990), Grimshaw (1995), Calvo and Akylas (1997)), v was set as -1. In that case, functions F and G depend on the slow wavenumber κ only (see Eqs. (5.116)–(5.117) and Calvo and Akylas (1997)). When $v \neq -1$, F and G would depend on both the slow and fast wavenumbers κ and k. However, notice that the dependence of F and G on the fast wavenumber k saturates when $|k| \gg 1$, because such dependence appears in forms like $\tanh(\pi k/2)$. Thus, when $|k| \gg 1$, F and G again depend on the slow wavenumber κ only. This fact would be sufficient to carry out the following analysis.

Since we anticipate pole singularities of $\hat{u}(k)$ and $\hat{w}(k)$ at $k_0 \approx \mp 1/2\epsilon$, this motivates us to solve Eqs. (5.113)–(5.114) in the large-k region of $|k| = O(1/\epsilon)$, i.e., $|\kappa| = O(1)$. In this region, Eqs. (5.113)–(5.114) can be simplified significantly. First, we simplify the nonlinear convolution terms $\widehat{u^2 w}$ and $\widehat{w^2 u}$. Using the convolution theorem of the Fourier transform (5.109), we have

$$\widehat{u^2 w}(\kappa) = \frac{1}{\epsilon^2}\int_{-\infty}^{\infty} \hat{w}(\kappa - \lambda)\left[\int_{-\infty}^{\infty} \hat{u}(\lambda - \lambda')\hat{u}(\lambda')d\lambda'\right] d\lambda. \tag{5.120}$$

Now we insert (5.118)–(5.119) into this equation and get

$$\begin{aligned}\widehat{u^2 w}(\kappa) = \frac{1}{8\epsilon^2}\int_{-\infty}^{\infty} &\text{sech}\frac{\pi(\kappa - \lambda)}{2\epsilon} G\left(\kappa - \lambda, \frac{\kappa - \lambda}{\epsilon}\right) \\ \times &\left[\int_{-\infty}^{\infty} \text{sech}\frac{\pi(\lambda - \lambda')}{2\epsilon}\text{sech}\frac{\pi\lambda'}{2\epsilon} F\left(\lambda - \lambda', \frac{\lambda - \lambda'}{\epsilon}\right) F\left(\lambda', \frac{\lambda'}{\epsilon}\right) d\lambda'\right] d\lambda.\end{aligned} \tag{5.121}$$

When $\epsilon \ll 1$ and $|\kappa| = O(1)$, the dominant contribution to the inner integral comes from the $0 < \lambda' < \lambda$ region (see also Akylas and Yang (1995)). In this region, the product of two sech terms in the inner integral is approximately $2\text{sech}(\pi\lambda/2\epsilon)$, which is a constant (here each sech function is simply exponential since its argument is large). In addition, due to the saturation property of functions F and G on the fast wavenumber k, these functions become dependent on the slow wavenumber κ only in the dominant contributions to this integral. Thus, this inner integral simplifies to

$$2\,\text{sech}\frac{\pi\lambda}{2\epsilon}\int_0^{\lambda} F(\lambda - \lambda')F(\lambda')d\lambda'. \tag{5.122}$$

Repeating the same procedure, the nonlinear convolution term in (5.121) then reduces to

$$\widehat{u^2 w}(\kappa) = \frac{1}{2\epsilon^2}\operatorname{sech}\frac{\pi\kappa}{2\epsilon}\int_0^\kappa G(\kappa-\lambda)\left[\int_0^\lambda F(\lambda-\lambda')F(\lambda')d\lambda'\right]d\lambda. \tag{5.123}$$

Similar expressions can be obtained for $\widehat{w^2 u}$ as well. When these reduced expressions are substituted into Eqs. (5.113)–(5.114), to the leading order in ϵ we get equations for $F(\kappa)$ and $G(\kappa)$ as

$$\frac{1}{2}\kappa^2(1+2\kappa)F(\kappa) = \int_0^\kappa G(\kappa-\lambda)\left[\int_0^\lambda F(\lambda-\lambda')F(\lambda')d\lambda'\right]d\lambda, \tag{5.124}$$

$$\frac{1}{2}\kappa^2(1-2\kappa)G(\kappa) = \int_0^\kappa F(\kappa-\lambda)\left[\int_0^\lambda G(\lambda-\lambda')G(\lambda')d\lambda'\right]d\lambda. \tag{5.125}$$

Notice that the velocity parameter v does not appear in these leading-order equations, indicating that the oscillation amplitudes to be derived below will not depend on v at the leading order.

Now we solve the simplified equations (5.124)–(5.125). The solution to these equations can be posed as a power series,

$$F(\kappa) = \sum_{n=0}^{\infty} a_n\kappa^n, \quad G(\kappa) = \sum_{n=0}^{\infty} b_n\kappa^n. \tag{5.126}$$

In view of the symmetry condition (5.115) and relations (5.118)–(5.119), we see that

$$G(\kappa) = F(-\kappa), \quad b_n = (-1)^n a_n. \tag{5.127}$$

When inserting the above two equations into the system (5.124)–(5.125), one gets the following recurrence relation for the coefficients a_n:

$$a_{n-1} + \frac{1}{2}a_n = \sum_{k=0}^{n}\left\{\sum_{l=0}^{k} l!(k-l)!a_l a_{k-l}\right\}\frac{(n-k)!}{(n+2)!}(-1)^{n-k}a_{n-k}. \tag{5.128}$$

Initial values of a_n can be obtained from Eq. (5.116) as $a_0 = 1$ and $a_1 = -3$. As $n \to \infty$, the nonlinear terms on the right-hand side of this recurrence relation become less important due to the fast decay of the coefficients $(n-k)!/(n+2)!$; thus we find that

$$a_n \to C(-2)^n, \quad n \to \infty, \tag{5.129}$$

where C is a constant. Our calculations give this constant as $C = 8.3571\ldots$. This constant was quoted as $C \approx 8.43$ in Wai et al. (1990) and as $C \approx 8.58$ in Grimshaw (1995) (the data-fitted value 13.24 in the former paper was for $\pi C/2$; see Grimshaw (1995)). Both of their values are inaccurate due to numerical issues.

From the coefficient formula (5.129), we see that the power series solution (5.126) for $F(\kappa)$ converges when $|\kappa| < 1/2$. In addition, this power series can be summed up, and its leading-order term near $\kappa = -1/2$ is

$$F(\kappa) \to \frac{C}{1+2\kappa}, \quad \kappa \to -\frac{1}{2}; \tag{5.130}$$

i.e., $F(\kappa)$ has a simple-pole singularity at $\kappa = -1/2$. Inserting this F expression into (5.118), we find that $\hat{u}(k)$ has a simple-pole singularity at $k = -1/2\epsilon$, and

$$\hat{u}(k) \to \frac{C}{2\epsilon} e^{-\pi/4\epsilon} \frac{1}{k + 1/2\epsilon}, \quad k \to -\frac{1}{2\epsilon}. \tag{5.131}$$

From earlier discussions (see Eq. (5.110)), we learned that the Fourier transform $\hat{u}(k)$ of $u(\theta)$ has a singularity at $k = k_0$. Recalling from (5.99) that $k = -1/2\epsilon$ is the leading-order term of k_0, the formula (5.131) then confirms this pole singularity in $\hat{u}(k)$. We also learned earlier that the residue of the pole singularity in $\hat{u}(k)$ gives $R \sin\sigma/\pi$, where R and σ are the amplitude and phase of cw tails in $u(\theta)$, see (5.106). Then from the residue of pole singularity in (5.131), we get the tail-amplitude formula

$$R = \frac{D}{\sin\sigma}, \qquad \text{where } D = \frac{\pi C}{2\epsilon} e^{-\pi/4\epsilon}. \tag{5.132}$$

Here σ is a free phase parameter. Notice that this tail amplitude R is indeed exponentially small in ϵ, as we have anticipated from earlier discussions. When $\sigma = \pm\pi/2$, the tail amplitude is D, which is minimal.

Now we show some numerical results to confirm these analytical calculations. These symmetric nonlocal waves in Eq. (5.96) can be easily obtained by the shooting method. For $\epsilon = 0.08$, $v = -1$, and $\sigma = \pi/2$, the solution profile is displayed in Fig. 5.6. This nonlocal wave contains a sech-like hump in the middle, and weak cw oscillations in the tails. Next, we quantitatively compare the tail amplitude formula (5.132) with direct numerical results at various values of ϵ. In these comparisons, we will take the oscillation phase $\sigma = \pi/2$; hence the theoretical value for the oscillation amplitude is $D(\epsilon)$. The numerically obtained oscillation amplitude $R(\epsilon, v)$ versus ϵ at two velocity values $v = \pm 1$ are plotted in Fig. 5.7(a). It is seen that in both cases, R decays very quickly as ϵ decreases to zero. In addition, when ϵ is fixed, the oscillation amplitude at $v = 1$ is smaller than that at $v = -1$. This figure also confirms our earlier statement that nonlocal waves in Eq. (5.96) exist for arbitrary values of the velocity v, not just for $v = -1$. To compare these numerical results with the analytical tail-amplitude formula $D(\epsilon)$ in Eq. (5.132), we plot in Fig. 5.7(b) the ratio of $R(\epsilon, v)/(\pi \exp(-\pi/4\epsilon)/2\epsilon)$ versus ϵ at three velocity values of $v = 0$ and ± 1. According to the analytical formula (5.132), this ratio should approach the theoretical limit $C = 8.3571\ldots$ as $\epsilon \to 0^+$ for all values of v (this C number is marked by an asterisk on the vertical axis). Fig. 5.7(b) shows that for all the three v values, this ratio indeed approaches C when $\epsilon \to 0^+$, confirming that the formula (5.132) gives the correct leading-order approximation for the oscillation amplitude $R(\epsilon, v)$.

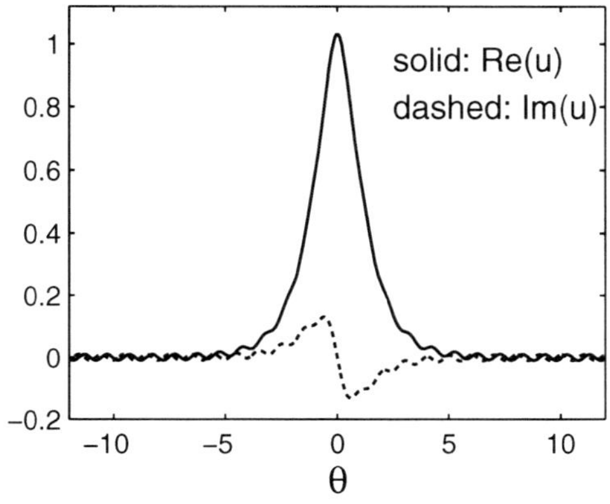

Figure 5.6. *A nonlocal wave in the NLS equation perturbed by third-order dispersion* (5.96), *with* $\epsilon = 0.08$, $v = -1$, *and oscillation phase* $\sigma = \pi/2$.

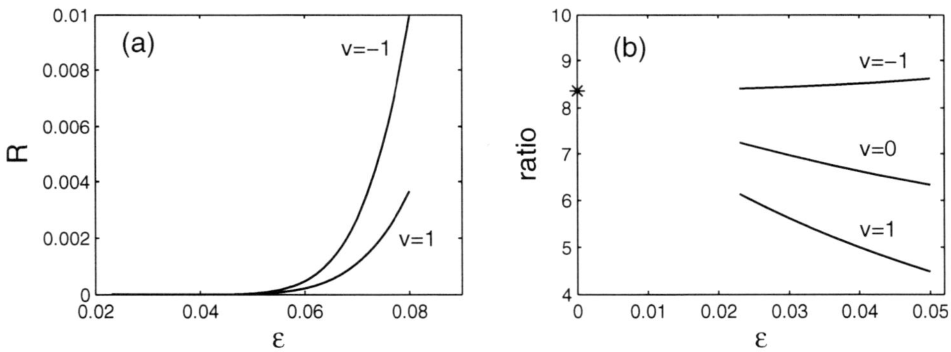

Figure 5.7. *Numerical results on oscillation amplitudes of nonlocal waves in Eq.* (5.96), *with oscillation phase* $\sigma = \pi/2$*:* (a) *oscillation amplitude* R *versus* ϵ *at two velocity values of* $v = \pm 1$*;* (b) *the ratio of* $R/(\pi \exp(-\pi/4\epsilon)/2\epsilon)$ *versus* ϵ *at three velocity values of* $v = 0$ *and* ± 1*; the theoretical limit* $C = 8.3571\ldots$ *as* $\epsilon \to 0^+$ *is also shown as an asterisk on the vertical axis for comparison.*

Radiation Solution

When an NLS soliton emits cw radiation under the effect of third-order dispersion, the cw radiation appears on only one side of the soliton (see Fig. 4.5 in Sec. 4.2.3). The reason is that the linear dispersion relation of Eq. (5.94) for the Fourier mode $e^{i(kx-\omega t)}$ is $\omega = k^2/2 + \epsilon k^3$. Thus, at the resonant wavenumber of the cw radiation $k_0 \approx -1/2\epsilon$, the group velocity is $\omega'(k_0) \approx 1/4\epsilon \gg 1$, meaning that the cw radiation will quickly spread along the positive

x-axis, but not along the negative x-axis. This situation is analogous to the radiation-shelf generation for a soliton in the perturbed complex modified KdV equation, which was analyzed in Sec. 4.4.

To construct a steady-state solution (5.95) in Eq. (5.94) which resembles this radiating solution, we can subtract the Fourier wave $Re^{ik_0\theta-i\sigma}$ from the symmetric nonlocal wave $u(\theta)$ and get a new function

$$u_a(\theta) = u(\theta) - Re^{ik_0\theta-i\sigma}. \tag{5.133}$$

Here R is the amplitude of cw tails in $u(\theta)$. Since R is exponentially small (see (5.132)), it is easy to check that this function $u_a(\theta)$ satisfies the steady-state equation (5.96) up to $O(R)$. Using the large-θ asymptotics (5.106) of $u(\theta)$ as well as the tail-amplitude formula (5.132), we see that the large-θ asymptotics of $u_a(\theta)$ is

$$u_a(\theta) \to \begin{cases} 0, & \theta \to -\infty, \\ 2De^{ik_0\theta+i\pi/2}, & \theta \to +\infty, \end{cases} \tag{5.134}$$

i.e., it decays to zero as $\theta \to -\infty$, and approaches a continuous wave as $\theta \to +\infty$. Thus this solution is asymmetric and resembles the radiating solution. Notice that the phase of its radiation shelf at $\theta \gg 1$ is $\sigma = \pi/2$, which is the phase of minimal tail amplitude in symmetric nonlocal waves, see Eq. (5.132). Meanwhile, the amplitude of this radiation shelf is twice the minimal tail amplitude D of symmetric nonlocal waves.

It must be pointed out that this "steady-state" asymmetric solution cannot be truly steady as it cannot exactly satisfy the steady-state equation (5.96); see Grimshaw (1995). Indeed, multiplying (5.96) by u^* and adding its complex conjugate, we find that

$$J = \epsilon v|u|^2 + \epsilon(|u|^2)_{\theta\theta} - 3\epsilon|u_\theta|^2 + \frac{1}{2}i(u_\theta u^* - u_\theta^* u) = \text{constant}. \tag{5.135}$$

For the solution $u_a(\theta)$ with the asymptotics (5.134), evaluation of J at $\theta \to -\infty$ gives $J = 0$, but evaluation of J at $\theta \to +\infty$ gives $J \approx -4D^2/\epsilon$, an obvious contradiction. What happens is that as $\theta \to +\infty$, the solution $u_a(\theta)$ would also contain an exponentially growing term $\nu e^{ik_-\theta}$ where $\nu = O(D^2/\epsilon^2)$; see Calvo and Akylas (1997). Since this growing term is of higher order than D, it was not accounted for in the leading-order formula (5.133) for $u_a(\theta)$. Due to this growing term, the asymmetric steady-state solution $u_a(\theta)$ is unbounded.

When an initial NLS (i.e., sech) soliton evolves under third-order dispersion perturbation, it can never reach this asymmetric steady-state solution since the sech soliton has a finite energy, while the energy of this asymmetric solution $u_a(\theta)$ is infinite. What will happen is that the sech soliton will continuously emit cw radiation of amplitude $2D$ (see

(5.134)) at the resonant wavenumber $k_0 \approx -1/2\epsilon$, and the front of this cw radiation will travel at the group velocity $\omega'(k_0) \approx 1/4\epsilon$ along the positive x-axis. The phase of this radiation shelf is $\pi/2$, which is the phase of minimal tail amplitude in symmetric stationary waves. Hence the radiation shelf selects its phase such that the radiation emission is minimal. Due to this transfer of energy into the cw radiation, the sech soliton will gradually decay. However, this soliton decay is very slow when $\epsilon \ll 1$. Indeed, the decay rate of the soliton's power (which is proportional to its amplitude) is approximately equal to the cw intensity $4D^2$ multiplying its group velocity $\omega'(k_0)$, which is about $\pi^2 C^2 \exp(-\pi/2\epsilon)/4\epsilon^3$ (Wai et al. (1990), Grimshaw (1995)). When $\epsilon = 0.04$, this decay rate is on the order of 10^{-11}, which is negligible. Thus sech solitons are very robust under the third-order dispersion perturbation.

From the above discussions, we learned that steady single-hump solitary waves cannot exist in the third-order NLS equation (5.94). However, steady double-hump solitary waves do exist in this equation. These double-hump solitary waves can be viewed as piecing together the asymmetric nonlocal wave $u_a(\theta)$ (which decays to zero as $\theta \to -\infty$) with its spatially reversed counterpart $u_a^*(-\theta + \Delta)$ (which decays to zero as $\theta \to +\infty$). Here Δ is the spacing between the two humps. Similarly, steady solitary waves with three or more humps exist in Eq. (5.94) as well. These multihump solitary waves were numerically reported by Akylas and Kung (1990) and Klauder et al. (1993), and then analytically constructed by Calvo and Akylas (1997). In their results, these solitary waves were found as isolated solutions existing at discrete values of the third-order dispersion coefficient ϵ. It was realized by Yang and Akylas (2003) that these solutions actually exist as continuous families for continuous ranges of ϵ. In addition, they exist inside the continuous spectrum of linear radiation modes in Eq. (5.94) and thus are examples of embedded solitons. Embedded solitons have many novel properties and will be discussed in the next section.

5.6 Embedded Solitons and Their Dynamics

From the previous section, it was seen that when the frequency of a "solitary" wave lies inside the continuous spectrum of linear radiation modes in a wave system, this nonlinear wave would be in resonance with the linear radiation modes. This resonance generically excites cw tails, and thus this wave would continuously leak energy. However, in some special situations, even though a solitary wave lies inside the continuous spectrum, it does not excite radiation modes; hence this solitary wave is nonradiating and can remain truly localized and stationary. Such solitary waves which reside inside the continuous spectrum of linear radiation modes are called embedded solitons (Yang et al. (1999)). These embedded solitons are in sharp contrast with traditional solitary waves which reside *outside* the continuous spectrum and can be termed gap solitons. Existence of embedded solitons has been reported in a wide variety of physical systems such as the fifth-order KdV equations (Kichenassamy and Olver (1992), Calvo and Akylas (1997), Champneys and Groves (1997), Yang (2001a), Tan et al. (2002)), the fourth-order NLS equations (Buryak (1995), Fujioka and Espinosa (1997)), the coupled KdV equations (Grimshaw and Cook (1996)), the second harmonic generation system (Yang et al. (1999), Pelinovsky and Yang (2002)), the massive Thirring model (Champneys et al. (1998), Champneys and Malomed (1999a)), the three-wave interaction system (Champneys and Malomed (1999b)), the third-order NLS equations (Akylas and Kung (1990), Klauder et al. (1993), Calvo and Akylas (1997), Gromov and Tyutin (1998), Gromov et al. (1999), Yang and Akylas (2003), Yang (2003), Elbert (2003), Pelinovsky and Yang (2005)), the coupled Bragg grating system (Mak et al. (2004)), and the discrete fourth-order NLS equation (Gonzalez-Perez-Sandia et al. (2004)); see also Champneys et al. (2001), Kivshar and Agrawal (2003) and the references therein. In addition, moving discrete breathers have been linked to embedded solitons (Pelinovsky and Rothos (2005)). Experimental observation of embedded solitons in a 2D photonic lattice system has also been reported (Wang et al. (2007)). Of these embedded solitons, some are isolated; i.e., they exist at discrete frequency values. Some others, on the other hand, exist as continuous families for a continuous range of frequency values (Gromov and Tyutin (1998), Yang and Akylas (2003), Yang (2003), Pelinovsky and Yang (2005)).

Due to the nature of embedded solitons residing inside the continuous spectrum of linear radiation modes, these solitons exhibit a number of novel dynamical properties which differ significantly from those of traditional gap solitons. In some systems, isolated and nonisolated embedded solitons are found to be semistable (Yang et al. (1999), Yang (2001a), Pelinovsky and Yang (2002), Tan et al. (2002), Yang and Akylas (2003)). That is, if the initial perturbed state has higher energy than the embedded soliton, then the solution can shed excess energy through cw radiation and relax back to this embedded soliton; but if the initial perturbed state has less energy than the embedded soliton, the solution would lose more and more cw radiation and gradually decay away. This semistability phenomenon would not happen for gap solitons. But in some other systems, isolated and nonisolated embedded solitons can be fully stable regardless of the initial perturbations (Yang (2001a), Yang (2003), Pelinovsky and Yang (2005)).

In this section, we discuss embedded solitons and their dynamics.

5.6.1 Isolated Embedded Solitons and Their Semistability

In some wave systems, embedded solitons exist at isolated frequency values. One physical model which gives rise to such isolated embedded solitons is the second harmonic generation (SHG) system with opposing dispersions and competing nonlinearities at fundamental and second harmonics. This model is

$$iU_t + U_{xx} + U^*V + \gamma_1(|U|^2 + 2|V|^2)U = 0, \tag{5.136}$$

$$iV_t - bV_{xx} + qV + \frac{1}{2}U^2 + \gamma_2(|V|^2 + 2|U|^2)V = 0, \tag{5.137}$$

where U and V are wave envelopes of the fundamental and second harmonic fields, $-b$ is the relative dispersion of the second harmonic, q is the frequency mismatch, and $\gamma_{1,2}$ are the Kerr coefficients (all variables have been nondimensionalized); see Agrawal (2007). This system is Hamiltonian if $\gamma_1 = \gamma_2$, and non-Hamiltonian otherwise. In our analysis below, γ_1 and γ_2 will be treated as arbitrary, and thus the analysis will hold in both Hamiltonian and non-Hamiltonian situations. In this general case, Eqs. (5.136)–(5.137) possess a conserved quantity

$$P = \int_{-\infty}^{\infty} (|U|^2 + 2|V|^2)dx, \tag{5.138}$$

which can be called the total power of this SHG system. This conserved quantity will play an important role in the following analysis.

Stationary solutions to the above SHG model are sought in the form

$$U(x,t) = u(x)e^{i\mu t}, \quad V(x,t) = v(x)e^{2i\mu t}, \tag{5.139}$$

where $u(x)$ and $v(x)$ are real functions which satisfy the following equations:

$$u_{xx} - \mu u + uv + \gamma_1(u^2 + v^2)u = 0, \tag{5.140}$$

$$-bv_{xx} + (q - 2\mu)v + \frac{1}{2}u^2 + \gamma_2(v^2 + u^2)v = 0, \tag{5.141}$$

and μ is a frequency constant.

Below, we consider only the case $b > 0$ and $\gamma_{1,2} < 0$. The continuous spectrum of radiation modes in this SHG system is the set of frequency values μ where linear waves $U \propto e^{ipx+i\mu t}$ or $V \propto e^{ipx+2i\mu t}$ with real values of p exist (the nonlinear terms in (5.136)–(5.137) are dropped in this consideration). In the present case, this continuous spectrum can be readily found to be

$$\mu \in (-\infty, 0] \cup [q/2, +\infty). \tag{5.142}$$

If μ lies in this continuous spectrum, the stationary solution u or v above generically cannot be localized. For instance, when $\mu > q/2 > 0$ and (u, v) are small at large distances, then symmetric stationary solutions would behave as

$$u(x) \to \alpha e^{-\sqrt{\mu}|x|}, \quad |x| \to \infty, \tag{5.143}$$

$$v(x) \to r \sin(k|x| + \sigma), \quad |x| \to \infty, \tag{5.144}$$

where $k=\sqrt{(2\mu-q)/b}$, and α, r, σ are constants. One can see that the v-component is nonlocal and contains a cw tail which extends to infinity. The amplitude of this tail is r, and its phase σ. The presence of this cw tail is due to the frequency μ of this stationary wave lying inside the continuous spectrum and thus in resonance with the linear radiation mode of wavenumber k given above. For this symmetric stationary solution (u, v),

$$u_x(0)=v_x(0)=0. \tag{5.145}$$

Since there are three parameters α, r, σ in the tails (5.143)–(5.144) which can be adjusted to meet the above symmetry condition, one of these three parameters is thus free, and this free parameter will be taken as the phase σ. For each given σ, one can use a simple shooting method to determine this symmetric stationary solution. Hence a continuous family of such solutions, parameterized by the phase σ of the v-component's cw tail, would be obtained. This situation is quite analogous to that in the third-order NLS equation of the previous section. To demonstrate these nonlocal stationary solutions, we take $b=q=1$, $\gamma_1=\gamma_2=-0.05$, and $\sigma=0$. The corresponding nonlocal solution at $\mu=0.9$ is displayed in Fig. 5.8(a). As μ changes inside the continuous spectrum $\mu\geq q/2=0.5$, the tail amplitude r will also vary. The dependence of r on μ is displayed in Fig. 5.8(b). As expected, this tail amplitude is nonzero for generic values of the frequency μ, hence the stationary solution is generically nonlocal.

However, Fig. 5.8(b) also shows that at the isolated frequency value $\mu_c=0.8114$ in the continuous spectrum $\mu\geq q/2=0.5$, the cw-tail amplitude r vanishes, and hence the corresponding solution (u_c, v_c) becomes localized! This localized solution is displayed in Fig. 5.8(c), and it is an isolated embedded soliton. It is isolated because there are no other embedded solitons existing at nearby frequency values. The existence of this embedded soliton is quite unusual and generally not anticipated.

Tail behaviors of nonlocal waves near the embedded soliton will prove important for our later analysis, so we describe them first. When $|\mu-\mu_c|\ll 1$, the tail amplitude r of these nonlocal waves is small, and so it can be expanded as

$$r(\mu,\sigma)=R(\sigma)(\mu-\mu_c)+O[(\mu-\mu_c)^2], \tag{5.146}$$

where $R(\sigma)$ is the slope of tail amplitude r at the embedded-soliton frequency μ_c and phase σ. Dependence of R on the phase σ can be explicitly determined, and it will be done later.

Existence of this embedded soliton inside the continuous spectrum is interesting enough. More interesting is how it evolves under perturbations. This evolution can be heuristically elucidated as follows. We have known that this embedded soliton is isolated. When it is perturbed, it will emit cw radiation due to resonance with the radiation modes, and its frequency will also shift due to this energy loss. In this process, the power P in

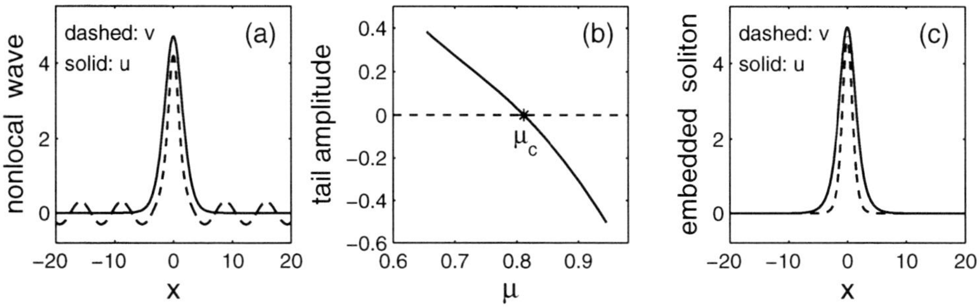

Figure 5.8. (a) *A nonlocal wave in the SHG model* (5.140)–(5.141) *at* $\mu = 0.9$ *and tail phase* $\sigma = 0$ *(other parameters are given in the text);* (b) *dependence of the tail amplitude r versus* μ *at* $\sigma = 0$*;* (c) *an embedded soliton at* $\mu_c = 0.8114$.

(5.138) is always conserved. Then, if the perturbed soliton initially has less power than the embedded soliton, because of power conservation, this perturbed soliton will never be able to relax back to the embedded soliton after radiation emission. Hence the solution will deviate further from the embedded soliton after the radiation loss, and its frequency will shift away from μ_c. In this process, the radiation shelf will intensify, and hence the perturbed soliton will decay away. On the other hand, if the perturbed soliton initially has higher power than the embedded soliton, the solution can shed excess power through cw radiation and approach the embedded soliton, and its frequency can move back toward μ_c. In this process, the radiation shelf will subside and gradually disappear. This is the novel semistability property of this isolated embedded soliton which has no counterpart in traditional gap solitons.

Below, we rigorously prove this semistability property by asymptotic methods. When this embedded soliton is perturbed, its solution can be expanded into the following perturbation series:

$$U(x,t) = \Big[u_c(x) + \epsilon\mu_1(T)u_1(x,t) + \epsilon^2 u_2(x,t) + O(\epsilon^3)\Big]\exp\left(i\int_0^t \mu\, d\tau\right), \tag{5.147}$$

$$V(x,t) = \Big[v_c(x) + \epsilon\mu_1(T)v_1(x,t) + \epsilon^2 v_2(x,t) + O(\epsilon^3)\Big]\exp\left(2i\int_0^t \mu\, d\tau\right), \tag{5.148}$$

$$\mu(T) = \mu_c + \epsilon\mu_1(T) + O(\epsilon^2), \tag{5.149}$$

where ϵ is a small parameter measuring the strength of the perturbation, and $T = \epsilon t$ is a slow time variable. When this expansion is substituted into Eqs. (5.136)–(5.137), we find that the first-order solution (u_1, v_1) satisfies the following equation:

$$(i\partial_t + L)\Phi_1 = F_0, \tag{5.150}$$

where

$$\Phi_1 = [u_1, v_1, u_1^*, v_1^*]^T, \quad F_0 = [u_c, 2v_c, -u_c, -2v_c]^T, \tag{5.151}$$

the superscript "T" represents vector transpose, and L is a linearization operator whose expression can be readily obtained. The operator L is eighth order. Its kernel contains two discrete (localized) eigenfunctions

$$\Psi_D^{(s)}(x) = F_0(x) \tag{5.152}$$

and

$$\Psi_D^{(a)}(x) = [u_c'(x), v_c'(x), u_c'(x), v_c'(x)]^T, \tag{5.153}$$

which are the Goldstein modes induced by the phase and translation invariances of the embedded soliton. Of these two discrete eigenfunctions, $\Psi_D^{(s)}(x)$ is symmetric, and $\Psi_D^{(a)}(x)$ antisymmetric. Their symmetries are marked by the superscripts "(s)" and "(a)" respectively. The kernel of L also contains four nonlocal but bounded real-valued continuous eigenfunctions of the form

$$\Psi_\pm^{(s)}(x) = \begin{bmatrix} \psi_\pm^{(s)}(x) \\ \phi_\pm^{(s)}(x) \\ \pm\psi_\pm^{(s)}(x) \\ \pm\phi_\pm^{(s)}(x) \end{bmatrix}, \quad \Psi_\pm^{(a)}(x) = \begin{bmatrix} \psi_\pm^{(a)}(x) \\ \phi_\pm^{(a)}(x) \\ \pm\psi_\pm^{(a)}(x) \\ \pm\phi_\pm^{(a)}(x) \end{bmatrix}, \tag{5.154}$$

where their symmetries are marked by the superscripts, and their normalized boundary conditions at large distances $|x| \gg 1$ are

$$\Psi_\pm^{(s)}(x) \to \begin{bmatrix} 0 \\ 1 \\ 0 \\ \pm 1 \end{bmatrix} \sin\left(k_c|x| + \sigma_\pm^{(s)}\right), \quad \Psi_\pm^{(a)}(x) \to \begin{bmatrix} 0 \\ 1 \\ 0 \\ \pm 1 \end{bmatrix} \sin\left(k_c x + \operatorname{sgn}(x)\sigma_\pm^{(a)}\right). \tag{5.155}$$

Here $k_c = \sqrt{(2\mu_c - q)/b}$, and $\sigma_\pm^{(s)}$, $\sigma_\pm^{(a)}$ are phase constants. For the embedded soliton shown in Fig. 5.8(c), we find numerically that

$$\sigma_+^{(s)} = 1.3270, \quad \sigma_-^{(s)} = 1.4007. \tag{5.156}$$

The values of $\sigma_\pm^{(a)}$ can also be obtained (not shown). The remaining two functions in the kernel of L are unbounded and thus not relevant to our analysis.

In addition to the above eigenfunctions, L also admits the following simple generalized eigenfunctions:

$$\Psi_G^{(s)}(x;\sigma) = \left[\frac{\partial u(x;\mu,\sigma)}{\partial\mu}, \frac{\partial v(x;\mu,\sigma)}{\partial\mu}, \frac{\partial u(x;\mu,\sigma)}{\partial\mu}, \frac{\partial u(x;\mu,\sigma)}{\partial\mu}\right]^T_{\mu=\mu_c}, \tag{5.157}$$

where

$$L\Psi_G^{(s)}(x;\sigma) = \Psi_D^{(s)}(x), \tag{5.158}$$

and σ is a free phase parameter. These generalized eigenfunctions are nonlocal and are induced by the free parameter μ in the two-parameter family $[u(x;\mu,\sigma), v(x;\mu,\sigma)]$ of nonlocal waves. Specifically, the generalized eigenrelation (5.158) can be derived by simply differentiating the stationary equations (5.140)–(5.141) with respect to μ. In view of the tail behaviors of stationary solutions (u,v) in Eqs. (5.144) and (5.146), we see that these generalized eigenfunctions have the following tail behaviors:

$$\Psi_G^{(s)}(x;\sigma) \to R(\sigma)\sin(k_c|x|+\sigma)[0,1,0,1]^T, \quad |x|\to\infty. \tag{5.159}$$

Since the SHG system (5.136)–(5.137) is not Galilean invariant (for $b \neq -1/2$), no more generalized eigenfunctions are expected for operator L.

It is noted that even though the number of these generalized eigenfunctions $\Psi_G^{(s)}(x;\sigma)$ appears to be infinite due to the free parameter σ, these functions are all linearly related with each other, and this fact will readily yield the explicit expression for the function $R(\sigma)$. To see this, we notice from Eq. (5.158) that for any two phase values σ and σ_0, the difference between their generalized eigenfunctions, $\Psi_G^{(s)}(x;\sigma) - \Psi_G^{(s)}(x;\sigma_0)$, lies in the kernel of operator L. By comparing the symmetry and boundary conditions of this difference function with the eigenfunctions in the kernel of L, it is clear that this difference function matches only $\Psi_+^{(s)}(x)$, and thus

$$\Psi_G^{(s)}(x;\sigma) - \Psi_G^{(s)}(x;\sigma_0) \propto \Psi_+^{(s)}(x). \tag{5.160}$$

Then by taking the boundary condition of this relation, we get

$$R(\sigma)\sin(k_c|x|+\sigma) - R(\sigma_0)\sin(k_c|x|+\sigma_0) \propto \sin(k_c|x|+\sigma_+^{(s)}). \tag{5.161}$$

By expanding each term into sine and cosine functions of the angle $k_c|x|+\sigma_+^{(s)}$ and equating coefficients of the term $\cos(k_c|x|+\sigma_+^{(s)})$, the formula for $R(\sigma)$ would be obtained. For convenience, let us take $\sigma_0 = \sigma_+^{(s)} + \frac{1}{2}\pi$. Then the formula for $R(\sigma)$ is

$$R(\sigma) = \frac{R(\sigma_+^{(s)} + \frac{1}{2}\pi)}{\sin(\sigma - \sigma_+^{(s)})}. \tag{5.162}$$

This formula shows that the minimal tail amplitude in stationary nonlocal waves v near the embedded soliton is

$$R_{min} = \left| R\left(\sigma_+^{(s)} + \frac{1}{2}\pi\right)\right|, \tag{5.163}$$

which occurs at two phase values

$$\sigma_{min} = \sigma_+^{(s)} \pm \frac{1}{2}\pi. \tag{5.164}$$

Notice that the above formula (5.162) closely resembles the tail-amplitude formula in Eq. (5.132) for nonlocal waves in the third-order NLS equation (see the previous section).

Now we solve the first-order equation (5.150). The most important feature of the first-order solution is that its v_1-component develops cw shelves with wavenumbers $\pm k_c$, which propagate to both $x = -\infty$ and $x = \infty$. The reason is that the kernel of L contains continuous eigenfunctions with oscillating tails $e^{\pm ik_c x}$ in their second and fourth components (see above). Thus, under localized forcing F_0, these oscillating tails will be excited. A completely analogous situation arises in the first-order equation of the soliton perturbation theory for the complex modified KdV equation, which was thoroughly analyzed in Sec. 4.4. This cw shelf generation is a natural consequence of a resonance between the embedded soliton and linear radiation modes. After excited, these cw shelves will propagate to $x = \pm\infty$ at their group velocities. The linear dispersion relation of the V equation (5.137), for Fourier modes $e^{i(px-\omega t)}$, is $\omega(p) = -bp^2 - q$. Since the wavenumbers of these cw shelves are $\pm k_c$, their group velocities are then $\omega'(\pm k_c) = \mp 2bk_c$. According to the Sommerfeld radiation condition, which says that these cw shelves must propagate from the source (F_0) region out to infinity, it is seen that in the $x \gg 1$ region (and when $t \gg 1$), the V component can contain only the $e^{-ik_c x}$ component whose group velocity is positive. Similarly, in the $x \ll -1$ region, the V-component can contain only the $e^{ik_c x}$-component whose group velocity is negative. Consequently, the first-order solution Φ_1 for $|x|, t \gg 1$ is

$$\Phi_1(x,t) \to \begin{pmatrix} 0 \\ h\, e^{-ik_c|x| + i\sigma_{rad}} \\ 0 \\ h\, e^{ik_c|x| - i\sigma_{rad}} \end{pmatrix}, \quad 0 \ll |x| \ll c_g t, \tag{5.165}$$

where $c_g = |\omega'(\pm k_c)| = 2bk_c$ is the magnitude of the two group velocities, σ_{rad} is the phase of the radiation shelf, and h is its amplitude. Here the shelf amplitudes at $x = \pm\infty$ are the same because the forcing term F_0 is symmetric in x. Note that an analogous radiation shelf arising in the soliton perturbation theory of the complex modified KdV equation was derived rigorously in Sec. 4.4.

The amplitude h and phase σ_{rad} of this radiation shelf are important for the quantitative analysis of the perturbed embedded soliton. For the present problem, these values turn out to be very easy to obtain, and one does not need to resort to adjoint continuous eigenfunctions of operator L (as one had to do in the complex modified KdV perturbation theory in Sec. 4.4). The key reason for this is that the first-order equation (5.150) here admits a simple inhomogeneous solution $\Psi_G^{(s)}(x;\sigma)$ in view of Eq. (5.158). Thus, by combining this inhomogeneous solution with homogeneous solutions in the kernel of L to match the Sommerfeld radiation condition (5.165), the values of h and σ_{rad} would be obtained. To proceed, we consider the $t \to \infty$ limit, where the first-order solution $\Phi_1(x,t)$ becomes stationary. In this case, the time derivative in Eq. (5.150) can be dropped. Since $\Psi_G^{(s)}(x;\sigma)$ is an inhomogeneous solution, the general stationary solution $\Phi_1(x)$ is then $\Psi_G^{(s)}(x;\sigma)$ plus the eigenfunctions in the kernel of L. Since the forcing term F_0 is symmetric, the radiation shelf in this stationary solution $\Phi_1(x)$ should be symmetric as well. Thus only symmetric continuous eigenfunctions in the kernel of L can appear in $\Phi_1(x)$. As a result, $\Phi_1(x)$ has the following expression:

$$\Phi_1(x) = \Psi_G^{(s)}(x;\sigma_0) + c_1\Psi_+^{(s)}(x) + ic_2\Psi_-^{(s)}(x), \tag{5.166}$$

where σ_0 is an arbitrary phase value, and c_1, c_2 are complex constants. Here the discrete eigenfunction $\Psi_D^{(s)}(x)$ in L's kernel is not included since this term can be removed by an $O(\epsilon)$ phase shift to the perturbative solution (5.147)–(5.148), and the discrete eigenfunction $\Psi_D^{(a)}(x)$ is not included since this term can be removed by an $O(\epsilon)$ position shift to the perturbative solution (5.147)–(5.148). For symmetric and real-valued perturbations, these phase and position shifts would vanish. Using the boundary conditions in (5.155) and (5.159), we find that when $|x| \to \infty$,

$$\Phi_1(x) \to \begin{pmatrix} 0 \\ R(\sigma_0)\sin(k_c|x|+\sigma_0) + c_1\sin(k_c|x|+\sigma_+^{(s)}) + ic_2\sin(k_c|x|+\sigma_-^{(s)}) \\ 0 \\ R(\sigma_0)\sin(k_c|x|+\sigma_0) + c_1\sin(k_c|x|+\sigma_+^{(s)}) - ic_2\sin(k_c|x|+\sigma_-^{(s)}) \end{pmatrix}. \tag{5.167}$$

This boundary condition must match the Sommerfeld radiation condition in Eq. (5.165). Hence the second and fourth components in the above boundary condition must be complex conjugate to each other, which yields that both c_1 and c_2 must be real. In addition, we must have

$$R(\sigma_0)\sin(k_c|x|+\sigma_0) + c_1\sin(k_c|x|+\sigma_+^{(s)}) + ic_2\sin(k_c|x|+\sigma_-^{(s)}) = h\,e^{-ik_c|x|+i\sigma_{rad}}. \tag{5.168}$$

The imaginary part of this relation is

$$c_2 \sin(k_c|x| + \sigma_-^{(s)}) = -h \sin(k_c|x| - \sigma_{rad}), \tag{5.169}$$

which yields

$$\sigma_{rad} = -\sigma_-^{(s)}, \quad c_2 = -h. \tag{5.170}$$

The real part of the relation (5.168) is

$$R(\sigma_0)\sin(k_c|x| + \sigma_0) + c_1 \sin(k_c|x| + \sigma_+^{(s)}) = h\cos(k_c|x| - \sigma_{rad}). \tag{5.171}$$

By expanding each term into sine and cosine functions of the angle $k_c|x| + \sigma_+^{(s)}$ and equating the coefficients of these functions, both h and c_1 would be obtained. For convenience, we take $\sigma_0 = \sigma_+^{(s)} + \frac{1}{2}\pi$. Then we find that

$$h = \frac{R(\sigma_+^{(s)} + \frac{1}{2}\pi)}{\cos(\sigma_-^{(s)} - \sigma_+^{(s)})}. \tag{5.172}$$

In generic cases, $\sigma_-^{(s)} - \sigma_+^{(s)} \neq 0$ or π, see (5.156). Then in view of Eq. (5.163), we see that $|h| > R_{min}$. In other words, when the perturbed embedded soliton in the SHG system (5.136)–(5.137) emits cw radiation, this radiation does not select the minimal tail amplitude of symmetric nonlocal waves to radiate. This contrasts the cw radiation in the NLS soliton under third-order dispersion perturbation, where the radiation amplitude was minimal of symmetric nonlocal waves (see previous section).

Emission of the radiation shelf in the first-order solution Φ_1 leaks energy from the perturbed embedded soliton and causes its frequency to shift. To determine how this perturbed soliton evolves under this radiation emission, several methods can be used (Yang (2001a), Pelinovsky and Yang (2002), Tan et al. (2002), Yang (2003)). The simplest method is to use conservation laws, which we will employ below (this method was also used in Sec. 4.4 in the treatment of a similar problem).

First, we substitute the perturbative solution (5.147)–(5.148) into the power conservation law (5.138). To $O(\epsilon^2)$, we get

$$\frac{d}{dt}\int_{-\infty}^{\infty}\Big[u_c^2 + \epsilon\mu_1 u_c(u_1 + u_1^*) + \epsilon^2\mu_1^2|u_1|^2 + \epsilon^2 u_c(u_2 + u_2^*) \\ + 2v_c^2 + 2\epsilon\mu_1 v_c(v_1 + v_1^*) + 2\epsilon^2\mu_1^2|v_1|^2 + 2\epsilon^2 v_c(v_2 + v_2^*)\Big]dx = 0. \tag{5.173}$$

When $t \gg 1$, the first-order solution (u_1, v_1) approaches a steady state (5.166) in the region $|x| \ll c_g t$. The second-order solution (u_2, v_2) approaches a steady state in this region as well. In the region $|x| \sim c_g t$, the v_1 solution contains a radiation shelf of height h which travels to $x = \pm\infty$ at the group velocity c_g. Using these facts, the above conservation law gives the dynamical equation for $\mu_1(T)$ as

$$\frac{d\mu_1}{dT}\int_{-\infty}^{\infty}\left[u_c(u_1+u_1^*)+2v_c(v_1+v_1^*)\right]dx = -4h^2 c_g \mu_1^2. \tag{5.174}$$

To calculate the integral on its left side, we use the steady-state solution (5.166). Taking $\sigma_0 = \frac{1}{2}\pi - \sigma_{rad}$ in this steady state and using the matching condition (5.171), we find that $c_1 = 0$; thus this steady state is

$$\Phi_1(x) = \Psi_G^{(s)}\left(x; \frac{1}{2}\pi - \sigma_{rad}\right) + ic_2\Psi_-^{(s)}(x). \tag{5.175}$$

Inserting this steady state into the above integral, and replacing μ_1 in (5.174) in favor of μ by Eq. (5.149), we finally obtain the dynamical equation for the frequency μ of the perturbed embedded soliton as

$$\frac{d\mu}{dt} = -\Gamma(\mu-\mu_c)^2, \tag{5.176}$$

where Γ is a constant whose expression is

$$\Gamma = \frac{2h^2 c_g}{\int_{-\infty}^{\infty}\left[u_c(x)\frac{\partial u(x;\mu,\sigma)}{\partial\mu} + 2v_c(x)\frac{\partial v(x;\mu,\sigma)}{\partial\mu}\right]_{\mu=\mu_c,\sigma=\frac{1}{2}\pi-\sigma_{rad}} dx}. \tag{5.177}$$

This dynamical equation is the normal form for evolution of the perturbed embedded soliton in the SHG system (5.136)–(5.137). The solution to this dynamical equation can be written out explicitly as

$$\mu(t) = \mu_c + \frac{\mu_0-\mu_c}{1+(\mu_0-\mu_c)\Gamma t}, \tag{5.178}$$

where $\mu_0 = \mu|_{t=0}$ is the perturbed soliton's initial frequency. Equation (5.176) and its solution (5.178) show that if $(\mu_0-\mu_c)\Gamma > 0$, then $\mu(t) \to \mu_c$, and the perturbed solution will approach the embedded soliton. In this process, the radiation shelf, whose height is proportional to $\mu-\mu_c$, will gradually disappear. However, if $(\mu_0-\mu_c)\Gamma < 0$, then $\mu(t)$ moves away from μ_c, and hence the perturbed solution will deviate apart from the embedded soliton. In this process, the radiation shelf will gradually intensify. This is the semistability property which was established heuristically earlier in this section.

Regarding the initial frequency value μ_0, it can be determined from the initial conditions $U(x,0)$ and $V(x,0)$, as will be briefly outlined below. Notice that the time-dependent solution $\Phi_1(x,t)$ of the first-order equation (5.150) is equal to the inhomogeneous solution $\Psi_G^{(s)}(x; \frac{1}{2}\pi - \sigma_{rad})$ plus the homogeneous solution (see Eq. (5.175)), and the homogeneous solution can be expanded into eigenfunctions of operator L. By setting $t = 0$ in this solution, a formula for $\Phi_1(x,0)$ would be obtained. Recall that $\Phi_1(x,0)$ can be obtained alternatively from the initial conditions and the expansions (5.136)–(5.137) as

$$\Phi_1(x,0) = \frac{1}{\mu_0 - \mu_c}\widehat{\Phi}_0(x), \tag{5.179}$$

where

$$\widehat{\Phi}_0(x) = \left[U(x,0) - u_c(x), V(x,0) - v_c(x), U^*(x,0) - u_c(x), V^*(x,0) - v_c(x)\right]^T.$$

To proceed further, we will use the symmetric discrete eigenfunction of L's adjoint operator L^A at the zero eigenvalue, which we denote as $\Psi_D^A(x)$. Here adjoint operators are defined under the inner product

$$\langle \mathbf{f}, \mathbf{g} \rangle = \int_{-\infty}^{\infty} \mathbf{f}^T(x)\,\mathbf{g}(x)\,dx. \tag{5.180}$$

If the SHG system (5.136)–(5.137) is Hamiltonian, i.e., $\gamma_1 = \gamma_2$, then $(JL)^A = JL$, where $J = \text{diag}(1,1,-1,-1)$, and hence $\Psi_D^A(x) = J\Psi_D^{(s)}(x)$. If the system is non-Hamiltonian, the structure of $\Psi_D^A(x)$ would still be the same as $J\Psi_D^{(s)}(x)$; i.e., $\Psi_D^A(x)$ is still real valued, its first and third components are equal to each other, and its second and fourth components are equal to each other. Now we take the inner product between $\Psi_D^A(x)$ and the formula of $\Phi_1(x,0)$. By utilizing the expression (5.179) as well as the orthogonality relations between $\Psi_D^A(x)$ and L's eigenfunctions, we get the formula for the initial frequency μ_0 as

$$\mu_0 = \mu_c + \frac{\langle \Psi_D^A(x), \widehat{\Phi}_0(x)\rangle}{\langle \Psi_D^A(x), \Psi_G^{(s)}(x;\pi/2-\sigma_{rad})\rangle}. \tag{5.181}$$

A similar formula for the initial velocity value in the perturbed embedded soliton in the general fifth-order KdV equation can be found in Tan et al. (2002). This μ_0 formula shows that, for two initial perturbations with opposite signs, i.e., $\widehat{\Phi}_0^{(1)}(x) = -\widehat{\Phi}_0^{(2)}(x)$, the corresponding values $\mu_0^{(1)} - \mu_c$ and $\mu_0^{(2)} - \mu_c$ would also be opposite. Thus the perturbed soliton would approach the embedded soliton in one case and decay in the other, indicating semistability. This completes the theoretical analysis on the evolution of the embedded soliton under small perturbations in the SHG system (5.136)–(5.137).

Now we numerically confirm the above theoretical analysis and semistability results. As before, we take $b = q = 1$ and $\gamma_1 = \gamma_2 = -0.05$ in the SHG model (5.136)–(5.137), where an embedded soliton exists at $\mu_c = 0.8114$ and is displayed in Fig. 5.8(c). For this embedded soliton, we numerically find from the formula (5.177) that $\Gamma = -0.586$. This embedded soliton is initially perturbed as

$$U(x,0) = u_c(x) + a_1 \text{sech} 2x, \quad V(x,0) = v_c(x) + a_2 \text{sech} 2x, \tag{5.182}$$

where a_1 and a_2 are small constants. When $a_1 = a_2 = 0.1$, numerical evolution of this perturbed soliton is displayed in Fig. 5.9(a, b). It is seen that the perturbed solution approaches the embedded soliton, and its frequency $\mu(t)$ approaches the embedded-soliton

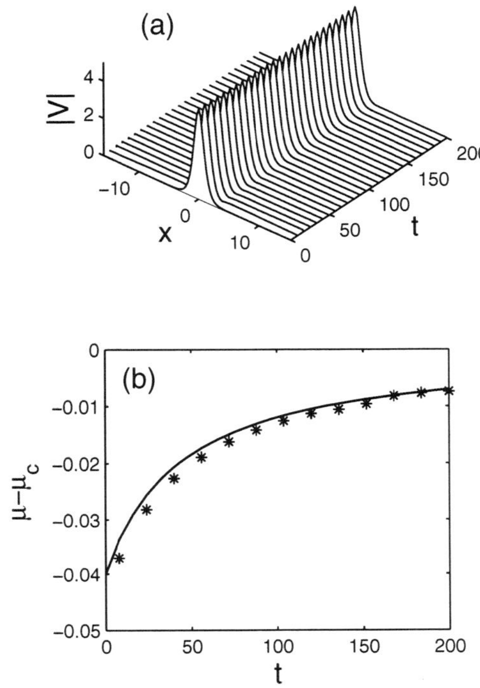

Figure 5.9. (a, b) *Stable evolution of a perturbed embedded soliton in the SHG system* (5.140)–(5.141)*, with the initial condition* (5.182) *and* $a_1 = a_2 = 0.1$*;* (c, d) *unstable evolution of a perturbed embedded soliton in the SHG system with the initial condition* (5.182) *and* $a_1 = a_2 = -0.1$. (a, c) *show evolutions of* $|V(x,t)|$, *and* (b, d) *show evolutions of the frequency deviation* $\mu(t) - \mu_c$*; stars are numerical values, and solid curves are the analytical formulae* (5.178).

frequency μ_c. If $a_1 = a_2 = -0.1$, numerical evolution of the perturbed soliton is displayed in Fig. 5.9(c, d). In this case, the perturbed solution deviates apart from the embedded soliton, and its frequency $\mu(t)$ moves away from μ_c. These two evolution simulations confirm the semistability results of the above analysis. To make quantitative comparisons, we first use the formula (5.181) to calculate μ_0 values for these two initial perturbations and find that $\mu_0 - \mu_c \approx -0.04$ for the former case and $\mu_0 - \mu_c \approx 0.04$ for the latter case. Inserting these μ_0 values and the above Γ value into the analytical frequency formula (5.178), we obtain the analytical frequencies for the two cases, plotted (as solid curves) in Fig. 5.9(b, d), alongside numerical values (as stars). Very good quantitative agreements can be observed for both cases.

In the above analysis, semistability of isolated embedded solitons was established for the SHG model (5.136)–(5.137). This analysis can be extended to other systems as well (see Yang (2001a), Tan et al. (2002) for instance). Generally speaking, if a wave system admits a nontrivial conserved quantity, then an isolated embedded soliton in such a system would be semistable under perturbations (regardless whether this system is Hamiltonian or not). In addition, the normal form for such perturbed embedded solitons would resemble Eq. (5.176).

5.6.2 Continuous Families of Embedded Solitons

For an embedded soliton with its frequency lying inside the system's continuous spectrum, the vanishing of resonant cw tails imposes a constraint on the soliton's frequency, which explains why the embedded soliton in the SHG system of the previous subsection was isolated (i.e., existing only at discrete frequency values). However, in some wave systems, nonlocal waves can have additional internal parameters (such as velocity) in addition to the frequency. Then the vanishing of resonant cw tails only imposes a relation between the frequency and the velocity. In such cases, embedded solitons could exist on continuous curves in the frequency-velocity parameter space. In other words, these embedded solitons would exist as continuous families. For such embedded solitons, the heuristic argument for semistability of isolated embedded solitons in the previous subsection would not apply. Indeed, when such a soliton is perturbed (regardless of the type of perturbations), it may, in principle, just shed some energy radiation and relax onto a nearby embedded soliton in the same solution family. If this happens, the embedded soliton would be nonlinearly stable. However, this stable evolution may not be realized, as the embedded soliton, when perturbed, may also deviate away from its solution family, resulting in nonlinear semistability or instability. As we will see below, all of these evolution scenarios are possible depending on the underlying wave equations, and thus stability of families of embedded solitons requires fresh and careful treatment.

One physical example where continuous families of embedded solitons exist is the third-order NLS equation

$$iU_t - iU_{xxx} + |U|^2U = 0, \tag{5.183}$$

which governs propagation of picosecond optical pulses at zero second-order dispersion point, where the third-order dispersion becomes dominant (see Sec. 1.3). Linear dispersion relation of this equation for Fourier modes $e^{i(kx+\mu t)}$ is $\mu = -k^3$. Thus the continuous spectrum of radiation modes for this equation is the entire μ-axis. As a result, any solitary wave in this equation, if one exists, is an embedded soliton.

We look for moving embedded solitons in this equation in the form

$$U(x,t) = \psi(\theta)e^{i\mu t}, \quad \theta = x - vt, \tag{5.184}$$

where μ and v are the frequency and velocity of this soliton. Note that Eq. (5.183) is not Galilean invariant, and thus the velocity v in this soliton is a nontrivial parameter. Inserting this form into (5.183), we obtain an ODE for $\psi(\theta)$ as

$$i(\psi_{\theta\theta\theta} + v\psi_\theta) + \mu\psi = |\psi|^2\psi. \tag{5.185}$$

Under variable scalings

$$\psi = v^{3/4}\Psi, \quad \theta = v^{-1/2}\Theta, \quad \mu = v^{3/2}\Omega, \tag{5.186}$$

Eq. (5.185) reduces to

$$i(\Psi_{\Theta\Theta\Theta} + \Psi_\Theta) + \Omega\Psi = |\Psi|^2\Psi, \tag{5.187}$$

which contains a single parameter Ω now. This equation does not admit single-hump embedded solitons. However, it admits an infinite number of double-hump embedded solitons at discrete parameter values $\Omega = \Omega_1, \Omega_2, \Omega_3, \ldots$, where

$$\Omega_1 = 0.8619, \quad \Omega_2 = 0.6959, \quad \Omega_3 = 0.6316, \quad \Omega_4 = 0.5939, \tag{5.188}$$

and so on (Akylas and Kung (1990), Klauder et al. (1993), Calvo and Akylas (1997), Yang and Akylas (2003)). The embedded solitons at Ω_1 and Ω_4 are displayed in Fig. 5.10. One can see that the $|\Psi(\Theta)|$ profiles of these solitons have two humps. In addition, the separation between these two humps increases with the decreasing value of Ω_n (i.e., with increasing index n).

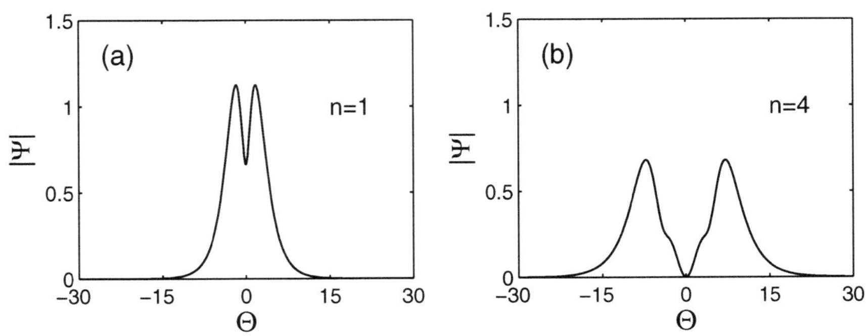

Figure 5.10. *Two embedded solitons in the normalized third-order NLS equation* (5.187) : (a) $\Omega = \Omega_1 = 0.8619$; (b) $\Omega = \Omega_4 = 0.5939$. *(After Yang and Akylas* (2003).*)*

Due to the variable scalings (5.186), infinite families of embedded solitons exist in the original equation (5.185). These solitons lie on the infinite number of curves

$$\mu = v^{3/2}\Omega_n, \quad n = 1,2,3,\ldots, \tag{5.189}$$

in the (v,μ) plane, and their profiles are the same as those shown in Fig. 5.10, except rescalings (5.186) along the θ and ψ directions. The power of these solitons in the nth family is

$$P_n(\mu) \equiv \int_{-\infty}^{\infty} |\psi_n(\theta;\mu)|^2 d\theta = b_n \mu^{2/3}, \tag{5.190}$$

where

$$b_n = \Omega_n^{-2/3} \int_{-\infty}^{\infty} |\Psi_n(\Theta)|^2 d\Theta. \tag{5.191}$$

Another third-order NLS equation which is closely related to (5.183) is

$$iU_t + U_{xx} + |U|^2 U = i\beta U_{xxx}, \tag{5.192}$$

where β is a real third-order dispersion coefficient. This equation governs propagation of picosecond optical pulses near zero second-order dispersion point, where the third-order dispersion becomes comparable with the second-order dispersion (see Sec. 1.3). Through a gauge transformation

$$U(x,t) = \frac{1}{\beta}\widehat{U}(\hat{x},\hat{t})e^{-\frac{1}{3}i\hat{x}+i\frac{1}{27}\hat{t}}, \quad \hat{x} = \frac{x}{\beta} + \frac{t}{3\beta^2}, \quad \hat{t} = \frac{t}{\beta^2}, \tag{5.193}$$

Eq. (5.192) becomes (5.183) for the scaled function $\widehat{U}(\hat{x},\hat{t})$, and thus these two equations share exactly the same properties (a related gauge transformation can be found in Yang and Akylas (2003)). When $\beta \ll 1$, we have shown in Sec. 5.5 that Eq. (5.192) does not admit single-hump embedded solitons. However, this equation admits continuous families of multihump embedded solitons as we have seen above.

Regarding the stability of these double-hump embedded solitons in the third-order NLS equations (5.183) and (5.192), it was shown by Yang and Akylas (2003) that they are linearly stable; i.e., their linearization spectrum does not contain unstable eigenvalues. Then how do they evolve nonlinearly under perturbations? Since these embedded solitons exist as continuous families with varying powers, one may suspect that these solitons, under *arbitrary* weak perturbations, may simply emit some radiation and adjust their shape to a nearby embedded soliton in the same solution family, as a regular soliton in the NLS equation would do under perturbations. This is not so, however. Numerical simulations by Yang and Akylas (2003) show that these double-hump embedded solitons are still semistable, just like isolated embedded solitons in the SHG system of the previous subsection. Specifically, these embedded solitons still evolve stably under perturbations of one sign and unstably under perturbations of the opposite sign. In other words, the freedom of continuous solution families with varying frequency and power is not sufficient to stabilize these embedded solitons.

Fully stable embedded solitons do exist, however. To demonstrate, we consider the generalized third-order NLS equation

$$U_t + U_{xxx} - i\alpha|U|^2U + \beta|U|^2U_x + \gamma\left(|U|^2\right)_x U = 0, \tag{5.194}$$

which arises in short-pulse propagation at zero second-order dispersion point in an optical fiber (see Sec. 1.3). This equation admits a one-parameter family of sech-shaped embedded solitons (Gromov and Tyutin (1998); see also Eq. (4.205) in Sec. 4.4):

$$U(x,t) = r\,\mathrm{sech}(rb\theta)e^{ik_*\theta - i\omega t}, \tag{5.195}$$

where

$$\theta = x - vt, \quad k_* = -\frac{\alpha}{2\gamma}, \quad b = \sqrt{\frac{\beta + 2\gamma}{6}} \tag{5.196}$$

and

$$v = b^2r^2 - 3k_*^2, \quad \omega = 2k_*(b^2r^2 + k_*^2). \tag{5.197}$$

Here the soliton amplitude r is arbitrary. In the frequency-velocity space (ω, v), these embedded solitons exist on a curve which is given parametrically by Eq. (5.197). When $\alpha, \gamma \ll 1$, Eq. (5.194) is a perturbed complex modified KdV equation. By a soliton perturbation theory, we proved in Sec. 4.4 of the previous chapter that this family of embedded solitons are nonlinearly fully stable (see Fig. 4.8). In other words, for arbitrary weak perturbations, a perturbed embedded soliton in this family will simply shed some radiation and relax onto a nearby embedded soliton in the same family. In the general case of $\alpha, \gamma = O(1)$, dynamics of these embedded solitons under perturbations can be analyzed by an asymptotic method (Pelinovsky and Yang (2005)), which is in the same spirit as the analysis of a perturbed embedded soliton in the SHG model of the previous subsection. Following this analysis, it was found that when the embedded soliton (5.195) is perturbed, its amplitude r and wavenumber k are governed by the following dynamical equations (Pelinovsky and Yang (2005)):

$$\frac{dr}{dt} = -\frac{1}{2}|h|^2b^3r^2(k - k_*)^2, \tag{5.198}$$

$$\frac{dk}{dt} = -\frac{1}{4}|h|^2b^5r^3(k - k_*), \tag{5.199}$$

where $h(r)(k - k_*)$ is the amplitude of the cw shelf which is emitted from the soliton. These dynamical equations are generalizations of similar equations (4.202)–(4.203) derived in Sec. 4.4 for $\alpha, \gamma \ll 1$. These equations admit a continuous family of fixed points, where $k = k_*$ and r is arbitrary. In addition, these fixed points are nonlinearly stable in Eqs. (5.198)–(5.199). This means that the continuous family of embedded solitons (5.195) are nonlinearly fully stable in the generalized third-order NLS equation (5.194).

Lastly, we would like to point out that solitons in many integrable equations are actually embedded solitons, as they reside in the continuous spectrum of radiation modes in those integrable equations. Examples include solitons in the fifth-order KdV hierarchy equation (Yang (2001a)), the complex modified KdV equation (Yang (2003)), the fourth-order NLS hierarchy equation, the Sasa–Satsuma equation, and many others. Embedded solitons in these equations always exist as continuous families with one or more nontrivial free parameters. In addition, they are fully stable due to the integrability of these equations. But such embedded solitons are degenerate due to the integrability. If the integrable equations are perturbed and become nonintegrable, those families of embedded solitons will reduce their number of free parameters by one. For instance, when the complex modified KdV equation is perturbed (see Eq. (5.194) and Sec. 4.4), its two-parameter family of solitons, parameterized by their amplitude and wavenumber (see (4.152)), reduces to a one-parameter family of embedded solitons (5.195). For another instance, when the fifth-order KdV hierarchy equation is perturbed, its one-parameter family of solitons (parameterized by its wave speed) disappears, except some isolated embedded solitons which persist at certain discrete speed values (Yang (2001a)).

5.7 Fractal Scattering in Collisions of Solitary Waves

One important and fascinating subject in nonintegrable equations is the collision between solitary waves. This collision arises in a wide range of physical situations. For instance, in soliton-based fiber communication systems, optical pulses traveling in different frequency channels pass through each other, giving rise to pulse collisions. In integrable systems, solitary waves (or more precisely, solitons) often collide with each other elastically, i.e., they emerge out of collision with their original shapes and speeds intact, and the only trace of collision is a position and phase shift in each soliton. The most familiar examples of such collisions occur in the KdV equation and the NLS equation. In some multicomponent integrable systems, the shapes of postcollision solitons may change due to polarization rotation or power transfer, such as in the Manakov equations and the focusing-defocusing coupled NLS equations (see Chapter 3). But in such cases, the speeds of solitons do not change, there is no energy radiation, and the collision process is still simple and can be well described by the inverse scattering theory and the N-soliton solutions.

In nonintegrable systems, however, collision of solitary waves can be much more complicated, and the collision outcome can depend on initial conditions in a sensitive fractal manner. This fractal-scattering phenomenon was first observed in kink-antikink collisions in sine-Gordon-type equations, such as the ϕ^4 model and a modified sine-Gordon model (Ablowitz et al. (1979), Campbell et al. (1983), Peyrard and Campbell (1983), Campbell et al. (1986), Anninos et al. (1991)). It was subsequently observed in collisions of kinks with delta-function defects in the sine-Gordon equation and the ϕ^4 model as well (Kivshar et al. (1991), Fei et al. (1992a, 1992b)). Then it was also observed in collisions between vector solitons in the coupled nonlinear Schrödinger equations (Yang and Tan (2000), Tan and Yang (2001a, 2001b)). What all these (nonintegrable) dispersive wave equations have in common is a shape-oscillation mode in their solitary waves. When the solitary waves first pass each other, this shape-oscillation mode is excited, which draws energy from the propagating waves. This energy transfer slows down the two waves and allows them to interact for some time. Under proper conditions, the energy in shape oscillations can return back to translations, which enables the two waves to escape. It is this resonant energy exchange between translations and shape oscillations which is responsible for fractal scatterings in these nonintegrable systems (Campbell et al. (1983)). To better understand these resonant collisions, reduced ODE models were constructed through the use of the variational method with an assumed solution ansatz (Anninos et al. (1991), Kivshar et al. (1991), Ueda and Kath (1990), Tan and Yang (2001b)). These ODE models were found to qualitatively capture the main features of resonant collisions in the underlying PDEs. Further analysis on these ODE models were performed by Goodman and Haberman (2004, 2005a, 2005b, 2007) and Goodman (2008), which shed more light on the resonant energy exchange inside these collisions.

In this section, we study the fractal scattering phenomenon in collisions of vector solitons in the coupled NLS equations

$$iA_t + A_{xx} + (|A|^2 + \beta|B|^2)A = 0, \tag{5.200}$$

$$iB_t + B_{xx} + (|B|^2 + \beta|A|^2)B = 0, \tag{5.201}$$

which govern optical pulse propagation in birefringent fibers and wavelength-division-multiplexed systems (see Sec. 1.3, or Menyuk (1987) and Agrawal (2007)). Here, β is the cross-phase-modulation coefficient. For simplicity, we will consider the collision between two orthogonally polarized and equal-amplitude vector solitons, whose initial conditions are

$$A(x,0) = \sqrt{2}\,\mathrm{sech}\left(x + \frac{1}{2}\Delta_0\right) e^{\frac{1}{4}iv_0x}, \tag{5.202}$$

$$B(x,0) = \sqrt{2}\,\mathrm{sech}\left(x - \frac{1}{2}\Delta_0\right) e^{-\frac{1}{4}iv_0x}. \tag{5.203}$$

Here v_0 is the relative velocity between the two initial solitons which will be called the *collision velocity*, and Δ_0 ($\gg 1$) is the initial soliton separation. In these initial conditions, the left soliton (5.202) is initially polarized in the A-component, and the right soliton (5.203) is initially polarized in the B-component. The collision results between these solitons do not depend on the specific value of Δ_0 as long as it is large enough. When $\beta = 0$ (the NLS case) or 1 (the Manakov case), this system is integrable, and the two solitons (5.202) and (5.203) will simply pass through each other without change in amplitude, velocity, or polarization (Zakharov and Shabat (1971), Manakov (1973)). For other values of β, this system is nonintegrable. This nonintegrable collision will be the focus of the study below.

5.7.1 PDE Simulation Results for Coupled NLS Equations

At each cross-phase-modulation coefficient β, Eqs. (5.200)–(5.201) were simulated with initial conditions (5.202)–(5.203) and $\Delta_0 = 20$, using the collision velocity v_0 as the control parameter. It was found that when β is very small (such as $\beta = 0.05$), the two solitons (5.202)–(5.203) pass through each other when v_0 is above a certain critical velocity v_c. In this case, the relative velocity between the two solitons after collision, which will be called the *exit velocity* and denoted by v_∞, is positive ($v_\infty > 0$). If v_0 is below this critical velocity v_c, the two solitons trap each other and form a bound state (Tan and Yang (2001b)). In this case, $v_\infty = 0$. As β increases, an interesting phenomenon occurs. That is, a sequence of reflection windows gradually appears just below the critical velocity v_c. When v_0 falls in one of these reflection windows, the two solitons are reflected back by each other after collision, and $v_\infty < 0$. This phenomenon is displayed in Fig. 5.11 at $\beta = 0.2$. Notice that these reflection windows are intertwined with trapping intervals. In addition, they are spaced closer and become narrower as the collision velocity v_0 approaches the critical velocity v_c. Furthermore, the quantity $(v_c^2 - v_n^2)^{-\frac{1}{2}}$, where v_n is the bottom point of the nth window, is almost a perfect linear function of the window index n, i.e.,

$$(v_c^2 - v_n^2)^{-\frac{1}{2}} = an + b, \tag{5.204}$$

where

$$a = 2.0588, \quad b = 0.0375. \tag{5.205}$$

When v_0 lies in these reflection windows, the two solitons first pass each other, then stay around and interact for some time, and then pass each other the second time and escape, see Fig. 5.11(b). So these reflection windows can be called two-pass windows. During interaction between the two passes, the solitons show a significant amount of width and amplitude oscillations. In different reflection windows of this sequence, the collision dynamics are similar, except that the interaction time between the two passes differs. It was found that this interaction time is a linear function of the window index n as well,

$$\omega T_n = 2n\pi + \rho, \tag{5.206}$$

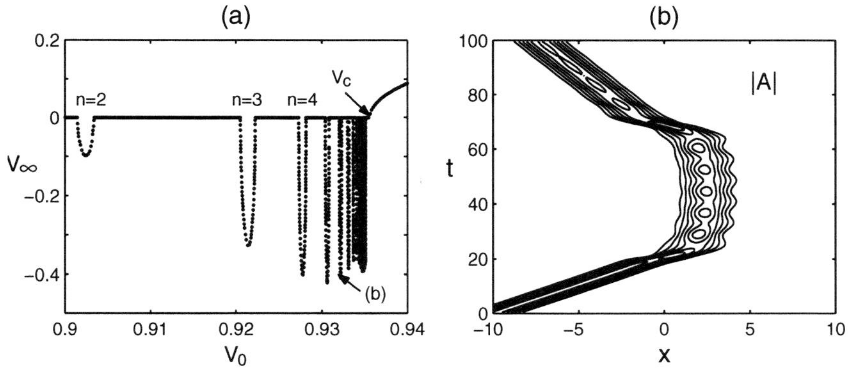

Figure 5.11. *Collision results in the coupled NLS equations* (5.200)–(5.201) *at* $\beta = 0.2$*:* (a) *exit-velocity versus collision-velocity graph;* (b) $|A|$ *contour with* v_0 *at the bottom of the reflection window marked in* (a)*; the* $|B|$ *contour is a mirror image of the* $|A|$ *contour around* $x = 0$*. (After Tan and Yang* (2001b)*.)*

where T_n is the time between the two passes when the collision velocity v_0 is at the bottom of the nth window, and

$$\omega = 0.7976, \qquad \rho = 0.9764. \tag{5.207}$$

As β increases from 0.2, the two-pass windows of Fig. 5.11 move closer to the critical velocity v_c. With further increase of β, windows of more complex collisions start to appear on the left side of the two-pass window sequence. In addition, these new windows start to develop their own window sequences inside the two-pass window sequence (Tan and Yang (2001b)). These developments subsequently lead to a fractal structure in the exit-velocity graph. This phenomenon is displayed in Fig. 5.12 at $\beta = 2/3$. First, one can notice that the exit-velocity graph now becomes much more complicated, with windows of transmission ($v_\infty > 0$), reflection ($v_\infty < 0$), and trapping ($v_\infty = 0$) all intertwined together. When one zooms into a small segment of this graph, one can get a structure which closely resembles the whole structure (see Fig. 5.12). In other words, this graph is a fractal. Collision dynamics on this fractal follows clear patterns, however. For example, at the three collision velocities marked in the upper row of Fig. 5.12, their collision dynamics are displayed in the lower row of the same figure. It is seen that in all three cases, the two solitons pass each other twice, but interaction times between the two passes are different. Hence the three reflection windows marked in Fig. 5.12 are counterparts of two-pass reflection windows at $\beta = 0.2$ in Fig. 5.11. If one zooms into other parts of the exit-velocity fractal in Fig. 5.12(a), the zoomed-in graph as well as the collision dynamics on it are related to the original graph and its collision dynamics as well (Yang and Tan (2001)).

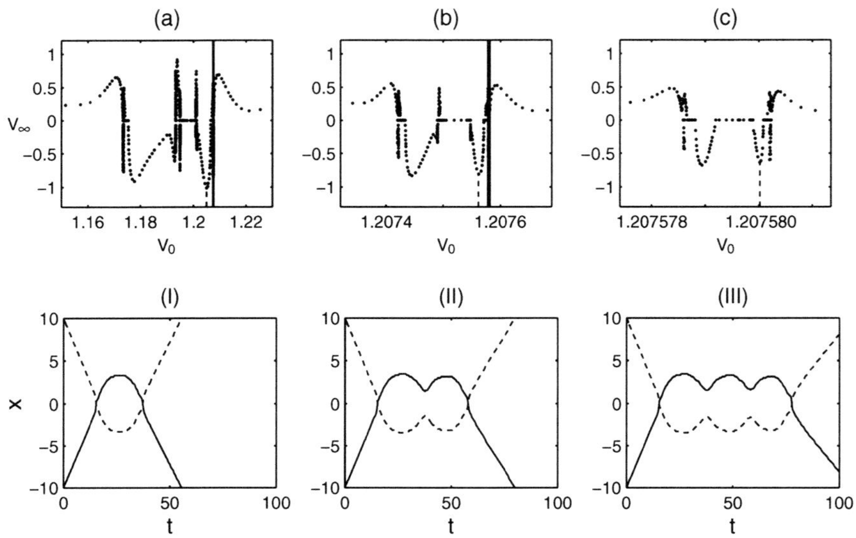

Figure 5.12. *Fractal structure and collision dynamics of vector solitons at* $\beta = 2/3$. *The upper row are the exit-velocity versus collision-velocity graphs, where* (b) *and* (c) *are successive amplifications of* (a) *in intervals marked by vertical solid lines. The lower row are collision dynamics with* v_0 *at bottoms of the reflection windows marked in the upper row. The curves plotted are positions of the maximum* $|A|$ *and* $|B|$ *amplitudes at each time (solid for* $|A|$ *and dashed for* $|B|$*). (After Yang and Tan* (2001).*)*

5.7.2 A Reduced ODE Model and Its Analysis

To better understand this fractal scattering in vector-soliton collisions, we analytically study this problem next. Due to the significant energy radiation emitted during these collisions, accurate or asymptotic analysis is hardly possible. Thus, we will resort to the variational method. A review of this method can be found in Malomed (2002).

In the variational method, solutions of the PDEs (5.200)–(5.201) are first formulated as solutions of the variational problem

$$\delta \int_a^b dt \int_{-\infty}^{\infty} dx\, \mathcal{L}(A, A^*, B, B^*) = 0, \tag{5.208}$$

where δ represents variation, the superscript "*" represents complex conjugation, and the Lagrangian density $\mathcal{L}$ is

$$\mathcal{L} = i(AA_t^* - A_t A^*) + i(BB_t^* - B_t B^*) + (2|A_x|^2 - |A|^4) + (2|B_x|^2 - |B|^4) - 2\beta|A|^2|B|^2. \tag{5.209}$$

Indeed, Eqs. (5.200)–(5.201) are nothing but the Euler–Lagrangian equations of the variational problem (5.208). To proceed further, one assumes a functional form (ansatz) for solutions $A(x,t)$ and $B(x,t)$, where the x-dependence is specified, but the form contains a number of parameters $\mathbf{p} = [p_1, p_2, \ldots, p_n]$ which are allowed to change with time, i.e.,

$$A(x,t) = A\,(x;\mathbf{p}(t)), \quad B(x,t) = B(x;\mathbf{p}(t)). \tag{5.210}$$

Inserting this solution ansatz into the variational problem (5.208) and performing integrations with respect to x, this variational problem then reduces to

$$\delta \int_a^b L[\mathbf{p}(t), \dot{\mathbf{p}}(t)]\, dt = 0, \tag{5.211}$$

where L is the Lagrangian

$$L = \int_{-\infty}^{\infty} \mathcal{L}(A, A^*, B, B^*)\, dx, \tag{5.212}$$

and $\dot{\mathbf{p}} = d\mathbf{p}/dt$. The Euler–Lagrangian equation for the reduced variational problem (5.211), i.e.,

$$\frac{d}{dt} L_{\dot{p}_j} - L_{p_j} = 0, \quad j = 1, \ldots, n, \tag{5.213}$$

then gives the evolution of parameters $\mathbf{p}(t)$ in the solution ansatz (5.210).

The success of the above variational method depends crucially on the proper choice of the solution ansatz (5.210). In addition, since energy radiation is ignored in this method, if radiation is significant in the underlying problem, the results from this method may become invalid.

Now we need to choose an ansatz for the solutions $A(x,t)$ and $B(x,t)$. Given the wide range of collision behaviors at different β values, selection of an ansatz suitable for all β values is very difficult. However, when β is small, a good and simple ansatz does exist. Our observation is that, at small β values, the collision is weak. Hence each of the solitons (5.202)–(5.203) still remains as a single pulse, whose shape remains close to its original one. But its width and amplitude can oscillate during collisions (see Fig. 5.11). In addition, the two solitons are related by the symmetry $B(x,t) = A(-x,t)$ due to the initial conditions (5.202)–(5.203). Thus we are motivated to select the following ansatz:

$$A(x,t) = \sqrt{2}\, a\, \mathrm{sech}\left(\frac{x + X/2}{w}\right) e^{i\left[\frac{1}{4}v(x+X/2) + \frac{b}{2w}(x+X/2)^2 + \sigma\right]}, \tag{5.214}$$

$$B(x,t) = \sqrt{2}\, a\, \mathrm{sech}\left(\frac{x - X/2}{w}\right) e^{i\left[-\frac{1}{4}v(x-X/2) + \frac{b}{2w}(x-X/2)^2 + \sigma\right]}, \tag{5.215}$$

where a, w, b, and σ are the two solitons' amplitude, width, chirp, and phase parameters, X is their separation distance, and v is their relative velocity. The chirp parameter b is introduced to induce width and amplitude oscillations in the solitons. Inserting this ansatz into the Lagrangian (5.212) and making some straightforward calculations, the Euler–Lagrangian equations (5.213) for the above parameters become (Ueda and Kath (1990), Tan and Yang (2001b))

$$\frac{d^2X}{dt^2} = \frac{16K\beta}{w^2}F'(\alpha), \tag{5.216}$$

$$\frac{d^2w}{dt^2} = \frac{16}{\pi^2 w^2}\left\{\frac{1}{w} - K - 3\beta K\left[\alpha F(\alpha)\right]'\right\}, \tag{5.217}$$

where

$$F(\alpha) = \frac{\alpha\cosh\alpha - \sinh\alpha}{\sinh^3\alpha} \tag{5.218}$$

and

$$\alpha = \frac{X}{w}, \qquad K = a^2 w = \text{constant}. \tag{5.219}$$

Corresponding to the initial solitons (5.202)–(5.203), the initial conditions for the above ODEs are

$$X(0) = \Delta_0, \;\; X'(0) = v_0, \;\; w(0) = 1, \;\; w'(0) = 0, \;\; K = 1. \tag{5.220}$$

Simulations of this reduced ODE model reveal that this model qualitatively captures the main features of soliton collisions in the PDEs (5.200)–(5.201). For instance, the simulation results of this ODE model at $\beta = 0.2$ are displayed in Fig. 5.13. It is seen that this model also exhibits a sequence of two-pass reflection windows, and collision dynamics in these windows resembles that in the PDE system. In addition, locations of these windows qualitatively follow the relation (5.204) as well. Furthermore, when β is small, the critical velocity v_c from this ODE model is quantitatively close to that in the PDE system. These agreements are encouraging, and they motivate us to further analyze this ODE model later in this subsection. But many differences also exist between the ODE and PDE systems. One difference is that at very small values of β (such as $\beta = 0.05$), this sequence of reflection windows persists in the ODE model, but disappears in the PDE system (Tan and Yang (2001b)). Another difference is that at small values of β where these two-pass reflection windows arise in both the PDE and ODE systems (such as $\beta = 0.2$), significant quantitative differences exist between the exit-velocity graphs of PDE simulations and ODE predictions, especially along the vertical axis (see Figs. 5.11 and 5.13). The third difference is that at small values of β where two-pass reflection windows exist in both the PDE and ODE systems

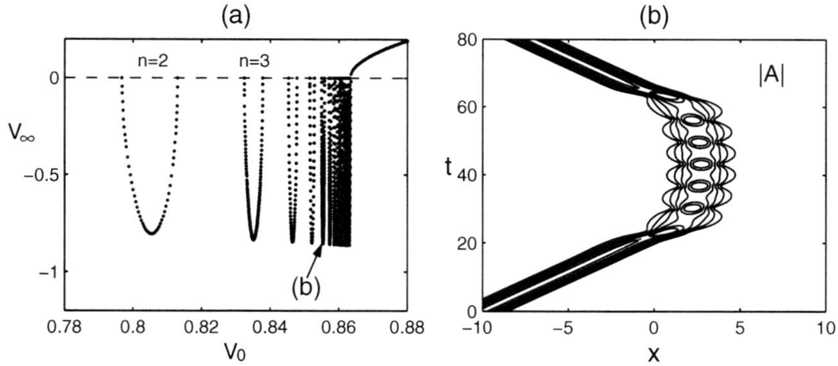

Figure 5.13. *Collision results in the reduced ODE model* (5.216)–(5.217) *at* $\beta = 0.2$*:* (a) *exit-velocity versus collision-velocity graph;* (b) $|A|$ *contour with* v_0 *at the bottom of the reflection window marked in* (a)*. (After Tan and Yang* (2001b).*)*

(such as $\beta = 0.2$), near each edge of such a reflection window, a sequence of three-pass transmission windows can be found in the ODE model, but such three-pass windows were not found in the PDE system (Tan and Yang (2001b)). The fourth difference is that, when β is not small, results from the ODE model and the PDE system show large deviations; hence the ODE model gives a poor description of the fractal structure in the PDE system at β values such as $\beta = 2/3$ (see Fig. 5.12). The reason for these differences at small β values is primarily due to the omission of energy radiation in the ODE model. This radiation is generated mostly when the two solitons pass through each other. It dissipates the translational motion of the solitons and makes their escape more difficult. Because of this radiation loss, certain windows in the ODE model (such as two-pass reflection windows at $\beta = 0.05$ and three-pass transmission windows at $\beta = 0.2$) did not materialize in the PDE system (Tan and Yang (2001b)). When β is not small, differences between the ODE model and the PDE system are partially due to the inappropriate ansatz (5.214)–(5.215), which did not take into account the soliton splitting which occurs during collisions in the PDE system (Yang and Tan (2001), Tan and Yang (2001b)). In view of these differences, results from the ODE model should be interpreted cautiously.

To shed more light on the collision dynamics, we analyze this reduced ODE model below along the lines of Goodman and Haberman (2005a). We will focus on the small-β case where the ODE model and the PDE system show reasonable agreement. In this case, the ODE model (5.216)–(5.217) can be further simplified. Indeed, since the collision between the two solitons is weak for small β, the width w undergoes only a small oscillation about its initial value $w = 1$. Thus we can write

$$w = 1 + W, \quad W \ll 1, \tag{5.221}$$

and partially linearize the ODEs (5.216)–(5.217) around $w = 1$. This w-linearization leads to the reduced system

$$\frac{d^2X}{dt^2} = 16\beta\left[F'(X)+G'(X)W\right], \tag{5.222}$$

$$\frac{d^2W}{dt^2} + \frac{16}{\pi^2}W = \frac{48\beta}{\pi^2}G(X), \tag{5.223}$$

where

$$G(X) = -[XF(X)]'. \tag{5.224}$$

With a further time rescaling,

$$\tau = 4\sqrt{\beta}\,t, \tag{5.225}$$

the above ODE system becomes

$$\ddot{X} = F'(X)+G'(X)W, \tag{5.226}$$

$$\ddot{W} + \lambda^2 W = \frac{3}{\pi^2}G(X), \tag{5.227}$$

with the fast frequency

$$\lambda = \frac{1}{\pi\sqrt{\beta}}. \tag{5.228}$$

Here, the dot represents the derivative to the rescaled time τ. This ODE system conserves the total energy

$$H = \left[\frac{1}{2}\dot{X}^2 - F(X)\right] + \frac{\pi^2}{6}\left(\dot{W}^2 + \lambda^2 W^2\right) - G(X)W, \tag{5.229}$$

where the first term is associated with the solitons' translational motion, the second term associated with the solitons' width oscillations, and the third term reflects the interaction between these two motions. Corresponding to the initial conditions (5.202)–(5.203) for the PDE system, the initial condition for this ODE system can be taken as

$$X \to -\infty, \quad \dot{X} \to V_0, \quad W \to 0, \quad \dot{W} \to 0, \quad \text{as } \tau \to -\infty, \tag{5.230}$$

where V_0 is the initial velocity in the scaled time τ, and it is related to the physical initial velocity v_0 by

$$v_0 = 4\sqrt{\beta}\,V_0 \tag{5.231}$$

in view of the time scaling (5.225). The reduced ODE system (5.226)–(5.227) is more amiable for analysis, and will be studied next.

First, notice that if W is held equal to zero, the phase portrait of Eq. (5.226), displayed in Fig. 5.14(a), shows three distinct types of orbits: closed orbits, corresponding to the two solitons bound together as a breather, unbounded orbits, corresponding to the solitons passing through each other, and the separatrix orbits, which form the boundary between the two regimes. The separatrix orbits are heteroclinic and connect degenerate saddle points at $X = \pm\infty$. Each orbit has a fixed energy

$$E = \frac{1}{2}\dot{X}^2 - F(X), \tag{5.232}$$

which is the translational-energy part of the whole system (5.226)–(5.227); see Eq. (5.229). This energy is negative on the closed orbits, positive on the unbounded orbits, and zero on the separatrix orbits. When W is allowed to vary, solutions may cross the separatrix orbits. We will show below that W remains $O(\sqrt{\beta})$ by variation of parameters. Thus, Eq. (5.226) can be studied by the perturbation method, with the W term as a small perturbation.

When β is small, interesting collision dynamics occurs at small initial velocities. Thus we assume $V_0 \ll 1$ below. In this case, the initial energy, $E_0 = V_0^2/2$, is small positive, and thus the orbit will remain close to the upper separatrix curve. Due to the perturbation term in Eq. (5.226), the energy of this orbit will decrease by an amount $\Delta E_1 < 0$ along the separatrix (see below). If the initial energy E_0 is larger than this energy loss $|\Delta E_1|$, the orbit will directly escape to $X = +\infty$, resulting in transmission of the two solitons. At the critical velocity V_c, $E_0 = |\Delta E_1|$. When V_0 is below this critical velocity, $E_0 < |\Delta E_1|$, and hence the energy of the orbit will become negative somewhere along the upper separatrix curve. As a result, the orbit cannot escape to $X = +\infty$. Instead, it will cross the upper separatrix curve, turn around, and move along the lower separatrix curve; see Fig. 5.14(a). The energy of the orbit along the lower separatrix also changes. But this energy change ΔE_2 can be either positive or negative depending on the time spent between the two passes where the orbit crosses $X = 0$ (see below). If the combined energy $E_0 + \Delta E_1 + \Delta E_2 > 0$, the orbit will cross the lower separatrix curve and escape to $X = -\infty$, resulting in reflection of the two solitons. This reflection is the strongest when $\Delta E_1 + \Delta E_2 = 0$, where the reflected solitons escape at exactly the same velocity as the initial solitons'. If $E_0 + \Delta E_1 + \Delta E_2 < 0$, the orbit cannot escape to $X = -\infty$ and has to turn around and move along the upper separatrix again, resulting in three passes between the two solitons.

Next, we analytically determine the critical velocities and locations of two-pass reflection windows in the ODE model (5.226)–(5.227).

Calculation of the Critical Velocity

To determine the critical velocity, we need to calculate the energy loss ΔE_1 as the orbit traverses the upper separatrix curve the first time. From Eq. (5.226) and the definition (5.232) of the energy, we have

$$\frac{dE}{d\tau} = \left(\ddot{X} - F(X)\right)\dot{X} = \dot{X}G'(X)W = \left(\frac{d}{d\tau}G(X(\tau))\right)W. \tag{5.233}$$

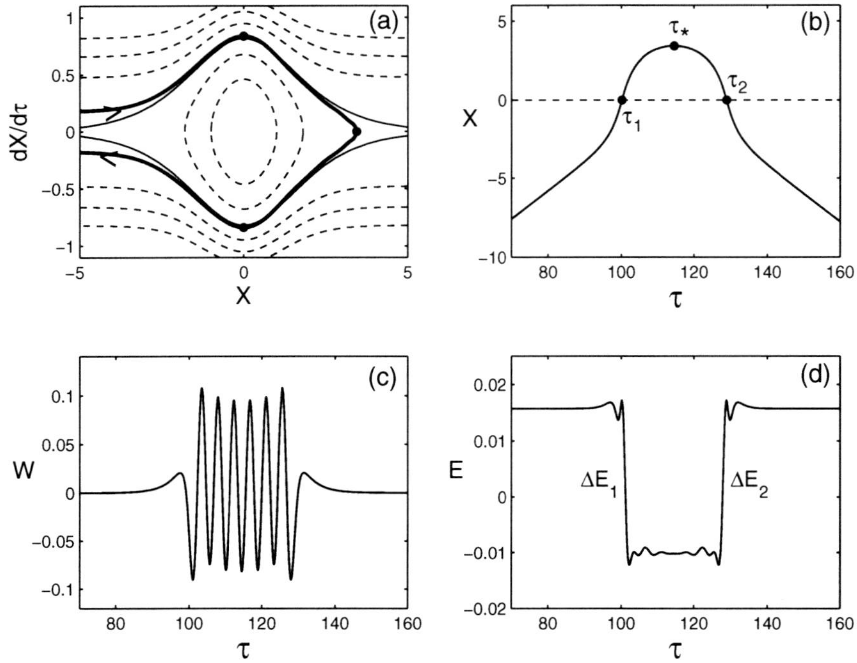

Figure 5.14. *Exact resonance in the ODE model* (5.226)–(5.227) *when* $\beta = 0.05$ *and* $V_0 = 0.1775$. *This* V_0 *value is at the bottom of the two-pass reflection window with* $n = 6$. (a) *the orbit in the phase space (thick solid line); the other curves are phase portraits of the unperturbed system* $\ddot{X} = F'(X)$, *where the thin solid lines are separatrix orbits;* (b) *the* $X(\tau)$ *graph;* τ_1 *and* τ_2 *are the times when* $X(\tau)$ *crosses zeros, and* τ_* *is the time when* $X(\tau)$ *is the largest;* (c) *the* $W(\tau)$ *graph, showing width oscillations between the two passes;* (d) *the energy* $E(\tau)$ *graph, showing energy loss during the first pass and energy gain during the second pass.*

Since the orbit $X(\tau)$ is near the upper separatrix curve, we will approximate $X(\tau)$ by $X_S(\tau - \tau_1)$ in the above equation, where $X_S(\tau)$ is the upper separatrix orbit which crosses zero at $\tau = 0$ (hence $X_S(\tau)$ is antisymmetric in τ), and τ_1 is the zero-crossing time of the orbit $X(\tau)$ (see Fig. 5.14). Under this approximation, the energy change along the upper separatrix is then

$$\Delta E_1 = \int_{-\infty}^{\infty} \left(\frac{d}{d\tau} G(X_S(\tau - \tau_1)) \right) W(\tau) d\tau = -\int_{-\infty}^{\infty} G(X_S(\tau - \tau_1)) \frac{dW(\tau)}{d\tau} d\tau. \quad (5.234)$$

This integral for the energy change along the separatrix is called the Melnikov integral in the literature. Given W's initial condition (5.230), we may solve the linear inhomogeneous equation (5.227) for $W(\tau)$ using variation of parameters and get

$$W(\tau) = \frac{3}{\pi^2\lambda}\left[\sin\lambda(\tau-\tau_1)\int_{-\infty}^{\tau-\tau_1} G(X_S(z))\cos\lambda z dz \right.$$
$$\left. - \cos\lambda(\tau-\tau_1)\int_{-\infty}^{\tau-\tau_1} G(X_S(z))\sin\lambda z dz\right]. \quad (5.235)$$

Here $X(\tau)$ has been approximated by $X_S(\tau-\tau_1)$ as well. This formula shows that $W(\tau)$ is $O(\lambda^{-1}) = O(\sqrt{\beta}) \ll 1$ for small β, which justifies our energy-change calculation above by the perturbation method. Inserting this $W(\tau)$ formula into (5.234) and making some simple calculations, we then obtain ΔE_1 as

$$\Delta E_1 = -\frac{3}{2\pi^2}\left|\int_{-\infty}^{\infty} G(X_S(\tau))e^{i\lambda\tau} d\tau\right|^2 = -\frac{3}{2\pi^2} I_c^2, \quad (5.236)$$

where

$$I_c(\beta) = \int_{-\infty}^{\infty} G(X_S(\tau))\cos\lambda\tau d\tau. \quad (5.237)$$

This formula shows that the energy drops when the orbit traverses along the upper separatrix curve the first time. The reason for it is that when $\tau \to -\infty$, $W \to 0$; hence this orbit initially has no width oscillations. But after the orbit moves over to the right end of the separatrix curve, width oscillations are excited, as Eq. (5.235) shows. Thus, due to conservation of the total energy (5.229), part of the translational energy E has transferred to width oscillations, resulting in a decrease in E.

At the critical velocity V_c, $\Delta E_1 = -E_0 = -V_c^2/2$, and hence

$$V_c = \frac{\sqrt{3}}{\pi} I_c. \quad (5.238)$$

Thus, calculation of V_c reduces to the calculation of the integral I_c. This integral can be readily computed numerically. When $\beta \ll 1$, it can also be determined asymptotically using complex analysis. After some calculations and converting the resulting V_c to the physical critical velocity v_c through the scaling (5.231), it is found that (Goodman and Haberman (2005a))

$$v_c = \frac{8\sqrt{3}}{5} e^{-T/\pi\sqrt{\beta}}\left[h_1\alpha\beta^{1/5} + h_2\alpha^2\beta^{2/5} + \cdots\right], \quad (5.239)$$

where

$$T \approx 2.1039, \quad h_1 \approx 2.1096, \quad h_2 \approx -0.6843, \quad \alpha \approx 2.0968.$$

Comparison between this critical-velocity formula from the ODE model and the numerical values from PDE simulations is made in Fig. 5.15(a), and good agreement can be seen.

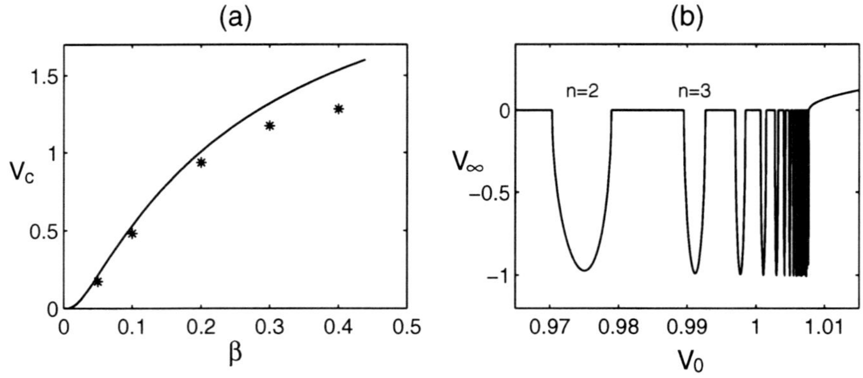

Figure 5.15. (a) *Graph of critical velocity* v_c *versus cross-phase-modulation coefficient* β*: the solid curve is the analytical formula* (5.239) *from the ODE model, and asterisks are numerical values from PDE simulations;* (b) *graph of the exit velocity versus initial velocity, obtained analytically from formulae* (5.239), (5.249), *and* (5.252) *for* $\beta = 0.2$.

Locations of Two-Pass Reflection Windows

Next, we calculate the locations of two-pass reflection windows, which lie below the critical velocity V_c. When the initial velocity falls in these reflection windows, the two solitons first pass each other, interact for some time, then pass each other again and escape. In the ODE system (5.226)–(5.227), the corresponding orbit is displayed in Fig. 5.14. Since the orbit $X(\tau)$ passes zero twice, energy also changes twice; see Fig. 5.14(d). The first energy change ΔE_1 has been obtained in Eq. (5.236). Now we calculate the second energy change ΔE_2 when the orbit passes zero the second time.

First of all, when the orbit moves near the upper separatrix to $X_{\max}$ (where $\tau = \tau_*$), the width function $W(\tau)$ in Eq. (5.235) becomes

$$W(\tau) \sim \frac{3I_c}{\pi^2 \lambda} \sin \lambda(\tau - \tau_1), \quad \tau \sim \tau_* \gg \tau_1. \tag{5.240}$$

When the orbit turns around and moves near the lower separatrix curve, the orbit $X(\tau)$ can be approximated by $X_S(\tau_2 - \tau)$, where τ_2 is the time when $X(\tau)$ crosses zero the second time (see Fig. 5.14(a, b)). Thus, the energy change ΔE_2 is

$$\Delta E_2 = -\int_{-\infty}^{\infty} G(X_S(\tau_2 - \tau)) \frac{dW(\tau)}{d\tau} d\tau. \tag{5.241}$$

The width function $W(\tau)$ during the second pass can still be calculated from Eq. (5.227) by the variation of parameters. But in view of the asymptotics (5.240) before the second pass, the formula for $W(\tau)$ now becomes

$$W(\tau)=\frac{3I_c}{\pi^2\lambda}\sin\lambda(\tau-\tau_1)+\frac{3}{\pi^2\lambda}\left[\sin\lambda(\tau-\tau_2)\int_{-\infty}^{\tau-\tau_2}G(X_S(z))\cos\lambda z\,dz\right.$$
$$\left.-\cos\lambda(\tau-\tau_2)\int_{-\infty}^{\tau-\tau_2}G(X_S(z))\sin\lambda z\,dz\right]. \tag{5.242}$$

Inserting this formula into (5.241), the energy change after the second pass ΔE_2 is then found to be

$$\Delta E_2=-\frac{3I_c^2}{\pi^2}\cos\lambda(\tau_2-\tau_1)-\frac{3I_c^2}{2\pi^2}. \tag{5.243}$$

Unlike the first energy change ΔE_1, this second energy change can be negative or positive, depending on the time $\tau_2-\tau_1$ spent between the two passes. The energy change ΔE_2 can be positive now, because the energy in width oscillations before the second pass can be transferred back to this translational energy E (recall that the total energy H in (5.229) is conserved). This precisely corresponds to the resonant energy exchange mechanism put forward by Campbell et al. to explain the appearance of reflection windows in ϕ^4-type equations (Campbell et al. (1983, 1986), Peyrard and Campbell (1983)). The strongest reflection (exact resonance) occurs when $\Delta E_1+\Delta E_2=0$, i.e., the two energy changes exactly cancel each other; hence all the width-oscillation energy is transferred back to the translational motion, and the solitons escape at the same velocity as the initial velocity. In this case, one easily finds from Eqs. (5.236) and (5.243) that $\cos\lambda(\tau_2-\tau_1)=-1$, and hence the condition for exact resonance is

$$\tau_2-\tau_1=\frac{(2n+1)\pi}{\lambda}, \tag{5.244}$$

where n is an integer. Physically, this n is the number of width peaks between the two passes; see Fig. 5.14(c). Recalling λ's expression (5.228) as well as the time rescaling (5.225), the exact-resonance condition in physical time is

$$t_2-t_1=\frac{\pi^2}{4}(2n+1). \tag{5.245}$$

This formula qualitatively agrees with the empirical formula (5.206), which was obtained from direct PDE simulations. Notice that this resonance condition is independent of β as long as β is small.

To determine the locations of reflection windows, we still need to relate the time $\tau_2-\tau_1$ with the initial velocity V_0. Notice from Fig. 5.14 that the solution between times τ_1 and τ_2 can be approximated by one half of the closed orbit of the unperturbed equation $\ddot{X}=F'(X)$ at the energy level $\mathcal{E}=E_0+\Delta E_1$ of time τ_*. The relation between the period $\mathcal{T}$ and energy $\mathcal{E}$ of the closed orbits in this unperturbed equation can be obtained either numerically or asymptotically. It is found that when $|\mathcal{E}|\ll 1$, the period $\mathcal{T}$ near the separatrix is given to the leading order by $\mathcal{T} = 2\pi/\sqrt{-2\mathcal{E}}$; see Goodman and Haberman (2005a). Thus, the leading-order approximation for $\tau_2 - \tau_1$ is

$$\tau_2 - \tau_1 = \frac{\pi}{\sqrt{-2\mathcal{E}}} = \frac{\pi}{\sqrt{V_c^2 - V_0^2}}. \tag{5.246}$$

Here, the formula $E_0 = V_0^2/2$ and the critical-velocity formulae (5.236)–(5.238) have been used. When this $\tau_2 - \tau_1$ formula is substituted into the exact-resonance relation (5.244), and the result is converted back to the physical velocity through the relation (5.231), we find that the initial velocity of exact resonance v_n is given by the formula

$$(v_c^2 - v_n^2)^{-1/2} = \frac{\pi}{4}(2n+1). \tag{5.247}$$

This resonant velocity v_n is at the bottom of the nth two-pass reflection window, and its formula qualitatively agrees with the empirical formula (5.204) obtained from direct PDE simulations as well.

Construction of the Exit-Velocity Graph

From the above analysis, we can further construct the exit-velocity graph when the initial velocity is above the critical velocity or is in one of the two-pass reflection windows. If $V_0 > V_c$, the solitons will directly escape after the first pass. The energy E_∞ after the escape is

$$E_0 + \Delta E_1 = E_\infty = \frac{1}{2}V_\infty^2, \tag{5.248}$$

where V_∞ is the exit velocity. Using the formula $E_0 = V_0^2/2$ as well as the formulae (5.236)–(5.238), and then converting the result into physical variables, we find that the formula for the physical exit velocity v_∞ is

$$v_\infty = \sqrt{v_0^2 - v_c^2}, \quad \text{when } v_0 > v_c. \tag{5.249}$$

When $V_0 < V_c$ and is in a two-pass reflection window, the two solitons pass each other twice and escape. The energy E_∞ after the two passes is

$$E_0 + \Delta E_1 + \Delta E_2 = E_\infty = \frac{1}{2}V_\infty^2. \tag{5.250}$$

Using the formulae (5.236), (5.243), and (5.244), this equation becomes

$$V_0^2 - 2V_c^2\left(1 + \cos\frac{\lambda\pi}{\sqrt{V_c^2 - V_0^2}}\right) = V_\infty^2. \tag{5.251}$$

Recalling λ's expression (5.228) and converting this equation into physical variables, we then find that when v_0 lies in the two-pass reflection windows, the physical exit velocity

v_∞ is given by the formula

$$v_\infty = -\sqrt{v_0^2 - 2v_c^2\left(1+\cos\frac{4}{\sqrt{v_c^2-v_0^2}}\right)}, \quad \text{when } v_0 < v_c. \tag{5.252}$$

If the initial velocity $v_0 < v_c$ and lies outside two-pass reflection windows, the quantity under the square root of the above formula would be negative, meaning that the solitons cannot escape after the first two passes and have to turn back and pass the third time. If we take $\beta = 0.2$, where $v_c = 1.008$ according to the formula (5.239), we obtain the exit-velocity graph from the formulae (5.249) and (5.252) shown in Fig. 5.15(b). It is seen that this analytical graph qualitatively agrees with the PDE simulation results in Fig. 5.11(a) and the ODE simulation results in Fig. 5.13(a).

The ODE model (5.222)–(5.223) and its subsequent analysis above correctly captures the main features of soliton collisions in the PDE system (5.200)–(5.201). More importantly, this analysis sheds much light on the resonant energy exchange between the translational motion and width oscillations of solitons during collisions and highlights the important role the collision time $t_2 - t_1$ plays in determining the amount of this energy exchange. The ansatz (5.214)–(5.215), which underlies the ODE model (5.222)–(5.223), is reasonable for small β where the collision is weak. When β is not small, fractal scatterings can arise in the PDE system (see Fig. 5.12). In this case, solitons can split during collisions, which makes the ansatz (5.214)–(5.215) invalid. Thus, the ODE model (5.222)–(5.223) is inadequate to describe fractal scatterings in vector-soliton collisions.

5.8 Fractal Scattering in Weak Interactions of Solitary Waves

It turns out that fractal scatterings can occur not only in collisions of solitary waves, but also in weak interactions of solitary waves if the nonlinear wave system is nonintegrable. Weak interactions between solitary waves arise when the two waves are well separated, and their relative velocities are small or zero. For example, in soliton-based fiber communication systems, two neighboring optical pulses traveling in the same frequency channel would experience this type of interactions. These interactions are caused by tail overlap between the two waves and are in general weak in strength. However, these weak interactions can sustain for a long time due to the small relative velocities between the two waves, and thus their accumulative effects can be very significant and highly nontrivial. Fractal scatterings in weak interactions were first reported by Dmitriev et al. (2001) in breathers of a weakly discrete sine-Gordon equation, and by Dmitriev and Shigenari (2002) in solitary waves of a weakly discrete NLS equation. Then it was also discovered by Zhu and Yang (2007) in a class of generalized NLS equations, and was found to exhibit a universal fractal structure regardless the specific forms of nonlinearity. To understand this universal fractal scattering in weak interactions, Zhu and Yang (2007) analyzed these generalized NLS equations by a perturbation method due to Karpman and Solovev (1981) and derived a simple and asymptotically accurate ODE model for weak interactions with arbitrary nonlinearities. After variable normalizations, this ODE model contains only a single constant parameter, which corresponds to different nonlinearities in the PDEs. This ODE model is a two degrees of freedom Hamiltonian system with highly coupled potentials, and its form is quite different from the variational ODE models derived previously for solitary wave collisions (Anninos et al. (1991), Kivshar et al. (1991), Ueda and Kath (1990), Tan and Yang (2001b)). Zhu et al. (2008a, 2008b) further analyzed this ODE model and derived a simple second-order map by using asymptotic methods near separatrix orbits. A remarkable feature of this map is that it does not contain any free parameters after various rescalings; thus it is universal for all weak interactions of solitary waves in the generalized NLS equations. This map was further investigated by Zhu et al. (2009) and was shown to give explicit analytical descriptions of the universal fractal structure in weak interactions, which agree very well with simulation results from the original PDEs and the reduced ODEs. After all this numerical and theoretical analysis, the physical mechanism for fractal scatterings in weak interactions also emerges. Based on the fact that this fractal scattering appears immediately when the generalized NLS equations are a weak perturbation of the NLS equation, the physical mechanism for fractal scattering is then traced to the sensitive dependence of PDE solutions near a second-order soliton of the NLS equation, whose zero in the spectral plane is double folded (degenerate). Under a small perturbation, this second-order soliton can either split into two escaping solitons or turn into a two-soliton bound state, which creates the sensitive dependence. This mechanism is very different from the resonant energy exchange mechanism for reflection windows and fractal scatterings in solitary wave collisions.

In this section, we consider fractal scatterings in weak interactions of the generalized NLS equation. This equation is of the form

$$iU_t + U_{xx} + F(|U|^2)U = 0, \tag{5.253}$$

where $F(\cdot)$ is a real-valued algebraic function with $F(0) = 0$. These equations govern various physical wave phenomena in nonlinear optics, fiber communications, and fluid dynamics (see Chapter 1, or Ablowitz and Segur (1981), Kivshar and Agrawal (2003)). Eq. (5.253) supports solitary waves of the form (see Eq. (5.7) in Sec. 5.1)

$$U(x,t) = u(\theta;\mu)e^{i\phi}, \tag{5.254}$$

where

$$\theta = x - \xi, \quad \xi = vt + x_0, \tag{5.255}$$

$$\phi = \frac{1}{2} v\theta + \sigma, \quad \sigma = \left(\mu + \frac{1}{4} v^2\right) t + \sigma_0, \tag{5.256}$$

$u(\theta;\mu)$ is a localized positive and symmetric function which satisfies the equation

$$u_{\theta\theta} - \mu u + F(u^2)u = 0, \tag{5.257}$$

and μ, v, x_0, σ_0 are four real constants which characterize this solitary wave. Physically, μ is the propagation constant which controls the wave's amplitude, v is the wave's velocity, x_0 is its initial position, and σ_0 is its initial phase. The asymptotic behavior of this solution at infinity is

$$u(\theta;\mu) \to c\, e^{-\sqrt{\mu}\,|\theta|}, \quad |\theta| \to \infty, \tag{5.258}$$

where c is the tail coefficient which depends on the nonlinear function F and the propagation constant μ. The power of this solitary wave is defined as

$$P(\mu) = \int_{-\infty}^{\infty} u^2(\theta;\mu)\, d\theta. \tag{5.259}$$

For some special nonlinearities F, analytical formulae for u, P, and c are available (see Sec. 5.1). In the general case, such analytical formulae may not exist but can be obtained numerically.

In weak interactions, two such solitary waves are initially well separated and have small relative velocities and amplitude differences. Then they would interfere with each other through tail overlapping. When time goes to infinity, they either separate from each other or form a bound state. The exit velocity, defined as $\Delta v_\infty = |v_2 - v_1|_{t\to\infty}$, depends on the initial conditions of the two waves.

5.8.1 PDE Simulation Results for Generalized NLS Equations

First, we numerically study weak interactions in the generalized NLS equation (5.253). For this purpose, we select two different nonlinearities,

$$F(|U|^2) = |U|^2 + \kappa |U|^4 \tag{5.260}$$

and

$$F(|U|^2) = |U|^2 + \rho |U|, \tag{5.261}$$

which are cubic quintic and quadratic cubic, respectively. Here κ and ρ are real parameters. To illustrate numerical results, we take $\kappa = 0.0003$ in (5.260) and $\rho = -0.0015$ in (5.261). The initial conditions are both taken as

$$x_{0,1} = -x_{0,2} = -5, \quad v_{0,k} = 0, \quad \mu_{0,k} = 1 \quad (k = 1,2), \tag{5.262}$$

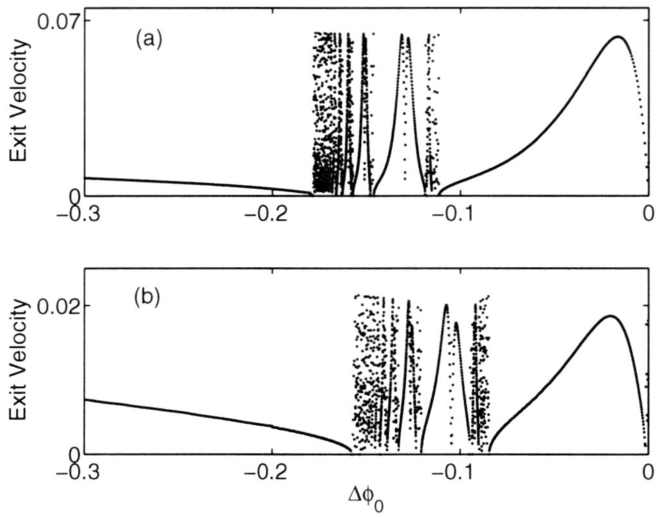

Figure 5.16. *Exit-velocity versus initial-phase-difference graphs for weak interactions in the generalized NLS equations* (5.253) *with initial conditions* (5.262)*:* (a) *cubic-quintic nonlinearity* (5.260) *with* $\kappa = 0.0003$*;* (b) *quadratic-cubic nonlinearity with* $\rho = -0.0015$.

$\phi_{0,1} = 0$, and the initial-phase difference $\Delta\phi_0 = \phi_{0,2} - \phi_{0,1}$ is used as the control parameter. The numerically obtained exit velocity Δv_∞ versus $\Delta\phi_0$ graphs for these two nonlinearities are plotted in Fig. 5.16(a, b), respectively. These graphs are symmetric in $\Delta\phi_0$, and thus only the $\Delta\phi_0 < 0$ portions are shown. It is seen that the two graphs are very similar. At small and large values of $|\Delta\phi_0|$, the graphs are smooth and simple. But in an intermediate interval of $|\Delta\phi_0|$, the graphs are not smooth and show intricate structures. This means that in this interval, the outcome of weak interactions is sensitive to the initial condition $\Delta\phi_0$. It turns out that these intricate structures are both fractals. To demonstrate, we take this nonsmooth portion of the graph in Fig. 5.16(a) and repeatedly zoom into it; the results are displayed in Fig. 5.17. From Fig. 5.17(a), one can see that this graph contains sequences of smooth double-peaked hills which cascade to the left and to the right. Between these hills are more delicate structures. When we zoom into one such structure between two hills (as marked in Fig. 5.17(a)), we got the zoomed-in graph displayed in Fig. 5.17(b). This graph is very similar to the original graph, except that the cascading directions of hills are reversed. When we further zoom into a structure between two hills in Fig. 5.17(b), the zoomed-in graph, shown in Fig. 5.17(c), is very similar to the graph in (b) as well, but with cascading directions of hills reversed as in the previous zoom. These repeated zooms clearly reveal the fractal nature of this graph. If we zoom into the graph of Fig. 5.16(b)

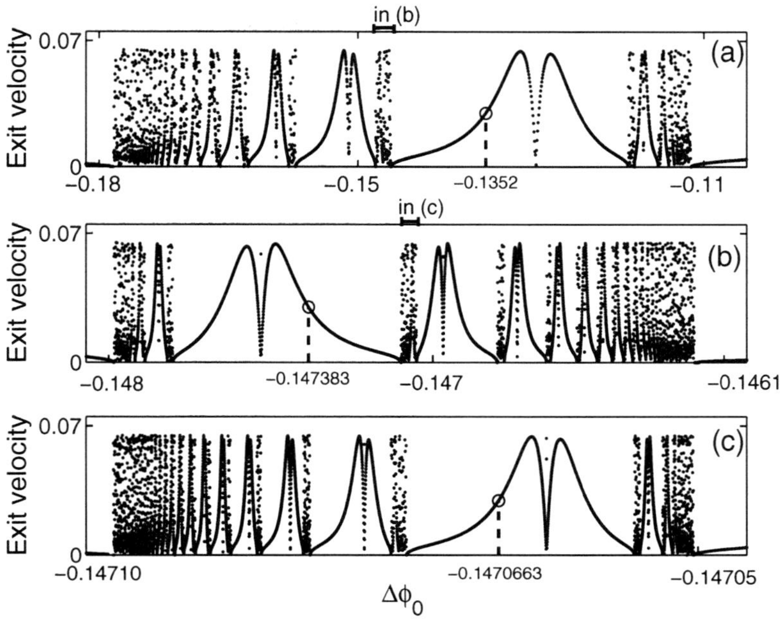

Figure 5.17. *The exit-velocity versus initial-phase-difference graphs of Fig.* 5.16(a) *and its two successive zoomed-in structures. The zoomed-in intervals are marked by upper bars in* (a) *and* (b). *Interaction dynamics at the circled points are shown in Fig.* 5.18.

for the quadratic-cubic nonlinearity, we will find that this graph is a fractal as well. In addition, this fractal is quantitatively similar to the one for the cubic-quintic nonlinearity in Fig. 5.16(a) after horizontal and vertical scalings. For these two nonlinearities in Fig. 5.16, the κ and ρ values were taken quite small, meaning that the generalized NLS equations (5.253) were weakly perturbed NLS equations. If the κ and ρ values are not small, these fractal structures will still persist, except that their locations on the $\Delta\phi_0$-axis will change (Zhu and Yang (2007)). Furthermore, if the initial condition is taken different from (5.262), say with different initial amplitudes $\mu_{0,1} \neq \mu_{0,2}$, fractal structures will persist as well (Zhu and Yang (2007)). From these findings, we conclude that fractal scattering is a common feature of weak interactions in the generalized NLS equations (5.253), and these fractal scatterings exhibit universal fractal structures. But this fractal scattering does not occur for every form of nonlinearity. For instance, when $\kappa < 0$ in the cubic-quintic nonlinearity (5.260), no fractal scattering will be found.

Dynamics of weak interactions on these fractals is an important issue. To illustrate this dynamics, we pick three $\Delta\phi_0$ values at the same relative positions of the fractal and

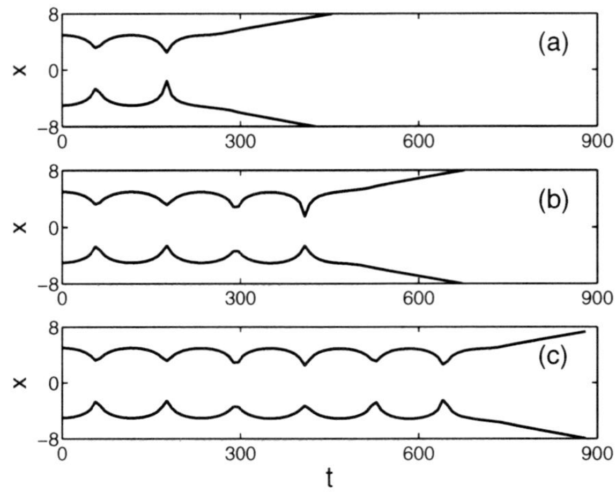

Figure 5.18. *Dynamics of weak interactions on the fractal of Fig.* 5.17. *The curves are the peak positions of the two solitons versus time at three values of* $\Delta\phi_0$ *(marked by circles in Fig.* 5.17*):* (a) -0.1352; (b) -0.147383; (c) -0.1470663.

its zoomed-in structures in Fig. 5.17 (as marked by circles), and the peak positions of these two solitons during weak interactions are displayed in Fig. 5.18. Several features can be observed. First, in all three cases, the two solitons stay apart from each other at all times, and hence their interaction is weak throughout the entire process. Second, in all three cases, positions of the two solitons oscillate several times before escape. The difference between the three cases is the number of oscillations before escape. This number of oscillations in the zoomed-in graph is two more than that in the original graph. Third, during their position oscillations, the widths of the two solitons do not oscillate. This contrasts collisions of vector solitons in Fig. 5.11 where width oscillations were significant during interactions. It also indicates that fractal scatterings in weak interactions have a different physical origin from the resonant energy exchange mechanism for solitary wave collisions. Fourth, during the weak interactions, there is very little energy radiation.

5.8.2 An Asymptotic ODE Model

In this subsection, we analytically study this universal fractal scattering of weak interactions in the generalized NLS equations (5.253) and derive an asymptotically accurate ODE model for it. In this study, the nonlinearity $F(|U|^2)$ in (5.253) can be arbitrary; i.e., Eq. (5.253) does not have to be close to the NLS equation. Our analysis follows Zhu and Yang (2007).

The basis of this analysis is the observation that during weak interactions, the two solitons are well separated, and there is very little energy radiation. Thus, the solution can be well approximated by the superposition of two individual solitons (5.254), but parameters in these solitons will slowly vary with time due to their interference. Dynamical equations for the soliton parameters can then be derived by a soliton perturbation theory, with their tail overlapping treated as a weak perturbation to each soliton. This is the same idea that was originally used by Karpman and Solovev (1981) for the study of weak interactions of NLS solitons (see Sec. 4.3). In the present case, since Eq. (5.253) is nonintegrable, and its nonlinearity is allowed arbitrary, there will be two twists in the execution of this idea. One is that a soliton perturbation theory for the nonintegrable equation (5.253) needs to be developed. The other one is that since the analytical expression for the solitary waves (5.254) is generally unavailable, new ways for calculating interaction integrals (which appear in this Karpman–Solovev-type analysis) need to be found.

Evolution of a Single Solitary Wave under Perturbations

First, we need to develop a perturbation theory for a single solitary wave (5.254) if Eq. (5.253) is under small perturbations. The perturbed equation can be written as

$$iU_t + U_{xx} + F(|U|^2)U = \epsilon G, \tag{5.263}$$

where function G is a perturbation term, and ϵ is a small parameter. Without perturbations ($\epsilon = 0$), the solitary wave (5.254) is an exact solution of Eq. (5.263), and its parameters v, μ, σ_0, x_0 are time independent. When the perturbation is present, these parameters will evolve slowly with time. The soliton perturbation theory for integrable equations is well known (see the previous chapter). For the nonintegrable equation (5.253), such a theory for the evolution of solitary-wave parameters on the slow time scale $T = \epsilon t$ can be similarly derived (Yang and Kaup (2000)). For this purpose, we first write the perturbed solution as

$$U(x,t) = \widehat{u}(\theta,t)e^{i\phi}, \tag{5.264}$$

where

$$\theta = x - \xi, \quad \xi = \int_0^t v\, d\tau + x_0, \tag{5.265}$$

and

$$\phi = \frac{1}{2}v\theta + \sigma, \quad \sigma = \int_0^t \left(\mu + \frac{1}{4}v^2\right) d\tau + \sigma_0. \tag{5.266}$$

Here v, μ, σ_0, x_0 are all functions of the slow time T. Substituting (5.264) into (5.263), we get the equation for $\widehat{u}$ as

$$\begin{aligned} i\widehat{u}_t + \widehat{u}_{\theta\theta} - \mu\widehat{u} + F(|\widehat{u}|^2)\widehat{u} = \epsilon G e^{-i\phi} - \epsilon\left(i\widehat{u}_\mu \mu_T - i\widehat{u}_\theta x_{0T}\right) \\ - \epsilon\,(v x_{0T}/2 - \theta v_T/2 - \sigma_{0T})\widehat{u}. \end{aligned} \tag{5.267}$$

We now expand $\widehat{u}(\theta,t)$ into a perturbation series

$$\widehat{u}(\theta,t)=u(\theta)+\epsilon\widehat{u}_1(\theta,t)+\epsilon^2\widehat{u}_2(\theta,t)+\cdots. \tag{5.268}$$

Inserting this expansion into (5.267), this equation at order ϵ^0 is satisfied automatically since $u(\theta)$ satisfies Eq. (5.257). At $O(\epsilon)$, the equation for $\widehat{u}_1$ can be written as

$$i\Psi_t+L\Psi=H, \tag{5.269}$$

where

$$\Psi=\begin{pmatrix}\widehat{u}_1+\widehat{u}_1^*\\ \widehat{u}_1-\widehat{u}_1^*\end{pmatrix},\quad L=\begin{pmatrix}0 & L_0\\ L_1 & 0\end{pmatrix}, \tag{5.270}$$

$$\begin{aligned}L_0&=\partial_{\theta\theta}-\mu+F(u^2),\\ L_1&=\partial_{\theta\theta}-\mu+F(u^2)+2u^2F'(u^2),\end{aligned} \tag{5.271}$$

and

$$H=\begin{bmatrix}Ge^{-i\phi}-G^*e^{i\phi}-2i(u_\mu\mu_T-u_\theta x_{0T})\\ Ge^{-i\phi}+G^*e^{i\phi}-(vx_{0T}-\theta v_T-2\sigma_{0T})u\end{bmatrix}. \tag{5.272}$$

Here the superscript "*" represents complex conjugation. The operator L has two eigenfunctions and two generalized eigenfunctions at the zero eigenvalue (see Sec. 5.2),

$$\Psi_{1d}=(0,u)^T,\quad \Psi_{2d}=(u_\theta,0)^T, \tag{5.273}$$

$$\Psi_{1g}=\left(u_\mu,0\right)^T,\quad \Psi_{2g}=(0,\theta u/2)^T, \tag{5.274}$$

and the eigenrelations are

$$L\Psi_{1d}=L\Psi_{2d}=0,\quad L\Psi_{1g}=\Psi_{1d},\quad L\Psi_{2g}=\Psi_{2d}. \tag{5.275}$$

In addition, this operator is self-adjoint under the inner product

$$\langle F_1,F_2\rangle\equiv\int_{-\infty}^{\infty}F_1^T\begin{pmatrix}0 & 1\\ 1 & 0\end{pmatrix}F_2\,d\theta, \tag{5.276}$$

where the superscript "T" represents transpose of a vector. To solve the first-order equation (5.269), we expand the inhomogeneous term H and solution Ψ into eigenfunctions of operator L (as was done in Sec. 4.1.2). It is easy to see that, in order for Ψ not to grow linearly with time, the coefficients of $\Psi_{1d},\Psi_{2d},\Psi_{1g}$, and Ψ_{2g} in the expansion of H must be zero. Since operator L is self-adjoint under the above inner product, eigenfunctions of L at different eigenvalues are orthogonal to each other. Because of this, these coefficients in

H's expansion can be readily found to be proportional to the inner products between H and these eigenfunctions. Then, in order for these coefficients to be zero, H must be orthogonal to $\Psi_{1d}, \Psi_{2d}, \Psi_{1g}$, and Ψ_{2g}, i.e.,

$$\langle H, \Psi_{1d} \rangle = \langle H, \Psi_{2d} \rangle = \langle H, \Psi_{1g} \rangle = \langle H, \Psi_{2g} \rangle = 0. \tag{5.277}$$

Evaluating integrals in these four orthogonality conditions, slow-time evolution equations for parameters v, μ, σ_0, x_0 will be obtained. These evolution equations can be written as

$$P\frac{dv}{dT} = 2\int_{-\infty}^{\infty} u_\theta (G^* e^{i\phi} + G e^{-i\phi})\, d\theta, \tag{5.278}$$

$$P_\mu \frac{d\mu}{dT} = \frac{1}{i}\int_{-\infty}^{\infty} u(G e^{-i\phi} - G^* e^{i\phi})\, d\theta, \tag{5.279}$$

$$P\frac{dx_0}{dT} = \frac{1}{i}\int_{-\infty}^{\infty} u\theta (G e^{-i\phi} - G^* e^{i\phi})\, d\theta, \tag{5.280}$$

$$P_\mu \left(\frac{v}{2}\frac{dx_0}{dT} - \frac{d\sigma_0}{dT} \right) = \int_{-\infty}^{\infty} u_\mu (G^* e^{i\phi} + G e^{-i\phi})\, d\theta. \tag{5.281}$$

Here, $P_\mu = P'(\mu)$. These equations will be used for the study of weak interactions of solitary waves below.

Weak Interactions of Two Solitary Waves

Now, we consider the weak interaction of two solitary waves in the generalized NLS equations (5.253). Here the tail overlapping between the two waves will be treated as a small perturbation, which causes the parameters of both waves to evolve on a slow time scale ϵt. The parameter ϵ is the magnitude of tail overlapping which is exponentially small with the spacing between the two waves. We will not introduce ϵ explicitly in the following analysis. To the leading order, the interacting solution is a superposition of two solitary waves,

$$U = U_1(x,t) + U_2(x,t), \tag{5.282}$$

where

$$U_k = u_k e^{i\phi_k}, \quad u_k = u(\theta_k; \mu_k), \quad k = 1, 2, \tag{5.283}$$

and the expressions for θ_k, ϕ_k are as in (5.265)–(5.266) but with indices $k = 1, 2$ added. By convention, we assign the initially left solitary wave with index $k = 1$, and the initially right solitary wave with index $k = 2$. To study the weak interaction between them, we require that they be both stable, and well separated, and that they have almost the same velocities and amplitudes. With the notation

$$\mu = (\mu_1 + \mu_2)/2, \quad v = (v_1 + v_2)/2, \tag{5.284}$$

and

$$\Delta\mu = \mu_2 - \mu_1, \ \ \Delta v = v_2 - v_1, \ \ \Delta\xi = \xi_2 - \xi_1, \quad \Delta\phi = \phi_2 - \phi_1, \tag{5.285}$$

these requirements are explicitly

$$P'(\mu) > 0, \quad |\Delta\mu\Delta\xi| \ll 1 \ll \mu\Delta\xi, \quad |\Delta v\Delta\xi| \ll 1. \tag{5.286}$$

Here, $P'(\mu) > 0$ is the Vakhitov–Kolokolov criterion for the linear stability of solitary waves in Eq. (5.253); see Sec. 5.3. The latter two requirements in (5.286) are the same as those made in Sec. 4.3 for the study of weak interactions of NLS solitons, and they are necessary for the asymptotic calculations below.

Now we insert the solution (5.282) into Eq. (5.253). By repeating similar calculations as in Sec. 4.3, we find that each solitary wave is governed by the following perturbed generalized NLS equations:

$$i\frac{\partial U_k}{\partial t} + \frac{\partial^2 U_k}{\partial \theta_k^2} + F(|U_k|^2)U_k = H_k, \quad k = 1,2, \tag{5.287}$$

where

$$H_k = -\left[F(|U_k|^2) + F'(|U_k|^2)|U_k|^2\right]U_{3-k} - F'(|U_k|^2)U_k^2 U_{3-k}^*. \tag{5.288}$$

In addition, under assumptions (5.286), the phase difference is

$$\Delta\phi \approx -v\Delta\xi/2 + \Delta\sigma, \tag{5.289}$$

which is independent of θ_k. Then, when the above soliton perturbation theory is applied to Eq. (5.287) and assumptions (5.286) are utilized, dynamical equations for $v_k, \mu_k, x_{k,0}$, and $\sigma_{k,0}$ will be obtained. For instance, the equation for μ_k is

$$P_\mu \frac{d\mu_k}{dt} = (-1)^k 2\sin\Delta\phi \int_{-\infty}^{\infty} uF(u^2)u^{(k)}d\theta, \tag{5.290}$$

where

$$u = u(\theta), \quad u^{(k)} = u\left(\theta + (-1)^k\Delta\xi\right). \tag{5.291}$$

Compared to the weak-interaction analysis of NLS solitons in Sec. 4.3, a new feature here is the calculation of interaction integrals in these dynamical equations. Since the nonlinearity

F is now arbitrary, and the explicit formula for $u(\theta)$ is unavailable, these integrals have to be calculated in different ways. Below we demonstrate how to calculate the integral in Eq. (5.290). First, we notice that the main contribution to this integral comes from the $\theta = O(1)$ region, where the function $u^{(k)}$ can be approximated through the asymptotics (5.258) as

$$u^{(k)} \approx c\, e^{(-1)^{k+1}\sqrt{\mu}\,\theta - \sqrt{\mu}\,\Delta\xi}. \tag{5.292}$$

Second, we notice that $u(\theta)$ satisfies Eq. (5.257), and thus $uF(u^2)$ in this integral can be replaced by $\mu u - u_{\theta\theta}$. Combining these two facts, we get

$$\int_{-\infty}^{\infty} uF(u^2)u^{(k)}d\theta = c\, e^{-\sqrt{\mu}\,\Delta\xi}\int_{-\infty}^{\infty}(\mu u - u_{\theta\theta})e^{\sqrt{\mu}\,\theta}d\theta. \tag{5.293}$$

Then performing integration by parts and utilizing the asymptotics (5.258), we finally obtain this integral as

$$\int_{-\infty}^{\infty} uF(u^2)u^{(k)}d\theta = 2\sqrt{\mu}\,c^2 e^{-\sqrt{\mu}\,\Delta\xi}. \tag{5.294}$$

Calculations of interaction integrals in the other dynamical equations for $v_k, x_{k,0}$, and $\sigma_{k,0}$ are similar (see Zhu and Yang (2007)). Using these integral formulae, the dynamical equations reduce to

$$P_\mu \frac{d\mu_k}{dt} = (-1)^k 4\sqrt{\mu}\,c^2 \sin(\Delta\phi)e^{-\sqrt{\mu}\,\Delta\xi}, \tag{5.295}$$

$$P\frac{dv_k}{dt} = (-1)^{k+1} 8\mu c^2 \cos(\Delta\phi)e^{-\sqrt{\mu}\,\Delta\xi}, \tag{5.296}$$

$$P\frac{dx_{k,0}}{dt} = -2D_1 \sin(\Delta\phi)e^{-\sqrt{\mu}\,\Delta\xi}, \tag{5.297}$$

$$P_\mu\left(\frac{v}{2}\frac{dx_{k,0}}{dt} - \frac{d\sigma_{k,0}}{dt}\right) = -2D_2\cos(\Delta\phi)e^{-\sqrt{\mu}\,\Delta\xi}, \tag{5.298}$$

where D_1 and D_2 are two constants. From these equations, we find that

$$\mu_t = v_t = 0, \tag{5.299}$$

$$\Delta\xi_{tt} = -\frac{16\mu c^2}{P}\cos(\Delta\phi)e^{-\sqrt{\mu}\,\Delta\xi}, \tag{5.300}$$

$$\Delta\phi_{tt} = \frac{8\sqrt{\mu}\,c^2}{P_\mu}\sin(\Delta\phi)e^{-\sqrt{\mu}\,\Delta\xi}, \tag{5.301}$$

and

$$\Delta v = \Delta\xi_t, \quad \Delta\mu = \Delta\phi_t. \tag{5.302}$$

These reduced ODEs are important results for the weak interaction of solitary waves in the generalized NLS equations (5.253). They show that the average propagation constant μ and the average velocity v do not change with time, but the separation $\Delta\xi$ and phase difference $\Delta\phi$ between the two solitons change according to Eqs. (5.300)–(5.301).

These dynamical equations can be further simplified by variable scalings. Introducing the notation and time scalings

$$\psi = \Delta\phi, \quad \zeta = -\sqrt{\mu}\,\Delta\xi, \quad \tau = \sqrt{16\mu^{3/2}c^2/P}\,t, \quad \varepsilon = P/(2\mu P_\mu) - 1, \tag{5.303}$$

these equations then reduce to

$$\begin{aligned} \zeta_{\tau\tau} &= \cos\psi\, e^{\zeta}, \\ \psi_{\tau\tau} &= (1+\varepsilon)\sin\psi\, e^{\zeta}. \end{aligned} \tag{5.304}$$

Equation (5.304) is the final ODE model we obtained for the analytical treatment of weak interactions in the generalized NLS equations (5.253). This model is asymptotically accurate when the two solitons are well separated, and is universal for the generalized NLS equations with arbitrary nonlinearities. It contains only a single parameter ε, which depends on the specific form of nonlinearity (as well as the time-invariant average propagation constant μ of the two waves). For the NLS equation

$$iU_t + U_{xx} + |U|^2U = 0, \tag{5.305}$$

which is obtained from (5.253) when $F(|U|^2) = |U|^2$, $P = 4\sqrt{\mu}$, and thus $\varepsilon = 0$. In this case, Eq. (5.304) reproduces the dynamical equations obtained in Sec. 4.3 for weak interactions of NLS solitons. If Eq. (5.253) is close to the NLS equation, then ε would be small. In particular, for both nonlinearities in Fig. 5.16, it is found that $\varepsilon = 0.001$. But for general types of nonlinearities, ε may not be small. This ODE model (5.304) is Hamiltonian for arbitrary values of ε. The conserved Hamiltonian energy is

$$H = \frac{1}{2}(\dot{\zeta}^2 - \dot{\psi}^2) - e^{\zeta}\cos\psi + \frac{\varepsilon}{2(1+\varepsilon)}\dot{\psi}^2. \tag{5.306}$$

Numerical simulations of this ODE model (5.304) reveal that this model gives a good quantitative description of weak interactions in the original PDE (5.253) for general types of nonlinearities. When $\varepsilon > 0$, this ODE model exhibits fractal scatterings; but when $\varepsilon < 0$, it does not show fractal scattering. This explains why in the PDE (5.253) fractal scatterings occur only for certain types of nonlinearities but not for others. When fractal scattering does occur, the fractal structure from this ODE model closely matches that from the original PDE. To illustrate, we take the cubic-quintic nonlinearity (5.260) with $\kappa = 0.0003$ and initial conditions (5.262) for the PDE as in Fig. 5.17. In the ODE model (5.304), the corresponding ε value is 0.001, and the corresponding initial condition is

$$\zeta_0 = -10, \quad \dot{\zeta}_0 = \dot{\psi}_0 = 0, \tag{5.307}$$

with $\psi_0 (= \Delta\phi_0)$ being the control parameter. For this case, exit-velocity fractals from PDE simulations and ODE predictions are both displayed in Fig. 5.19(a, b). It is seen that the ODE graph not only captures all the main features of the PDE graph, but also gives a good quantitative prediction for the PDE graph. The main difference between the two graphs is that, at the middle of each hill, the ODE graph exhibits peaks of infinite height, but the PDE graph dips down instead. In those parameter regions, the two waves come together and interact strongly, which makes the reduced ODE model (5.304) and its prediction invalid.

5.8.3 A Universal Separatrix Map

In order to analytically explain the onset of fractal scattering in the ODE model (5.304) when $\varepsilon > 0$, we will investigate this model asymptotically for $|\varepsilon| \ll 1$ in this subsection. This case corresponds to weakly perturbed NLS equations such as those in Figs. 5.16 and 5.17. We will show that when $\varepsilon = 0$, Eq. (5.304) possesses a family of separatrix orbits. Then, in the perturbed system (5.304) with $\varepsilon \neq 0$, when the initial condition lies near those separatrix orbits, sensitive dependence to initial conditions can arise, and fractal scatterings can appear. To analytically explain this fractal scattering, we will derive a separatrix map for the orbits on a Poincaré section and show that this map exhibits a universal fractal when $\varepsilon > 0$ but not when $\varepsilon < 0$. This analysis will also be based on Melnikov-type integrals near separatrix orbits, similar to that in the study of resonant reflection windows in solitary wave collisions in the previous section. However, since the unperturbed system (5.304) is now four dimensional and strongly coupled, this analysis will contain a number of new features. For example, since the separatrix orbits here are not unique, one now needs to find a way to select the separatrix orbit which best approximates the perturbed solution. The derivation below follows Zhu et al. (2008b).

To study Eq. (5.304), we first define the following quantities:

$$E = \frac{1}{2}(\dot{\zeta}^2 - \dot{\psi}^2) - e^{\zeta}\cos\psi, \tag{5.308}$$

$$M = \dot{\zeta}\dot{\psi} - e^{\zeta}\sin\psi, \tag{5.309}$$

$$W = -\frac{1}{C}\operatorname{arccoth}\frac{\dot{\zeta} + i\dot{\psi}}{2C}, \tag{5.310}$$

and

$$C = \sqrt{(E + iM)/2}. \tag{5.311}$$

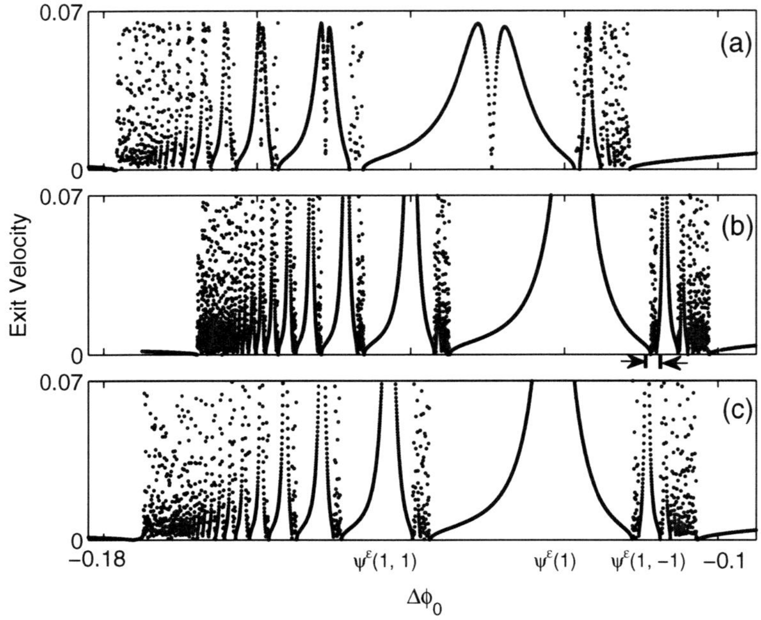

Figure 5.19. *Comparisons of exit-velocity fractals in weak interactions of solitary waves in Eq.* (5.253) *for the cubic-quintic nonlinearity* (5.260) *with* $\kappa = 0.0003$ *and initial conditions* (5.262)*:* (a) *PDE simulation results;* (b) *simulation results from the ODE model* (5.304)*;* (c) *simulation results from the separatrix map* (5.355)–(5.356)*. The marked segment in* (b) *is amplified in Fig.* 5.23*, where the solution dynamics is shown. Labels* $\psi^{\varepsilon}(1), \psi^{\varepsilon}(1,1), \ldots$ *are locations of singular peaks in* (c)*, which correspond to intersections of the initial-value curve* (5.360) *with the singular curves* $\gamma(1), \gamma(1,1), \ldots$ *in the map's fractal in Fig.* 5.22 *(from Zhu et al.* (2009) *after an error in figure* (b) *is corrected).*

Here, arccoth(z) is the inverse hyperbolic cotangent function, which can be expressed in terms of logarithmic functions as $\text{arccoth}(z) = \ln[(z+1)/(z-1)]/2$. The functions $\sqrt{\cdot}$ and arccoth$(\cdot)$ in the above definitions are multivalued. To make them unique, we choose $\text{Im}(\sqrt{\cdot}) \geq 0$ and $\text{Im}(\text{arccoth}(\cdot)) \in [0, \pi)$ at the initial time $\tau = \tau_0$, and we require these functions to be continuous at later times.

Integrable Solutions and Separatrix Orbits

When $\varepsilon = 0$, Eq. (5.304) is integrable (see Sec. 4.3). By defining

$$Y = \zeta + i\psi, \tag{5.312}$$

Eq. (5.304) simplifies to

$$Y_{\tau\tau} = e^Y. \tag{5.313}$$

Multiplying this equation by $\dot{Y}$ and integrating once, we get

$$\dot{Y}^2 - 2e^Y = 4C_0^2, \tag{5.314}$$

where C_0 is the value of C at the initial time $\tau = \tau_0$. Notice that E and M are the real and imaginary parts of the quantity $(\dot{Y}^2 - 2e^Y)/2$; hence E and M are two conserved quantities of this integrable system, and they will be called the energy and momentum, respectively. In addition to E and M, one can verify that for this integrable system, $\dot{W} = 1$, and hence $\text{Im}(W)$ is the third conserved quantity.

When Eq. (5.314) is combined with (5.313), we get

$$\ddot{Y} = -\frac{1}{2}(4C_0^2 - \dot{Y}^2). \tag{5.315}$$

This equation is the same as Eq. (4.142) in the previous chapter, and hence its analytical solution has been given by Eq. (4.145) there. For the convenience of the following analysis, we will express this solution in another form. First, the solution to Eq. (5.315) can be written as

$$\dot{Y}(\tau) = -2C_0 \coth C_0(\tau - \tau_0 + W_0), \tag{5.316}$$

where W_0 is the value of W at $\tau = \tau_0$. Utilizing the initial condition $Y(\tau_0) = Y_0$ as well as the relation

$$\dot{Y}_0^2 - 2e^{Y_0} = 4C_0^2 \tag{5.317}$$

obtained by setting $\tau = \tau_0$ in (5.314), and recalling properties of hyperbolic functions, we then find that the solution to (5.316) can be written as

$$Y(\tau) = \ln\left[2C_0^2 \text{csch}^2 C_0(\tau - \tau_0 + W_0)\right], \tag{5.318}$$

where $\text{csch}(z) = 1/\sinh(z)$ is the hyperbolic cosecant function. Using the relation (5.317) and properties of hyperbolic functions, it is easy to check that this solution and the solution given in Eq. (4.145) of the previous chapter are equivalent. The asymptotic behavior of the solution (5.318) can be easily determined. If we let $C_0 = a + ib$, where a, b are the real and imaginary parts of C_0, then if $a \neq 0$,

$$Y(\tau) \to -2|a|\tau - \text{sgn}(a) 2b\tau i, \quad \tau \to \infty. \tag{5.319}$$

We see that ζ escapes to $-\infty$ with exit velocity $2|a|$, and thus we call these orbits escape orbits. For these orbits, the sign of $\dot{\psi}_\infty$ is determined only by the sign of M since $\text{sgn}(\dot{\psi}_\infty) = -\text{sgn}(ab) = -\text{sgn}(M)$. Physically, the sign of $\dot{\psi}_\infty$ is related to the final power distribution between the two solitary waves after interactions. If $a = 0$ but $b \neq 0$, i.e., $E < 0$ and $M = 0$, then the orbits are periodic orbits with period

$$T_p = \pi/|b| = \sqrt{2}\pi|E|^{-1/2}. \tag{5.320}$$

If $a = b = 0$, i.e., $E = M = 0$, then the right-hand side of Eq. (5.314) is zero. Integrating this first-order equation, we readily find that the solution can be written as

$$Y_s(\tau) = -\ln\left[i\sigma A^{-\frac{1}{2}} + (\tau - \tau_M)/\sqrt{2}\right]^2, \tag{5.321}$$

where $A = e^{\zeta_M}$, ζ_M is the maximum of ζ, τ_M is the time when $\zeta = \zeta_M$, and σ is the sign of $\dot{\psi}$ at $\tau = \tau_M$. This solution can also be obtained by taking $C_0 \to 0$ in the general solution (5.318). Escape orbits are unbounded, while periodic orbits are bounded. The orbits (5.321) separate the unbounded orbits from bounded ones, and hence they are the separatrix orbits. On these orbits,

$$\zeta \to -\infty, \quad \dot{\zeta} \to 0, \quad \psi \to 0, \quad \dot{\psi} \to 0, \quad \text{as } \tau \to \pm\infty. \tag{5.322}$$

Notice that these separatrix orbits form a continuous family since their maximum ζ_M is arbitrary. In addition, σ can be ± 1.

It is noted that the solution (5.318), for general values of E and M, can develop finite-time singularities for certain initial conditions. Singularity would occur if at certain time $\tau_c > \tau_0$,

$$C_0(\tau_c - \tau_0 + W_0) = n\pi i, \quad n = 0, \pm 1, \pm 2, \dots. \tag{5.323}$$

By splitting this complex relation into the real and imaginary parts, and recalling our choices for branches of $\sqrt{\cdot}$ and $\text{arccoth}(\cdot)$ below Eq. (5.310), we find that when $a \neq 0$, the condition for singular solutions is

$$S(\tau = \tau_0) = 2k, \quad k = 0, 1, 2, \dots, \tag{5.324}$$

where the function $S(\tau)$ is defined as

$$S = \frac{2|C|^2 \text{Im}(W)}{\pi \text{Re}(C)}. \tag{5.325}$$

If $a = 0$, the condition for singular solutions is $\text{Im}(W_0) = 0$. Physically, at the time of singularity τ_c, the two solitary waves strongly collide, and thus τ_c is the collision time. Let us call the initial values of the integrable system (5.304) which satisfy the above singularity conditions *singularity points*. Then if the initial condition is near one of those singularity points, the two waves will come close to each other and strongly interact. If the initial condition is not near those singularity points, the two waves will remain apart from each other for all times, and their interaction will be weak throughout. These singularity points are important, since it was observed from numerical simulations that fractal scatterings appear near these singularity points when $0 < \varepsilon \ll 1$ (Zhu and Yang (2007)). More will be said about this later in this section.

Perturbed Solutions

Next, we study solutions in the perturbed system (5.304) where $\varepsilon \ll 1$. To motivate our analysis, we first show a typical numerical solution. For this purpose, we take $\varepsilon = 0.001$, which corresponds to both the cubic-quintic and quadratic-cubic nonlinearities in the PDE simulations of Fig. 5.16. The initial condition is taken as (5.307) and $\psi_0 = -0.1081$, which corresponds to the initial condition (5.262) in the PDE simulations. Time evolutions of the solution (ζ, ψ) are plotted in Fig. 5.20. We see that ζ undergoes five oscillations and then escapes to $-\infty$. Each oscillation corresponds to a "bouncing" motion between the two solitons, similar to that found in fractal scatterings of soliton collisions (see Fig. 5.12). Each bounce occurs at a local maximum of ζ, where the two waves are closest and most energy exchange between them takes place. After a bounce, the two waves retreat from each other, but may turn around and bounce again. Each local minimum of ζ (where separation between the two waves is locally the largest) is called a saddle approach. Saddle approaches create sensitive dependence on initial conditions, and sequences of saddle approaches (repeated bounces) lead to fractal scattering. For each orbit, we use indices $n = 0, 1, 2, \ldots$ to label its saddle-approach sequence. All variables with subscript n represent values of these variables at the nth saddle approach. Conditions for the existence of saddle approaches are that

$$\dot{\zeta} = 0, \qquad \ddot{\zeta} = \cos\psi\, e^{\zeta} > 0. \tag{5.326}$$

Here Eq. (5.304) has been used for the second condition. Thus in view of H's definition (5.306) as well as $\varepsilon \ll 1$, the Hamiltonian H must be negative at a saddle approach. The same holds for E as well.

Next we examine evolutions of E, M, and W. Evolutions of E and M are also shown in Fig. 5.20. It is seen that E and M are initially small and remain small throughout interactions. They stay roughly constant in vicinities of saddle approaches, but undergo significant changes near bouncing points. This is understandable, since interactions are the weakest at maximal separations (saddle approaches) and the strongest at minimal separations (bouncing points). From one saddle approach to the next, the net change of E is approximately zero, but the net change of M is significant and is comparable to M itself. Notice also that during interactions, E remains negative, which is expected in view of the conditions for saddle approaches above. But M can change the sign. The evolution of $\mathrm{Im}(W)$ is similar to that of M, but $\mathrm{Re}(W)$ approximately grows linearly with time since $\dot{W} = 1$ in the integrable case.

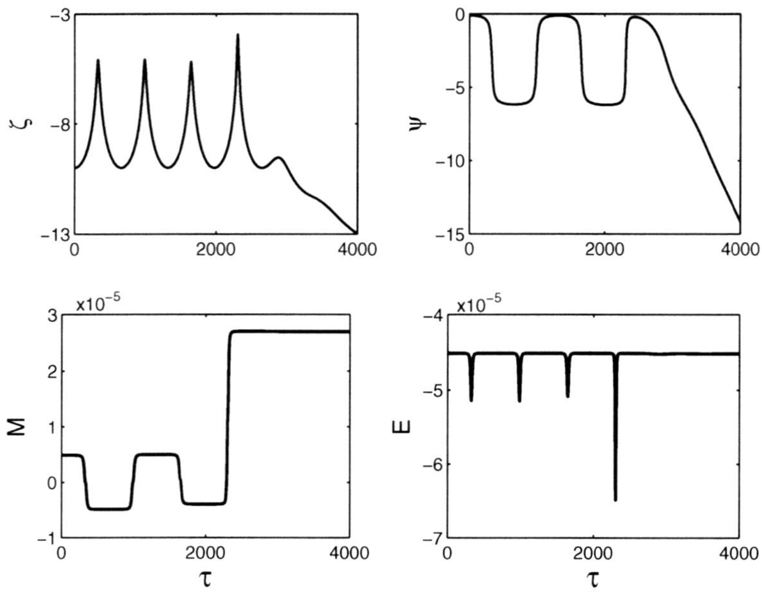

Figure 5.20. *Evolutions of ζ, ψ as well as E, M in the perturbed system* (5.304) *with $\varepsilon = 0.001$. The initial condition is $\zeta_0 = -10$, $\dot{\zeta}_0 = \dot{\psi}_0 = 0$, and $\psi_0 = -0.1081$. (After Zhu et al.* (2008b).*)*

From the above observations, we see that if we have a way of determining the net changes ΔE, ΔM, and $\Delta \mathrm{Im}(W)$ from one saddle approach to another, then we would be able to completely determine the final state of the orbit. Indeed, since almost all orbits eventually escape, i.e., $\zeta \to -\infty$ when $\tau \to \infty$, then by setting $\tau = \infty$ in the definitions (5.306) and (5.309) we get

$$\dot{\zeta}_\infty^2 - \frac{\dot{\psi}_\infty^2}{1+\epsilon} = 2H, \quad \dot{\zeta}_\infty \dot{\psi}_\infty = M_\infty. \tag{5.327}$$

Then the exit velocity $|\dot{\zeta}_\infty|$ can be obtained as

$$|\dot{\zeta}_\infty| = \sqrt{H + \sqrt{H^2 + M_\infty^2/(1+\varepsilon)}}. \tag{5.328}$$

Since H is always conserved, $|\dot{\zeta}_\infty|$ is then determined only by M_∞. The sign of $\dot{\zeta}_\infty$ is always negative since ζ escapes to $-\infty$ when $\tau \to \infty$. After $\dot{\zeta}_\infty$ is known, the power-distribution parameter $\dot{\psi}_\infty$ can be derived from Eq. (5.327) accordingly.

In order to calculate changes in E, M, and $\mathrm{Im}(W)$, we first need to find a good approximation for the perturbed solution. For this purpose, we notice that in weak interactions, E and M are initially small since $\zeta_0 \ll -1, \dot{\zeta}_0, \dot{\psi}_0 \ll 1$ (see (5.308)–(5.309)). Because $\varepsilon \ll 1$,

E and M will remain small during later evolutions. Thus we expect the perturbed solution in (5.304) to be near a separatrix solution (5.321). Specifically, between two adjacent saddle approaches, there is a separatrix orbit (5.321) near which the perturbed solution lies. Notice that the separatrix orbits (5.321) form a continuous family whose maximum ζ_M is a free parameter. Notice also from Fig. 5.20(a) that in the perturbed solution, the ζ-maxima between saddle approaches can change from one segment to another. Thus, the maximum ζ_M of the separatrix orbit which approximates the perturbed solution will also vary from one segment to the next. The determination of ζ_M in a separatrix orbit to approximate each segment of the perturbed solution is a new feature of this analysis.

To calculate the net changes ΔE, ΔM, and $\Delta \mathrm{Im}(W)$ between saddle approaches, we first determine how M, E, and W evolve with time. In the integrable case, M and E are conserved, and $\dot{W} = 1$. In the perturbed case, we find from (5.304) that

$$\frac{dM}{d\tau} = \varepsilon e^{\zeta} \sin\psi\, \dot{\zeta}, \tag{5.329}$$

$$\frac{dE}{d\tau} = -\varepsilon e^{\zeta} \sin\psi\, \dot{\psi}, \tag{5.330}$$

$$\frac{dW}{d\tau} = -\frac{\dot{E} + i\dot{M}}{2(E + iM)} W + D, \tag{5.331}$$

where

$$D = -\frac{(\dot{\zeta} + i\dot{\psi})(\dot{E} + i\dot{M})}{2e^{\zeta + i\psi}(E + iM)} + 1 + \varepsilon \sin^2\psi + i\varepsilon \sin\psi \cos\psi. \tag{5.332}$$

If the quantities E_n, M_n, W_n at the nth saddle approach are given, then we can calculate these quantities at the $(n+1)$st saddle approach by integrating the above dynamical equations. Recall that saddle approaches exist only when $E_n < 0$. Thus in these calculations, E_n will be required to be always negative. Our calculations will be performed in the asymptotic regime $\varepsilon \ll 1, E_n \ll 1$, and $M_n \ll 1$. Because $\varepsilon \ll 1$, the perturbed solution is close to the unperturbed solution between the two saddle approaches. Because $E_n, M_n \ll 1$, the solutions (both perturbed and unperturbed) will be close to a separatrix orbit. The outline of our calculations is the following. We first derive changes in the momentum M and energy E from one saddle approach to the next by Melnikov integrals along a separatrix orbit (5.321). The parameters of this separatrix orbit are determined by requiring it to have the same ζ-maximum point as the unperturbed solution (5.318). To find the change in W, we use the method of variation of parameters to integrate the linear inhomogeneous equation (5.331) from one saddle approach to the next. The time between the two saddle approaches, which comes into the calculations, is not simple in general, so we approximate it under the additional assumption that $M_n/E_n \ll 1$. In this case, the unperturbed orbit is approximately a periodic orbit with period (5.320), and this period gives the needed time.

It turns out that this additional assumption of $M_n/E_n \ll 1$ is not just for the convenience of the analysis. In the parameter regions of fractal scatterings, this condition actually must be satisfied in most parts of the weak-interaction process. The reason is that this condition guarantees the integrable solution (5.318) to be quasi-periodic; hence the two solitons will repeatedly bounce off each other and create a sequence of saddle approaches, and these saddle approaches are critical for the sensitive dependence of solutions and the fractal structures. In weak interactions of Fig. 5.20, the ratio M_n/E_n is indeed small (on the order of 0.1) before the fourth (last major) bounce. After the fourth bounce, M/E is not small. But now, M and E do not change much, so there is no need to do calculations in that region. Thus this assumption of $M_n/E_n \ll 1$ is satisfied for this example. Similarly, in the fractal regions of Fig. 5.19, this assumption is satisfied as well. In some other fractal structures which have been reported for different initial conditions, M/E may not be small initially (see Zhu and Yang (2007), Zhu et al. (2008b)). But it becomes small after the first major bounce and then remains small for the rest of the interactions. Thus the assumption of $M_n/E_n \ll 1$ is appropriate for fractal scatterings in weak interactions.

Derivation of the Separatrix Map

We now start our calculations by integrating Eqs. (5.329)–(5.331) from one saddle approach to the next. Let us consider M first. From (5.329), we have

$$\Delta M_n = M_{n+1} - M_n = \int_{\tau_n}^{\tau_{n+1}} \varepsilon\, e^{\zeta} \sin\psi\, \dot{\zeta}\, d\tau, \tag{5.333}$$

where τ_n and τ_{n+1} are the times of the nth and $(n+1)$st saddle approaches. Since $\varepsilon \ll 1$, the perturbed solution inside the integrand will be approximated by the unperturbed solution. Thus we obtain

$$\Delta M_n = \int_{\tau_n}^{(\tau_{n+1})_u} \varepsilon\, e^{\zeta_u} \sin\psi_u\, \dot{\zeta}_u\, d\tau. \tag{5.334}$$

Here the variables with subscript "u" represent the unperturbed solution. Since E_n, $M_n \ll 1$, we can further approximate the unperturbed solution in the integrand of (5.334) by a separatrix solution (5.321). This separatrix solution is not unique since it contains parameters σ, ζ_M, and τ_M. We now show how to select these parameters so that the resulting separatrix orbit best approximates the unperturbed solution. We cannot ask the separatrix orbit to start at the same nth saddle approach since ζ of the separatrix orbit does not have local minima. Notice that most contributions to the integral of (5.334) come from the ζ-maximum region; thus it is natural to ask the separatrix orbit to have the same ζ-maximum point as the unperturbed solution. In addition, since σ is the sign of $\dot{\psi}$ at the ζ-maximum in the separatrix orbit, it is the sign of $\dot{\psi}_u$ at the ζ-maximum. A valuable feature of our system (5.304) is that the unperturbed equations admit explicit closed-form solutions; see (5.318).

This makes it easy to implement the above requirements. These requirements uniquely determine the separatrix orbit (5.321) and lead to the following conditions:

$$\sigma_n = \operatorname{sgn}\left\{\operatorname{Im}\left[-2C_n\coth(C_n(\tau_u^* + W_n))\right]\right\}, \tag{5.335}$$

$$A_n = 2\left|C_n^2\operatorname{csch}^2\left[C_n(\tau_u^* + W_n)\right]\right|. \tag{5.336}$$

Here $\tau_n + \tau_u^*$ is the time where ζ_u reaches its maximum. However, we still do not know τ_u^* due to the complexity of the integrable solution (5.318). This is where the assumption $M_n/E_n \ll 1$ comes in. Due to this assumption, we can expand the general solution as a series in M_n/E_n. From its leading-order term, which is a periodic solution with period (5.320), we find the leading-order approximation for τ_u^* as

$$\tau_u^* = \frac{1}{2}T_p = \frac{\pi}{\sqrt{2|E_n|}}. \tag{5.337}$$

To clearly illustrate these ideas, we plot the perturbed solution, the unperturbed solution, and the separatrix solution together in Fig. 5.21. This figure was generated with typical numerical values in fractal regions when $\varepsilon \ll 1$. From this figure, we see that (i) the unperturbed solution is very close to the perturbed solution; (ii) the separatrix solution is very close to the unperturbed solution near the ζ-maximum. Away from the ζ-maximum, the separatrix orbit and the other two orbits start to deviate from each other. But in those regions, the contribution to the integral of (5.334) is negligible since ζ is large negative. Thus our separatrix-orbit approximation for the unperturbed solution is valid.

When the separatrix solution (5.321) is substituted into (5.334), we readily find that

$$\Delta M_n = \varepsilon \int_{-\infty}^{+\infty} e^{\zeta_s} \sin\psi_s \, \dot{\zeta}_s \, d\tau = \frac{1}{2}\varepsilon\pi\sigma_n A_n, \tag{5.338}$$

where σ_n and A_n are given by (5.335)–(5.336). Note that in the near-separatrix approximation, the saddle approaches τ_n and τ_{n+1} are asymptotically large, so that τ_n and τ_{n+1} are approximated by $-\infty$ and $+\infty$ in the above calculations, as is usual for Melnikov-type integrals.

By similar calculations from (5.330) and utilizing the symmetry properties of separatrix solutions (5.321), we find that

$$\Delta E_n = \varepsilon \int_{-\infty}^{+\infty} e^{\zeta_s} \sin\psi_s \, \dot{\psi}_s \, d\tau = 0. \tag{5.339}$$

In other words, the energy does not change from one saddle approach to another. This is precisely what Fig. 5.20 shows. This fact is rather dramatic, and it contributes to the rather simple map we will obtain later in this section.

The ΔM_n formula (5.338) depends on W_n (in view of (5.336)), and thus we also need to calculate ΔW_n in order to continue the iteration. Notice that Eq. (5.331) for W is an

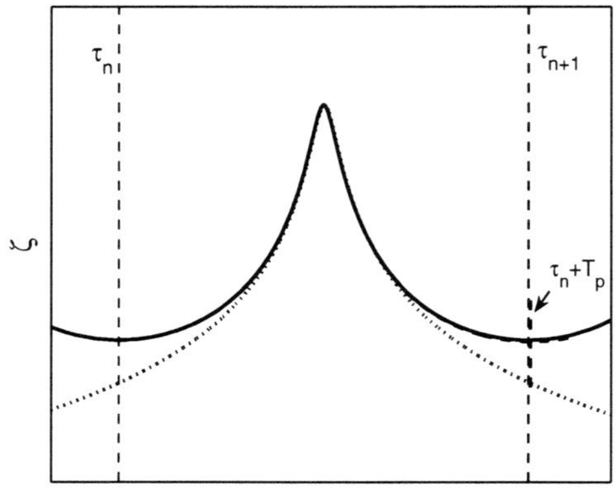

Figure 5.21. *Plots of the perturbed solution $\zeta(\tau)$ (solid line), the unperturbed solution (dashed line), as well as the separatrix solution (dotted line) in our analysis. The solid and dashed lines are almost indistinguishable.*

inhomogeneous linear ODE. Recalling the definition (5.311) of C, we see that a homogeneous solution of this ODE is $W = 1/C(\tau)$. Thus by the method of variation of parameters, we can integrate this inhomogeneous ODE from τ_n to τ_{n+1} and get

$$W_{n+1} = W_n \frac{C_n}{C_{n+1}} + \frac{\int_{\tau_n}^{\tau_{n+1}} \sqrt{E(\tau)+iM(\tau)}\, D(\tau)d\tau}{\sqrt{E_{n+1}+iM_{n+1}}}. \tag{5.340}$$

Since $E, M \ll 1$, the perturbed solution lies near the separatrix orbit (5.321) whose parameters are specified by (5.335)–(5.336). Thus we can approximate $\zeta + i\psi$ in the formula (5.332) of $D(\tau)$ by this separatrix solution Y_s. On the separatrix orbit, $\dot{Y}_s e^{-Y_s} = -(\hat{\tau} + i\alpha)$, where $\hat{\tau} = \tau - \tau_n - T_p/2$, and $\alpha = \sigma_n\sqrt{2/A_n}$. Here the approximation (5.337) for the time of the separatrix peak has been used. Then using integration by parts, we find that to the leading order in ε,

$$W_{n+1} = W_n \frac{C_n}{C_{n+1}} + \frac{(\hat{\tau}+i\alpha)\sqrt{E+iM}\,\big|_{\tau_n}^{\tau_{n+1}}}{\sqrt{E_{n+1}+iM_{n+1}}}. \tag{5.341}$$

Recalling the definition (5.311) for C as well as the expressions above for $\hat{\tau}$ and α, the equation for W_{n+1} becomes

$$W_{n+1} = W_n \frac{C_n}{C_{n+1}} + \frac{\pi}{\sqrt{2|E_n|}}\left(1+\frac{C_n}{C_{n+1}}\right) + i\sqrt{2}\sigma_n A_n^{-\frac{1}{2}}\left(1-\frac{C_n}{C_{n+1}}\right), \tag{5.342}$$

where σ_n and A_n are given by (5.335)–(5.336).

This map for W_n, as well as the one for M_n in (5.338), is not simple. It turns out that, under the same assumption $M_n/E_n \ll 1$ we have made above, these maps can be greatly simplified. First, due to (5.339), $E_n = E_0$ for any n. Thus we will use E_0 to replace E_n. Next, due to the assumption $M_n/E_n \ll 1$, we can expand C_n/C_{n+1} asymptotically as

$$\frac{C_n}{C_{n+1}} = 1 - \frac{M_{n+1} - M_n}{2E_0} i. \tag{5.343}$$

In addition, the third term on the right-hand side of (5.342) is of higher order in M_n/E_0 and can be neglected. Thus, to leading order in M_n/E_0,

$$W_{n+1} = W_n \left(1 - \frac{M_{n+1} - M_n}{2E_0} i\right) + \frac{\pi}{\sqrt{2|E_0|}} \left(2 - \frac{M_{n+1} - M_n}{2E_0} i\right). \tag{5.344}$$

Notice that at the first (zeroth) saddle approach, $\dot{\zeta}_0 = 0$, and $C_0 = \sqrt{E_0/2} + O(M_n/E_0)$. In addition, $|\dot{\psi}_0| \leq \sqrt{2|E_0|}$ in view of (5.308) and (5.326). Thus from (5.310) and recalling our choices for branches of $\sqrt{\cdot}$ and $\operatorname{arccoth}(\cdot)$ below (5.310), the leading-order approximation for $\operatorname{Re}(W_0)$ under the assumption $M_n/E_0 \ll 1$ is that

$$\operatorname{Re}(W_0) = -\pi/\sqrt{2|E_0|}. \tag{5.345}$$

Then the general formula for W_n is

$$W_n = \frac{\pi}{\sqrt{2|E_0|}} \left(2n - 1 - \widehat{S}_n \frac{M_n}{2E_0} i\right), \tag{5.346}$$

where

$$\widehat{S}_{n+1} M_{n+1} = 2n M_{n+1} - (2n - \widehat{S}_n) M_n. \tag{5.347}$$

Indeed, by inserting (5.346) and (5.347) into (5.344) and neglecting second-order terms in M_n/E_0 due to the assumption $M_n/E_n \ll 1$ and (5.339), we see that Eq. (5.344) is satisfied. Notice from (5.346) that

$$\widehat{S}_n = -\frac{\operatorname{Im}(W_n) 2E_0 \sqrt{2|E_0|}}{\pi M_n}. \tag{5.348}$$

Recalling the definition (5.325) for $S(\tau)$, the above expression of $\widehat{S}_n$ is simply the leading-order term of $S_n = S(\tau_n)$ under the assumption $M_n/E_n \ll 1$. Thus we will treat $\widehat{S}_n = S_n$ below by neglecting the higher-order difference between them.

Substituting formula (5.346) into (5.335)–(5.337) and keeping only the leading-order terms with respect to M_n/E_0, we find that

$$\sigma_n = \operatorname{sgn}\left(\frac{4\sqrt{2}|E_0|}{(2n - S_n)\pi} \frac{E_0}{M_n}\right) = -\operatorname{sgn}[(2n - S_n) M_n], \tag{5.349}$$

$$A_n = \frac{16|E_0|}{(2n - S_n)^2 \pi^2} \frac{E_0^2}{M_n^2}. \tag{5.350}$$

Here we have used the asymptotic expansions of $\coth(n\pi i+z)=\coth(z)\to 1/z$ and $\text{csch}^2(n\pi i+z)=\text{csch}^2(z)\to 1/z^2$ as $z\to 0$ for any integer of n. Introducing a new variable

$$Q_n \equiv (2n-S_n)M_n, \tag{5.351}$$

then after substituting (5.349) and (5.350) into (5.338) and utilizing (5.347), we finally obtain the following simplified map:

$$M_{n+1}=M_n-\text{sgn}(Q_n)\frac{8|E_0|^3\varepsilon}{\pi Q_n^2}, \tag{5.352}$$

$$Q_{n+1}=Q_n+2M_{n+1}, \tag{5.353}$$

with initial conditions M_0 and Q_0, where $Q_0=-S_0M_0$ in view of (5.351). Here the initial value S_0 is calculated from (5.325) and (5.310). This map was derived near the separatrix orbits (5.321) and is thus called the separatrix map. It is asymptotically accurate when $\varepsilon\ll 1$, $M_n, E_n\ll 1$, and $M_n/E_n\ll 1$.

The separatrix map (5.352)–(5.353) can be further simplified. Introducing normalized variables

$$G=\frac{8|E_0|^3\varepsilon}{\pi}, \quad m_n=G^{-1/3}M_n, \quad q_n=G^{-1/3}Q_n, \tag{5.354}$$

this map then becomes

$$m_{n+1}=m_n-\frac{\text{sgn}(\varepsilon q_n)}{q_n^2}, \tag{5.355}$$

$$q_{n+1}=q_n+2m_{n+1}. \tag{5.356}$$

This is a simple but important second-order area-preserving map. It does not have any parameters in it (except the sign of ε). This universal map governs weak two-wave interactions in the generalized NLS equations (5.253).

It is noted that the separatrix map (5.355)–(5.356) somewhat resembles the standard map

$$m_{n+1}=m_n+K\sin q_n, \tag{5.357}$$

$$q_{n+1}=q_n+m_{n+1}, \tag{5.358}$$

where q_n is taken modulo 2π, and K is a constant. This map is also area preserving and is a classic example which exhibits Hamiltonian chaos when $K>0$ (Chirikov (1979), Lichtenberg and Lieberman (1992)). As we will show below, the separatrix map (5.355)–(5.356) exhibits intricate dynamics when $\epsilon>0$ as well, but its dynamics is qualitatively very different from that in the standard map. By eliminating the variable m_n, the map (5.355)–(5.356) can be rewritten as a single equation for q,

$$q_{n+1}-2q_n+q_{n-1}+\frac{2\text{sgn}(\varepsilon)}{|q_n|^3}q_n=0, \tag{5.359}$$

which is a stationary discrete NLS equation with an unusual nonlinear term.

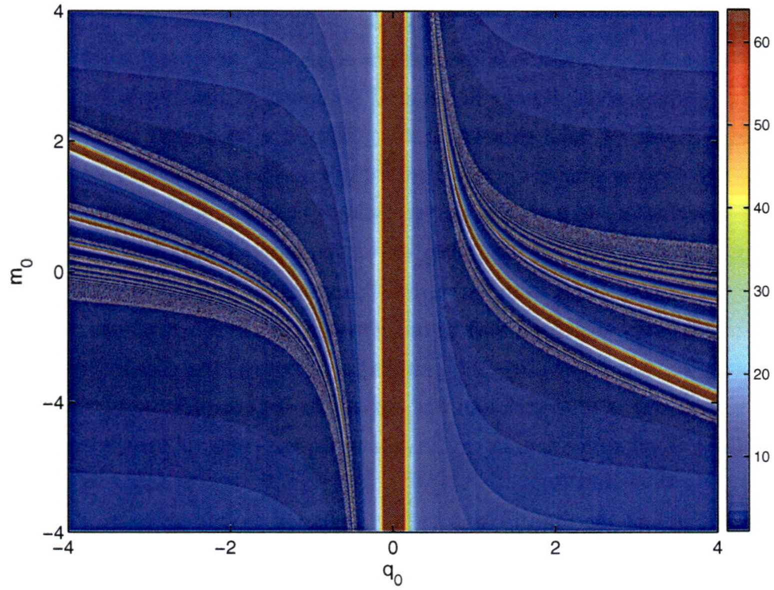

Figure 5.22. *Graph of* $|m_\infty|$ *in the initial-value space* (q_0, m_0) *for the separatrix map* (5.355)–(5.356) *with sgn*$(\varepsilon) = 1$. *Colors represent value levels of* $|m_\infty|$. *(After Zhu et al.* (2009).*)*

5.8.4 Fractal in the Separatrix Map

When $\text{sgn}(\varepsilon) = 1$, the separatrix map (5.355)–(5.356) exhibits a fractal structure in the graph of $|m_\infty|$ over the initial-condition space (q_0, m_0). This fractal is displayed in Fig. 5.22. It provides the key for the understanding of universal fractal structures in weak interactions in the generalized NLS equations (5.253).

The geometric structure of this fractal in the separatrix map can be completely ascertained. It contains an infinite number of curvy red stripes of varying thickness (where $|m_\infty|$ is high). In the middle of each red stripe, there is a curve where $|m_\infty| = \infty$. On this singular curve, $q_n = 0$ for some $n \geq 0$, which causes m_{n+1} and the subsequent m values to be infinite. These singular curves together form the backbone of the map's fractal. Each singular curve γ is characterized by a unique finite binary sequence $\mathbf{a} = \text{sgn}(q_0, q_1, \ldots, q_{n-1})$, where $q_n = 0$. Hence it can be denoted by $\gamma(\mathbf{a})$. For example, $\gamma(1)$ and $\gamma(-1)$ are located in the middle of the thickest red stripes in the right and left half planes of the fractal in Fig. 5.22, respectively. It is easy to see that for any binary sequence $\mathbf{a}$, $\gamma(\mathbf{a})$ and $\gamma(-\mathbf{a})$ are antisymmetric to each other about the origin, and thus we consider only singular curves on the right half plane with $\mathbf{a} = (1, a_1, \ldots, a_{n-1})$ below. First, the singular curves $\{\gamma(1), \gamma(1,1), \gamma(1,1,1), \ldots\}$ form the primary cascading sequence from the left to the right. This sequence accumulates to the limit curve $\gamma(1,1,1,1,\ldots)$, which is the right boundary of the fractal region. This primary sequence of singular curves defines the overall geometry of the map's fractal in Fig. 5.22. Next, on the left side of each primary curve, there is a secondary structure which contains infinite and closely spaced curves. Inside each secondary structure of a primary curve $\gamma(\mathbf{a})$, there is a secondary sequence of curves $\{\gamma(\mathbf{a},-1),\gamma(\mathbf{a},-1,-1),\gamma(\mathbf{a},-1,-1,-1),\ldots\}$ which cascades to the left. This secondary curve sequence defines the overall geometry of the secondary structure. Near each curve in this secondary sequence (and on its right-hand side), there is a tertiary structure, and so on. This pattern in the structure continues indefinitely. Every time one zooms into a higher-order structure, the cascading direction of its higher-order sequence is reversed, and the side of the higher-order structure relative to its associated lower-level curve (either on the left or on the right) is also reversed (Zhu et al. (2009)).

The dynamics of the map is intimately connected to this geometric structure of the fractal in Fig. 5.22. It turns out that for all initial conditions (q_0,m_0) in the same red stripe of the fractal, their orbits are closely related. Specifically, in a red stripe of an arbitrary binary sequence $\mathbf{a}=(a_0,a_1,a_2,\ldots,a_{n-1})$, all orbits are very close to each other up to (q_n,m_n). In particular, the signs $\mathrm{sgn}(q_0,q_1,\ldots,q_{n-1})$ are all the same and given by this binary sequence $\mathbf{a}$. Beyond the point (q_n,m_n), m_k quickly approaches a constant m_∞ whose value is given by the color bar in Fig. 5.22, and q_k approaches $\pm\infty$. When (q_0,m_0) is on the singular curve of the red stripe, $|m_k|=\infty$ $(k>n)$.

The map's fractal in Fig. 5.22 provides the key for a complete understanding of fractal structures in weak interactions of solitary waves. Furthermore, it enables us to make quantitative predictions for those fractal structures in the PDE (5.253) and the ODE model (5.304). For this purpose, notice that for any initial conditions of the PDE or ODE with a control parameter (such as the initial-phase difference $\Delta\phi_0$), the corresponding initial conditions for the map (5.355)–(5.356) form a continuous curve in the (q_0,m_0) space. The intersection of this initial-condition curve with the map's fractal in Fig. 5.22 then gives the map's prediction for m_∞, which in turn gives the map's prediction for the fractal structure in the exit-velocity graph of the PDE or ODE through Eq. (5.328). As an example, we consider the cubic-quintic nonlinearity (5.260) and the initial conditions (5.262) as in Fig. 5.17. The corresponding initial conditions for the ODE are (5.307), and for the map are

$$q_0=m_0=-\frac{1}{2}\pi^{1/3}\tan\psi_0\cdot\varepsilon^{-1/3}. \tag{5.360}$$

Here $S_0=-1$, and $\cos\psi_0>0$ due to the requirement of $E<0$ in the text below Eq. (5.326). These initial conditions form a straight line of slope one in Fig. 5.22. Using the m_∞ values on this straight line as well as Eq. (5.328), and scaling the result back to physical units, the exit-velocity graph predicted from the map is displayed in Fig. 5.19(c). This graph is also a fractal, and it closely matches those of the PDE and the ODE in (a, b) of that same figure. In particular, the peaks of infinite height in Fig. 5.19(c), marked by $\psi^{\epsilon}(\mathbf{a})$, correspond to the intersections between this initial-condition line (5.360) and the singular curves $\gamma(\mathbf{a})$ in the map's fractal; the hills in Fig. 5.19(c) correspond to red stripes in the map's fractal; the left boundary of the fractal region $\phi_L = -0.175$ in Fig. 5.19(c) corresponds to the intersection between the initial-condition line (5.360) and the limit curve $\gamma(1,1,1,1,\ldots)$ of the primary sequence in the map's fractal; and the right boundary of the fractal region $\phi_R = -0.103$ in Fig. 5.19(c) corresponds to the intersection between the initial-condition line (5.360) and the limit curve $\gamma(1,-1,-1,-1,\ldots)$ of the secondary sequence of $\gamma(1)$ in the map's fractal.

From the dynamics of the map on its fractal in Fig. 5.22, we can obtain the dynamics of the map in the exit-velocity fractal of Fig. 5.19(c). Then using the correspondence of exit-velocity fractals from the map, the ODE and the PDE in Fig. 5.19, we can ascertain the dynamics of weak interactions in the ODE and the original PDE for $\Delta\phi_0$ (i.e., ψ_0) anywhere in their exit-velocity fractals. First, we consider the ODE dynamics for ψ_0 in its exit-velocity fractal of Fig. 5.19(b). Each ψ_0 value in this ODE's fractal lies on a hill with a singular peak in its middle. This singular peak corresponds to $\psi^{\epsilon}(\mathbf{a})$ in the map's exit-velocity fractal in Fig. 5.19(c). Suppose the binary sequence associated with this singular peak is $\mathbf{a} = (a_0, a_1, a_2, \ldots, a_{n-1})$. When ψ_0 is at this singular peak in the ODE's fractal, the $\zeta(t)$ solution oscillates n times (i.e., the two waves bounce with each other n times), and then approaches infinity at a finite time t_c and terminates. The $-\dot{\psi}$ solution exhibits spikes at the n bounce points before the singularity time t_c, whose sign sequence is given by $\mathbf{a}$. Thus, the physical meaning of the binary sequence $\mathbf{a}$ of a singular peak in the exit-velocity fractal of the ODE is that it gives the sign sequence of the $-\dot{\psi}(t)$ solution at the bounce points before the collision time. The reason is that the sign of q_k is the same as the sign of $-\Delta M_k$ in view of Eq. (5.355), which is then the same as the sign of $-\sigma_k$ in view of Eq. (5.338), which in turn is the same as the sign of $-\dot{\psi}$ at the kth bounce point in view of the text below Eq. (5.321). When ψ_0 is at other points of the hill with this binary sequence $\mathbf{a}$, the ODE solutions are very close to this singular solution of the singular peak up to the singularity time t_c. At $t \approx t_c$, the two waves are the closest (i.e., ζ is maximal). Beyond this time, all $\zeta(t)$ solutions go to $-\infty$, while all $\dot{\psi}(t)$ solutions approach constants $\dot{\psi}_\infty$ which have the same sign on the same side of the hill but opposite sign between the two sides of the hill. To illustrate the above general results, the ODE dynamics on a hill of binary sequence $\mathbf{a} = (1,-1,1)$ in the exit-velocity fractal of Fig. 5.19(b) is displayed in Fig. 5.23(a, b).

Next we discuss the PDE dynamics on a hill of an arbitrary binary sequence $\mathbf{a} = (a_0, a_1, a_2, \ldots, a_{n-1})$ in the exit-velocity fractal of Fig. 5.19(a). For all $\Delta\phi_0$ values on the hill, the PDE solutions are very close to each other up to the time when the two waves are the closest to each other. This time can be called the "collision" time, and is almost the same for the entire hill. Before this collision time, the two waves bounce n times. At each bounce

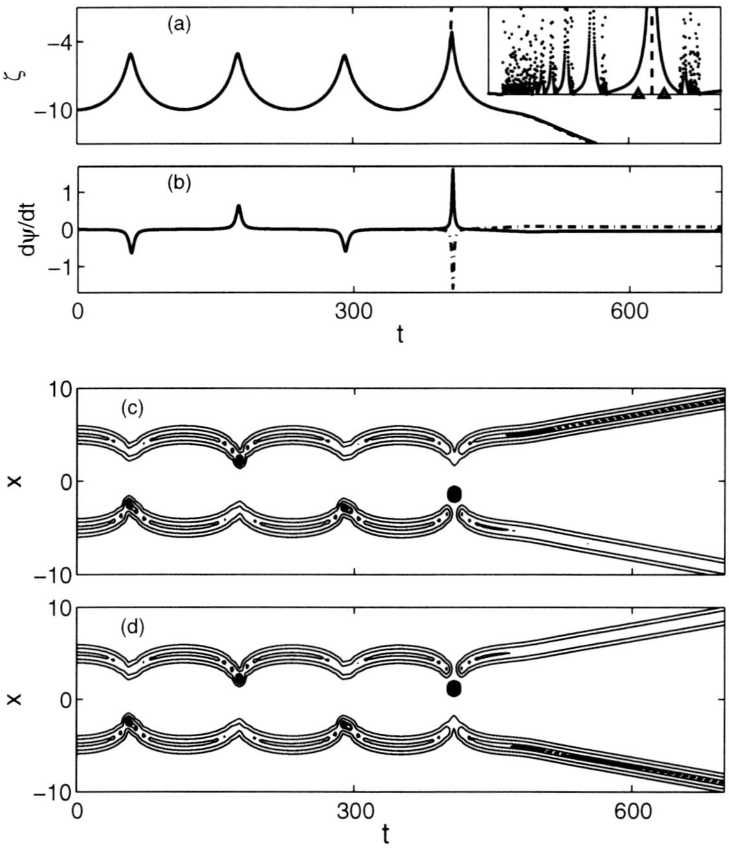

Figure 5.23. *Dynamics of ODE and PDE solutions in the exit-velocity fractals of Fig.* 5.19. *The inset in* (a) *is amplification of the segment marked in the ODE fractal of Fig.* 5.19(b)*, where the widest hill is the one of binary sequence* $\mathbf{a} = (1, -1, 1)$*. The ODE solutions at three marked points* $\psi_a < \psi_b < \psi_c$ *in the inset (by triangles and a vertical dashed line) are plotted as dash-dotted, dashed, and solid lines in* (a, b)*. Here,* $\psi_a = -0.10822$*,* $\psi_b = -0.108167$*, and* $\psi_c = -0.10812$*. PDE solutions at points corresponding to* ψ_a *and* ψ_c *in the PDE fractal of Fig.* 5.19(a) *are shown in* (c, d)*, respectively (*$\Delta\phi_0 = -0.118228$ *in* (c) *and* -0.118136 *in* (d)*). (After Zhu et al.* (2009)*.)*

point, the left wave has higher (or lower) amplitude if the corresponding digit in the binary sequence $\mathbf{a}$ is 1 (or -1). Thus the physical meaning of the binary sequence $\mathbf{a}$ of a hill in the exit-velocity fractal of the PDE is that it gives the sequence of relative amplitudes between the two waves at the bounce points before the collision time, with digit 1 (-1) meaning that the left wave is higher (lower). For most of the points on the hill (except a small portion in the middle), the two waves separate from each other after the collision time, and the exiting waves have opposite relative amplitudes on the two sides of the hill. The specific relative amplitudes of the two waves after escape are determined by the sign of $-\dot{\psi}_\infty$. The left wave will have higher (lower) amplitude if $-\dot{\psi}_\infty$ is positive (negative). As an example, the PDE dynamics on the hill of binary sequence $\mathbf{a} = (1, -1, 1)$ in the exit-velocity fractal of Fig. 5.19(a) is displayed in Fig. 5.23(c, d). Using the above general results, we can now understand the PDE dynamics in Fig. 5.18. Indeed, it is easy to see that the hill of the circled $\Delta\phi_0$ value in Fig. 5.17(a) corresponds to the binary sequence $\mathbf{a} = 1$, the hill of the circled $\Delta\phi_0$ value in Fig. 5.17(b) corresponds to the binary sequence $\mathbf{a} = (1, 1, -1)$, and the hill of the circled $\Delta\phi_0$ value in Fig. 5.17(c) corresponds to the binary sequence $\mathbf{a} = (1, 1, -1, 1, -1)$. These correspondences explain the number of position oscillations seen in Fig. 5.18.

By now, we have linked all fractal structures in weak interactions of the PDEs (5.253) to this unique fractal of the map in Fig. 5.22. Through this connection, we can determine how those PDE fractals change as the nonlinearities and initial conditions vary. To demonstrate, we consider the cubic-quintic nonlinearity (5.260) and the initial conditions (5.262) as before, but we allow the quintic coefficient κ to vary now. This varying κ causes the parameter ε to change. To determine the fractals in the PDEs for varying values of ϵ, we notice that the map's $|m_\infty|$ fractal at the intersection with the initial-condition line (5.360) on the right half plane lies between two accumulation points, $(q_a, q_a) = (0.741, 0.741)$ and $(q_b, q_b) = (1.271, 1.271)$ on singular curves $\gamma(1, -1, -1, -1, \ldots)$ and $\gamma(1, 1, 1, 1, \ldots)$, respectively. In view of the connection (5.360) between (q_0, m_0) and ψ_0, the corresponding ψ_0 values of these two accumulation points are

$$\phi_L = -\arctan(2q_b\pi^{-1/3}\varepsilon^{1/3}), \quad \phi_R = -\arctan(2q_a\pi^{-1/3}\varepsilon^{1/3}). \tag{5.361}$$

These ϕ_L and ϕ_R values are the left and right boundaries of the map's exit-velocity fractal on the negative $\Delta\phi_0$-axis (see Fig. 5.19(c)), and they are the map's predictions for the fractal regions in the PDEs. These formulae show that $\psi_{L,R} \to 0$ as $\varepsilon \to 0^+$, which means that the fractal region of the PDE shrinks to $\Delta\phi_0 = 0$ as $\varepsilon \to 0^+$, and its shrinking rate is proportional to $\varepsilon^{1/3}$ for $\varepsilon \ll 1$. This is precisely what happens. To illustrate, we choose two κ values 0.0029 and 0.0003 in the cubic-quintic nonlinearity (5.260), which correspond to $\varepsilon = 0.01$ and 0.001, respectively. Fractal structures of the PDEs for these two κ values are displayed in Fig. 5.24(a, b). It is seen that the fractal region indeed shrinks to $\Delta\phi_0 = 0$ as ε decreases to zero. We further recorded the ϕ_L and ϕ_R values in the PDE fractals at a number of ε points, and the data is plotted in Fig. 5.24(c). The theoretical formulae (5.361) for ϕ_L and ϕ_R are also plotted for comparison. It is seen that the PDE values and the map's analytical predictions agree very well, confirming the asymptotic accuracy of the separatrix map (5.355)–(5.356).

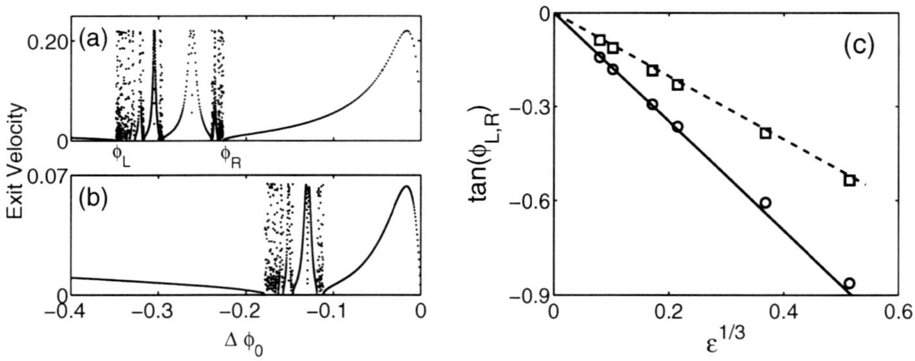

Figure 5.24. (a, b) *Exit-velocity versus initial-phase-difference graphs of the PDE for the cubic-quintic nonlinearity* (5.260) *with* $\kappa = 0.0029$ *and* 0.0003 *(corresponding to* $\varepsilon = 0.01$ *and 0.001), respectively;* (c) *plots of the left and right boundaries* ϕ_L *and* ϕ_R *of the fractal region versus* ε *from PDE simulations (circles and squares) and map predictions (solid and dashed lines). (After Zhu et al.* (2008a).*)*

Lastly, we note that when $\epsilon < 0$, the separatrix map (5.355)–(5.356) does not have a fractal structure in the graph of $|m_\infty|$ over the (q_0, m_0) plane (Zhu et al. (2009)). This explains why the ODE model (5.304) and the corresponding PDE system do not exhibit fractal scatterings when $\epsilon < 0$.

5.8.5 Physical Mechanism for Fractal Scatterings in Weak Interactions

After all this theoretical analysis, we now discuss the physical mechanism which causes fractal scatterings in weak interactions of solitary waves. From the analysis on the ODE model (5.304), we see that this fractal scattering is caused by the perturbed solutions (with $\varepsilon \neq 0$) lying near the separatrix orbits of the integrable system ($\varepsilon = 0$). It is well known that trajectories near separatrix orbits are sensitive to small perturbations, and this is the origin of these fractal scatterings. From the point of view of the PDE (5.253), $\varepsilon = 0$ corresponds to the NLS equation, and the separatrix orbits (5.321) correspond to second-order solitons of the NLS equation whose zero in the spectral plane is double folded (these second-order solitons are also called double-pole solitons in the literature). These second-order solitons can be viewed as the limit of a two-soliton solution whose two simple zeros approach each other in the spectral plane. Analytical expressions for these second-order solitons can be found in Gagnon and Stiévenart (1994). When the velocity of this solution is normalized to zero by Galilean transformation (see (5.7)), and its amplitude at $t = \pm\infty$ normalized to $\sqrt{2}$

through dilation transformation $t \to \alpha^2 t, x \to \alpha x, U \to \alpha^{-1} U$, this second-order soliton in the NLS equation (5.305) can be written as

$$U(x,t) = 2\sqrt{2}\,\frac{(x+c)\sinh x - (1+2it)\cosh x}{\cosh^2 x + (x+t)^2 + 4t^2}\, e^{it}, \tag{5.362}$$

where c is a free real parameter (this solution can also be obtained from the general formulae for higher-order solitons in Shchesnovich and Yang (2003a, 2003b)). When $t \to \pm\infty$, these solutions approach two single solitons of equal amplitude and phase, and the distance between them approaches infinity logrithmically. These characteristics closely match the separatrix orbits (5.321) and their large-time asymptotics (5.322). The sensitivity of trajectories near separatrix orbits (5.321) in the perturbed ODE (5.304) with $|\varepsilon| \ll 1$ has its counterpart in the PDE (5.253) as well. Here $|\varepsilon| \ll 1$ corresponds to the PDE (5.253) being close to the NLS equation. In this case, when the initial condition of the PDE is two well-separated solitary waves with approximately the same velocities and amplitudes (as in weak interactions), this initial condition is then close to the second-order NLS soliton (5.362). Due to degeneracy of the double zero in this second-order soliton, when it is perturbed, this double zero can split either vertically or horizontally in the spectral plane. Horizontal splitting gives rise to a pair of simple zeros with unequal real parts; hence the solution will split into two escaping single NLS solitons. Vertical splitting, on the other hand, gives rise to a pair of simple zeros with equal real parts; hence the solution will form a two-soliton bound state. Since which way this splitting will occur depends sensitively on perturbations, the PDE solutions near second-order NLS solitons are then sensitive to perturbations as well. This is the physical origin of fractal scattering in weak interactions in the PDEs (5.253).

One important feature of fractal scattering in weak interactions is that this fractal scattering appears only in certain portions of the initial-condition space. Indeed, for the initial condition (5.262) or its ODE counterpart (5.307), the interval of fractal scattering shrinks to $\Delta\phi_0 = 0$ when $\varepsilon \to 0^+$ (see Fig. 5.24). This phase value $\Delta\phi_0 = 0$ is a singularity point in the integrable ODE (5.304). At a singularity point, the two waves in the NLS equation will come together and collide with each other. For other initial conditions, it was found that the fractal regions also shrink to singularity points of the integrable ODE when $\varepsilon \to 0^+$ (Zhu and Yang (2007), Zhu et al. (2009)). Recall that for weak interactions $E_0 \ll 1, M_0 \ll 1$ hold for all initial conditions. Thus, all initial conditions lie near the separatrix orbits of the integrable ODE. Then a natural question is, why does sensitive dependence and fractal scattering in the perturbed system appear only near the singularity points when $\varepsilon \ll 1$? The reason is that when the initial condition is not near a singularity point, the two waves stay far apart and quickly escape from each other in the integrable system. Then if $\varepsilon \ll 1$, the perturbed solution behaves similarly, and hence no sensitive dependence can arise. Only when the initial condition lies near a singularity point can the two waves come close to each other, which allows the sensitive dynamics near separatrix orbits to play out (see Fig. 5.20). In the PDE language, this means that fractal scattering in weak interactions of the perturbed NLS equation can arise only when the initial conditions are near singularity points where two NLS solitons would collide with each other.

5.9 Transverse Instability of Solitary Waves

One-dimensional solitary waves, when viewed in two spatial dimensions, become soliton stripes which are uniform along the second dimension (called transverse direction). Then an important question is whether these soliton stripes are stable against transverse perturbations. The transverse stability of 1D NLS solitons was first analyzed by Zakharov and Rubenchik (1974). By using a regular perturbation expansion in small transverse wavenumbers, the authors showed that 1D NLS solitons were unstable to transverse long-wave perturbations for both signs of transverse dispersion. In addition, the unstable eigenvalues were limited to a finite interval of transverse wavenumbers. The Zakharov–Rubenchik analysis is valid for small transverse wavenumbers. When the transverse wavenumber is moderate or large, transverse instability was numerically investigated by Cohen et al. (1976), Saffman and Yuen (1978), Anderson et al. (1979), and Deconinck et al. (2006). For small transverse wavenumbers, the numerical results agree well with the Zakharov–Rubenchik analysis. In addition, when the longitudinal and transverse dispersions have the same sign (the elliptic case), the numerical results also confirm the Zakharov–Rubenchik prediction of wavenumber cutoff for transverse instability; see also Janssen and Rasmussen (1983) for further analysis near this cutoff. But when the longitudinal and transverse dispersions have the opposite sign (the hyperbolic case), the numerical results reveal that the prediction of wavenumber cutoff for transverse instability is incorrect, as additional bifurcations of complex eigenvalues arise which persist for all large transverse wavenumbers (Anderson et al. (1979), Deconinck et al. (2006)). Some reviews on transverse instability can be found in Kuznetsov et al. (1986) and Kivshar and Pelinovsky (2000). Below, we examine transverse stability of 1D solitary waves in generalized NLS equations for both signs of transverse dispersion. We will present the Zakharov–Rubenchik analysis for long transverse waves, and numerical results for moderate and short transverse waves.

Consider the $(2+1)$-dimensional generalized NLS equation

$$iU_t + U_{xx} + \alpha U_{yy} + F(|U|^2)U = 0, \tag{5.363}$$

where x is the longitudinal direction, y is the transverse direction, $\alpha = \pm 1$ is the transverse dispersion coefficient, and $F(\cdot)$ is a real-valued function with $F(0) = 0$. When $\alpha = 1$, the longitudinal and transverse dispersions have the same sign, and it will be called the elliptic case. When $\alpha = -1$, the longitudinal and transverse dispersions have the opposite sign, and it will be called the hyperbolic case. This equation models the propagation of a monochromatic laser beam through a nonlinear medium (where $\alpha = 1$; see Chapter 1) as well as evolution of a gravity-capillary wave packet in deep water (where $\alpha = \pm 1$); see Ablowitz and Segur (1981). This equation admits transversely uniform solitary waves

$$U(x,y,t) = u(x)e^{i\mu t}, \tag{5.364}$$

where $u(x)$ is a localized positive function with a single hump, and μ is the propagation constant. The function $u(x)$ satisfies the ODE

$$u_{xx} - \mu u + F(u^2)u = 0. \tag{5.365}$$

Now we consider the stability of this 1D solitary wave in the presence of transverse perturbations. As we did in Secs. 5.2 and 5.3, we perturb this solitary wave (5.364) by normal modes as

$$U(x,y,t) = e^{i\mu t}\left\{u(x) + [v(x)+w(x)]e^{\lambda t + iky} + \left[v^*(x) - w^*(x)\right]e^{\lambda^* t - iky}\right\}, \tag{5.366}$$

where $v, w \ll 1$ are small perturbations, and k is the transverse wavenumber. Substituting this perturbed solution into (5.363) and linearizing, we obtain the following linear-stability eigenvalue problem:

$$(L_0 - \alpha k^2)w = -i\lambda v, \qquad (L_1 - \alpha k^2)v = -i\lambda w, \tag{5.367}$$

where

$$L_0 = \partial_{xx} - \mu + F(u^2), \tag{5.368}$$

$$L_1 = \partial_{xx} - \mu + F(u^2) + 2u^2 F'(u^2). \tag{5.369}$$

This eigenvalue problem is the counterpart of Eq. (5.19) in Sec. 5.2. From (5.365), we see that

$$L_0 u = 0, \qquad L_1 u_x = 0. \tag{5.370}$$

Thus when $k = 0$, the eigenvalue problem (5.367) has a zero eigenvalue with two corresponding discrete eigenfunctions

$$v_0 = 0, \qquad w_0 = u, \tag{5.371}$$

and

$$v_0 = u_x, \qquad w_0 = 0, \tag{5.372}$$

which are symmetric and antisymmetric, respectively. These are the Goldstein modes induced by phase and translation invariances of 1D solitons in Eq. (5.363). When $k \neq 0$, this zero eigenvalue will bifurcate out. Below we will calculate this eigenvalue bifurcation for $k \ll 1$. In these calculations, the following two relations will be utilized:

$$L_0(xu) = 2u_x, \qquad L_1 u_\mu = u. \tag{5.373}$$

The first relation can be verified directly by noting the form of L_0 and the fact of $L_0 u = 0$, and the second relation can be obtained by differentiating (5.365) with respect to μ.

Before this bifurcation calculation, we should notice some general properties of this eigenvalue problem (5.367). Since $u(x)$ has no zeros, from (5.370) and the properties of Schrödinger operators, L_0 is semi-negative-definite. Hence when $\alpha > 0$, $L_0 - \alpha k^2$ is negative definite for $k \neq 0$. Consequently, the eigenvalue equation (5.367) can be rewritten as

$$(L_1 - \alpha k^2)v = -\lambda^2 (L_0 - \alpha k^2)^{-1} v. \tag{5.374}$$

By taking the inner product of this equation with v, we see that λ^2 is always real. Thus when $\alpha > 0$ (elliptic case), any eigenvalue λ of (5.367) is either real or purely imaginary. As a consequence, unstable eigenvalues can only bifurcate out from zero. Because of this, wavenumbers k_b of instability bifurcations can be easily determined (see below). When $\alpha < 0$ (hyperbolic case), however, complex eigenvalues λ can arise. These complex eigenvalues can bifurcate out from the imaginary axis through Hopf bifurcations, or from the real axis when two real eigenvalues collide. Bifurcation points of complex eigenvalues are difficult to predict analytically. This is why the hyperbolic case is harder to analyze, and numerical computations become more important.

5.9.1 Instability of Long Transverse Waves

Now we calculate bifurcations of the zero eigenvalue when $k \ll 1$, where the transverse waves are long. In this case, the eigenmodes admit the following regular perturbation expansions in k:

$$v = v_0 + kv_1 + k^2 v_2 + \cdots, \tag{5.375}$$

$$w = w_0 + kw_1 + k^2 w_2 + \cdots, \tag{5.376}$$

$$-i\lambda = k\lambda_1 + k^2\lambda_2 + \cdots. \tag{5.377}$$

For the symmetric eigenmode which bifurcates from (5.371), we have $v_0 = 0$ and $w_0 = u$. Substituting the above expansion into (5.367), at $O(k)$, we get

$$L_1 v_1 = \lambda_1 u, \tag{5.378}$$

whose solution is

$$v_1 = \lambda_1 u_\mu \tag{5.379}$$

in view of (5.373). Here the homogeneous term in proportion to u_x is not included in v_1 since we are seeking symmetric eigenfunctions. At $O(k)$, we also get from (5.367) that $L_0 w_1 = 0$, whose solution can be taken as $w_1 = 0$ (since its homogeneous solution in proportion to u can be absorbed into w_0). At $O(k^2)$, we get from (5.367) that

$$L_0 w_2 = \lambda_1 v_1 + \alpha u. \tag{5.380}$$

Taking the inner product of this equation with the homogeneous solution u and noticing that L_0 is self-adjoint, we see that the solvability condition for w_2 is that the inhomogeneous term $\lambda_1 v_1 + \alpha u$ be orthogonal to u. Then recalling the expression (5.379) as well as the perturbation expansion of λ, we find that the leading-order expression for λ^2 of the symmetric eigenmode is

$$\lambda_s^2 = \frac{2P(\mu)}{P'(\mu)}\alpha k^2, \tag{5.381}$$

where $P(\mu) = \int u^2(x;\mu)dx$ is the power of the 1D solitary wave $u(x)$.

To seek the antisymmetric eigenmode which bifurcates from (5.372), we let $v_0 = u_x$ and $w_0 = 0$ in the expansions (5.375)–(5.376). Substituting these expansions into (5.367), at $O(k)$ we get

$$L_0 w_1 = \lambda_1 u_x, \qquad L_1 v_1 = 0, \tag{5.382}$$

whose solutions are

$$w_1 = \frac{1}{2}\lambda_1 x u, \qquad v_1 = 0. \tag{5.383}$$

At $O(k^2)$, we get

$$L_1 v_2 = \lambda_1 w_1 + \alpha u_x. \tag{5.384}$$

The solvability condition of this equation is that the inhomogeneous term $\lambda_1 w_1 + \alpha u_x$ be orthogonal to the homogeneous solution u_x, from which the formula for λ_1 would be obtained. Then recalling the perturbation expansion of λ, we find the leading-order expression for λ^2 of the antisymmetric eigenmode as

$$\lambda_a^2 = -\frac{4\langle u_x, u_x\rangle}{P}\alpha k^2. \tag{5.385}$$

According to the Vakhitov–Kolokolov stability criterion (see Sec. 5.3), the solitary wave (5.364) is stable to longitudinal (x-direction) perturbations if and only if $P'(\mu) > 0$. In this case, we see from (5.381) and (5.385) that when $\alpha = 1$, the symmetric mode is unstable, while when $\alpha = -1$, the antisymmetric mode is unstable. Thus the 1D solitary wave (5.364) is always transversely unstable for both signs of the transverse dispersion coefficient α.

5.9.2 Instability of Short Transverse Waves

When k is not small, the above eigenvalue formulae (5.381) and (5.385) will not apply. In addition, new unstable eigenvalues may appear. To determine all unstable eigenvalues for the whole range of transverse wavenumbers, numerical methods can be used. For each k value, the whole spectrum of the eigenvalue problem (5.367) can be computed by the Fourier collocation method as explained in Secs. 2.4.3 and 7.3. Below we apply this numerical method to the (2+1)-dimensional NLS equation

$$iU_t + U_{xx} + \alpha U_{yy} + |U|^2 U = 0, \tag{5.386}$$

where $F(|U|^2) = |U|^2$ in (5.363), and we give a complete description of transverse instabilities of NLS solitons for both the elliptic and hyperbolic cases. Without loss of generality, we take $\mu = 1$ in (5.364); then the 1D NLS soliton is

$$u(x) = \sqrt{2}\,\mathrm{sech} x. \tag{5.387}$$

The Elliptic Case

In this case, $\alpha = 1$, and thus the nonlinearity is self-focusing along both the x and y directions. Since λ is either real or purely imaginary in this case (see above), the bifurcation points k_b of transverse instabilities occur only when $\lambda = 0$, i.e.,

$$(L_0 - k_b^2)w = (\partial_{xx} - 1 - k_b^2 + 2\,\mathrm{sech}^2 x)w = 0 \tag{5.388}$$

or

$$(L_1 - k_b^2)w = (\partial_{xx} - 1 - k_b^2 + 6\,\mathrm{sech}^2 x)v = 0. \tag{5.389}$$

The first equation has a bounded (symmetric) solution only when $k_b = 0$, while the second equation has bounded symmetric and antisymmetric solutions when $k_b = \pm\sqrt{3}$ and 0, respectively. When combined with the above general bifurcation calculations at $k \ll 1$, we see that the unstable transverse modes are symmetric and limited in the interval $0 < |k| < \sqrt{3}$. When $k \ll 1$, the asymptotic formula (5.381) reduces to

$$\lambda_s^2 \to 4k^2, \qquad k \to 0. \tag{5.390}$$

Between $0 < k^2 < 3$, the transverse stability spectrum as well as accurate eigenvalues can be obtained numerically. The full stability spectrum when $k = 1$ is displayed in Fig. 5.25(a). It is seen that all eigenvalues are real or purely imaginary, in agreement with the above general analysis. There is one pair of real eigenvalues, one stable and the other one unstable. There is also a pair of internal modes. The continuous spectrum consists

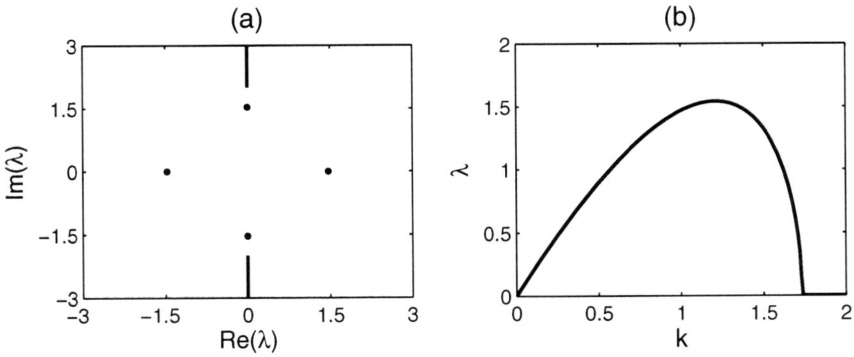

Figure 5.25. (a) *Transverse stability spectrum of an NLS-soliton stripe with* $\mu = 1$ *and transverse wavenumber* $k = 1$ *in the elliptic case* ($\alpha = 1$)*;* (b) *curve of unstable eigenvalue* λ_s *versus* k.

of two segments on the imaginary axis: $\pm i[1 + k^2, \infty)$, or $\pm i[2, \infty)$ since $k = 1$ here. As k varies, the curve of the unstable eigenvalue $\lambda_s(k)$ is as plotted in Fig. 5.25(b). It matches the asymptotic formula (5.390) for $k \ll 1$, and the instability is indeed restricted to the finite interval of $|k| \in [0, \sqrt{3}]$, as expected.

How does the solution evolve after the onset of transverse instability? To investigate this question, we add to the NLS soliton (5.387) 1% random noise perturbation along both the longitudinal and transverse directions, and then simulate its subsequent nonlinear evolution by the numerical methods described in Sec. 7.1. A typical time evolution result is shown in Fig. 5.26. It is seen that the initially uniform soliton stripe breaks up into a sequence of lumps. This instability is called neck-type instability in the literature. After breakup, the lumps collapse, where the widths of the lumps shrink to zero, and their peaks go to infinity. Wave collapse is a typical phenomenon in the 2D NLS equation. It will be examined in more detail in the next section.

The Hyperbolic Case

Unlike the elliptic case, transverse instability in the hyperbolic case ($\alpha = -1$) has caused some confusion in the literature; see Zakharov and Rubenchik (1974), Saffman and Yuen (1978), Ablowitz and Segur (1979), Anderson et al. (1979), Pelinovsky (2001). For a long time, it was believed that transverse instability in this case was also limited to a finite range of wavenumbers, and short transverse waves were stable. The situation was finally clarified by Deconinck et al. (2006), who showed numerically that short transverse waves were also unstable due to oscillatory instabilities (caused by complex eigenvalues). Below we describe transverse instabilities in this hyperbolic case.

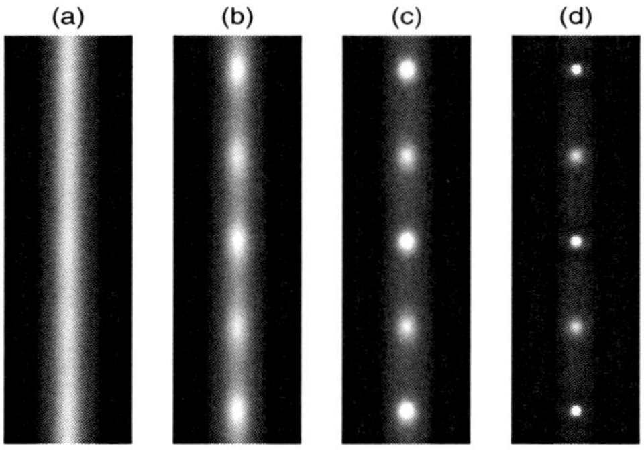

Figure 5.26. *Transverse instability and subsequent evolution of an NLS-soliton stripe in the elliptic case.* (a) *is a perturbed initial state, and* (b, c, d) *are its subsequent time evolutions.*

First, we present the whole transverse stability spectra at two transverse wavenumbers, $k = 0.5$ and 1.5, and the results are shown in Fig. 5.27(a, b). At $k = 0.5$, the discrete spectrum contains a pair of real eigenvalues and a quadruple of complex eigenvalues. But at $k = 1.5$, the discrete spectrum contains two quadruples of complex eigenvalues. The appearance of complex eigenvalues in the transverse stability of the hyperbolic case is an important feature which was not known until the numerical work of Anderson et al. (1979).

As k continuously varies from small to large values, various eigenvalue bifurcations take place. All these bifurcations as well as the complete branches of unstable eigenvalues on the k interval are depicted in Fig. 5.27(c, d). At $k = 0$, a real unstable eigenvalue $\lambda_a(k)$ of odd symmetry bifurcates out from the Goldstein mode (5.372), and its leading-order expression is given by (5.385). This eigenvalue increases with k until $k = 0.8$, after which it decreases. At $k = 1$, another real unstable eigenvalue $\lambda_2(k)$ of odd symmetry bifurcates out from zero with eigenfunction $v = 0$, $w = \tanh(x)$. At this k number, the upper and lower edges of the continuous spectrum (see Fig. 5.27(a)) merge together at the origin, which induces this pair of real eigenvalues $\pm\lambda_2(k)$ to bifurcate out. When $k \to 1^+$, $\lambda_2(k) \to 2\sqrt{2}(k-1)^{1.5}$ to the leading order (Deconinck et al. (2006)). This second eigenvalue $\lambda_2(k)$ increases with k until $k = 1.08$, where it collides with the first eigenvalue $\lambda_a(k)$ of the same parity on the real axis. Upon collision, these two real eigenvalues leave the real axis and bifurcate into a pair of complex eigenvalues. As k increases further, these complex eigenvalues always persist, and their growth rates (real parts) decrease exponentially with k. Thus, unlike the elliptic case, short transverse waves are unstable in the hyperbolic case due to oscillatory (complex-eigenvalue) instabilities.

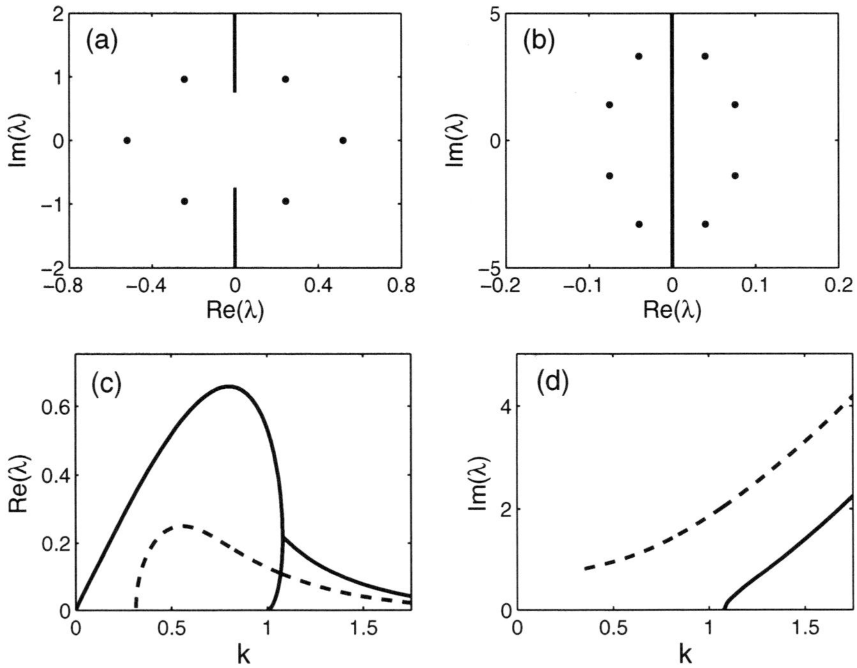

Figure 5.27. (a, b) *Transverse stability spectra of an NLS-soliton stripe with* $\mu = 1$ *in the hyperbolic case* ($\alpha = -1$)*:* (a) $k = 0.5$*;* (b) $k = 1.5$*;* (c, d) *all unstable eigenvalues* λ *in the* k *interval:* (c) $\mathrm{Re}(\lambda)$ *versus* k*;* (d) $\mathrm{Im}(\lambda)$ *versus* k.

In addition to the unstable transverse modes described above, there is another pair of unstable complex eigenvalues which bifurcate out due to a Hopf bifurcation at $k = 0.31$, where two internal modes collide and leave the imaginary axis. This pair of unstable eigenvalues persist for all $k > 0.31$, but their growth rates are always less than the unstable eigenvalues described above.

In this hyperbolic case, since the most unstable eigenmodes are antisymmetric, this antisymmetric instability leads to a bending of the soliton stripe along the transverse direction and induces snake-type instability. After the onset of this instability, since there are no fully localized solitons nor wave collapse in this case, the solution eventually has to decay away. This evolution scenario is confirmed in Fig. 5.28. Here the initial condition is an NLS-soliton stripe perturbed by 1% random noise along the longitudinal and transverse directions; see Fig. 5.28(a). Over time, a snaking instability develops; see Fig. 5.28(b). Afterwards, this snaking pattern breaks up into arrays of filaments (see Fig. 5.28(c)), which then gradually disappear; see Fig. 5.28(d).

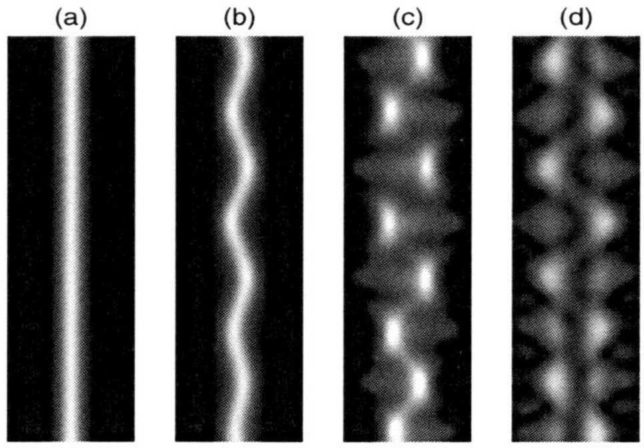

Figure 5.28. *Transverse instability and subsequent evolution of an NLS-soliton stripe in the hyperbolic case.* (a) *is a perturbed initial state, and* (b, c, d) *are its subsequent time evolution.*

5.9.3 Experimental Demonstrations of Transverse Instabilities

Experimental observation of transverse instability and breakup of uniform soliton stripes due to neck-type instability in the elliptic case was performed by Mamaev et al. (1996). In the experiment, the authors used a 10 mW beam from a He-Ne laser (wavelength 0.6328μm) to create a stripe beam with a controlled waist. The beam was directed into a photorefractive crystal of SBN:60. It was polarized along the crystal c-axis (to take advantage of the largest component of the electro-optic tensor of SBN), and propagated in the plane perpendicular to the c-axis. The crystal measured 10 mm along the direction of propagation. The elliptical beam waist measured about 15 μm along the c-axis and 2 mm along the vertical (transverse) direction. A variable DC voltage was applied along the crystal c-axis to create self-focusing nonlinearity. Fig. 5.29 shows the input (a) and output (b–f) beams for different values of the applied voltage, i.e., different values of the nonlinearity. Without the DC voltage, the output beam spreads due to diffraction (Fig. 5.29(b)). As the nonlinearity increases, the beam starts to self-focus (Fig. 5.29(c)) and form a 1D soliton which is uniform along the vertical direction (Fig. 5.29(d)). A further increase of nonlinearity leads to transverse instability, and the stripe beam breaks up into a periodic sequence of filaments (Fig. 5.29(e,f)).

It should be pointed out that the mathematical model for the above experiment is not exactly the $(2+1)$-dimensional generalized NLS equation (5.363) (with $\alpha = 1$), since the photorefractive nonlinearity in the experiment is strongly anisotropic. Using a more sophisticated mathematical model which incorporates anisotropy (Zozulya and Anderson (1995)), Mamaev et al. (1996) found that the transverse instability of neck type also occurs, in agreement with the experiments; see also Infeld and Lenkowska–Czerwinska (1997). Note also that the photorefractive nonlinearity is saturable rather than cubic. Under saturable

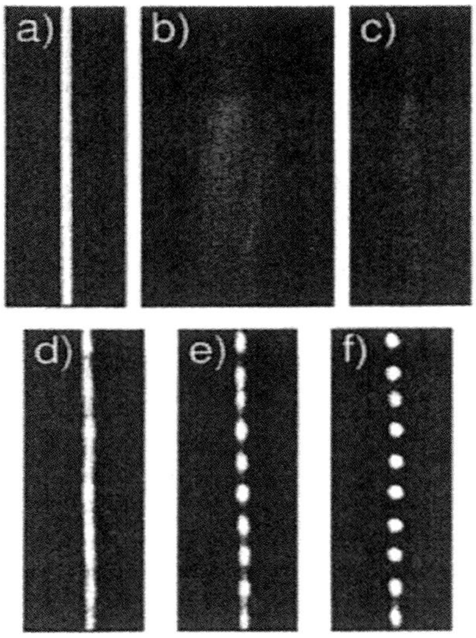

Figure 5.29. *Experimental observation of transverse instability of a uniform soliton stripe in a photorefractive medium:* (a) *input beam, and* (b–f) *output beams at different values of the applied DC voltage* 0, 600, 1000, 1560, *and* 2200 V, *respectively. (After Mamaev et al.* (1996).)

nonlinearity, wave collapse is suppressed. Thus after the stripe beam breaks up into filaments, each filament forms a stable photorefractive soliton instead of collapsing.

5.10 Wave Collapse in the Two-Dimensional NLS Equation

A striking phenomenon in the 2D NLS equation is that, for certain initial conditions (which can be very smooth), the solution can develop spikes of infinite amplitude in finite time. This phenomenon is called wave collapse. It was first observed numerically by Kelley (1965). Then it was proved analytically by Vlasov et al. (1971), who showed that wave collapse always occurs if the Hamiltonian energy of the initial wave is negative. Thus, negative Hamiltonian is a sufficient condition for wave collapse. A necessary condition for collapse was given by Weinstein (1983), who showed that wave collapse would not occur if the power P of the initial wave is below a critical power P_c, where P_c is the power of the ground-state solitary waves in the 2D NLS equation. The singularity structure of the solution near the collapse point has been analyzed by many authors, and the solution has been shown to asymptotically approach a quasi-self-similar radially symmetric profile (a comprehensive review of these studies can be found in Sulem and Sulem (1999)). Recent optical experiments on wave collapse have confirmed this universal collapsing profile;

see Moll et al. (2003). In this section, we describe the phenomenon of wave collapse in the 2D NLS equation.

The 2D NLS equation is

$$iU_t + U_{xx} + U_{yy} + |U|^2 U = 0. \tag{5.391}$$

This equation is Hamiltonian, and its Hamiltonian energy

$$H = \int_{-\infty}^{\infty} \left[|U_x|^2 + |U_y|^2 - \frac{1}{2}|U|^4 \right] dxdy \tag{5.392}$$

is a conserved quantity. The power

$$P = \int_{-\infty}^{\infty} |U|^2 dxdy \tag{5.393}$$

is another conserved quantity in this equation. This equation also admits a variance identity (see Vlasov et al. (1971))

$$Q_{tt} = 8H, \qquad Q \equiv \int_{-\infty}^{\infty} (x^2 + y^2)|U|^2 dxdy, \tag{5.394}$$

which can be easily verified. This identity, called the virial theorem, plays a critical role in the study of wave collapse in Eq. (5.391). Since H is conserved, this virial theorem yields

$$Q(t) = 4Ht^2 + Q'(0)t + Q(0). \tag{5.395}$$

If $H < 0$, then this formula shows that Q would become zero at a finite time $t = t_f$. But since Q is strictly nonnegative, $Q = 0$ would imply that U is zero everywhere except at the origin $(x, y) = (0,0)$. Then due to power conservation, the solution must develop a singularity at the origin, and hence wave collapse occurs. Note that the time t_f of vanishing Q is only an upper bound for the time of wave collapse. The actual collapse could occur earlier and at different locations other than the origin.

The 2D NLS equation (5.391) admits a family of ground-state solitary waves

$$U(x, y, t) = u(r)e^{i\mu t}, \tag{5.396}$$

where $r = \sqrt{x^2 + y^2}$ is the polar radius, and $u(r)$ is a positive and monotonically decaying function which satisfies the equation

$$u'' + \frac{1}{r}u' - \mu u + u^3 = 0 \tag{5.397}$$

and the boundary conditions $u'(0) = 0$, $u(\infty) = 0$. These ground-state solutions were first reported by Chiao et al. (1964) and are sometimes called Townes solutions. These solutions are given by

$$u(r; \mu) = \sqrt{\mu}\, u_1(\sqrt{\mu}\, r), \tag{5.398}$$

where $u_1(r)$ is the solution of Eq. (5.397) with $\mu = 1$. Thus, the power of this family of ground states is

$$P(\mu) = 2\pi \int_0^\infty r u^2(r) dr = 2\pi \int_0^\infty r u_1^2(r) dr \equiv P_c, \tag{5.399}$$

which is a constant. Here $P_c \approx 11.70$. The Hamiltonian H of these ground states is obviously zero due to the virial theorem (5.394) and the fact that Q is time independent for these states. According to the Vakhitov–Kolokolov stability criterion in Sec. 5.3, these ground states are *linearly* stable. However, if one takes the initial condition as the ground state $u(r)$ multiplying a constant,

$$U_0(x, y) = (1+\epsilon)u(r), \tag{5.400}$$

then when $\epsilon > 0$, it is easy to show that $H < 0$, and thus wave collapse occurs. When $-1 < \epsilon < 0$, then $H > 0$, $Q(0) > 0$, and $Q'(0) = 0$; thus $Q(t)$ will monotonically increase to infinity, indicating that the solution will spread out to large distances. In view of the power conservation, the solution then will disperse and decay over time. These two evolution scenarios indicate that the ground states are *nonlinearly* unstable.

An important property of the 2D NLS equation (5.391), first noted by Talanov (1970), is that it is invariant under the "pseudoconformal" transformation

$$\hat{x} = \frac{x}{t_* - t}, \quad \hat{y} = \frac{y}{t_* - t}, \quad \hat{t} = \frac{1}{t_* - t}, \tag{5.401}$$

$$\widehat{U} = (t_* - t) e^{\frac{ir^2}{4(t_* - t)}} U(x, y, t), \tag{5.402}$$

where t_* is any real constant. A direct consequence of this transformation is that, from each stationary ground-state solution (5.396), one can construct a self-similar collapsing solution

$$U(x, y, t) = \frac{1}{t_* - t} u\left(\frac{r}{t_* - t}\right) e^{i \frac{\mu - r^2/4}{t_* - t}}, \tag{5.403}$$

where $t_* > 0$ is the time of collapse. Notice that this collapsing solution has critical power P_c. More interestingly, from direct calculations or the virial theorem (5.394), we can find its Hamiltonian to be

$$H = \frac{1}{2}\pi \int_0^\infty r^3 u^2(r) dr, \tag{5.404}$$

which is positive. This means that a solution can also collapse when $H > 0$. Thus, $H < 0$ is a sufficient, but not necessary, condition for wave collapse. In the pseudoconformal transformation (5.401)–(5.402), if we let $t_* < 0$, then we can construct a self-similar decaying solution from the stationary ground-state solution. This decaying solution can be written as

$$U(x, y, t) = \frac{1}{t_* + t} u\left(\frac{r}{t_* + t}\right) e^{-i \frac{\mu - r^2/4}{t_* + t}}, \tag{5.405}$$

where $t_* > 0$. This decaying solution also has critical power P_c and positive Hamiltonian H, just like the collapsing solution (5.403).

On wave collapse, a necessary condition for collapse was given by Weinstein (1983), which says that wave collapse cannot occur if the initial power P_0 of the solution is below the critical power P_c of the ground states. For example, for the initial condition (5.400), when $-1 < \epsilon < 0$, $P_0 < P_c$, and hence the solution does not collapse, as we have mentioned above.

When a solution collapses, an important question is the rate of blowup and solution profile during its collapse. Since the ground states (5.396) are nonlinearly unstable to small perturbations, in view of the pseudoconformal transformation (5.401)–(5.402), the self-similar collapsing solutions (5.403) are then unstable as well. Thus these special collapsing solutions cannot describe wave collapse for generic initial conditions (especially those with negative Hamiltonian energies). Indeed, accurate numerical computations and asymptotic analysis reveal that, close to wave collapse, the solution blows up at the rate of $[\ln|\ln(t_* - t)|/(t_* - t)]^{1/2}$, which is quite different from the blowup rate of $1/(t_* - t)$ exhibited in the special collapsing solutions (5.403). Numerical computations and asymptotic analysis also reveal that the solution profile close to collapse is radially symmetric and quasi-self-similar, and it resembles the Townes profile at the center of collapse. A comprehensive review on this issue can be found in Sulem and Sulem (1999).

It is noted that wave collapse also occurs in certain other equations such as the 3D NLS equation (Sulem and Sulem (1999)), the Benney–Roskes (also called Davey–Stewartson) equation (see Benney and Roskes (1969), Davey and Stewartson (1974), Ablowitz et al. (2005)), the 1D generalized NLS equation with dual-power-law nonlinearity (Pelinovsky et al. (1996)), and others. It can even occur in integrable equations, as Fig. 2.5 in Sec. 2.10 has shown. But this collapse is suppressed for certain types of nonlinearities (such as saturable nonlinearity).

Chapter 6

Nonlinear Wave Phenomena in Periodic Media

In previous chapters, we have shown nonlinear wave phenomena occurring in homogeneous media, where the corresponding model equations all have space-independent coefficients. In this chapter, we discuss nonlinear wave phenomena in periodic media, where the corresponding model equations have spatially periodic coefficients now. The periodic media fundamentally change the dispersion relations, and thus induce a wide variety of novel nonlinear phenomena which have no counterparts in homogeneous media. These new phenomena attracted a lot of interest in nonlinear optics and Bose–Einstein condensates and were heavily studied in the past decade.

First we describe the physical situations where nonlinear wave propagation in periodic media arises. In Sec. 1.2, we have shown that when a light beam travels in a Kerr medium, if the refractive index of the medium exhibits periodic transverse variation, then under certain approximations, propagation of this light beam is described by the NLS equation with an external potential (1.43). Physical examples include photonic crystal fibers with weak index variation (Argyros et al. (2005)), laser-written waveguide arrays (Szameit et al. (2006)), and planar etched waveguide arrays on AlGaAs substrates (Eisenberg et al. (1998), Iwanow et al. (2004)). After variable normalizations, this model equation reduces to

$$iU_z + U_{xx} + U_{yy} - V(x, y)U + \sigma |U|^2 U = 0, \tag{6.1}$$

where U is the envelope function of the light beam, z is the distance of propagation, (x, y) is the transverse plane, $V(x, y)$ is a periodic potential function, and σ is the sign of nonlinearity. In most Kerr media, the nonlinearity is self-focusing, thus $\sigma = 1$. Equation (6.1) is a model equation for nonlinear wave propagation in periodic media which exhibits Kerr (cubic) nonlinearity.

In photorefractive crystals, the nonlinearity is of saturable type (see Sec. 1.2). If an ordinarily polarized lattice beam with a periodic transverse intensity pattern is launched into such a crystal and made uniform along the propagation direction, then the propagation of an extraordinarily polarized probe beam is governed by a generalized NLS equation (1.48) with saturable nonlinearity and a lattice potential (when the anisotropic property of photorefractive nonlinearity is neglected). After variable normalizations, this governing equation becomes

$$iU_z + U_{xx} + U_{yy} - \frac{E_0}{1 + I_L(x,y) + |U|^2} U = 0, \tag{6.2}$$

where U is the envelope function of the light beam, z is the distance of propagation, (x, y) is the transverse coordinate, E_0 is the applied DC field, and $I_L(x, y)$ is the lattice intensity function which is periodic in x and y. When $E_0 > 0$, the nonlinearity is of self-focusing type, while when $E_0 < 0$, the nonlinearity is of self-defocusing type. Equation (6.2) is a model equation for nonlinear wave propagation in periodic media which exhibits saturable nonlinearity.

Another physical situation of nonlinear wave phenomena in periodic media is the Bose–Einstein condensates. In recent experiments, the condensate was loaded into an optical lattice (see Morsch and Oberthaler (2006) for a review). This lattice forms a periodic potential for the condensate and can be used to manipulate the behavior of the condensate in various interesting ways. In this case, the Gross–Pitaevskii equation (1.62), after variable normalizations, becomes exactly the same as Eq. (6.1) above (see Dalfovo et al. (1999), Morsch and Oberthaler (2006)). Here, U represents the macroscopic condensate wavefunction, $V(x, y)$ is the periodic potential formed by the optical lattice, z is time, and $\sigma = -\text{sgn}(g)$ is the opposite sign of the s-wave scattering length g of the atoms in the condensate. For some atoms such as ^{87}Rb and ^{23}Na, the interaction between atoms is repulsive, and hence $\sigma = -1$ (Anderson et al. (1995), Davis et al. (1995)). For some other atoms such as ^{85}Rb and ^{7}Li, the interaction between atoms is attractive, and hence $\sigma = 1$ (Bradley et al. (1995), Cornish et al. (2000)). The effective interaction between atoms in a condensate can also be tuned by Feshbach resonance from repulsive to attractive, and vise versa (Cornish et al. (2000)).

In the sections below, we will focus on the Kerr model (6.1). Here, σ is allowed to be both positive and negative; i.e., we will consider both self-focusing and self-defocusing nonlinearities in optical media (or both attractive and repulsive atomic interactions in Bose–Einstein condensates). For simplicity, we will take the potential function $V(x, y)$ to be proportional to $\sin^2 x$ in one dimension and $\sin^2 x + \sin^2 y$ in two dimensions. For other periodic potentials, if they are symmetric in x and y, then our analytical results below still hold. When the wave amplitude U is small, the saturable model (6.2) is similar to the Kerr model (6.1), and thus both models would give similar results. At higher amplitudes, results from the two models may deviate.

6.1 One-Dimensional Gap Solitons Bifurcated from Bloch Bands and Their Stability

In this section, we study the Kerr model equation (6.1) in one dimension. This 1D model is

$$iU_t + U_{xx} - V(x)U + \sigma|U|^2U = 0, \tag{6.3}$$

where

$$V(x) = V_0 \sin^2 x, \tag{6.4}$$

V_0 is the strength of the periodic potential, and $\sigma = \pm 1$ is the sign of nonlinearity. Solitary waves in this equation are sought in the form

$$U(x,t) = u(x)e^{-i\omega t}, \tag{6.5}$$

where $u(x)$ is a real-valued amplitude function which satisfies the equation

$$u_{xx} - V(x)u + \omega u + \sigma u^3 = 0, \tag{6.6}$$

and ω is the wave's frequency. To determine these solitary waves, one needs to understand the linear dispersion relation of Eq. (6.6) for infinitesimal waves u. The fundamental property of a periodic medium is that its dispersion relation exhibits Bloch bands and bandgaps. These Bloch bands and bandgaps play a key role in the solution dynamics of Eq. (6.3), and thus will be analyzed first below.

6.1.1 Bloch Bands and Bandgaps

When function $u(x)$ is infinitesimal, Eq. (6.6) becomes a linear equation,

$$u_{xx} - V(x)u + \omega u = 0. \tag{6.7}$$

Bounded solutions of this linear equation are called Bloch modes, and the corresponding frequencies ω form Bloch bands. This linear 1D equation is equivalent to Mathieu's equation. Its solution is

$$p(x;\omega) = e^{ikx}\widehat{p}(x;\omega), \tag{6.8}$$

where $\widehat{p}(x;\omega)$ is periodic with the same period π as the potential $V(x)$, and $\omega = \omega(k)$ is the 1D dispersion relation. These Bloch waves and the dispersion relation can be computed effectively by a Fourier collocation method (see Skorobogatiy and Yang (2009) for a description of this method and a sample MATLAB code). The dispersion diagram for $V_0 = 6$ is shown in Fig. 6.1(a). It is seen that this dispersion diagram contains a number of Bloch bands,

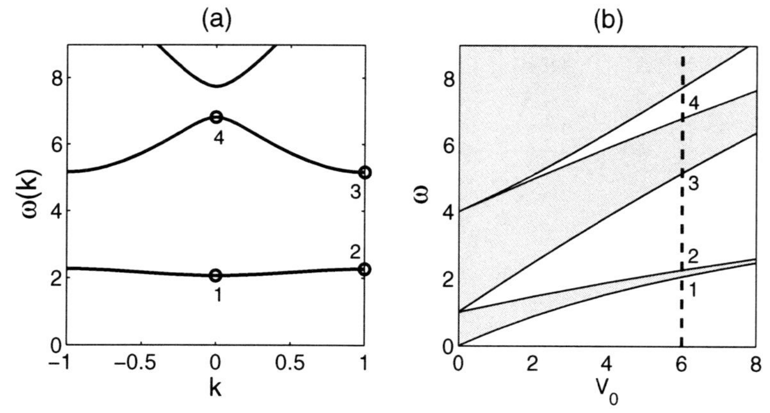

Figure 6.1. (a) *Dispersion curves of the* 1*D equation* (6.7) *with* $V_0 = 6$*;* (b) *Bloch bands (shaded regions) and bandgaps at various values of potential levels* V_0 *in Eq.* (6.7). *Bloch modes at the marked points "1–4" are displayed in Fig.* 6.2.

separated by gaps between them. These gaps are called bandgaps. The gap below the lowest Bloch band is called the semi-infinite gap. The higher gaps are called consecutively as the first gap, the second gap, and so on. When ω falls in these gaps, linear Bloch-wave propagation is forbidden. At the edges of Bloch bands, $k = 0$ or 1. When $k = 0$, the Bloch wave (6.8) is π-periodic, while when $k = 1$, the Bloch wave is 2π-periodic. The first four Bloch waves $p(x; \omega_k)$ at the lowest four band edges $\omega = \omega_k, 1 \leq k \leq 4$, are displayed in Fig. 6.2. Notice that these Bloch waves at band edges are all real valued. In addition, they are either symmetric or antisymmetric in x. At other values of the potential strength V_0, the bandgap structure of this 1D equation (6.7) is shown in Fig. 6.1(b). It is seen that in this 1D case, bandgaps appear at any nonzero potential strength V_0. In addition, the number of bandgaps is infinite.

The above Bloch solutions at band edges have infinitesimal amplitudes. When amplitudes of these solutions increase, these Bloch waves will localize and form solitary-wave structures. The corresponding frequency ω then moves from band edges into bandgaps. In the next section, we analyze how solitary waves bifurcate out from Bloch-band edges by multiscale perturbation methods.

6.1.2 Envelope Equations of Bloch Waves

In this section, we develop an asymptotic theory to analyze small-amplitude solitary waves bifurcating from edges of Bloch bands in Eq. (6.6). These solitary waves are Bloch-wave packets, and we will derive their envelope equations below. This derivation is similar to that developed by Baizakov et al. (2002) and Pelinovsky et al. (2004).

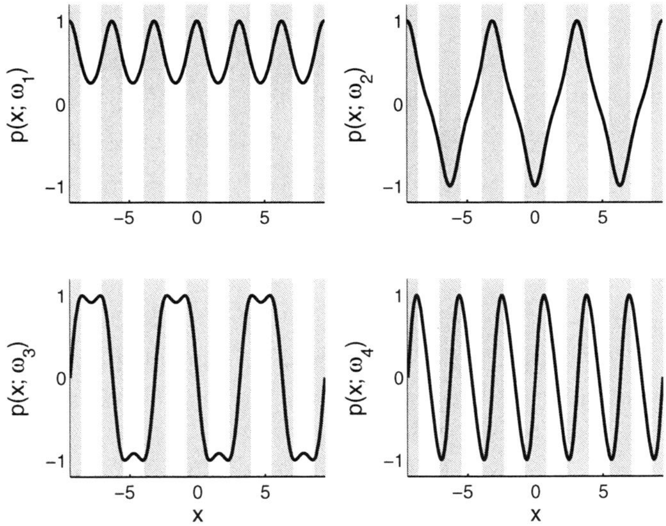

Figure 6.2. *One-dimensional Bloch waves in Eq.* (6.7) *at the lowest four edges of Bloch bands marked from* 1 *to* 4 *in Fig.* 6.1. *Shaded regions represent lattice sites, i.e., regions of low potential values* $V(x)$.

Let us consider a band edge ω_0, whose corresponding Bloch mode is $p(x;\omega_0)$. Below we denote this Bloch mode simply as $p(x)$. Note that $p(x)$ is either symmetric or antisymmetric in x, i.e.,

$$p(-x) = \pm p(x), \tag{6.9}$$

since the potential $V(x)$ is symmetric in x. When the solution $u(x)$ of Eq. (6.6) is infinitesimally small, this solution at the band edge is then just proportional to this Bloch wave $p(x)$. When $u(x)$ is small but not infinitesimal, the solution $u(x)$ then becomes a Bloch-wave packet and can be expanded into a multiscale perturbation series:

$$u = \epsilon u_0 + \epsilon^2 u_1 + \epsilon^3 u_2 + \cdots, \tag{6.10}$$

$$\omega = \omega_0 + \tau\epsilon^2, \tag{6.11}$$

where $\epsilon \ll 1$ is the small amplitude of this wave,

$$u_0 = A(X)p(x) \tag{6.12}$$

is the leading-term Bloch-wave packet, $X = \epsilon x$ is the slow spatial variable of the envelope function A, and $\tau = \pm 1$. Substituting the above expansions into Eq. (6.6), the equation at $O(\epsilon)$ is automatically satisfied. At $O(\epsilon^2)$, the equation for u_1 is

$$\frac{\partial^2 u_1}{\partial x^2} - V(x)u_1 + \omega_0 u_1 = -2\frac{\partial^2 u_0}{\partial x \partial X}. \tag{6.13}$$

Here the partial derivative ∂_{xx} on the left side of the equation is with respect to the fast variable x only. The homogeneous equation of (6.13) has a periodic solution $p(x)$. In order for the inhomogeneous equation (6.13) to admit a periodic solution in the fast variable x, its inhomogeneous term must be orthogonal to its homogeneous solution $p(x)$ over a period 2π, i.e.,

$$\int_0^{2\pi} \frac{\partial^2 u_0}{\partial x \partial X} p(x) dx = 0. \tag{6.14}$$

Here the integration is with respect to the fast variable x only. This condition is called the Fredholm condition in the literature. Here the integration length is 2π rather than π since the homogeneous solution $p(x)$ may have period 2π; see Fig. 6.2. Recalling the u_0 solution (6.12), it is easy to check that the above Fredholm condition is satisfied automatically; thus we can find an x-periodic solution to Eq. (6.13) as

$$u_1 = \frac{dA}{dX} \nu(x), \tag{6.15}$$

where $\nu(x)$ is a periodic solution of equation

$$\nu_{xx} - V(x)\nu + \omega_0 \nu = -2p_x. \tag{6.16}$$

At $O(\epsilon^3)$, the equation for u_2 is

$$\frac{\partial^2 u_2}{\partial x^2} - V(x)u_2 + \omega_0 u_2 = -\left(2\frac{\partial^2 u_1}{\partial x \partial X} + \frac{\partial^2 u_0}{\partial X^2} + \tau u_0 + u_0^3\right). \tag{6.17}$$

Substituting the formulae (6.12) and (6.15) for u_0 and u_1 into this equation, we get

$$-\left\{\frac{\partial^2 u_2}{\partial x^2} - V(x)u_2 + \omega_0 u_2\right\} = \frac{d^2 A}{dX^2}\left[2\nu'(x) + p(x)\right] + A^3 p^3(x) + \tau A\, p(x). \tag{6.18}$$

Before applying the Fredholm condition to this inhomogeneous equation, we notice the following identity:

$$\int_0^{2\pi} \left[2\nu'(x) + p(x)\right] p(x) dx = D \int_0^{2\pi} p^2(x) dx, \tag{6.19}$$

where

$$D \equiv \frac{1}{2}\frac{d^2\omega}{dk^2}\bigg|_{\omega=\omega_0} \tag{6.20}$$

is a second-order dispersion coefficient at the band edge ω_0. Identity (6.19) can be verified by expanding the linear Bloch wave solution (6.8) of Eq. (6.7) into a power series of $k - k_0$ around the band edge $\omega = \omega_0$ (where $k = k_0$), pursuing the expansion to second order in $k - k_0$, and then applying the Fredholm condition to the second-order linear inhomogeneous equation (Pelinovsky et al. (2004), Skorobogatiy and Yang (2009)). Utilizing this identity, the Fredholm condition for Eq. (6.18) leads to the following stationary NLS equation for the envelope function A:

$$D\frac{\partial^2 A}{\partial X^2} + \tau A + \sigma\alpha A^3 = 0, \tag{6.21}$$

where the nonlinear coefficient α is

$$\alpha = \frac{\int_0^{2\pi} p^4(x)\,dx}{\int_0^{2\pi} p^2(x)\,dx}, \tag{6.22}$$

which is always positive. A localized solution in this stationary NLS equation (6.21) exists only when

$$\text{sgn}(D) = \text{sgn}(\sigma) = -\text{sgn}(\tau). \tag{6.23}$$

The condition $\text{sgn}(D) = -\text{sgn}(\tau)$ simply means that the frequency ω in (6.11) always lies outside the Bloch bands (i.e., inside bandgaps). Hence these solitary waves $u(x)$ are often called gap solitons. The condition $\text{sgn}(D) = \text{sgn}(\sigma)$ shows that the nonlinearity and dispersion must be of the same sign. Recall from the dispersion curves in Fig. 6.1(a) that at the lower edge of each Bloch band, $D > 0$, and at the upper edge of each Bloch band, $D < 0$. Thus under self-focusing nonlinearity ($\sigma = 1$), gap solitons will bifurcate out from every lower band edge, and under self-defocusing nonlinearity ($\sigma = -1$), gap solitons will bifurcate out from every upper band edge. The appearance of gap solitons under self-defocusing nonlinearity is a distinctive phenomenon of periodic media which cannot occur in homogeneous media. Under the above conditions, the localized solution in Eq. (6.21) then is

$$A(x) = \sqrt{\frac{2}{|\alpha|}}\,\text{sech}\frac{X}{\sqrt{|D|}}. \tag{6.24}$$

6.1.3 Locations of Envelope Solutions

The envelope equation (6.21) is translation invariant. Thus any spatial translation $A(X - X_0)$ of its solution would still be a solution, where X_0 is an arbitrary real constant. However, only when X_0 takes some special values can the perturbation series solution (6.10) truly satisfy the original equation (6.6). The reason is that X_0 must satisfy a certain additional constraint. This constraint is exponentially small in ϵ. Thus it cannot be captured by the power series expansion (6.10), but needs to be calculated separately. The analysis below follows Pelinovsky et al. (2004), but with certain corrections.

First we derive the constraint for the envelope solution. Multiplying Eq. (6.6) by u_x, integrating from $-\infty$ to $+\infty$, and using integration by parts, we get the following constraint on a true solitary wave solution $u(x)$ to Eq. (6.6):

$$\int_{-\infty}^{+\infty} V'(x)u^2(x)dx = 0. \tag{6.25}$$

Substituting the perturbation expansion (6.10) of the solution $u(x)$ into the above equation and utilizing the solutions $u_0, u_1, \ldots$ obtained earlier, this constraint becomes

$$\begin{aligned} I(x_0) \equiv \epsilon^2 \int_{-\infty}^{+\infty} A^2(X - X_0)V'(x)p^2(x)dx \\ + 2\epsilon^3 \int_{-\infty}^{+\infty} A(X - X_0)A'(X - X_0)V'(x)p(x)\nu(x)dx + \cdots = 0. \end{aligned} \tag{6.26}$$

Here $X_0 = \epsilon x_0$ is the center position of the envelope solution A. The integrals in this constraint can be easily calculated. Let us take the first integral as an example. Since $V'(x)$ is antisymmetric and $p^2(x)$ is symmetric, and both functions are π-periodic, their product $V'(x)p^2(x)$ then is antisymmetric and π-periodic. Thus, this product has the following Fourier series expansion:

$$V'(x)p^2(x) = \sum_{m=1}^{\infty} c_m \sin 2mx, \tag{6.27}$$

where $c_1, c_2, \ldots$ are the Fourier coefficients. When this Fourier expansion is substituted into the first integral in (6.26), every Fourier mode in this expansion contributes an exponentially small term to this integral, and the exponential rate of decay of these terms is larger for higher values of m. Keeping only the leading-order exponentially small term from the first Fourier mode ($m = 1$), the first integral in Eq. (6.26) then becomes

$$\epsilon^2 c_1 \int_{-\infty}^{+\infty} A^2(X - X_0) \sin 2x \, dx.$$

Shifting the x variable in this integral and recalling the formula of $A(X)$ in (6.24), this integral can be simplified to $W_1 \sin 2x_0$, where

$$W_1 = \frac{2\epsilon c_1}{|\alpha|} \int_{-\infty}^{+\infty} \operatorname{sech}^2 \frac{X}{\sqrt{|D|}} \cos \frac{2X}{\epsilon} dX = \frac{4\pi c_1 |D|}{|\alpha|} \operatorname{csch} \frac{\pi |D|^{1/2}}{\epsilon}, \tag{6.28}$$

and $\operatorname{csch} x = 1/\sinh x$ is the hyperbolic cosecant function. Notice that $W_1 = O(e^{-\pi\sqrt{|D|}/\epsilon})$ and thus is exponentially small in ϵ. Repeating the same calculation to the other terms in (6.26), we find that all those terms are of the form $W_n \sin 2x_0$, where W_n is an x_0-independent constant whose expression is analogous to that in (6.28). It is important to notice that all terms in the constraint (6.26) are of the same order in ϵ, even though one may be tempted to think the second and higher terms are of higher order in ϵ than the first term. Let us demonstrate this fact by the second term in (6.26). Rewriting $2A(X - X_0)A'(X - X_0)$ in this second term as $d\{A^2(X-X_0)\}/dX$, recalling $X=\epsilon x$, and performing integration by parts, this second term then becomes

$$-\epsilon^2 \int_{-\infty}^{\infty} A^2(X-X_0)[V'(x)p(x)\nu(x)]'\,dx,$$

which is clearly of the same order in ϵ as the first term in (6.26). Because of this, all the coefficients W_n are of the same order in ϵ. This fact was missed by Pelinovsky et al. (2004). Collecting all terms of the same form $W_n \sin 2x_0$ in (6.26), this constraint then becomes

$$I(x_0)=\left(\sum_{n=1}^{\infty} W_n\right)\sin 2x_0 = 0. \tag{6.29}$$

This constrain would be satisfied if

$$\sin 2x_0 = 0. \tag{6.30}$$

Thus, the envelope solution A can be centered at two locations:

$$x_0 = 0,\ \pi/2. \tag{6.31}$$

In the first case ($x_0=0$), the peak of the envelope function A is located at a lattice site (a potential minimum), and the corresponding soliton is called an on-site soliton. In the second case ($x_0=\pi/2$), the peak of the envelope is located between two lattice sites (a potential maximum), and the corresponding soliton is called an off-site soliton.

It should be pointed out that the constraint (6.25) is only a necessary but not sufficient condition for gap solitons. Thus the above analysis does not guarantee true gap solitons would exist at the two locations (6.31). Numerical results, however, do confirm that gap solitons indeed exist at $x_0=0$ and $\pi/2$; see below and Pelinovsky et al. (2004).

6.1.4 Families of Gap Solitons Bifurcated from Band Edges

The above analysis predicts two families of gap solitons bifurcating from every Bloch-band edge under either self-focusing or self-defocusing nonlinearity. It also gives the leading-order expressions for these solitons when the frequency ω is near the band edge (where the wave amplitude is small). Away from the band edge, this asymptotic analysis breaks down. In this section, we numerically determine these gap solitons both near the band edge and away from it. In these computations, we set the potential strength $V_0=6$ as before.

Numerically, we indeed find two families of gap solitons (on-site and off-site) bifurcating from each band edge. The two soliton families from the lower and upper edges $\omega_{1,2}$ of the first Bloch band are displayed in Figs. 6.3 and 6.4. Fig. 6.3 shows the power curves of these solitons, with the power defined as $P=\int |u|^2 dx$, while Fig. 6.4 displays typical

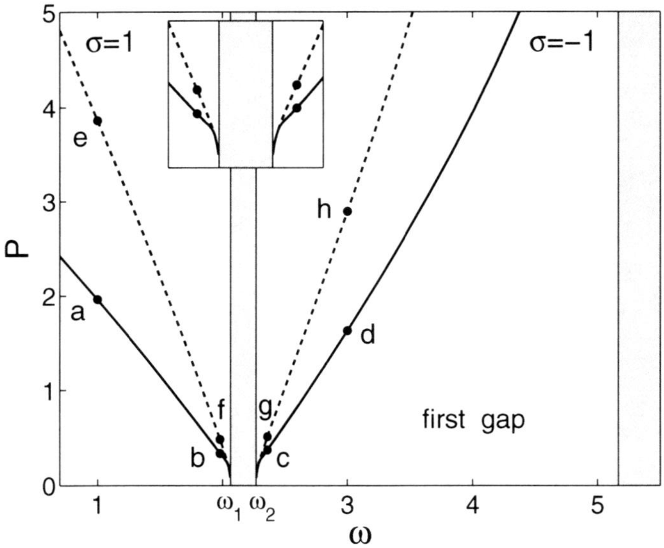

Figure 6.3. *Power curves of on-site and off-site 1D solitons bifurcated from lower and upper edges of the first Bloch band under self-focusing and self-defocusing nonlinearities, respectively ($V_0 = 6$). Solid curves: on-site solitons; dashed curves: off-site solitons. The inset shows power curves near band edges. Soliton profiles at the marked points are shown in Fig. 6.4.*

soliton profiles near and far away from the band edges. From Fig. 6.3, we see that two power curves bifurcate out from each band edge, with the lower curve for on-site solitons, and the upper curve for off-site solitons. Hence, at the same frequency value, the off-site soliton has higher power than the on-site soliton. Notice that the two soliton families in the semi-infinite gap exist under self-focusing nonlinearity ($\sigma = 1$), and the other two soliton families in the first gap exist under self-defocusing nonlinearity ($\sigma = -1$), which is consistent with the analytical results above.

Now we examine soliton profiles in these solution families. Inside the semi-infinite gap (under self-focusing nonlinearity), solitons are all positive (i.e., all intensity peaks are in phase), which can be seen in Fig. 6.4(a, b, e, f). On-site solitons (see (a, b)) have a single intensity peak residing at a lattice site (potential minimum), while off-site solitons (see (e, f)) have two equal-intensity peaks residing at two adjacent lattice sites. These off-site solitons are sometimes called in-phase dipole solitons since their two equal-intensity peaks have the same phase. Near the lower edge ω_1, these solitons develop in-phase tails on both sides, and their amplitudes decrease, see (b, f). When $\omega \to \omega_1$, these solitons approach the Bloch wave of Fig. 6.2(a) with infinitesimal amplitude. In the first bandgap (under self-defocusing

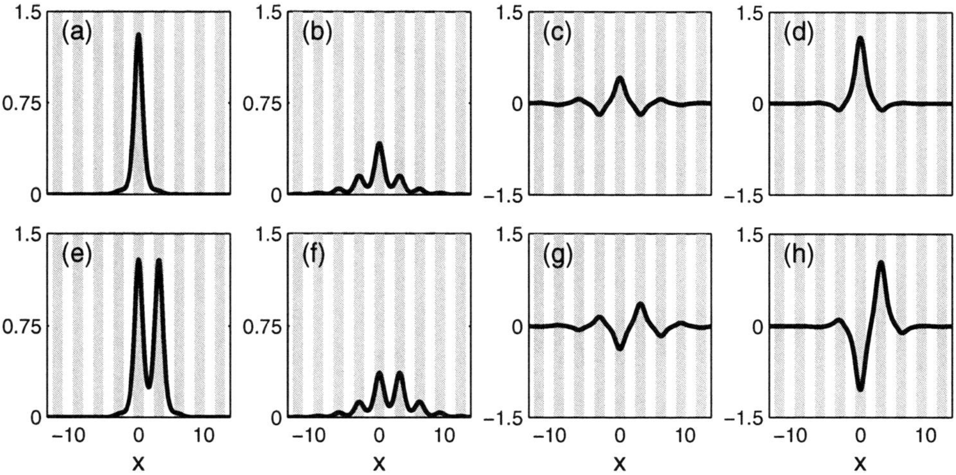

Figure 6.4. *Soliton profiles at the marked points in Fig.* 6.3. (a, b, e, f)*: solitons in the semi-infinite gap under self-focusing nonlinearity.* (c, d, g, h)*: solitons in the first gap under self-defocusing nonlinearity. Upper row: on-site solitons; lower row: off-site solitons. The vertical gray stripes represent lattice sites (low potentials).*

nonlinearity), adjacent peaks of solitons are out of phase; see Fig. 6.4(c, d, g, h). On-site solitons here (see (c, d)) have a single intensity peak at a lattice site, flanked by out-of-phase tails. Off-site solitons (see (g, h)) have two equal-intensity and out-of-phase peaks at two adjacent lattice sites, and are sometimes called out-phase dipole solitons. The out-phase structure of neighboring peaks in these gap solitons originates from that of the Bloch wave at the upper band edge ω_2 (see Fig. 6.2(b)), from which these gap solitons bifurcate out. Solitons bifurcated from higher bands have more complex spatial structures and will not be shown here.

6.1.5 Stability of Gap Solitons Bifurcated from Band Edges

Now we examine the linear stability of these gap solitons which bifurcate from band edges. For the two solution families in the semi-infinite gap (under self-focusing nonlinearity), solitons are all positive; see Fig. 6.4(a, b, e, f). But the generalized Vakhitov–Kolokolov criterion in Sec. 5.3.2 does not give direct stability information since the number of positive eigenvalues in the linear operator L_1 is not known analytically. For the solution families in the first and higher gaps, solitons are not sign definite (see Fig. 6.4(c, d, g, h)), and thus the generalized Vakhitov–Kolokolov criterion in Sec. 5.3.2 does not apply at all. This means that alternative analytical methods or numerical methods are necessary to determine the stability of gap solitons. This question was studied by Pelinovsky et al. (2004). Through a number of approximations, these authors derived an analytical formula for the small eigenvalues which bifurcate out from zero when the gap soliton bifurcates out from a band edge. This formula was expressed through the derivative of the function $I(x_0)$ which was defined in Eq. (6.26). Mistaking the first term as the leading order term in the series of (6.26), these authors obtained a more explicit formula for these small eigenvalues. This eigenvalue formula shows that on-site solitons near band edges are linearly stable, and off-site solitons near band edges are linearly unstable, which qualitatively agrees with their numerical results (no quantitative comparison was made between analytical and numerical results in this work). What was not realized by Pelinovsky et al. (2004) is that all terms in the series of (6.26) for $I(x_0)$ are of the same order in ϵ (see Sec. 6.1.3). Thus their eigenvalue formula from the first term of (6.26) is erroneous. If one attempts to sum up all terms in the series of (6.26), the algebra involved in it is formidable, if not impossible. Even if the sum of all those terms in (6.26) can be obtained, the eigenvalue formula by Pelinovsky et al. (2004) is still questionable since certain approximations they made were not justified.

Below, we study the linear stability of on-site and off-site gap solitons by numerical methods. First, we perturb these solitons as

$$U(x,t) = \left\{u(x) + [v(x) + w(x)]e^{\lambda t} + [v^*(x) - w^*(x)]e^{\lambda^* t}\right\} e^{-i\omega t}, \tag{6.32}$$

where (v, w) are infinitesimal normal-mode perturbations, λ is the mode's eigenvalue, and the superscript "*" represents complex conjugation. Substituting this perturbed solution into Eq. (6.3) and linearizing, we find that these normal modes satisfy the following linear eigenvalue problem (see Secs. 5.2 and 5.3):

$$L_0 w = -i\lambda v, \quad L_1 v = -i\lambda w, \tag{6.33}$$

where

$$L_0 = \partial_{xx} + \omega - V(x) + \sigma u^2, \tag{6.34}$$

$$L_1 = \partial_{xx} + \omega - V(x) + 3\sigma u^2. \tag{6.35}$$

This linear eigenvalue problem can be solved numerically (see Chapter 7). To illustrate, we consider on-site and off-site solitons in the semi-infinite and first gaps, which are shown in Figs. 6.3 and 6.4. Stability spectra for these solitons at $\omega = 1$ in the semi-infinite gap and $\omega = 3$ in the first gap (see Fig. 6.4(a,e,d,h)) are displayed in Fig. 6.5. It is seen that both on-site solitons have no unstable eigenvalues and thus are linearly stable. But both off-site solitons contain a real positive eigenvalue and thus are linearly unstable. We have also computed the stability spectra for other solitons on these four solution branches and obtained similar results. That is, both branches of on-site solitons are all linearly stable, and both branches of off-site solitons are all linearly unstable (due to presence of a positive eigenvalue).

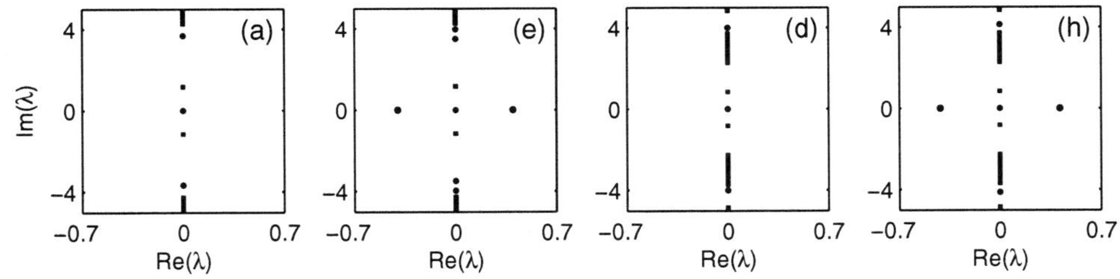

Figure 6.5. *Stability spectra for the on-site and off-site solitons shown in Fig.* 6.4(a, e, d, h), *respectively.*

Now we examine the magnitude of this unstable positive eigenvalue in off-site solitons. To illustrate, we consider the branch of off-site solitons in the semi-infinite gap with $V_0 = 6$ and $\sigma = 1$ (see Fig. 6.3). The unstable positive eigenvalue λ is numerically computed for this entire off-site soliton branch and is plotted in Fig. 6.6(a). It is seen that λ bifurcates out from zero when the soliton bifurcates out from the band edge $\omega_1 = 2.063182$. In addition, this positive eigenvalue persists for the entire branch, indicating that this entire branch of off-site solitons is linearly unstable. To determine the functional dependence of λ on ω near the band edge ω_1, we replot this λ in Fig. 6.6(b) in logarithmic scale versus $1/\epsilon$, where $\epsilon = \sqrt{\omega_1 - \omega}$. We find that

$$\lambda \to \frac{C}{\sqrt{\epsilon}} e^{-\pi\sqrt{D}/2\epsilon}, \quad \epsilon \to 0, \tag{6.36}$$

where D is the dispersion coefficient defined in (6.20), and $C \approx 5.38$. This formula is also plotted in Fig. 6.6(b), and it matches the numerical eigenvalues very well. This formula shows that λ is exponentially small near the band edge. We have also examined the unstable positive eigenvalue for the branch of off-site solitons in the first gap with $\sigma = -1$ and $V_0 = 6$ (see Fig. 6.3), and obtained results similar to that in Fig. 6.6(a, b). Specifically, a positive eigenvalue λ bifurcates out from zero when the soliton bifurcates out from the band edge $\omega_2 = 2.266735$, and it persists for the entire solution branch. In addition, near the band edge ω_2, the λ formula has the same form as (6.36), except that $\epsilon = \sqrt{\omega - \omega_2}$, and the constant C takes on a different value.

Due to linear instability, off-site solitons, when perturbed, will break up and evolve into other solution states. Now we numerically examine the final state of off-site solitons following their breakup. For this purpose, we initially perturb these solitons by a weak phase gradient as

$$U(x,0) = u(x)e^{i\beta x}, \tag{6.37}$$

where $u(x)$ is the off-site soliton, and $\beta \ll 1$ is a phase-gradient parameter. Our numerics show that when the off-site soliton has higher amplitude (away from band edges), under the

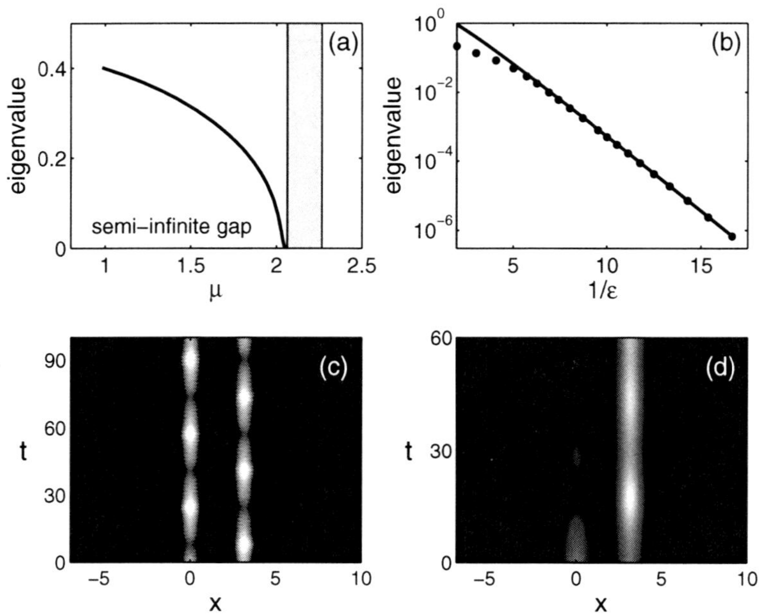

Figure 6.6. (a, b) *Unstable eigenvalues of off-site solitons in the semi-infinite gap with* $V_0 = 6$ *and* $\sigma = 1$ *(see Fig.* 6.3*). In* (b), $\epsilon = \sqrt{\omega_1 - \omega}$; *the dots are numerical eigenvalues, and the solid line is the formula* (6.36). (c, d) *Instability developments of high- and low-amplitude off-site solitons in Fig.* 6.4(e, f), *respectively.*

above phase-gradient perturbation, it will evolve into an oscillating state, where the energy periodically shifts between two adjacent lattice sites. An example is shown in Fig. 6.6(c), where the off-site soliton is taken as the high-amplitude one in Fig. 6.4(e) in the semi-infinite gap, and $\beta = 0.1$ in the perturbation (6.37). But when the off-site soliton has lower amplitude (near band edges), then under the phase-gradient perturbation (6.37), it will shift all its energy into one lattice site and evolve into an on-site soliton. An example of this development is shown in Fig. 6.6(d), where the soliton is taken as the low-amplitude one in Fig. 6.4(f) in the semi-infinite gap, and $\beta = 0.1$. In both evolution scenarios, the symmetry of the underlying off-site soliton is broken. Thus this instability is a symmetry-breaking instability.

From the above numerical results (as well as those in Pelinovsky et al. (2004)), we can draw general conclusions on the stability of on-site and off-site gap solitons. All off-site solitons are linearly unstable due to the presence of a positive eigenvalue, which causes symmetry-breaking instability. Near the band edge ω_n of soliton bifurcation, this unstable eigenvalue is exponentially small, and its analytical expression is given by (6.36), where $\epsilon = \sqrt{|\omega - \omega_n|}$, D is the dispersion coefficient at this band edge, and C is a constant whose value depends on the underlying band edge. The origin of this unstable eigenvalue is the following (see Pelinovsky et al. (2004)). Near a band edge, the gap soliton is a low-amplitude Bloch-wave packet. The envelope of this wave packet is governed by Eq. (6.21). This envelope equation is both phase and translation invariant; thus the linearization spectrum of its envelope solution contains a zero eigenvalue of multiplicity four, two induced by phase invariance and the other two induced by translation invariance. In the full lattice equation (6.3), however, the translation invariance is destroyed, and thus the two multiplicities of the zero eigenvalue associated with spatial translations must bifurcate out. For on-site solitons, this zero eigenvalue bifurcates along the imaginary axis, thus creating no instability. But for off-site solitons, this bifurcation is along the real axis, which then creates a positive eigenvalue, as we have seen above.

It is interesting to note that the eigenvalue formula (6.36) has the same form as the unstable eigenvalue of elevation waves in the fifth-order KdV equation; see Calvo et al. (2000). The unstable eigenvalue in that work was calculated by exponential asymptotics techniques. Using similar techniques, the eigenvalue formula (6.36) for the present lattice problem was also derived analytically by Hwang et al. (2010), and their analytical formula fully agrees with the numerical results.

Regarding on-site gap solitons, they generally do not possess real positive eigenvalues, and thus may be linearly stable; see Fig. 6.5(a, d). However, it should be cautioned that these solitons may suffer oscillatory instabilities due to complex eigenvalues (Pelinovsky et al. (2004)). When that happens, on-site solitons will also become unstable. However, these oscillatory instabilities, even if they exist, are often very weak when the on-site solitons reside in lower bandgaps (Pelinovsky et al. (2004)).

6.2 One-Dimensional Gap Solitons Not Bifurcated from Bloch Bands

The previous section showed that from each edge of a Bloch band, only two families of gap solitons bifurcate out. It turns out that in addition to these two solution families, the lattice equation (6.3) also admits infinite families of gap solitons which do not bifurcate from band edges. The power curves of these solitons always exhibit multibranches which do not touch band edges. Solitons on the upper branches are always linearly unstable, but solitons on the lowest branch can be stable. In this section, we consider several such soliton families and determine their stability properties.

Stable Dipole Solitons and Their Solution Families

In the previous section, we showed that off-site solitons are of dipole type and linearly unstable (see Figs. 6.3–6.6). It turns out that stable dipole solitons do exist in the lattice equation (6.3), and they do not bifurcate from band edges. To illustrate, we still take $V_0 = 6$.

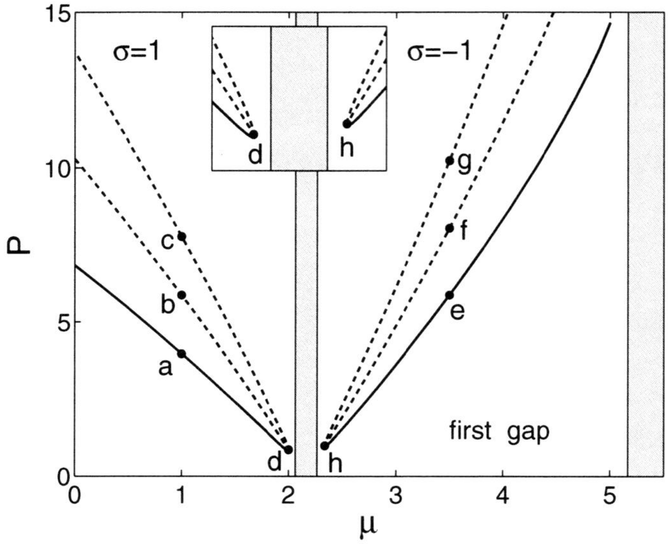

Figure 6.7. *Power curves of dipole-soliton families which do not bifurcate from band edges ($V_0 = 6$). The left curves are in the semi-infinite gap with self-focusing nonlinearity ($\sigma = 1$), and the right curves are in the first gap with self-defocusing nonlinearity ($\sigma = -1$). The dashed parts are unstable branches. The inset shows power curves near band edges. Soliton profiles and their stability spectra at the marked points are shown in Figs.* 6.8 *and* 6.9.

In the semi-infinite gap under self-focusing nonlinearity, we can find a family of solutions which are displayed in Figs. 6.7 and 6.8. The power curve of this family exhibits three branches which are connected near the Bloch band (but not touching it). Solitons on the lower branch have an out-phase dipole profile; see Fig. 6.8(a). For convenience, we call these solitons in $(1, -1)$ configuration. On the middle branch, however, the solitons develop an extra intensity peak at the left side, and turn into $(1, 1, -1)$ configuration; see Fig. 6.8(b). On the upper branch, the solitons develop extra intensity peaks on both sides and turn into $(1, 1, -1, -1)$ configuration; see Fig. 6.8(c). These three branches are connected near the Bloch band, where the solitons have lower amplitude and are broader; see Fig. 6.8(d). Notice that the lower and upper branches are antisymmetric, while the middle branch is asymmetric.

In the first gap under self-defocusing nonlinearity, we can also find a family of solutions which do not bifurcate from band edges. Its power curve, displayed in Fig. 6.7, also contains three branches. Solutions on the lower branch have an in-phase dipole structure, i.e., in $(1, 1)$ configuration (see Fig. 6.9(e)), while solutions on the middle and upper branches are

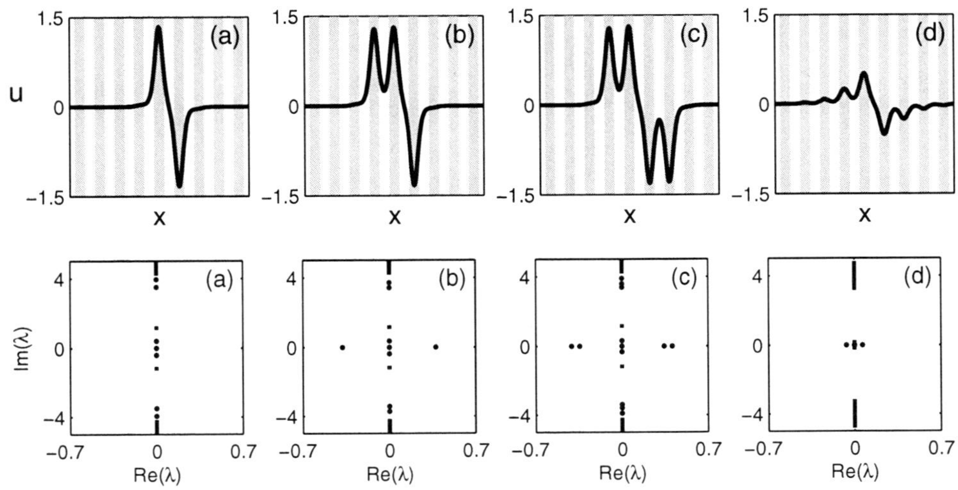

Figure 6.8. *Upper row: soliton profiles at the marked points in the semi-infinite gap of Fig.* 6.7 *($\sigma = 1$). Lower row: stability spectra of the solitons in the upper row.*

in $(-1, 1, 1)$ and $(-1, 1, 1, -1)$ configurations, respectively (see Fig. 6.9(f, g)). These three branches are also connected near the Bloch band, where the solitons are low and broad; see Fig. 6.9(h). For this solution family, the lower and upper branches are symmetric, and the middle branch is asymmetric.

We have examined the stability properties of these dipole-soliton families and found that their lower branches are linearly stable, while their middle and upper branches are linearly unstable. To demonstrate, the stability spectra for solitons in the upper rows of Figs. 6.8 and 6.9 are displayed in the lower rows of those same figures. For both families, the spectra of solitons on the lower branch do not contain any unstable eigenvalue, and are thus stable. The spectra of solitons on the middle and upper branches, however, contain one and two positive eigenvalues, respectively, and are thus unstable. What happens here is that, as the soliton moves from the lower branch to the branching point (where the three branches meet), when it crosses the power minimum point, a positive (unstable) eigenvalue bifurcates out, and hence the soliton becomes linearly unstable. This change of stability at the power minimum point should occur in view of the general result in Sec. 5.4. This positive eigenvalue, after its creation, persists as the soliton moves onto the middle and upper branches. At the branching point (see Figs. 6.8(d) and 6.9(h)), another eigenvalue bifurcates out from the origin. Along the upper branch, this eigenvalue bifurcates along the real axis, hence creating a second positive (unstable) eigenvalue. Along the middle and lower branches, however, this eigenvalue bifurcates along the imaginary axis, and hence no second unstable eigenvalue appears. These eigenvalue bifurcations explain why solitons on

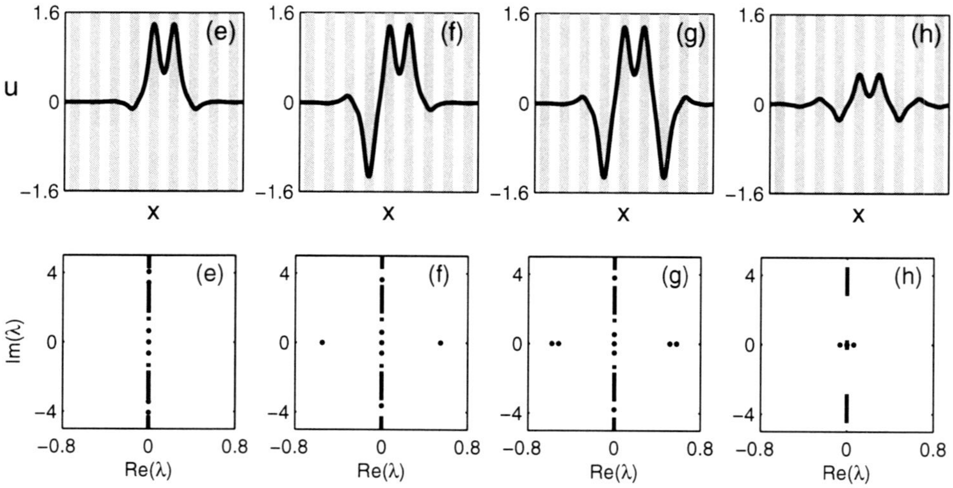

Figure 6.9. *Upper row: soliton profiles at the marked points in the first gap of Fig. 6.7 ($\sigma = -1$). Lower row: stability spectra of the solitons in the upper row.*

the middle branch contain only one positive eigenvalue, while solitons on the upper branch contain two positive eigenvalues.

Stable Truncated-Bloch-Wave Solitons and Their Solution Families

At every edge of a Bloch band, an infinitesimal spatially periodic Bloch wave exists (see Figs. 6.1 and 6.2). When ω moves away from the edge (toward the left under self-focusing nonlinearity and toward the right under self-defocusing nonlinearity), this infinitesimal (linear) Bloch wave bifurcates out into a finite-amplitude (nonlinear) Bloch wave which is also spatially periodic (Alexander et al. (2006), Wang et al. (2009)). This nonlinear Bloch wave exists not only when ω lies in a bandgap, but also when ω lies in a Bloch band. When ω lies inside a bandgap, we can truncate this nonlinear periodic Bloch wave to a finite number of intensity peaks and obtain a corresponding solitary wave. This resulting solitary wave is called a truncated-Bloch-wave soliton (Wang et al. (2009)). Such a soliton was first reported experimentally by Anker et al. (2005) and theoretically investigated by Alexander et al. (2006) and Wang et al. (2009). To illustrate this type of solitons, we consider self-defocusing nonlinearity and take $V_0 = 6$. A truncated-Bloch-wave soliton in the first gap is displayed in Fig. 6.10(b). This soliton is truncated from the nonlinear Bloch wave which bifurcates out from the lower edge of the first Bloch band $\omega_1 = 2.063182$, and hence its seven intensity peaks are all in phase, similar to the linear Bloch wave at edge ω_1; see Fig. 6.2(a). This soliton, once in existence, will generate its own solution family. The power curve of

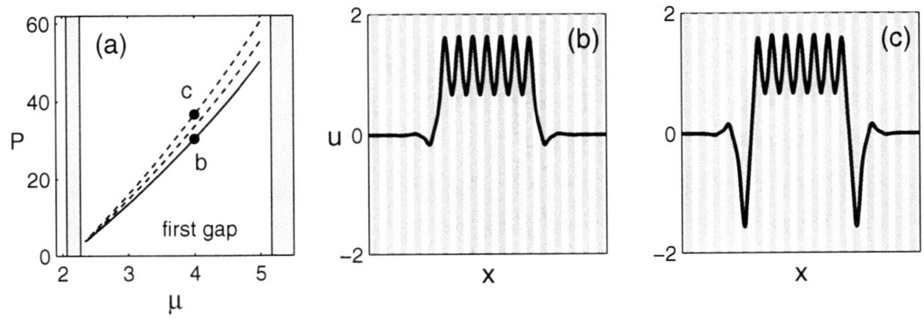

Figure 6.10. (a) *Power curve for a family of truncated-Bloch-wave solitons in the first gap under self-defocusing nonlinearity* ($\sigma = -1$)*; the dashed parts are unstable branches.* (b, c) *Soliton profiles at the marked points on the lower and upper branches in* (a). *(After Wang et al.* (2009).*)*

this family is shown in Fig. 6.10(a). This curve also has triple branches. While solutions on the lower branch have seven in-phase intensity peaks in $(1,1,1,1,1,1,1)$ configuration (see Fig. 6.10(b)), solutions on the middle branch develop one more intensity peak at one edge of the soliton and turn into $(-1,1,1,1,1,1,1,1)$ configuration, and solutions on the upper branch develop two more intensity peaks at both edges and turn into $(-1,1,1,1,1,1,1,1,-1)$ configuration; see Fig. 6.10(c). Regarding the stability of this solution family, the lower branch is found to be linearly stable, while the middle and upper branches are linearly unstable (Wang et al. (2009)). These behaviors are analogous to those of the in-phase dipole family under self-defocusing nonlinearity as shown in Figs. 6.7 and 6.9. In fact, the in-phase dipole soliton in Fig. 6.9(e) is also a truncated-Bloch-wave soliton, where the nonlinear Bloch wave is truncated to two intensity peaks instead of seven. Thus these two soliton families are closely related and hence share similar properties. One can carry this connection even further. That is, we can also view the on-site gap soliton in Fig. 6.4(d) as the simplest truncated-Bloch-wave soliton, where the nonlinear Bloch wave is truncated to a single intensity peak. Under this viewpoint, truncated-Bloch-wave solitons then provide a link between infinitely extended nonlinear Bloch waves and the simplest on-site gap solitons (Alexander et al. (2006)). If the nonlinear Bloch wave in the first gap is truncated to other numbers of intensity peaks, the corresponding solitons and their solution families can be found as well (Wang et al. (2009)).

The results above and in the previous section enable us to develop some intuition on the existence and stability of gap solitons in the lattice equation (6.3). On existence, these results suggest that for both self-focusing and self-defocusing nonlinearities, one can find gap solitons with an arbitrary arrangement of intensity peaks and phase structures. For instance, one should be able to find solitons in $(1,1,1)$, $(1,1,1,-1)$, or $(1,-1,1,-1)$

configurations. These intuitions are indeed correct, if the lattice potential is sufficiently deep. For instance, with $V_0 = 6$, we have found all these soliton structures for both self-focusing and self-defocusing nonlinearities. Each of these solitons induces its own solution families, and thus infinite families of gap solitons exist in the lattice equation (6.3). Except for the on-site and off-site soliton families considered in the previous section, none of the other soliton families bifurcates from band edges, and thus their power curves always have multiple branches. For each family, while the middle and upper branches of the power curve are always linearly unstable, the lowest branch can be stable.

The above stability results allow us to develop some intuition on the stability of gap solitons as well. Notice that under self-focusing nonlinearity ($\sigma = 1$), out-phase dipoles $(1, -1)$ and $(-1, 1)$ (see Fig. 6.8(a)) are stable, while in-phase dipoles $(1, 1)$ and $(-1, -1)$ (see Fig. 6.4(e)) are unstable. Under self-defocusing nonlinearity ($\sigma = -1$), the situation is just the opposite: in-phase dipoles are stable, while out-phase dipoles are unstable (see Figs. 6.4(h) and 6.9(e)). These stability results on dipoles can help us predict the stability of gap solitons in more general configurations. Specifically, if a soliton contains an in-phase dipole such as $(1, 1)$ under self-focusing nonlinearity, or out-phase dipole such as $(1, -1)$ under self-defocusing nonlinearity, then this soliton is expected to be linearly unstable. For example, for the $(1, 1, -1)$ soliton in Fig. 6.8(b) under self-focusing nonlinearity, since it contains an in-phase dipole $(1, 1)$, it is then expected to be unstable, in agreement with its spectrum in Fig. 6.8. For another example, the $(-1, 1, 1)$ soliton in Fig. 6.9(f) contains an out-phase dipole $(-1, 1)$ under self-defocusing nonlinearity and thus is expected to be unstable, again in agreement with its spectrum in Fig. 6.9. On the other hand, the $(1, -1, 1, -1)$ soliton can be expected to be stable under self-focusing nonlinearity since it comprises only (stable) out-phase dipoles; and the $(1, 1, 1, 1)$ soliton can be expected to be stable under self-defocusing nonlinearity since it comprises only (stable) in-phase dipoles. These expectations have been verified for deep potentials (such as $V_0 = 6$). We note in passing that the above intuition on gap-soliton stability in the lattice equation (6.3) under self-focusing nonlinearity is consistent with the analytical results by Pelinovsky et al. (2005) on soliton stability in the discrete NLS equation under self-focusing nonlinearity in the weak intersite coupling limit (which corresponds to deep potentials in the lattice equation (6.3)).

The above results and intuitions on the existence and stability of gap solitons hold only when the potential is sufficiently deep. In shallow potentials, the lower branches of many solution families can become unstable as well due to oscillatory instabilities (Wang et al. (2009)). What is more, many solution families can even totally disappear if the potential is too shallow (Wang et al. (2009)).

Lastly, we point out that the gap solitons considered in this section are relatively simple. There are also many other gap solitons which are related to edges of higher Bloch bands and thus have more complicated profiles. Such examples can be found in Efremidis and Christodoulides (2003), Pelinovsky et al. (2004), and Zhang and Wu (2009).

6.3 Two-Dimensional Gap Solitons Bifurcated from Bloch Bands

In this section, we study gap solitons in two spatial dimensions. Many of the results for the 1D case above can be carried over to the 2D case. In such cases, our description will be brief. However, many new phenomena also arise in two dimensions, such as a much wider variety of gap-soliton structures, which often have no counterparts in one dimension. Such results will be described in more detail.

The model equation (6.1) in two dimensions is

$$iU_t + U_{xx} + U_{yy} - V(x,y)U + \sigma |U|^2 U = 0, \tag{6.38}$$

where $\sigma = \pm 1$ is the sign of the cubic nonlinearity,

$$V(x,y) = V_0 \left(\sin^2 x + \sin^2 y\right) \tag{6.39}$$

is the 2D periodic potential, and V_0 is the depth of this potential. Notice that the spatial periods of this potential along the x and y directions are the same. This type of potential is sometimes called a square potential in the literature. This potential is separable in x and y, which makes the theoretical analysis a little easier. Similar analysis can be repeated for other types of periodic potentials and nonlinearities with minimal changes. Solitary waves of Eq. (6.38) are sought in the form

$$U(x,y,t) = u(x,y)e^{-i\omega t}, \tag{6.40}$$

where amplitude function $u(x,y)$ is a solution of the following equation:

$$u_{xx} + u_{yy} - [F(x) + F(y)]u + \omega u + \sigma |u|^2 u = 0, \tag{6.41}$$

$$F(x) = V_0 \sin^2 x, \tag{6.42}$$

and ω is the wave's frequency.

In the rest of this section, we will determine solitary waves in Eq. (6.41) which are bifurcated from Bloch bands. To do so, we first need to understand Bloch bands and bandgaps of this 2D equation.

6.3.1 2D Bloch Bands and Bandgaps

When the solution $u(x,y)$ is infinitesimal, Eq. (6.41) becomes a linear equation,

$$u_{xx} + u_{yy} - [F(x) + F(y)]u + \omega u = 0. \tag{6.43}$$

Since the periodic potential in this equation is separable, its Bloch solutions and Bloch bands can be constructed from solutions of the 1D equation. Specifically, the 2D Bloch solution

$u(x, y)$ of (6.43) and its frequency ω can be written in a separable form,

$$u(x, y; \omega) = p(x; \widehat{\omega}_a)\, p(y; \widehat{\omega}_b), \quad \omega = \widehat{\omega}_a + \widehat{\omega}_b, \tag{6.44}$$

where $p(x; \widehat{\omega})$ is a Bloch solution of the linear 1D equation (6.7) at frequency $\widehat{\omega}$. Using the 1D results of Sec. 6.1 and the above connection (6.44) between the 1D and 2D Bloch solutions, we can construct the dispersion surfaces and bandgap structures of the 2D problem (6.43). In particular, the 2D Bloch modes are of the form

$$u(x, y; \omega) = e^{ik_x x + ik_y y}\, \widehat{p}\,[x; \widehat{\omega}(k_x)]\, \widehat{p}\,[y; \widehat{\omega}(k_y)], \tag{6.45}$$

where k_x, k_y are wavenumbers in the first Brillouin zone $-1 \le k_x, k_y \le 1$, $\widehat{p}(x; \widehat{\omega})$ is a π-periodic function in x, $\widehat{\omega}(k)$ is the 1D dispersion relation, and

$$\omega = \widehat{\omega}(k_x) + \widehat{\omega}(k_y) \tag{6.46}$$

is the 2D dispersion relation. This 2D dispersion relation at the potential depth $V_0 = 6$ is shown in Fig. 6.11(a). This figure contains a number of dispersion surfaces whose ω values form the 2D Bloch bands. Between these surfaces, two bandgaps can be seen. At other potential depth V_0, the 2D bandgap structure is summarized in Fig. 6.11(b). This figure reveals that, unlike the 1D case, there is only a finite number of bandgaps in the 2D problem at any given potential depth V_0. In addition, bandgaps appear only when V_0 is above a certain threshold (which is approximately 1.4 here). As the potential gets deeper, so does

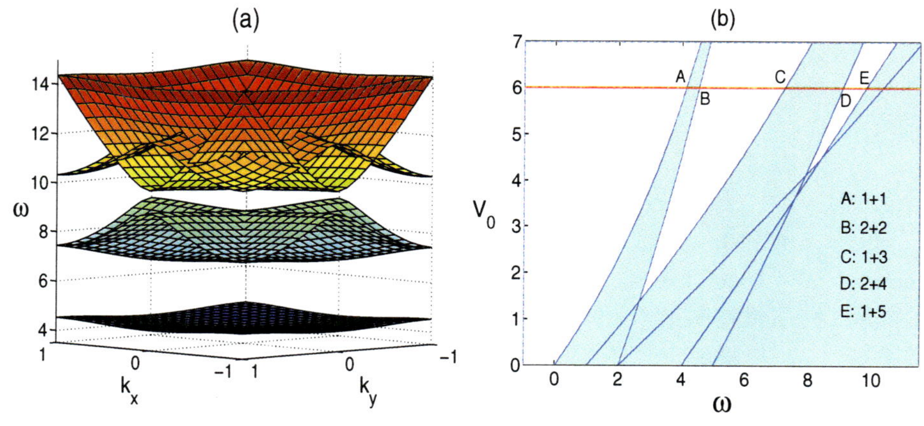

Figure 6.11. (a) *Dispersion surfaces of the linear 2D equation* (6.43) *at potential depth* $V_0 = 6$; (b) *2D bandgap structure for various potential depths* V_0. *Letters A, B, C, D, E mark the edges of Bloch bands at* $V_0 = 6$. *Insets like "A:* $1+1$*" are explained in the text. (After Shi and Yang* (2007).*)*

the number of bandgaps. At $V_0 = 6$, two bandgaps exist. The edges of these Bloch bands are marked in Fig. 6.11(b) by letters A, B, C, D, E, respectively.

The locations of band edges in the first Brillouin zone are important, as they determine the symmetry properties of the corresponding Bloch modes. In the first Brillouin zone, band edges A and B in Fig. 6.11(b) are located at $(k_x, k_y) = (0,0)$ and $(1,1)$, respectively. These two locations are the center and corner of the Brillouin zone and are called Γ and M points in the literature. At these two points, only a single Bloch wave exists. Band edges C and D, however, are located at $(k_x, k_y) = (0,1)$ and $(1,0)$, which are on the edges of the Brillouin zone and are called X points in the literature. At these X points, two linearly independent Bloch waves exist. Bloch solutions at these band edges are $u(x,y) = p(x;\widehat{\omega}_1)p(y;\widehat{\omega}_1)$ for A, $p(x;\widehat{\omega}_2)p(y;\widehat{\omega}_2)$ for B, $p(x;\widehat{\omega}_1)p(y;\widehat{\omega}_3)$ and $p(y;\widehat{\omega}_1)p(x;\widehat{\omega}_3)$ for C, $p(x;\widehat{\omega}_2)p(y;\widehat{\omega}_4)$ and $p(y;\widehat{\omega}_2)p(x;\widehat{\omega}_4)$ for D, where $\widehat{\omega}_k (1 \le k \le 4)$ are the lowest four 1D band edges (see Fig. 6.1), and $p(x;\widehat{\omega}_k)$ are the corresponding 1D Bloch waves at these edges (see Fig. 6.2). For convenience, we denote edge A as "$1+1$", B as "$2+2$," C as "$1+3$," and D as "$2+4$." The Bloch modes at A, B have the 90° rotational symmetry, i.e., $u(x,y) = u(y,x)$. Adjacent peaks in the Bloch mode of A are in-phase, while those of B are out of phase. The Bloch modes $p(x;\widehat{\omega}_1)p(y;\widehat{\omega}_3)$ and $p(x;\widehat{\omega}_2)p(y;\widehat{\omega}_4)$ at C, D do not have this 90° rotational symmetry however. Since the square lattice $V(x,y)$ in (6.39) admits this rotational symmetry, this means that the companion modes $p(y;\widehat{\omega}_1)p(x;\widehat{\omega}_3)$ and $p(y;\widehat{\omega}_2)p(x;\widehat{\omega}_4)$, which are 90° rotations of $p(x;\widehat{\omega}_1)p(y;\widehat{\omega}_3)$ and $p(x;\widehat{\omega}_2)p(y;\widehat{\omega}_4)$, are Bloch modes at C, D as well. The coexistence of several linearly independent Bloch modes at a band edge is a new feature in two spatial dimensions, and it has important implications for soliton bifurcations in the next subsection.

6.3.2 Envelope Equations of 2D Bloch Waves

In this subsection, we study bifurcations of small-amplitude Bloch-wave packets from edges of 2D Bloch bands and derive their envelope equations. The main difference between 1D and 2D cases is that, because there can be two linearly independent Bloch modes at a 2D band edge, solitons can bifurcate from a linear combination of them. This leads to coupled envelope equations for the two Bloch modes and a wider variety of soliton structures that often have no counterparts in 1D. Our derivation will follow Shi and Yang (2007). In certain special cases, envelope equations similar to (6.54)–(6.55) but without the γ terms were also derived by Brazhnyi et al. (2006, 2007).

Let us first consider a generic 2D band edge ω_0, where two linearly independent Bloch modes coexist. At such an edge one can write $\omega_0 = \widehat{\omega}_{0,1} + \widehat{\omega}_{0,2}$, where $\widehat{\omega}_{0,1}$ and $\widehat{\omega}_{0,2}$ are two 1D band edges with $\widehat{\omega}_{0,1} \neq \widehat{\omega}_{0,2}$. The corresponding 2D Bloch modes are $p_1(x)p_2(y)$ and $p_1(y)p_2(x)$, where $p_n(x) = p(x;\widehat{\omega}_{0,n})$ is the 1D Bloch mode at the 1D edge $\widehat{\omega}_{0,n}$. When the solution $u(x,y)$ of Eq. (6.41) is infinitesimal, this solution (at the band edge) is a linear superposition of these two Bloch modes. When $u(x,y)$ is small but not infinitesimal, this solution then becomes a combination of these two Bloch-wave packets and can be expanded into a multiscale perturbation series,

$$u = \epsilon u_0 + \epsilon^2 u_1 + \epsilon^3 u_2 + \cdots, \tag{6.47}$$

$$\omega = \omega_0 + \tau\epsilon^2, \tag{6.48}$$

where $\epsilon \ll 1$ is the small amplitude of this solitary wave,

$$u_0 = A_1(X,Y)p_1(x)p_2(y) + A_2(X,Y)p_2(x)p_1(y) \tag{6.49}$$

is the leading-order two-Bloch-wave packets, $X = \epsilon x, Y = \epsilon y$ are slow spatial variables of envelope functions A_1 and A_2, and $\tau = \pm 1$. The derivation of envelope equations for A_1 and A_2 below is analogous to that for the 1D case, and thus will only be sketched. Substituting the above expansions into the 2D lattice equation (6.41), this equation at $O(\epsilon)$ is automatically satisfied. At $O(\epsilon^2)$, the equation for u_1 is

$$\frac{\partial^2 u_1}{\partial x^2} + \frac{\partial^2 u_1}{\partial y^2} - [F(x)+F(y)]u_1 + \omega_0 u_1 = -2\left(\frac{\partial^2 u_0}{\partial x\partial X} + \frac{\partial^2 u_0}{\partial y\partial Y}\right). \tag{6.50}$$

It is easy to verify that the solution to this equation is

$$u_1 = \frac{\partial A_1}{\partial X}\nu_1(x)p_2(y) + \frac{\partial A_1}{\partial X}\nu_2(y)p_1(x) + \frac{\partial A_2}{\partial X}\nu_2(x)p_1(y) + \frac{\partial A_2}{\partial X}\nu_1(y)p_2(x), \tag{6.51}$$

where $\nu_n(x)$ is a periodic solution of the 1D equation

$$\nu_{n,xx} - F(x)\nu_n + \widehat{\omega}_{0,n}\nu_n = -2p_{n,x}, \quad n = 1,2. \tag{6.52}$$

At $O(\epsilon^3)$, the equation for u_2 is

$$\begin{aligned}&\frac{\partial^2 u_2}{\partial x^2} + \frac{\partial^2 u_2}{\partial y^2} - [F(x)+F(y)]u_2 + \omega_0 u_2 \\ &\quad = -\left(2\frac{\partial^2 u_1}{\partial x\partial X} + 2\frac{\partial^2 u_1}{\partial y\partial Y} + \frac{\partial^2 u_0}{\partial X^2} + \frac{\partial^2 u_0}{\partial Y^2} + \tau u_0 + |u_0|^2 u_0\right).\end{aligned} \tag{6.53}$$

Substituting the formulae (6.49) and (6.51) for u_0 and u_1 into the right-hand side of this equation, utilizing the Fredholm condition, (which requires the inhomogeneous term in (6.53) to be orthogonal to the homogeneous solutions $p_1(x)p_2(y)$ and $p_1(y)p_2(x)$ over 2π period in x and y), and recalling the relation (6.19), the following coupled nonlinear equations for envelope functions A_1 and A_2 are obtained (Shi and Yang (2007)):

$$\begin{aligned}D_1\frac{\partial^2 A_1}{\partial X^2} + D_2\frac{\partial^2 A_1}{\partial Y^2} + \tau A_1 + \sigma\Big[\alpha|A_1|^2 A_1 + \beta\left(A_1^* A_2^2 + 2A_1|A_2|^2\right) \\ + \gamma\left(|A_2|^2 A_2 + A_2^* A_1^2 + 2A_2|A_1|^2\right)\Big] = 0,\end{aligned} \tag{6.54}$$

$$D_2\frac{\partial^2 A_2}{\partial X^2}+D_1\frac{\partial^2 A_2}{\partial Y^2}+\tau A_2+\sigma\Big[\alpha|A_2|^2A_2+\beta\left(A_2^*A_1^2+2A_2|A_1|^2\right)$$
$$+\gamma\left(|A_1|^2A_1+A_1^*A_2^2+2A_1|A_2|^2\right)\Big]=0. \qquad (6.55)$$

Here, the superscript "*" represents complex conjugation,

$$D_n \equiv \frac{1}{2}\,\widehat{\omega}''(k)\big|_{\widehat{\omega}=\widehat{\omega}_{0,n}} \qquad (6.56)$$

is the 1D second-order dispersion at 1D edge $\widehat{\omega}_{0,n}$, and

$$\alpha=\frac{\int_0^{2\pi}\int_0^{2\pi} p_1^4(x)p_2^4(y)\,dxdy}{\int_0^{2\pi}\int_0^{2\pi} p_1^2(x)p_2^2(y)\,dxdy}, \qquad (6.57)$$

$$\beta=\frac{\int_0^{2\pi}\int_0^{2\pi} p_1^2(x)p_2^2(x)p_1^2(y)p_2^2(y)\,dxdy}{\int_0^{2\pi}\int_0^{2\pi} p_1^2(x)p_2^2(y)\,dxdy}, \qquad (6.58)$$

$$\gamma=\frac{\int_0^{2\pi}\int_0^{2\pi} p_1^3(x)p_2(x)p_2^3(y)p_1(y)\,dxdy}{\int_0^{2\pi}\int_0^{2\pi} p_1^2(x)p_2^2(y)\,dxdy} \qquad (6.59)$$

are nonlinear coefficients. Notice that α and β are always positive, but γ may be positive, negative, or zero. For instance, at edges C and D of Fig. 6.11, $\gamma=0$. But at edge E, $\gamma<0$.

At special 2D band edges where only a single Bloch mode exists (such as points A and B in Fig. 6.11), the envelope equation for this single Bloch mode would be simpler. In this case, the single Bloch mode has the form $p_1(x)p_1(y)$ with frequency $\omega_0=2\widehat{\omega}_{0,1}$, where $p_1(x)=p(x;\widehat{\omega}_{0,1})$ is the 1D Bloch wave at edge $\widehat{\omega}_{0,1}$. The leading-order solution now becomes

$$u_0=A_1(X,Y)p_1(x)p_1(y), \qquad (6.60)$$

and the envelope equation for $A_1(X,Y)$ can be readily found to be

$$D_1\left(\frac{\partial^2 A_1}{\partial X^2}+\frac{\partial^2 A_1}{\partial Y^2}\right)+\tau A_1+\sigma\alpha_0|A_1|^2A_1=0, \qquad (6.61)$$

where D_1 is as given in Eq. (6.56), and the nonlinear coefficient α_0 is

$$\alpha_0=\frac{\int_0^{2\pi}\int_0^{2\pi} p_1^4(x)p_1^4(y)\,dxdy}{\int_0^{2\pi}\int_0^{2\pi} p_1^2(x)p_1^2(y)\,dxdy}. \qquad (6.62)$$

For this single Bloch-wave packet, an envelope equation similar to (6.61) has also been derived by Baizakov et al. (2002).

Similarly to the 1D case, envelope solutions (A_1,A_2) must be centered at certain special positions in the lattice owing to constraints that are the counterparts of (6.25) for the

1D case. Following similar calculations as in one dimension, we can show that the center (x_0, y_0) of envelope solutions can only be located at four possible positions (Shi and Yang (2007)):

$$(x_0, y_0) = (0,0), \left(0, \frac{\pi}{2}\right), \left(\frac{\pi}{2}, 0\right), \left(\frac{\pi}{2}, \frac{\pi}{2}\right). \tag{6.63}$$

Solitary waves centered at $(x_0, y_0) = (0,0)$ of minimum potential are called on-site solitons, while those centered at other three positions are called off-site solitons.

Envelope equations (6.54)–(6.55) and (6.61) admit localized solutions only when their coefficients satisfy certain conditions. Thus, soliton bifurcations at 2D band edges are possible only under such conditions as well. For instance, at edge A in Fig. 6.11, $D_1 > 0$. Thus soliton bifurcation is possible only when $\sigma = 1$, i.e., the nonlinearity is self-focusing. The situation at edge C is similar. But situations at edges B and D are just the opposite, i.e., soliton bifurcations at these edges are possible only under self-defocusing nonlinearity.

6.3.3 Families of 2D Gap Solitons Bifurcated from Band Edges

Envelope equations (6.54)–(6.55) and (6.61) admit various types of solutions, and each envelope solution generates four families of gap solitons corresponding to the four envelope locations (6.63). Thus, many soliton families can bifurcate out from each edge of a 2D Bloch band. Among these solution families, off-site soliton families are always linearly unstable, just like the 1D case. Because of this, we will only consider on-site soliton families in the rest of this subsection.

At the band edge A in Fig. 6.11, the envelope equation is the scalar 2D NLS equation (6.61) with $D_1 > 0$. Under self-focusing nonlinearity ($\sigma = 1$), this envelope equation admits a positive and radially symmetric ground-state solution $A_1(X, Y)$, whose corresponding gap-soliton solution $u_0(x, y)$ in (6.60) is a positive wave packet (since the Bloch wave at edge A is positive everywhere as well). From this positive wave packet, a family of positive on-site solitons bifurcates out in the semi-infinite gap. The power curve of this solution family is displayed on the left side of Fig. 6.12. Here the power is defined as $P = \int |u|^2 dxdy$ as before. A distinctive feature of this power curve is that it is nonmonotonic. It has a nonzero minimum, below which solitons do not exist. This contrasts the 1D case where solitons exist at all power levels (see Fig. 6.3). It should be noted that as ω approaches the edge A, the power P approaches a finite value, *not* infinity (see the inset in this power figure). In addition, the power P *linearly* approaches this finite value on the band edge. Near edge A, the soliton is a slowly modulated positive Bloch-wave packet described by the analysis in the previous subsection. This profile is displayed in Fig. 6.12(b). It has one main peak at a lattice site, flanked by in-phase tails on all four sides. Away from this edge, the soliton becomes more localized, and its tails gradually disappear; see Fig. 6.12(a). This family

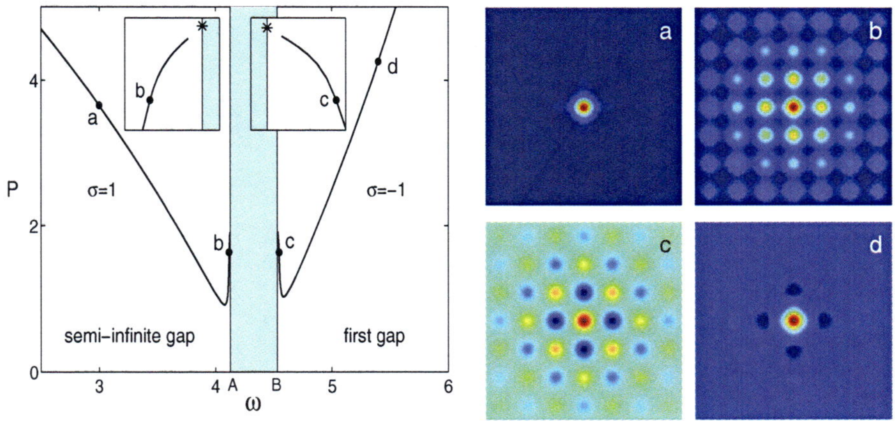

Figure 6.12. *Left: power curves of on-site solitons bifurcated from edge points "A, B" of the first Bloch band under self-focusing and self-defocusing nonlinearities, respectively ($V_0 = 6$). The insets show power curves near band edges, where asterisks are the limit power values on the edges. Right: soliton profiles at the marked points in the left figure. (After Shi et al.* (2008).*)*

of solitons has been investigated both theoretically and experimentally by Fleischer et al. (2003b), Yang and Musslimani (2003), and Efremidis et al. (2003).

At band edge B, the envelope equation is the scalar 2D NLS equation (6.61) with $D_1 < 0$, which admits a positive ground-state solution under *self-defocusing* nonlinearity ($\sigma = -1$). The corresponding gap soliton is a sign-changing wave packet (since the Bloch wave at edge B is sign changing). From this wave packet, a family of on-site solitons bifurcates out in the first gap. The power curve of this soliton family is also displayed at the left side of Fig. 6.12. Similar to the soliton family bifurcated from edge A, this solution family also has a nonzero minimum power. In addition, the power remains finite as ω approaches edge B. Near the edge, the profile of the soliton is displayed in Fig. 6.12(c). It has one main peak at a lattice site, flanked by out-of-phase tails on all sides. Away from the edge, the soliton becomes more localized; see Fig. 6.12(d). This solution family has been investigated theoretically and experimentally by Fleischer et al. (2003b), Efremidis et al. (2003), and Lou et al. (2007).

At edge C, there are two linearly independent Bloch waves, and their envelope equations are (6.54)–(6.55) with $D_1, D_2 > 0$, $D_1 \neq D_2$, and $\gamma = 0$. This coupled system admits several types of envelope solutions under self-focusing nonlinearity. One of them is $A_1 \neq 0, A_2 = 0$, i.e., the soliton is a wave packet of a single Bloch wave. In this case, the A_1 equation (6.54) is a single 2D NLS equation with different dispersion coefficients along the X and Y directions. Thus, it admits a positive elliptical-shaped envelope soliton.

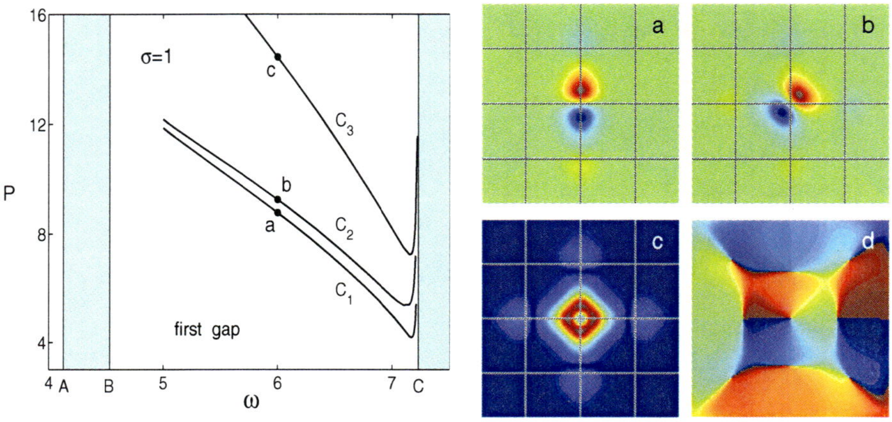

Figure 6.13. *Left: power curves of on-site solitons bifurcated from edge C of the second Bloch band under self-focusing nonlinearity* ($V_0 = 6$). *Right:* (a, b) *soliton profiles at the marked points "a,b" in the left figure;* (c, d) *amplitude and phase distributions of the soliton at point "c" of the left figure. The grids in* (a, b, c) *are the lattice sites (of minimal potentials). (After Shi et al.* (2008).*)*

Bifurcating from the resulting Bloch-wave packet, a family of on-site gap solitons is then generated from edge C. We denote this solution family as the C_1 family, and its power curve is displayed at the left side of Fig. 6.13. At $\omega = 6$ away from the edge C (see point "a" on the power curve), the soliton in this family is displayed in Fig. 6.13(a). This soliton contains two equal-intensity peaks aligned along the vertical direction, and the two peaks are out of phase with each other. However, different from the dipole solitons considered earlier (such as Fig. 6.8(a)), the two peaks of the dipole here are located at a *single* lattice site, *not* at two different lattice sites. Thus we can call this soliton a single-site dipole soliton. Near edge C, the soliton in this C_1 family becomes broad along the vertical direction, but remains relatively narrow along the horizontal direction. Because of this, it was called a reduced-symmetry soliton by Fischer et al. (2006), who first reported this type of solutions.

At edge C, envelope equations (6.54)–(6.55) also admit other solutions. One of them is $A_1 > 0, A_2 > 0$. These envelope solutions are both ellipse shaped but stretched along orthogonal directions (Shi and Yang (2007)). In this case, the soliton consists of wave packets of both Bloch waves. From this solution, a family of on-site gap solitons bifurcates out from edge C. We denote this solution family as the C_2 family, and its power curve is also displayed at the left side of Fig. 6.13. At $\omega = 6$ (see point "b" on the power curve), the soliton in this family is displayed in Fig. 6.13(b). This soliton is also an out-phase dipole residing at a single lattice site, similar to the one in Fig. 6.13(a). However, this single-site dipole is now aligned along the diagonal direction. Near edge C, the soliton in this C_2

family becomes broad along both the horizontal and vertical directions. This C_2 family of solutions was first reported by Shi and Yang (2007).

Another envelope solution admitted at edge C is where $A_1 > 0$, and A_2 is purely imaginary. In this case, these envelopes of the two Bloch waves have a $\pi/2$ phase delay. The family of on-site gap solitons bifurcating from these solutions is denoted as the C_3 family, and its power curve is also displayed at the left side of Fig. 6.13. Unlike the (real-valued) solitons in the C_1 and C_2 families, solitons in the C_3 family are complex valued. At $\omega = 6$ (see point "c" on the power curve), this complex soliton is displayed in Fig. 6.13(c, d). Its intensity field consists of a ring at a lattice site, and weaker tails in the far distance. The phase plot shows that when winding around this ring, the phase of the soliton increases by 2π. Thus this soliton has a vortex structure with a unit topological charge. These single-site vortex solitons, for saturable nonlinearity, were first reported theoretically by Manela et al. (2004), and then experimentally observed by Bartal et al. (2005).

At other band edges (such as D, E in Fig. 6.11), various families of gap solitons bifurcate out under either self-focusing or self-defocusing nonlinearity as well. See Shi and Yang (2007) for details.

6.4 Stability of 2D Gap Solitons Bifurcated from Bloch Bands

Stability of 2D gap solitons in the previous section is an important issue. In 1D, on-site gap solitons can be linearly stable both near band edges and away from them (see Sec. 6.1.5). The situation turns out to be different in two dimensions. One may notice that the power curves of these 2D soliton families all have a nonzero power minimum (see Figs. 6.12 and 6.13). Thus, in view of the results in Sec. 5.4, solitons on one side of the power minimum must be linearly unstable. But we do not know yet which side of solitons are unstable. The Vakhitov–Kolokolov stability criterion is not applicable either since these gap solitons are sign-indefinite in general.

In this section, we determine stability properties of 2D gap solitons bifurcated from Bloch bands. As in one dimension, off-site 2D solitons are all linearly unstable. Thus we only consider on-site solitons below. First, we will show analytically that for solitons near band edges, if $\sigma P'(\omega) > 0$, then these solitons are linearly unstable due to the presence of a positive eigenvalue. For self-focusing nonlinearity ($\sigma = 1$), solitons with power slope $P'(\omega) > 0$ are linearly unstable. This power-slope condition is the same as that in the Vakhitov–Kolokolov criterion of Sec. 5.3 (notice the notational difference of $\omega = -\mu$, where μ is the propagation constant in Sec. 5.3, thus $P'(\omega) = -P'(\mu)$). But for self-defocusing nonlinearity ($\sigma = -1$), solitons with power slope $P'(\omega) < 0$ are linearly unstable. This power-slope condition becomes the opposite of that in the Vakhitov–Kolokolov criterion. Thus our instability condition $\sigma P'(\omega) > 0$ is an extension and modification of the Vakhitov–Kolokolov criterion to sign-indefinite solitons. When combining this stability condition with the power curves in Figs. 6.12 and 6.13, we see that these 2D on-site solitons near Bloch bands are all linearly unstable. This means that solitons on the Bloch-band side of the power minimum point are linearly unstable. Hence solitons on the other side of the power minimum point away from Bloch bands could be linearly stable (if no other instabilities arise). Numerically, we will show that on-site solitons bifurcating from edges A, B of the first band (see Fig. 6.12) are indeed linearly stable on the side of the power minimum point away from the edges. But solitons bifurcating from edge C of the second band (see Fig. 6.13) remain unstable away from the edge due to additional instabilities (the same holds for solitons bifurcating from edges D and E as well).

6.4.1 Analytical Calculations of Eigenvalue Bifurcations near Band Edges

In this subsection, we analytically calculate eigenvalues of on-site 2D solitons near band edges and show that positive (unstable) eigenvalues appear when $\sigma P'(\omega) > 0$. Near a band edge, these solitons are low-amplitude Bloch-wave packets whose envelope equations are (6.54)–(6.55) or (6.61). These envelope equations are translation invariant (along both x and y directions), phase invariant, and their envelope solitons have a constant power independent of the frequency. Due to these properties, the linearization spectrum of these envelope solitons has a zero eigenvalue of multiplicity eight: two associated with phase invariance, four associated with translation invariance, and the remaining two associated with constant power. In the full lattice equation (6.38), the phase invariance persists, but the translation invariances are broken, and the power of gap solitons is no longer a constant (see Figs. 6.12 and 6.13). As a result, in the linearization spectrum of a gap soliton near the band edge, the zero eigenvalue only has multiplicity two (induced by phase invariance), and the other six multiplicities of the zero eigenvalue have to bifurcate out. The four multiplicities of the zero eigenvalue associated with spatial translations bifurcate out along the imaginary axis for on-site solitons, thus not creating instabilities (for off-site solitons, this bifurcation is along the real axis and hence symmetry-breaking instability is generated). These translation-induced eigenvalue bifurcations are analogous to the 1D case, and the bifurcated eigenvalues are exponentially small in the soliton amplitude. The bifurcation of the remaining two multiplicities of the zero eigenvalue associated with the envelope-soliton's constant power is new and has no counterpart in the 1D case. This eigenvalue bifurcation can create instabilities to on-site 2D solitons and will be calculated analytically below. We will show that this bifurcation is along the real axis and creates instability when

$\sigma P'(\omega) > 0$. A surprising feature of this bifurcation is that the bifurcated eigenvalue is algebraically small, rather than exponentially small, in the soliton amplitude. This motivates us to use a power series expansion to calculate this eigenvalue. This calculation follows Shi et al. (2008) (except a notational change of $\omega = -\mu$).

For simplicity, we consider real-valued gap solitons that are single-Bloch-wave packets near band edges; i.e., the A_2 term in Eq. (6.49) is absent. These solitons include the ones bifurcating from edges A and B, as well as the C_1 family bifurcating from the edge C, among others (see Figs. 6.12 and 6.13). Near the band edge ω_0, this single-Bloch-wave packet soliton is given by

$$u = \epsilon u_0 + \epsilon^2 u_1 + \epsilon^3 u_2 + \epsilon^4 u_3 + \cdots, \tag{6.64}$$

$$\omega = \omega_0 + \tau \epsilon^2, \tag{6.65}$$

where $\epsilon \ll 1$ is the small amplitude of the soliton, $p_1(x)p_2(y)$ is a Bloch wave at edge ω_0, the leading-order term u_0 is a slowly varying packet of this single Bloch wave,

$$u_0 = A(X,Y)p_1(x)p_2(y), \tag{6.66}$$

$X = \epsilon(x - x_0), Y = \epsilon(y - y_0)$, and (x_0, y_0) is the center position of the envelope function. For on-site solitons, $(x_0, y_0) = (0,0)$. In (6.66), we denoted the envelope function as $A(X,Y)$ rather than $A_1(X,Y)$ of Eq. (6.49) for simplicity, and the other notations are the same as those in Sec. 6.3.2. Here the 1D Bloch functions $p_1(x)$ and $p_2(x)$ can be the same such as at edges A and B. They can also be different such as at edges C and D. For real-valued gap solitons u, the envelope function $A(X,Y)$ is also real and satisfies the stationary 2D NLS equation

$$D_1 \frac{\partial^2 A}{\partial X^2} + D_2 \frac{\partial^2 A}{\partial Y^2} + \tau A + \sigma \alpha A^3 = 0. \tag{6.67}$$

When $p_1(x) = p_2(x)$, then $D_1 = D_2$; hence this equation reduces to (6.61). When $p_1(x) \neq p_2(x)$, this equation is a reduction of the coupled envelope equations (6.54)–(6.55) by taking $A_2 = 0$, which is possible at band edges where $\gamma = 0$ (such as edges C and D). The envelope equation (6.67) admits a positive and single-humped ground-state solution $A(X,Y) > 0$, which induces families of gap solitons such as those bifurcated from edges A and B in Fig. 6.12 and the C_1 family bifurcated from the edge C in Fig. 6.13. Stability of such gap solitons near band edges will be analyzed below.

Before we conduct the stability analysis of such real-valued 2D gap solitons, we need some information on these solitons in their asymptotic expansion (6.64). To obtain such information, we first define the following linear operator:

$$L_0 = \frac{\partial^2}{\partial x^2} + \frac{\partial^2}{\partial y^2} - [F(x) + F(y)] + \omega + \sigma u^2(x,y). \tag{6.68}$$

This operator is the counterpart of operator (5.69) for the generalized NLS equations with potentials (5.64). Then the solitary wave equation (6.41) becomes

$$L_0 u(x,y) = 0. \tag{6.69}$$

Since the solution $u(x,y)$ contains both fast and slow spatial variables (x,y) and (X,Y), in our multiscale asymptotic analysis below it is necessary to separate derivatives to these fast and slow variables in L_0, so that L_0 is rewritten as

$$L_0 = M_0 + \epsilon M_1 + \epsilon^2 M_2 + \epsilon^2 \sigma (u_0 + \epsilon u_1 + \cdots)^2, \tag{6.70}$$

where

$$M_0 = \frac{\partial^2}{\partial x^2} + \frac{\partial^2}{\partial y^2} - [F(x) + F(y)] + \omega_0, \tag{6.71}$$

$$M_1 = 2\left(\frac{\partial^2}{\partial x \partial X} + \frac{\partial^2}{\partial y \partial Y}\right), \tag{6.72}$$

$$M_2 = \frac{\partial^2}{\partial X^2} + \frac{\partial^2}{\partial Y^2} + \tau. \tag{6.73}$$

Here the partial derivatives to x and y in M_0 and M_1 are with respect to the fast variables x and y only. Substituting the expansions (6.64) and (6.70) into the solitary wave equation (6.69), we get the following equations for the solutions u_n at various orders of ϵ:

$$\epsilon^1 : \; M_0 u_0 = 0, \tag{6.74}$$

$$\epsilon^2 : \; M_0 u_1 = -M_1 u_0, \tag{6.75}$$

$$\epsilon^3 : \; M_0 u_2 = -M_1 u_1 - (M_2 + \sigma u_0^2) u_0, \tag{6.76}$$

$$\epsilon^4 : \; M_0 u_3 = -M_1 u_2 - (M_2 + 3\sigma u_0^2) u_1, \tag{6.77}$$

$$\epsilon^5 : \; M_0 u_4 = -M_1 u_3 - (M_2 + 3\sigma u_0^2) u_2 - 3\sigma u_0 u_1^2. \tag{6.78}$$

The first-order equation (6.74) is satisfied automatically. The second-order equation (6.75) admits a periodic solution in fast variables (x,y),

$$u_1 = \frac{\partial A}{\partial X} \nu_1(x) p_2(y) + \frac{\partial A}{\partial Y} p_1(x) \nu_2(y), \tag{6.79}$$

where $\nu_n(x)$ is a solution of Eq. (6.52). To make $\nu_n(x)$ unique, we require its symmetry to be the same as that of $p_n'(x)$, i.e., opposite of that of $p_n(x)$. The Fredholm condition on the third-order equation (6.76) gives the envelope equation (6.67).

The power curve of gap solitons near band edges will prove to be important for our stability analysis. This power is

$$P(\omega) = \int_{-\infty}^{\infty} \int_{-\infty}^{\infty} |u|^2 dx dy. \tag{6.80}$$

Near a band edge, this power function can be calculated analytically. Using the perturbation series (6.64) as well as the expressions (6.66) and (6.79) for u_0 and u_1, it is easy to find that (Shi et al. (2008))

$$P(\omega) = P_0 + O(\epsilon^2) = P_0 + O(|\omega - \omega_0|), \tag{6.81}$$

where

$$P_0 = G\int_{-\infty}^{\infty}\int_{-\infty}^{\infty} A^2(X,Y)dXdY = \frac{C_0 G\sqrt{D_1 D_2}}{\alpha} \tag{6.82}$$

is the limit power on the band edge,

$$G = \frac{1}{\pi^2}\int_0^{\pi}\int_0^{\pi} p_1^2(x)p_2^2(y)dxdy \tag{6.83}$$

is the average value of the squared Bloch-wave function $p_1^2(x)p_2^2(y)$, and

$$C_0 = 11.70 \tag{6.84}$$

is the power of the ground-state soliton in the 2D potentialless NLS equation with unit dispersion and nonlinearity coefficients. This power formula shows that the limit power P_0 on the band edge is finite, and the power curve near the band edge is a linear function of $\omega - \omega_0$, both in agreement with the numerical power curves in Figs. 6.12 and 6.13.

Now we study the linear stability of these gap solitons $u(x,y)$. For this purpose, we perturb these solitons by normal modes as

$$\begin{aligned} U(x,y,t) = e^{-i\omega t}\Big\{u(x,y) + \Big[v(x,y) + i\lambda^{-1}w(x,y)\Big]e^{\lambda t} \\ + \Big[v^*(x,y) - [i\lambda^{-1}w(x,y)]^*\Big]e^{\lambda^* t}\Big\}, \end{aligned} \tag{6.85}$$

where $v, w \ll 1$ are normal-mode perturbations, and λ is the mode's eigenvalue. Here we introduced a factor $i\lambda^{-1}$ in front of w, so that the resulting eigenvalue problem (6.86) involves λ^2 only. This treatment has been used in Sec. 5.4 in the study of stability switching at a power extremum. Substituting the perturbed solution (6.85) into the original evolution equation (6.38) and neglecting higher-order terms in (v,w), we obtain the linear eigenvalue problem

$$L_0 w = -\lambda^2 v, \quad L_1 v = w, \tag{6.86}$$

where L_0 has been defined in (6.68), and L_1 is defined as

$$L_1 = \frac{\partial^2}{\partial x^2} + \frac{\partial^2}{\partial y^2} - [F(x) + F(y)] + \omega + 3\sigma u^2(x,y). \tag{6.87}$$

This operator L_1 is the counterpart of (5.70) for the generalized NLS equations with potentials (5.64). Since this eigenvalue problem is linear, we will treat its eigenfunctions (v,w) as $O(1)$ quantities below. In addition, we have the relation (6.69) as well as the relation

$$L_1 u_\omega = -u. \tag{6.88}$$

This second relation is obtained by taking the partial derivative of (6.69) with respect to ω. Furthermore, by taking the inner products of (6.86)'s first equation with u and the second equation with u_ω, and recalling the self-adjoint properties of operators L_0, L_1 as well as the relations (6.69) and (6.88), we see that for nonzero eigenvalues λ, functions u and v are orthogonal to each other, and u_ω and w are orthogonal to each other, i.e.,

$$\langle u, v\rangle = 0 \tag{6.89}$$

and

$$\langle u_\omega, w\rangle = 0. \tag{6.90}$$

Here the inner product is defined as

$$\langle f, g\rangle = \int_{-\infty}^{\infty}\int_{-\infty}^{\infty} f(x,y)\, g(x,y)\, dx dy. \tag{6.91}$$

In our multiscale perturbation analysis below, we need to separate the partial derivatives in L_1 into those with respect to the fast and slow variables as we have done for the operator L_0. In doing so we get

$$L_1 = M_0 + \epsilon M_1 + \epsilon^2 M_2 + 3\epsilon^2 \sigma (u_0 + \epsilon u_1 + \cdots)^2, \tag{6.92}$$

which is the analogue of (6.70) for L_0.

Our objective here is to determine discrete eigenvalues λ and eigenfunctions (v,w) for low-amplitude 2D gap solitons near edges of Bloch bands. Before detailed calculations, we first lay out our plan. These discrete eigenvalues are small since the soliton's amplitude is low. Then in view of Eqs. (6.69) and (6.86), we see that w should be proportional to u to the leading order. Indeed, we will find that $w \sim \epsilon u = O(\epsilon^2)$; see (6.132). Correspondingly, $v \sim -\epsilon u_\omega$ in view of (6.86) and (6.88). It is important to recognize that near band edges, $u \sim \epsilon A(X,Y) p_1(x) p_2(y)$, and thus $v \sim -\epsilon u_\omega = O(\epsilon^2 A_\omega) = O(A_\tau) = O(1)$, which is two orders larger than w. But w cannot be exactly equal to ϵu, because $w = \epsilon u$ does not satisfy the orthogonality condition (6.90) in view that the power curve's slope is not zero near band edges (see Figs. 6.12 and 6.13). Calculation of the higher-order correction to the leading term ϵu of w is a key step in our analysis. Through systematic perturbative calculations, we will show that the higher-order correction to w is given by (6.132), where the function ζ is proportional to λ^2. Then by inserting this w formula into the orthogonality relation (6.90), the expression for λ^2 will be obtained.

Now we start to calculate the eigenvalue λ and eigenfunction (v, w) for low-amplitude real-valued 2D gap solitons near band edges. For this purpose, we expand this eigenfunction and eigenvalue into the following power series of ϵ:

$$v = v_0 + \epsilon v_1 + \epsilon^2 v_2 + \cdots, \tag{6.93}$$

$$w = w_0 + \epsilon w_1 + \epsilon^2 w_2 + \cdots, \tag{6.94}$$

$$\lambda^2 = \epsilon \eta_1 + \epsilon^2 \eta_2 + \cdots. \tag{6.95}$$

Inserting these expansions and those of L_0 and L_1 into the eigenvalue problem (6.86), at $O(1)$, we get

$$M_0 w_0 = 0, \tag{6.96}$$

$$M_0 v_0 = w_0. \tag{6.97}$$

From (6.96), we see that $w_0 = h_0(X, Y) p_1(x) p_2(y)$. The Fredholm condition for (6.97) says that w_0 must be orthogonal to the homogeneous Bloch solution $p_1(x) p_2(y)$ over 2π period in x and y, i.e.,

$$\int_0^{2\pi} \int_0^{2\pi} w_0\, p_1(x) p_2(y)\, dx dy = 0. \tag{6.98}$$

Here the integration is with respect to the fast variables x and y only. From this Fredholm condition and the w_0 formula, we see that $h_0(X, Y) = 0$, and thus $w_0 = 0$. Then from (6.97), we get

$$v_0 = \phi(X, Y) p_1(x) p_2(y). \tag{6.99}$$

At $O(\epsilon)$, we get

$$M_0 w_1 = -\eta_1 v_0, \tag{6.100}$$

$$M_0 v_1 = w_1 - M_1 v_0. \tag{6.101}$$

Applying the Fredholm condition to Eq. (6.100), we see that $\eta_1 = 0$, and hence $w_1 = h_1(X, Y) p_1(x) p_2(y)$. Applying the Fredholm condition to (6.101) and noticing that $M_1 v_0$ and $p_1(x) p_2(y)$ have the opposite symmetry in fast (x, y) variables and hence the integral of their product is zero over one period of $p_1(x)$ and $p_2(y)$, we find that $h_1(X, Y) = 0$, thus $w_1 = 0$. In this case, the solution v_1 to (6.101) is

$$v_1 = \frac{\partial \phi}{\partial X} \nu_1(x) p_2(y) + \frac{\partial \phi}{\partial Y} p_1(x) \nu_2(y). \tag{6.102}$$

Here we did not add into v_1 the homogeneous solution $h(X,Y)p_1(x)p_2(y)$ of (6.101) because it turns out from later calculations that such a homogeneous term must be zero, and so it is left out at this early stage in order to simplify the analysis.

At $O(\epsilon^2)$, we get

$$M_0 w_2 = -\eta_2 v_0, \tag{6.103}$$

$$M_0 v_2 = w_2 - M_1 v_1 - (M_2 + 3\sigma u_0^2) v_0. \tag{6.104}$$

Applying the Fredholm condition to (6.103), we get $\eta_2 = 0$. Thus

$$w_2 = \psi(X,Y)p_1(x)p_2(y). \tag{6.105}$$

The Fredholm condition for (6.104) is

$$\int_0^{2\pi}\int_0^{2\pi}\left[w_2 - M_1 v_1 - (M_2 + 3\sigma u_0^2)v_0\right] p_1(x)p_2(y)\,dx\,dy = 0. \tag{6.106}$$

Utilizing the expressions (6.99), (6.102), and (6.105) for (v_0, v_1, w_2) as well as the relation (6.19), the above Fredholm condition reduces to

$$\mathcal{L}_1 \phi = \psi, \tag{6.107}$$

where

$$\mathcal{L}_1 \equiv D_1 \frac{\partial^2}{\partial X^2} + D_2 \frac{\partial^2}{\partial Y^2} + \tau + 3\sigma\alpha A^2. \tag{6.108}$$

Next we proceed to higher orders of ϵ. We will see that the equations for the next few higher-order terms in w are decoupled from the higher-order terms in v, because λ^2 turns out to be high order in ϵ, or $O(\epsilon^6)$ to be specific. Thus we will consider higher-order terms in w only. At $O(\epsilon^3)$, we have

$$M_0 w_3 = -\eta_3 v_0 - M_1 w_2. \tag{6.109}$$

Its Fredholm condition shows that $\eta_3 = 0$ (since the integral of $M_1 w_2$ multiplying $p_1(x)p_2(y)$ is zero over one period of $p_1(x)$ and $p_2(y)$). In this case, the solution w_3 is then

$$w_3 = \frac{\partial \psi}{\partial X} \nu_1(x)p_2(y) + \frac{\partial \psi}{\partial Y} p_1(x)\nu_2(y). \tag{6.110}$$

Here we did not add into w_3 the homogeneous solution $h(X,Y)p_1(x)p_2(y)$ of (6.109) either because such a term turns out to be zero as in the v_1 case.

At $O(\epsilon^4)$, we have

$$M_0 w_4 = -\eta_4 v_0 - M_1 w_3 - (M_2 + \sigma u_0^2) w_2. \tag{6.111}$$

Utilizing the expressions (6.99), (6.105), and (6.110) for (v_0, w_2, w_3) as well as the relation (6.19), the Fredholm condition for (6.111) reduces to

$$\mathcal{L}_0\psi = -\eta_4\phi, \tag{6.112}$$

where

$$\mathcal{L}_0 \equiv D_1\frac{\partial^2}{\partial X^2} + D_2\frac{\partial^2}{\partial X^2} + \tau + \sigma\alpha A^2. \tag{6.113}$$

Now it is important to recognize that (see Sec. 5.3), the two equations (6.107) and (6.112) form the eigenvalue problem for the linear stability of ground-state envelope solitons $Q(X,Y,T) = A(X,Y)e^{-i\tau T}$ in the 2D envelope NLS equation

$$iQ_T + D_1 Q_{XX} + D_2 Q_{YY} + \sigma\alpha|Q|^2 Q = 0. \tag{6.114}$$

The linear-stability spectrum of these ground-state solitons in this 2D constant-coefficient NLS equation is well known. This spectrum contains a single discrete eigenvalue which is zero. Thus $\eta_4 = 0$. Corresponding to this zero eigenvalue, there are four discrete eigenfunctions which are induced by the phase and translation invariances of Eq. (6.114) as well as its ground states having a constant power (the algebraic multiplicity of this zero eigenvalue is eight). The constant-power-induced eigenfunction is

$$\phi = -A_\tau, \quad \psi = A. \tag{6.115}$$

As we have explained at the beginning of this subsection, we will calculate eigenvalue bifurcations from this eigenmode below. Note that this eigenmode exists only in two spatial dimensions because only 2D envelope solitons in the envelope NLS equation (6.114) have constant power. This mode does not exist in other spatial dimensions (such as in one or three dimensions) because powers of envelope solitons in such cases are not constant. Because of this, the calculation of eigenvalue bifurcation from this eigenmode below has no counterpart in 1D and 3D lattice equations.

With the above choice (6.115), and in view of the expressions (6.99), (6.102), (6.105), and (6.110) for (v_0, v_1, w_2, w_3) as well as the expressions (6.66) and (6.79) for (u_0, u_1), we see that

$$v_0 = -u_{0\tau}, \quad v_1 = -u_{1\tau}, \tag{6.116}$$

$$w_0 = w_1 = 0, \quad w_2 = u_0, \quad w_3 = u_1. \tag{6.117}$$

In addition,

$$\eta_1 = \eta_2 = \eta_3 = \eta_4 = 0. \tag{6.118}$$

Next we continue to pursue higher-order terms in the power series expansions of the eigenfunction w and eigenvalue λ^2. The function w_4 satisfies Eq. (6.111). Utilizing (6.76), (6.117), and (6.118), we find that

$$w_4 = u_2 + \zeta(X,Y)p_1(x)p_2(y). \tag{6.119}$$

Here the inclusion of the homogeneous solution $\zeta(X,Y)p_1(x)p_2(y)$ in w_4 is necessary, and in fact crucial, in our analysis. To obtain the equation for w_5, we expand the eigenvalue problem (6.86) to $O(\epsilon^5)$ and get

$$M_0 w_5 = -\eta_5 v_0 - M_1 w_4 - (M_2 + \sigma u_0^2)w_3 - 2\sigma u_0 u_1 w_2. \tag{6.120}$$

Inserting the expressions (6.99), (6.117), and (6.119) for (v_0, w_2, w_3, w_4) into the above equation and utilizing the Fredholm condition on the u_3 equation (6.77), one can readily show that the Fredholm condition for (6.120) is satisfied only when $\eta_5 = 0$. In this case, by utilizing the u_3 equation (6.77), the solution w_5 can be found to be

$$w_5 = u_3 + \frac{\partial \zeta}{\partial X} \nu_1(x)p_2(y) + \frac{\partial \zeta}{\partial Y} p_1(x)\nu_2(y). \tag{6.121}$$

Here the homogeneous solution of (6.120) of the form $h(X,Y)p_1(x)p_2(y)$ is not included in (6.121), because it will not enter the leading-order asymptotic expression for the eigenvalue λ^2.

Next we proceed to the equation for w_6, whose Fredholm condition will determine the function $\zeta(X,Y)$ in the expression (6.119) of w_4. The equation for w_6 can be obtained by expanding the eigenvalue problem (6.86) to $O(\epsilon^6)$, and we get

$$M_0 w_6 = -\eta_6 v_0 - M_1 w_5 - (M_2 + \sigma u_0^2)w_4 - 2\sigma u_0 u_1 w_3 - \sigma(2u_0 u_2 + u_1^2)w_2. \tag{6.122}$$

Utilizing the expressions (6.99), (6.117), (6.119), and (6.121) of $(v_0, w_2, w_3, w_4, w_5)$, and the Fredholm condition on the u_4 equation (6.78) as well as the relation (6.19), we find that the Fredholm condition for (6.122) reduces to

$$\mathcal{L}_0 \zeta = \eta_6 A_\tau. \tag{6.123}$$

This equation determines the function ζ. Notice that A satisfies Eq. (6.67), i.e., $\mathcal{L}_0 A = 0$, and thus (6.123) has a homogeneous solution A. This homogeneous solution is orthogonal to the inhomogeneous term $\eta_6 A_\tau$ since the power of envelope solitons A in Eq. (6.67) is independent of the frequency parameter τ. Thus Eq. (6.123) admits a localized solution ζ.

The solution ζ to Eq. (6.123) can be derived more explicitly as follows. By differentiating equation (6.67) with respect to τ, we find that

$$\mathcal{L}_1 A_\tau = -A. \tag{6.124}$$

When this relation is combined with Eq. (6.123), we get

$$\mathcal{L}_1 \mathcal{L}_0 \zeta = -\eta_6 A. \tag{6.125}$$

By introducing normalized variables

$$\widehat{X} = \left|\frac{\tau}{D_1}\right|^{1/2} X, \quad \widehat{Y} = \left|\frac{\tau}{D_2}\right|^{1/2} Y, \quad \widehat{A} = \left|\frac{\alpha}{\tau}\right|^{1/2} A, \tag{6.126}$$

we find that the function ζ can be expressed as

$$\zeta = -\frac{\eta_6}{|\tau|\sqrt{\alpha|\tau|}} \widehat{\zeta}(\widehat{X}, \widehat{Y}), \tag{6.127}$$

where $\widehat{\zeta}(\widehat{X}, \widehat{Y})$ is a radially symmetric solution of the following parameter-free equation:

$$\widehat{\mathcal{L}}_1 \widehat{\mathcal{L}}_0 \widehat{\zeta}(\widehat{X}, \widehat{Y}) = \widehat{A}. \tag{6.128}$$

Here $\widehat{A}(\widehat{X}, \widehat{Y})$ is the unique ground-state solution of the following parameter-free 2D NLS equation:

$$\widehat{\mathcal{L}}_0 \widehat{A}(\widehat{X}, \widehat{Y}) = 0, \tag{6.129}$$

and $\widehat{\mathcal{L}}_0$, $\widehat{\mathcal{L}}_1$ are the normalized linear operators

$$\widehat{\mathcal{L}}_0 \equiv \frac{\partial^2}{\partial \widehat{X}^2} + \frac{\partial^2}{\partial \widehat{Y}^2} - 1 + \widehat{A}^2, \tag{6.130}$$

$$\widehat{\mathcal{L}}_1 \equiv \frac{\partial^2}{\partial \widehat{X}^2} + \frac{\partial^2}{\partial \widehat{Y}^2} - 1 + 3\widehat{A}^2. \tag{6.131}$$

To summarize the above results, we insert the expressions (6.117) and (6.119) for (w_2, w_3, w_4) into the perturbation series expansion (6.94) and find that the asymptotic formula for the eigenfunction w is

$$w = \epsilon\left[u + \epsilon^3 \zeta(X,Y) p_1(x) p_2(y) + O(\epsilon^4)\right], \tag{6.132}$$

where ζ is given by (6.127). Notice that $w = O(\epsilon^2)$ since $u = O(\epsilon)$.

Now we insert the above perturbation-series solution of w into the orthogonality relation (6.90). To the leading order, we get

$$\left\langle u_\omega,\ u + \epsilon^3 \zeta(X,Y) p_1(x) p_2(y)\right\rangle = 0. \tag{6.133}$$

Notice that

$$\langle u_\omega, u\rangle = \frac{1}{2} P'(\omega), \tag{6.134}$$

where P is the power of the soliton. By taking the leading-order term ϵu_0 for u and using the relation (6.65) between ω and τ, we find that to the leading order,

$$\begin{aligned}\left\langle u_\omega, \epsilon^3 \zeta(X,Y) p_1(x) p_2(y) \right\rangle &= \epsilon^2 \langle A_\tau(X,Y) p_1(x) p_2(y), \zeta(X,Y) p_1(x) p_2(y) \rangle \\ &= G \int_{-\infty}^{\infty} \int_{-\infty}^{\infty} A_\tau(X,Y) \zeta(X,Y) dX dY, \end{aligned} \tag{6.135}$$

where G is defined in Eq. (6.83). Here the second equality can be obtained by expanding $p_1^2(x) p_2^2(y)$ in the integrand into Fourier series. The constant term in this Fourier series makes a dominant contribution to the integral, which is given at the end of (6.135). The other harmonics in this Fourier series make exponentially small contributions to the integral, which are thus omitted. To evaluate the last integral in (6.135), we see from (6.124) and (6.126) that

$$A_\tau = -\frac{\sigma}{\sqrt{\alpha|\tau|}} \widehat{\mathcal{L}}_1^{-1} \widehat{A}(\widehat{X}, \widehat{Y}). \tag{6.136}$$

Here the fact of $\mathrm{sgn}(D_1) = \mathrm{sgn}(D_2) = \sigma$ has been used in order for the envelope solution A in Eq. (6.67) to be localized. Then using the normalized variables (6.126) as well as the above expression (6.136) for A_τ and the expression (6.127) for ζ, and recalling the condition of $\mathrm{sgn}(\tau) = -\sigma$ for the envelope solution A in Eq. (6.67) to be localized, one finds that

$$\int_{-\infty}^{\infty} \int_{-\infty}^{\infty} A_\tau(X,Y) \zeta(X,Y) dX dY = \frac{\eta_6 C_1 \sqrt{D_1 D_2}}{\alpha \tau^3}, \tag{6.137}$$

where

$$C_1 \equiv -\int_{-\infty}^{\infty} \int_{-\infty}^{\infty} \widehat{\zeta}(\widehat{X}, \widehat{Y}) \widehat{\mathcal{L}}_1^{-1} \widehat{A}(\widehat{X}, \widehat{Y}) d\widehat{X} d\widehat{Y} = 0.8684. \tag{6.138}$$

Inserting the expressions (6.134), (6.135), and (6.137) into (6.133), we obtain the coefficient η_6 as

$$\eta_6 = -\frac{P'(\omega_0) \alpha \tau^3}{2 C_1 G \sqrt{D_1 D_2}}. \tag{6.139}$$

Substituting this expression into the expansion (6.95) of λ^2 and recalling the relation (6.65) as well as the power formula (6.82), we finally obtain the leading-order asymptotic expression for λ^2 as

$$\lambda^2 = -\frac{C_0}{2C_1} \frac{P'(\omega_0)}{P(\omega_0)} (\omega - \omega_0)^3, \tag{6.140}$$

where $P(\omega_0) = P_0$ is the limit power value at the band edge ω_0 (see (6.82)), and constants (C_0, C_1) are given in (6.84) and (6.138). This asymptotic formula is our final analytical result on the eigenvalue bifurcation from zero near band edges. This formula has important consequences as we will elaborate below.

First this formula shows that λ^2 is always real. If λ^2 is negative, then λ is purely imaginary and thus stable, while if λ^2 is positive, then λ is purely real and unstable. In view of (6.65) and the existence condition of envelope solitons in Eq. (6.67), we see that $\text{sgn}(\omega-\omega_0) = -\text{sgn}(\sigma)$, and thus the formula (6.140) also shows that

$$\text{sgn}(\lambda^2) = \text{sgn}[\sigma P'(\omega_0)]. \tag{6.141}$$

This relation tells us that for self-focusing nonlinearity ($\sigma = 1$), the eigenvalue λ is unstable if $P'(\omega_0) > 0$, i.e., the slope of the power curve at the band edge is positive. This is analogous to the Vakhitov–Kolokolov stability criterion in Sec. 5.3 (after the notational difference of $\omega = -\mu$ is accounted for). Notice that the gap solitons here are sign-indefinite in general; thus our stability criterion for self-focusing nonlinearity is an extension of the Vakhitov–Kolokolov criterion to sign-indefinite solitons. But for self-defocusing nonlinearity ($\sigma = -1$), this eigenvalue λ is unstable if $P'(\omega_0)$ is negative. This is just the opposite of what the Vakhitov–Kolokolov criterion says. Thus our stability criterion for self-defocusing nonlinearity is a modification of the Vakhitov–Kolokolov criterion. These results reveal the important role the sign of nonlinearity plays in the connection between stability and the power slope.

The above conditions for linear instability can be reduced into an equivalent but simpler condition. Notice that under self-focusing nonlinearity, gap solitons bifurcate out from the left edge of a Bloch band. Thus when $P'(\omega_0) > 0$, the powers of gap solitons near the band edge are lower than the limit power P_0 on the band edge. Under self-defocusing nonlinearity, gap solitons bifurcate out from the right edge of a Bloch band. Thus when $P'(\omega_0) < 0$, the powers of gap solitons near the band edge are also lower than the limit power P_0 on the band edge. Consequently, our results above for linear instability of 2D gap solitons for both self-focusing and self-defocusing nonlinearities can be condensed into the following simple criterion: *Two-dimensional gap solitons near band edges are unstable if their powers are lower than the limit powers on the band edges.*

It is noted that these stability results were derived for real-valued 2D gap solitons bifurcating from a single Bloch mode at a band edge. Thus these results can be directly applied to gap solitons near band edges A and B in Fig. 6.12 and the C_1 family of gap solitons near edge C in Fig. 6.13. From the power curves in those figures, which show that powers near band edges are all lower than those on the edges, we immediately conclude that these gap solitons near band edges are all linearly unstable. The above stability analysis can also be extended to other 2D gap solitons, such as the C_2 and C_3 families bifurcated from edge C in Fig. 6.13. In view of the power curves in Fig. 6.13, solitons in C_2 and C_3 families near edge C are expected to be linearly unstable as well.

Formula (6.140) also tells us that the unstable eigenvalue λ is proportional to $|\omega - \omega_0|^{3/2}$. Recalling Eq. (6.65), we see that $|\omega-\omega_0|^{3/2}$ is proportional to ϵ^3, where ϵ is the amplitude of the gap soliton. Thus λ is proportional to the cubic power of the soliton's amplitude. This contrasts the eigenvalues bifurcating from translation-invariance-induced eigenmodes of envelope solitons near band edges, which are exponentially small in ϵ instead (see Sec. 6.1.5).

Formula (6.140) holds for both on-site and off-site 2D gap solitons near band edges. For off-site 2D solitons, they also suffer symmetry-breaking instabilities with exponentially small eigenvalues, just as in the 1D case (see Sec. 6.1.5). But for on-site 2D solitons, the eigenvalue bifurcation from formula (6.140) is the only source of instability which has no counterpart in one dimension.

The linear instability obtained above is a width instability. With the onset of this instability, the perturbed soliton contracts or expands depending on initial perturbations. This instability is of the same type as the Vakhitov–Kolokolov instability shown in Fig. 5.3 and the instability near a power extremum analyzed in Sec. 5.4. The reason for this being a width instability is quite similar to that for an instability near a power extremum in Sec. 5.4. Specifically, when the leading-order unstable eigenfunctions (v, w) in (6.116) and (6.132) are scaled back to small perturbations and substituted into the perturbed solution (6.85), since $\lambda > 0$ for the unstable mode, the perturbed solution is approximately a gap soliton with its frequency ω shifted (upward or downward), and the shift amount exponentially increases with time. Since the frequency controls the width of the gap soliton, this instability is hence a width instability. Numerical demonstration of this width instability will be made in the next subsection (see Fig. 6.16).

6.4.2 Numerical Stability Results

In this subsection, we numerically study stability properties of on-site 2D gap solitons bifurcating from band edges. In particular, we will numerically confirm the eigenvalue bifurcation formula (6.140) near band edges. In addition, we will determine the stability of gap solitons away from band edges.

We first consider on-site solitons bifurcating from edges A and B as shown in Fig. 6.12. From their power curves and the stability results of the previous subsection, these solitons near band edges are linearly unstable. Then, in view of the stability switching at a power extremum in Sec. 5.4, this unstable eigenvalue in (6.140) will become stable when ω moves to the other side of the power minimum away from the edges; hence solitons on that side can become stable (if no additional instabilities arise). Numerically, we have determined the stability spectra for the four gap solitons shown in Fig. 6.12(a–d), and these spectra are displayed in Fig. 6.14. We see that for both solution families bifurcated from edges A and B, solitons near band edges indeed possess a positive (unstable) eigenvalue and are thus unstable; see Fig. 6.14(b, c). But solitons on the other side of the power minimum do

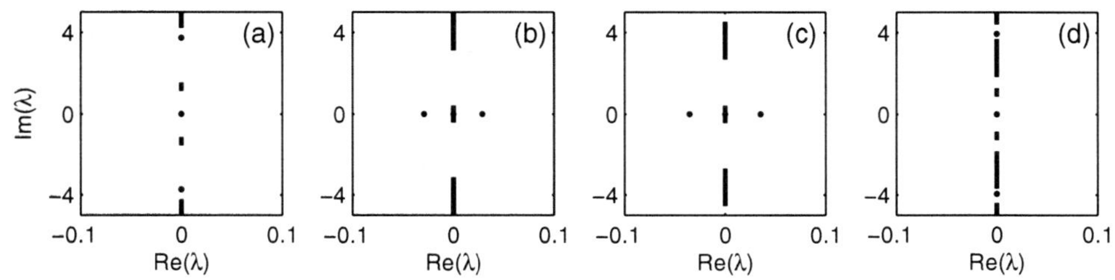

Figure 6.14. *Stability spectra of on-site 2D solitons bifurcated from edges A and B in Fig.* 6.12. *The letters* "a–d" *correspond to the same ones in Fig.* 6.12.

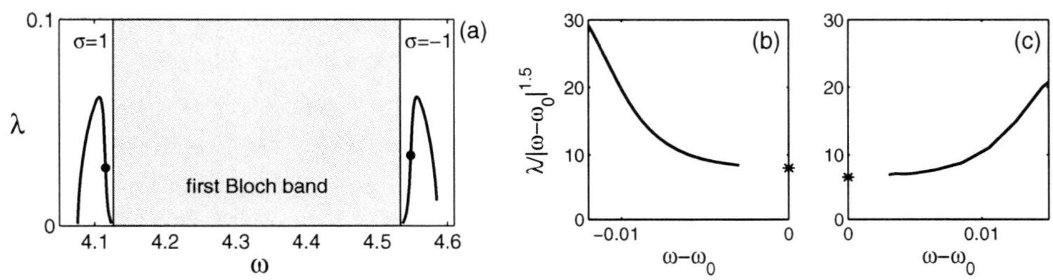

Figure 6.15. (a) *Positive eigenvalues of on-site 2D solitons near band edges A (left) and B (right) in Fig.* 6.12. (b, c) *the ratio* $\lambda/|\omega-\omega_0|^{3/2}$ *versus* $\omega-\omega_0$ *for these unstable eigenvalues in* (a); (b) *is for edge A* ($\omega_0=4.1264$)*, and* (c) *for edge B* ($\omega_0=4.5335$)*. The marker* "$*$" *represents the theoretical limit ratio* β *in* (6.142). *(After Shi et al.* (2008).*)*

not possess any unstable eigenvalue, and are thus linearly stable; see Fig. 6.14(a, d). These results qualitatively agree with our analysis above. Numerically, we have also tracked the positive eigenvalue in these solitons near band edges as ω moves, and the eigenvalue graphs are displayed in Fig. 6.15(a). For both solution families, these unstable eigenvalues bifurcate out from zero at band edges and return back to zero exactly at the power minimum of the respective solution family. To quantitatively verify the analytical eigenvalue formula (6.140), we plot in Fig. 6.15(b, c) the ratio $\lambda/|\omega-\omega_0|^{3/2}$ versus $\omega-\omega_0$ for both families. According to (6.140), as $\omega\to\omega_0$,

$$\frac{\lambda}{|\omega-\omega_0|^{3/2}}\to\beta,\qquad \beta\equiv\left[\frac{C_0}{2C_1}\frac{|P'(\omega_0)|}{P(\omega_0)}\right]^{1/2}. \tag{6.142}$$

Numerically, we indeed see that for both families, $\lambda/|\omega-\omega_0|^{3/2}$ approaches the theoretical β value of the respective family (marked by asterisks in Fig. 6.15(b, c)) when $\omega\to\omega_0$, which confirms the asymptotic formula (6.140).

For the unstable on-site solitons near band edges, how do they evolve under perturbations? We have shown at the end of the previous subsection that this instability is a width instability. To demonstrate this, we consider the unstable soliton $u(x,y)$ in Fig. 6.12(c), which is near edge B under self-defocusing nonlinearity. This soliton is initially perturbed as

$$U(x,y,0) = (1+\delta)\,u(x,y), \tag{6.143}$$

where $\delta \ll 1$ is a perturbation constant. When $\delta = 0.1$, this perturbed state is shown in Fig. 6.16(a). By numerically simulating the original 2D lattice equation (6.38) from this initial condition, the solution at $t = 40$ is shown in Fig. 6.16(b). It is seen that the initially broad solution has contracted to a narrow one. This wave focusing under self-defocusing nonlinearity ($\sigma = -1$) is a unique feature in a periodic medium which has no counterpart in a homogeneous medium. The peak amplitude of the solution versus time is plotted in Fig. 6.16(d). It is seen that this peak amplitude oscillates and slowly approaches a final equilibrium value. If $\delta < 0$ in (6.143), the evolution is quite different. In this case, we take $\delta = -0.1$. Then the corresponding solution at $t = 50$ is shown in Fig. 6.16(c), and the peak amplitude versus time is shown in Fig. 6.16(d). We see that the solution now steadily expands and decays. These instability developments confirm that the instability of this gap soliton is indeed a width instability, similar to the Vakhitov–Kolokolov instability shown in Fig. 5.3. This is not surprising, as the stability criterion we derived in the previous subsection is indeed an extension and modification of the Vakhitov–Kolokolov criterion to gap solitons. Notice that the early stages of these instability developments also resemble the collapse or decay of the ground-state soliton under perturbations in the constant-coefficient 2D NLS equation (see Sec. 5.10). This resemblance originates from the fact that the envelope of

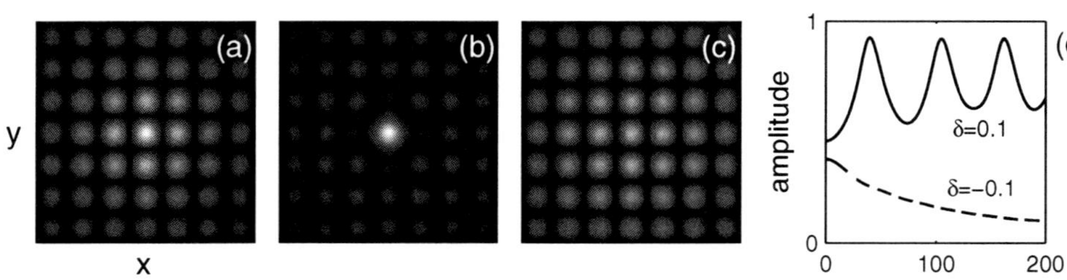

Figure 6.16. *Instability development of the on-site gap soliton near edge B under self-defocusing nonlinearity; (see Fig.* 6.12(c)*).* (a, b) *Solutions* $|U|$ *at* $t = 0$ *and* 40 *for perturbation* (6.143) *with* $\delta = 0.1$*;* (c) *solution* $|U|$ *at* $t = 50$ *for perturbation* (6.143) *with* $\delta = -0.1$*;* (d) *peak amplitudes of the solutions versus time.*

the 2D gap soliton near band edges is indeed governed by a constant-coefficient 2D NLS equation; see (6.67).

Next we consider on-site solitons bifurcating from the edge C, which were shown in Fig. 6.13. For the C_1 family which bifurcates from a single Bloch-wave packet, the stability result of the previous subsection directly applies. Since the slope of its power curve is positive near edge C, these solitons near band edges are thus also linearly unstable due to width instability. At the other side of the minimum-power point away from edge C, this width instability is absent. Then are solitons on that side linearly stable as in the solution families from edges A and B? The answer is negative, because the solitons there suffer an additional oscillatory instability caused by complex eigenvalues. To demonstrate, the stability spectrum for the soliton in the C_1 family at $\omega = 6$ (see Fig. 6.13(a)) is displayed in Fig. 6.17(a). It is seen that the spectrum contains a quadruple of complex eigenvalues which make this soliton unstable. We have also tracked this quadruple of complex eigenvalues as ω moves and found that these eigenvalues persist for the entire C_1 family. When this solution family bifurcates out from edge C, this quadruple of complex eigenvalues bifurcate out from $\pm i(\omega_C - \omega_A)$ on the imaginary axis, where ω_A and ω_C are the ω values of band edges A and C. The origin of this eigenvalue bifurcation is the following. For any gap soliton of frequency ω (with $V_0 = 6$ as in Fig. 6.13), by taking the limit $|x|, |y| \to \infty$ to the eigenvalue problem (6.86) and considering the resulting eigenvalue equations for $v + i\lambda^{-1} w$ and $v - i\lambda^{-1} w$, we can find that the continuous spectrum of this gap soliton is a collection of line segments,

$$\pm i\,[\omega - \omega_B, \omega - \omega_A], \quad \pm i\,[\omega - \omega_D, \omega - \omega_C], \quad \pm i(-\infty, \omega - \omega_E]$$

on the imaginary axis, where $\omega_A, \ldots, \omega_E$ are the frequency values at band edges A to E, respectively. Then, at edge C (where $\omega = \omega_C$), the continuous-spectrum segments

$$\pm i\,[\omega_C - \omega_B, \omega_C - \omega_A] = \pm[2.6957i, 3.1028i]$$

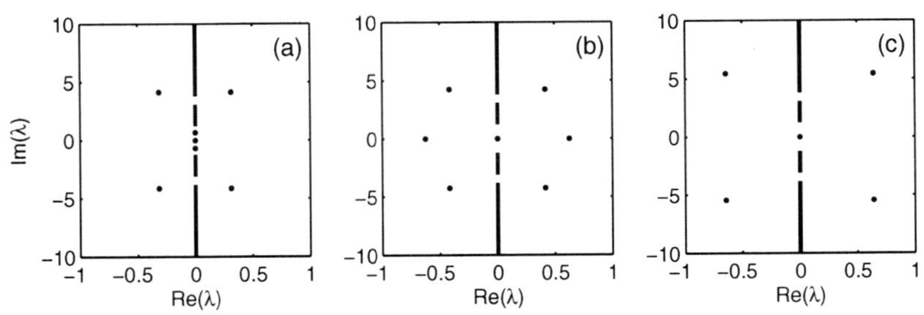

Figure 6.17. *Stability spectra of on-site 2D solitons bifurcated from edge C in Fig.* 6.13. *The letters* a–c *correspond to the same ones in Fig.* 6.13.

are embedded inside the other continuous-spectrum segments

$$\pm i(-\infty, \omega_C - \omega_E] = \pm[2.5799i, \infty i).$$

Hence $\pm i(\omega_C - \omega_A)$ are the edges of embedded continuous-spectrum segments. When solitons in the C_1 family bifurcate out from edge C, a quadruple of complex eigenvalues bifurcate out from these embedded continuum-segment edges. This complex-eigenvalue bifurcation occurs simultaneously with the real-eigenvalue bifurcation of the previous subsection at edge C. When ω moves away from the C edge, while the real unstable eigenvalue becomes stable at the minimum-power point, the complex eigenvalues persist for the entire solution family, making them all unstable.

Now we consider the C_2 and C_3 soliton families bifurcated from edge C. These two families bifurcate not from a single Bloch-wave packet, but from a combination of two Bloch-wave packets. Thus the stability analysis of the previous subsection does not directly apply. Numerically we have found that these two families are all linearly unstable as well. To demonstrate, the stability spectra for the solitons in these families at $\omega = 6$ (see Fig. 6.13(b, c)) are displayed in Fig. 6.17(b, c). For the soliton in the C_2 family, the spectrum contains a pair of real eigenvalues and a quadruple of complex eigenvalues. For the soliton in the C_3 family, the spectrum contains a quadruple of complex eigenvalues. These unstable eigenvalues persist for the entire solution families, and thus these soliton families are all unstable, just like the C_1 family.

From the above numerical stability results, one may conclude that of all on-site gap solitons bifurcated from band edges, only those bifurcated from the lowest edges A and B of the first Bloch band can be stable (when they are not near the edges). The solitons bifurcated from higher edges are generally all unstable. Notice from Fig. 6.12 that the intensity peaks of solitons from edges A and B are all located precisely at the lattice sites (i.e., potential minima), which makes their stability possible. However, the intensity peaks of solitons from higher edges are generally not located at lattice sites (see Fig. 6.13), which makes them prone to instability.

6.5 Two-Dimensional Gap Solitons Not Bifurcated from Bloch Bands

As in the 1D case, many families of 2D gap solitons do not bifurcate from band edges. Two such families and their stability properties will be described below.

The first family we present is a family of vortex solitons which reside in the semi-infinite gap under self-focusing nonlinearity ($\sigma = 1$). At the potential depth $V_0 = 6$, this solution family is displayed in Fig. 6.18. The power curve of this family has double branches

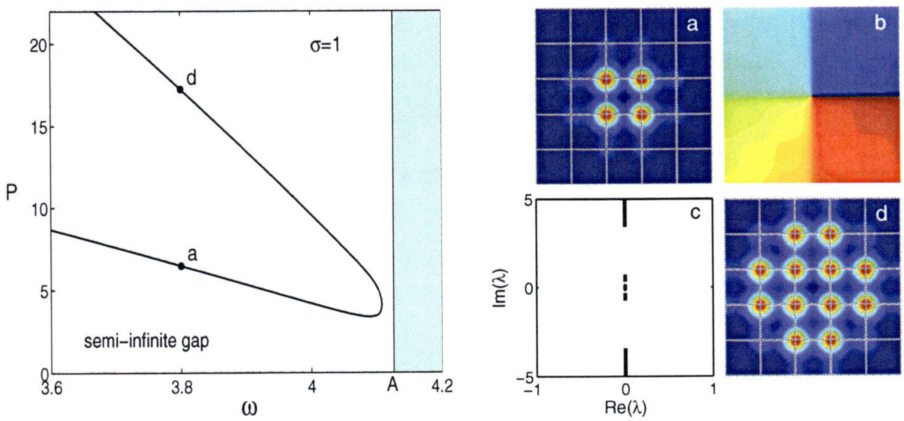

Figure 6.18. *Left: power curve of off-site vortex solitons (with charge one) in the semi-infinite gap under self-focusing nonlinearity* ($V_0 = 6$). *Right:* (a, b) *amplitude and phase distributions of the vortex soliton at point "a" of the lower branch in the left figure;* (c) *stability spectrum of the vortex soliton in* (a)*;* (d) *amplitude distribution of the vortex soliton at point "d" of the upper branch in the left figure (the phase structure is similar to* (b)*). The grids in* (a, d) *are lattice sites. (After Wang and Yang* (2008).*)*

and does not touch the band edge A. The solution at $\omega = 3.8$ on the lower branch is displayed in Fig. 6.18(a, b). This solution is complex valued and has four intensity peaks which are located at four adjacent lattice sites in a square configuration. The center of this soliton is between the four lattice sites (off-site). When winding around this soliton center, the phase of the solution increases by 2π, which gives a vortex phase structure (with unit topological charge). In particular, the phases of the four intensity peaks differ by $\pi/2$ between each other. Hence this soliton is a unit-charge off-site vortex soliton. This type of solitons was first theoretically reported by Yang and Musslimani (2003), see also Baizakov et al. (2003) for report of an on-site vortex soliton whose center is located on a lattice site. They were subsequently observed by Neshev et al. (2004) and Fleischer et al. (2004) in photorefractive crystals. As ω gets near the band edge A, this soliton becomes less localized. When ω moves to the upper branch, this soliton evolves into a twelve-peak structure in a cross configuration (see Fig. 6.18(d)), but its phase structure remains the same as Fig. 6.18(b) on the lower branch (Wang and Yang (2008)). Regarding the stability of these vortex solitons, it was found by Yang and Musslimani (2003) that the ones on the lower branch are linearly stable in deep potentials. To demonstrate, the stability spectrum for the vortex soliton in Fig. 6.18(a, b) is displayed in (c) of the same figure. This spectrum does not contain any unstable eigenvalues, and thus this vortex soliton is stable. Vortex solitons on the upper branch, however, are linearly unstable. An intuitive explanation for this is that, on the upper branch, the additional intensity peaks which developed have roughly the same phase as the original intensity peaks which are adjacent to them. From the stability results of 1D gap solitons in Sec. 6.2, we learned that under self-focusing nonlinearity, two adjacent intensity peaks with the same phase are linearly unstable. This phenomenon holds for two dimensions as well; see Yang et al. (2004b). Thus the vortex solitons on the upper branch are expected to be unstable. This family of vortex solitons resides in the semi-infinite gap under self-focusing nonlinearity. Under self-defocusing nonlinearity, similar vortex solitons with four intensity peaks in a square configuration exist in the first bandgap. Such gap vortex solitons have been studied theoretically and experimentally by Ostrovskaya and Kivshar (2004), Sakaguchi and Malomed (2004), Wang and Yang (2008), and Song et al. (2008).

The second gap-soliton family we present is a family of real-valued solitons in the first gap under self-defocusing nonlinearity. On the lower branch, these solitons contain four adjacent intensity peaks which are in phase with each other and form a number-7 configuration. The power curve of this solution family is displayed at the left side of Fig. 6.19, and the 7-shaped soliton at $\omega = 6$ of the lower branch is shown in Fig. 6.19(a, b). This soliton is real valued; hence its phase is either zero or π as shown by two different colors in (b). The four intensity peaks of the soliton are in phase as the phase diagram in (b) shows. The most important property of this 7-shaped soliton is that it is linearly

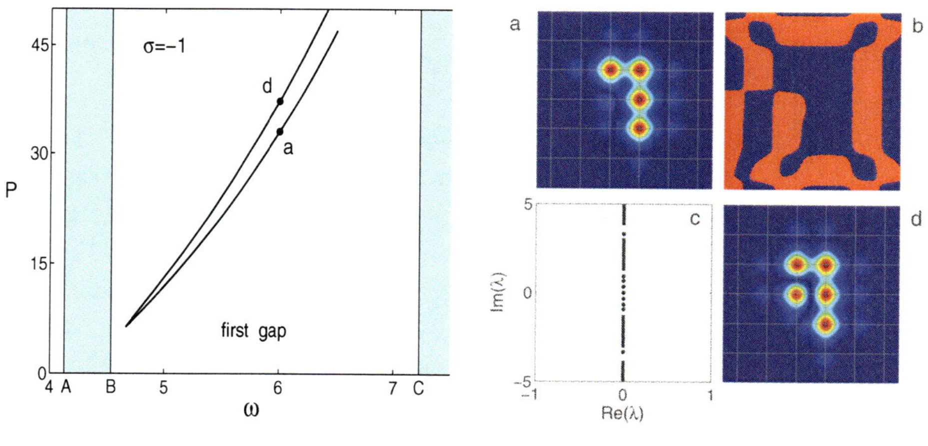

Figure 6.19. *Left: power curve of the 7-shaped soliton family in the first gap under self-defocusing nonlinearity ($V_0 = 6$). Right:* (a, b) *amplitude and phase distributions of the 7-shaped soliton at point "a" of the lower branch in the left figure;* (c) *stability spectrum of the 7-shaped soliton in* (a); (d) *amplitude distribution of the soliton at point "d" of the upper branch in the left figure (the phase structure is similar to* (b)*).*

stable. This is demonstrated by the stability spectrum of this soliton, which is displayed in Fig. 6.19(c). From the intuition we obtained on 1D gap solitons (see Sec. 6.2), under self-defocusing nonlinearity, in-phase intensity peaks at lattice sites tend to be stable. The stability of this 7-shaped in-phase soliton shows that this intuition holds for two dimensions as well. But the stability of this 7-shaped soliton is even more remarkable, as it disproves a conventional wisdom, which says that solitons with peculiar geometries (such as those with corners) are always unstable. Regarding the other solitons on this lower branch, they are found to be stable as well (except near the turning point of the power curve). On the upper branch, the solitons develop an extra intensity peak which is out of phase with its two neighbors; see Fig. 6.19(d). These solitons are found to be linearly unstable. This is not surprising, as these solitons contain out-phase intensity peaks, which are expected to be linearly unstable under self-defocusing nonlinearity from our knowledge on 1D gap solitons.

The 7-shaped solitons in Fig. 6.19 indicate that the 2D lattice can support a great variety of gap solitons in almost arbitrary configurations. In earlier work (Yang et al. (2004b), Alexander et al. (2006)), dipole, quadrupole, and truncated-Bloch-wave solitons in two dimensions have been reported. Fig. 6.19 shows that the shapes of 2D gap solitons actually can be much more flexible than those reported. More importantly, these flexible soliton structures can also be stable. Indeed, one can use the stable 7-shaped soliton in Fig. 6.19(a) as a building block to construct other stable shapes of solitons. For instance, by extending the two horizontal intensity peaks rightward, we can get a T-shaped soliton. If we further extend the three vertical intensity peaks upward, then we can get a cross-shaped soliton. These extended soliton structures are still linearly stable if their intensity peaks are all in phase under self-defocusing nonlinearity.

6.6 Experimental Results

In the past ten years, many experiments have been performed on nonlinear wave phenomena in periodic media, mostly in spatial optics and Bose–Einstein condensates. The experiments in spatial optics were often performed in waveguide arrays or photorefractive crystals. The waveguide arrays could be created by etching an AlGaAs substrate (Eisenberg et al. (1998)) or by high-intensity laser writing (Szameit et al. (2006)). The nonlinearity in these waveguide arrays is cubic, as in the model (6.1), and of self-focusing type ($\sigma = 1$). In photorefractive crystals, the periodic medium could be created by interfering pairs of laser beams (Fleischer et al. (2003b), Neshev et al. (2003)), or by using amplitude masks (Martin et al. (2004), Chen and Yang (2007)). The nonlinearity in photorefractive crystals is of saturable type (as in the model (6.2)), and can be either self-focusing or self-defocusing depending on the sign of the voltage applied to the crystal (Singh et al. (1995), Segev et al. (1996)). In Bose–Einstein condensates, an optical lattice can be created by interfering pairs of laser beams. This optical lattice forms a periodic potential in the condensate. The nonlinear interaction between atoms of the condensate is cubic and is described by the Gross–Pitaevskii equation (1.62), or its nondimensionalized version (6.1). This interaction can be attractive ($\sigma = 1$) or repulsive ($\sigma = -1$), depending on the type of atoms in the condensate.

The first experimental observation of solitons in a 1D periodic medium was performed by Eisenberg et al. (1998) in planar etched waveguide arrays. In the experiment, light was coupled to the central waveguide. At low power, light spread out to many waveguides due to linear diffraction. At high power, however, light was localized close to the input waveguide and formed a lattice soliton. This soliton is the counterpart of the on-site 1D solitons in the semi-infinite gap under self-focusing nonlinearity as shown in Figs. 6.3 and 6.4. Subsequently, Fleischer et al. (2003a) observed on-site 1D lattice solitons in photorefractive crystals for both self-focusing and self-defocusing nonlinearities, and Neshev et al. (2003) observed both on-site and off-site 1D lattice solitons in photorefractive crystals for self-focusing nonlinearity. The latter's experimental results are shown in Fig. 6.20. In this experiment, a photonic lattice (with spacing $16\,\mu$m) was first created in the crystal by laser interference. Then a probe beam was launched into this lattice, and its center was either on a lattice site or between lattice sites. The applied voltage was positive, thus creating a self-focusing (saturable) nonlinearity. At low intensities of the probe beam, diffraction resembling Bloch waves of the periodic potential was observed, and the diffraction pattern was almost independent of the initial position of the probe beam relative to the lattice (Fig. 6.20(b)). However, when the intensity of the probe beam was increased, two distinct states were formed (Fig. 6.20(c)). When the maximum of the probe beam was centered between lattice sites, an off-site soliton was generated (Fig. 6.20(c), middle). However, when the probe beam was centered on a lattice site, an on-site soliton was formed (Fig. 6.20(c), right). These solitons are the counterparts of the on-site and off-site 1D solitons in the semi-infinite gap as shown in Figs. 6.3 and 6.4. Experimentally, it was also observed that the off-site soliton was unstable and transformed into a nonsymmetric structure as a result of small perturbations of the beam position. This phenomenon coincides with the theoretical result in Sec. 6.1 that off-site gap solitons are linearly unstable due to symmetry-breaking instability.

In Bose–Einstein condensates (BECs), the nonlinear interaction is repulsive ($\sigma = -1$) for many atoms (such as ^{87}Rb). The first BEC experiment on the observation of gap solitons in a 1D optical lattice under repulsive atom-atom interaction was made by Eiermann et al. (2004), and their experimental result is shown in Fig. 6.21. The experiment was performed with ^{87}Rb BEC. The atoms were first loaded into a crossed light beam dipole trap. Then a periodic potential was ramped up by a standing light wave of wavelength $\lambda = 783\,$nm. This

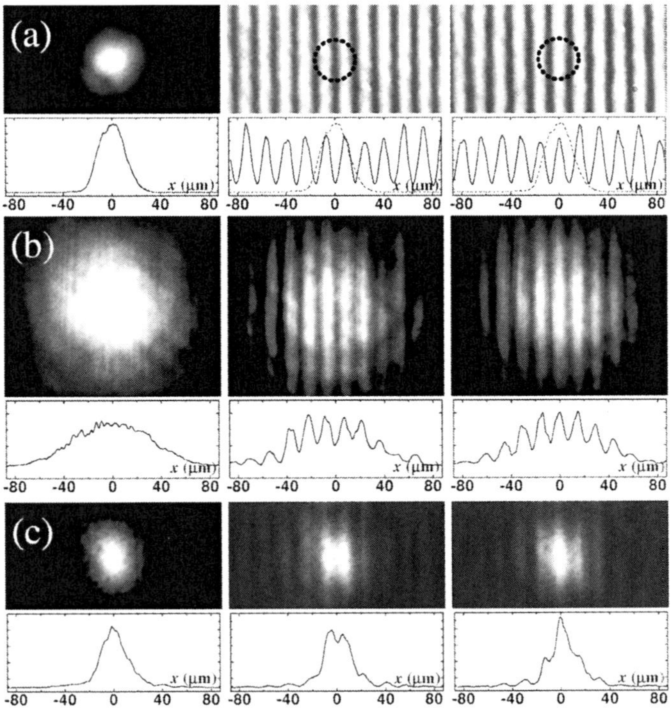

Figure 6.20. *Experimental observation of on-site and off-site* 1*D lattice solitons (with self-focusing nonlinearity):* (a) *input beam and optical lattice (power,* 23 μW*);* (b) *output probe beam at low power (*$2 \times 10^{-3}\,\mu$W*);* (c) *localized states at high power (*$87 \times 10^{-3}\,\mu$W*). Left: vibrating piezoelectric translational mirror; middle: off-site excitation; right: on-site excitation. Electric field,* 3600 V/cm*; lattice period,* 16 μm. *(After Neshev et al.* (2003).*)*

periodic potential was weak since the nondimensionalized potential depth as in the model equations (6.3)–(6.4) was $V_0 \approx 0.7$. Here the potential depth was nondimensionalized by the recoil energy $E_r = (\hbar^2/2m)(\pi^2/d^2)$, where $\hbar$ is the Planck constant, m is the mass of the atom, and d is the period of the optical lattice; see Eq. (1.62). Next, the atomic cloud was released into the 1D horizontal waveguide by switching off one dipole trap beam. Subsequently, the atomic ensemble was prepared to match the Bloch-wave packet near the upper edge of the first Bloch band (see ω_2 in Fig. 6.1) by accelerating the periodic potential. The time evolution of the resulting atomic wave packet was shown in Fig. 6.21. It is seen that after approximately 25 ms, a small peak was formed which does not change in shape or amplitude. This peak is a gap soliton under repulsive atom-atom interaction. The width of this matter-wave soliton was about $\sim$6 μm, and it contained approximately 250 atoms.

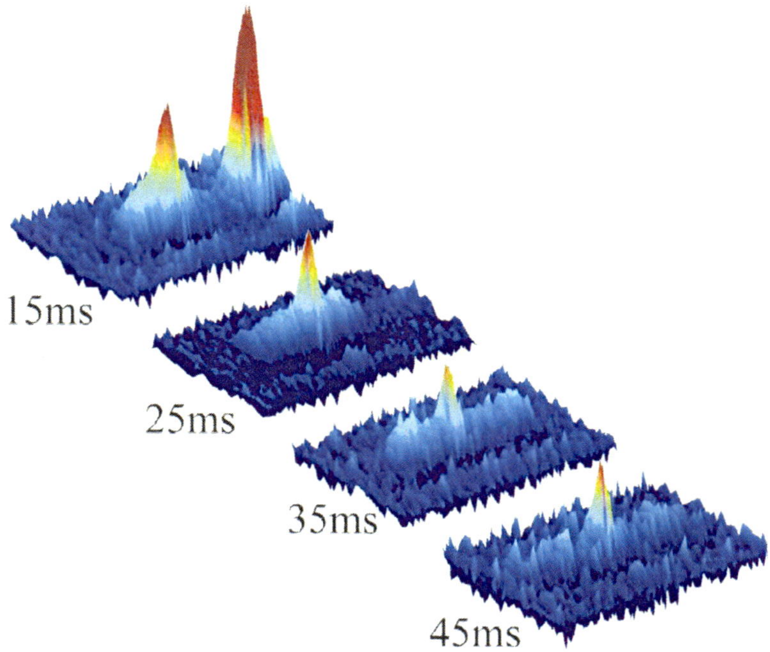

Figure 6.21. *Observation of BEC gap solitons in a quasi-*1*D weak optical lattice with repulsive atom-atom interaction. The atomic density deduced from absorption images is shown for different propagation times. After approximately* 25 ms, *a stationary gap soliton with approximately* 250 *atoms is formed. Excessive atoms are radiated and disperse over time. Soliton width* $\sim 6\,\mu$m*; lattice period* $\sim 0.4\,\mu$m. *(After Eiermann et al.* (2004).*)*

Notice that the period of the optical lattice was $\lambda/2$, which was about $0.4\,\mu$m. Thus the observed gap soliton was a Bloch-wave packet whose theoretical counterpart is the 1D gap solitons near the upper edge of the first Bloch band under self-defocusing nonlinearity as shown in Figs. 6.3 and 6.4.

One-dimensional truncated Bloch-wave solitons (as those shown in Fig. 6.10) were also observed first in BEC experiments by Anker et al. (2005), and their experimental results are shown in Fig. 6.22. The experiment was performed with ^{87}Rb BECs which exhibits repulsive atom-atom interaction. These ^{87}Rb atoms were loaded into a 1D optical lattice in the same way as was done in the experiment of Fig. 6.21. However, the lattice potential in this experiment was much deeper than that in the experiment of Fig. 6.21, since the depth of the potential was $V_0 \sim 10\,E_r$, where E_r is the recoil energy. The number of atoms in the experiment was adjusted from 2000 ± 200 to 5000 ± 600. Initially, both wave packets expanded. At $t \sim 35$ ms, the wave packet with higher initial atomic density developed steep edges and stopped expanding, and a truncated Bloch-wave soliton akin to Fig. 6.10(b)

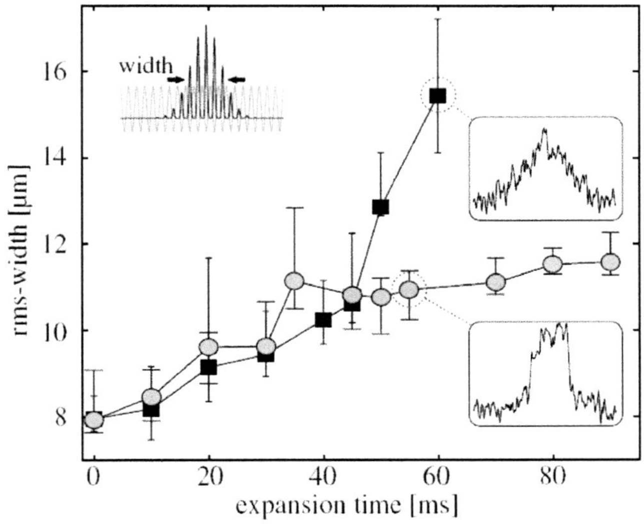

Figure 6.22. *Observation of* 1*D truncated-Bloch-wave solitons with Bose-condensed* ^{87}Rb *atoms. The dynamics of the wave packet width along the periodic potential is shown for two different initial atom numbers. By increasing the number of atoms from* 2000 ± 200 *(squares) to* 5000 ± 600 *(circles), the repulsive atom-atom interaction leads to the stopping of the global expansion of the wave packet. The insets show that the wave packet remains almost Gaussian in the diffusive (linear) regime but develops into a truncated-Bloch-wave soliton with steep edges in the self-trapping (nonlinear) regime. Lattice period* $\sim 0.4\,\mu$m. *(After Anker et al.* (2005).*)*

formed (see inset in Fig. 6.22). The root-mean-square (rms) width of this soliton was about $12\,\mu$m, while the period of the optical lattice was about $0.4\,\mu$m. Thus this observed soliton had many more intensity peaks than the one shown in Fig. 6.10(b). In contrast, the wave packet with the lower initial atomic density continued to expand with a Gaussian shape, and this was due to the linear evolution of the wave packet at low atom densities.

In 2D periodic media, the first experimental observation of gap solitons was performed by Fleischer et al. (2003b) in photorefractive crystals. In this experiment, a 2D photonic lattice was first created in the crystal by interfering two pairs of laser beams and was ordinarily polarized. Then an extraordinarily polarized Gaussian probe beam was launched into the photonic lattice. At low power, the probe beam always diffracted and scattered. But at higher power, the probe beam self-localized into a 2D gap soliton for both self-focusing and self-defocusing (saturable) nonlinearities. These solitons are the counterparts of on-site 2D gap solitons bifurcated from edges A and B of the first Bloch band under self-focusing and

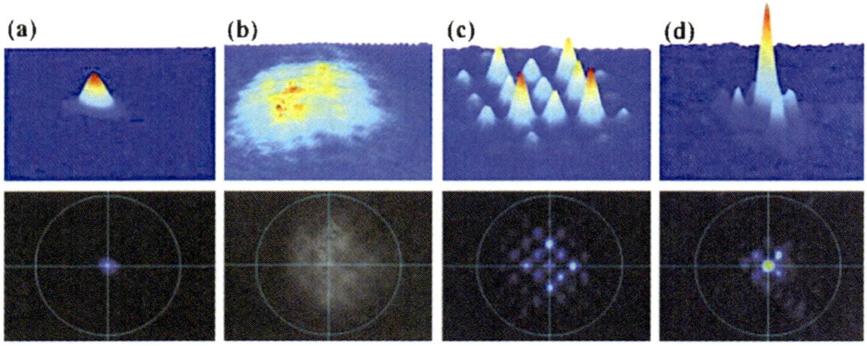

Figure 6.23. *Experimental demonstration of 2D lattice solitons in a photorefractive crystal and photonic lattice under self-focusing nonlinearity.* (a) *Input,* (b) *diffraction output without lattice,* (c) *diffraction with lattice at low voltage (*900 V/cm*), and* (d) *2D soliton at high voltage (*3000 V/cm*). Top: 3D intensity plots; bottom: 2D transverse patterns. Lattice period:* 20 μm*; crystal length:* 8 mm. *(After Martin et al.* (2004).*)*

self-defocusing nonlinearities as shown in Fig. 6.12. Subsequently, observation of on-site 2D gap solitons in photorefractive crystals under self-focusing nonlinearity was also performed by Martin et al. (2004), and their experimental result is shown in Fig. 6.23. In this experiment, a 2D square lattice was created by passing an ordinarily polarized beam through an amplitude mask. This mask provided spatial periodic modulation to the otherwise uniform beam (modulation period 20 μm). This lattice beam from the mask remained nearly invariant as it propagated in the crystal. Then an extraordinarily polarized probe beam was also launched into the crystal and propagated collinearly with the lattice. In addition, a uniform incoherent background beam was used as "dark illumination." The amount of (saturable) nonlinearity experienced by the probe beam was tuned by the applied DC voltage to the crystal. At low voltage (900 V/cm), diffraction of the probe beam was observed (Fig. 6.23(c)), which shows that most of the energy flows from the center towards the diagonal directions of the lattice. But at higher voltage (3000 V/cm), a lattice soliton was observed (Fig. 6.23(d)), with most of its energy concentrated in the center and the four neighboring sites along the principal axes of the lattice. This observed soliton is the counterpart of on-site 2D gap solitons bifurcated from edge A of the first Bloch band under self-focusing nonlinearity as shown in Fig. 6.12. The observation of 2D gap solitons bifurcated from edge B of the first band under self-defocusing nonlinearity was also performed by Lou et al. (2007) in photorefractive crystals.

Two-dimensional gap solitons bifurcated from the lower edge (i.e., edge C) of the second Bloch band under self-focusing nonlinearity (see Fig. 6.13) have also been observed by Bartal et al. (2005) and Fischer et al. (2006) in photorefractive crystals. The experiment

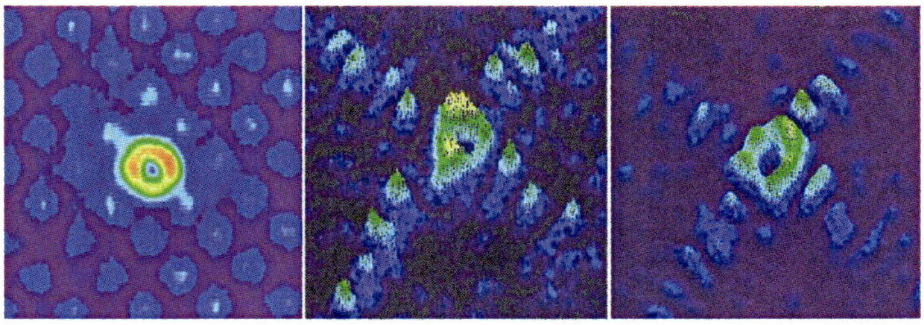

Figure 6.24. *Experimental observation of C_3-family vortex solitons bifurcated from edge C of the second Bloch band under self-focusing nonlinearity (see Fig.* 6.13*). Left: intensity distribution of the input vortex-ring beam photographed (for size comparison) on the background of the optically induced lattice; middle: output intensity distribution of a low-intensity ring beam experiencing linear diffraction in the lattice; right: output intensity distribution of a high intensity ring beam, which has evolved into a C_3-family vortex soliton. Lattice spacing:* 13 μm*; crystal length:* 5 mm*. (After Bartal et al.* (2005)*.)*

of Bartal et al. (2005) was performed in a 5 mm long photorefractive crystal. Two pairs of plane waves interfered to optically induce a square lattice (with spacing 13 μm) inside the crystal. These plane waves were ordinarily polarized and remained invariant along the propagation direction. The probe beam was polarized extraordinarily so that it experienced both the photonic lattice and the photorefractive (saturable) nonlinearity. The input beam was generated by reflecting the probe beam off a vortex mask of unit topological charge and then imaging the ring beam onto the crystal input face. The width of the input ring beam was comparable to the size of a single waveguide and was launched around a single waveguide. The intensity pattern of the initial ring entering the lattice is shown in Fig. 6.24 (left panel). The middle panel of this figure shows the output (linear) diffraction pattern of a low-intensity beam after 5 mm propagation in the lattice, exhibiting preferential diffraction along the lattice axes but retaining the size of the central ring. As the probe beam intensity was increased to one-half of the lattice peak intensity, the input beam reshapes and forms a soliton (Fig. 6.24 (right panel)). The phase structure of this soliton was probed by interference with a weakly diverging Gaussian beam. It was found that the simple singly charged vortex structure of the input beam evolved into a more complex phase structure. This soliton is the counterpart of the vortex soliton of C_3 family shown in Fig. 6.13(c).

Two-dimensional gap solitons of C_1 family from edge C of the second Bloch band under self-focusing nonlinearity (see Fig. 6.13) were observed by Fischer et al. (2006), and their experimental result is shown in Fig. 6.25. This experiment was performed in a 20 mm long photorefractive crystal. A 2D photonic lattice (spacing 23 μm) was created by interfering two pairs of coherent ordinary polarized beams. Another extraordinary polarized

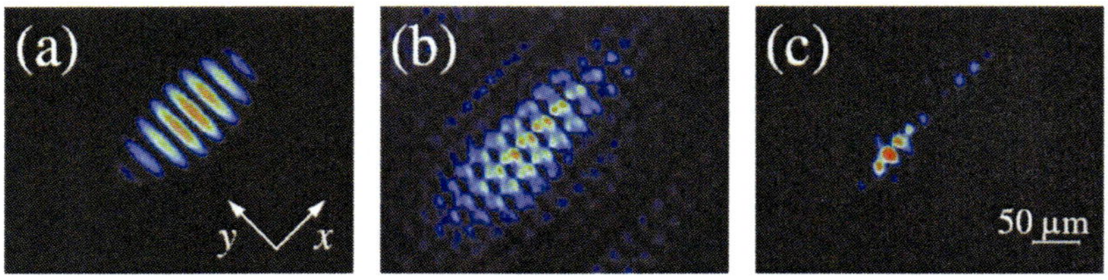

Figure 6.25. *Experimental observation of C_1-family gap solitons bifurcated from edge C of the second Bloch band under self-focusing nonlinearity (see Fig.* 6.13). (a) *Input beam profile;* (b) *output beam at low power;* (c) *localized state at high power. Lattice spacing:* 23 μm; *crystal length:* 20 mm. *(After Fischer et al.* (2006).*)*

beam was first directed onto a programmable phase modulator to allow for phase and amplitude engineering. The beam produced by the modulator was then also launched onto the front face of the crystal and combined with the lattice beam. In order to match the field profile of the linear Bloch state at edge C (see Fig. 6.13), a broad input beam with several out-of-phase peaks was used (Fig. 6.25(a)). To compensate for the anisotropic diffraction, the input beam was made elongated along the x direction. Such a broad beam allowed for spectrally pure excitation of the Bloch state from edge C of the second band, which could then be moved adiabatically into the gap as the beam self-focused, thus avoiding radiation into other spectral bands. At low intensities (20 nW; see Fig. 6.25(b)), the beam diffracted linearly, and its width increased more in the x than in the y direction despite the fact that the input size along the x direction was larger. The spatial structure of the output beam represents the characteristic fine structure of the Bloch wave from edge C of the second band, with out-of-phase lobes along the x direction as confirmed by interferometric measurements. At higher laser powers, the beam focused until a strongly localized state was formed at about 230 nW (Fig. 6.25(c)), confined to three lattice sites in the x direction and one in the y direction. This localized state is the counterpart of the 2D gap soliton of C_1 family under self-focusing nonlinearity as shown in Fig. 6.13. Recall from Sec. 6.4 that a broad 2D Bloch-wave packet near band edges is linearly unstable due to width instability, and the packet would contract to more localized states if the input power exceeds the threshold for the soliton formation (see Fig. 6.16). This width instability explains the focusing of the broad input beam at high powers in Fig. 6.25. Notice that the strongly localized soliton in Fig. 6.25 also traversed along the x direction, thus it evolved between on-site and off-site states. The soliton in Fig. 6.13(a), on the other hand, is just an on-site state.

One may recall from Sec. 6.4.2 that gap solitons bifurcated from edge C of the second Bloch band are all linearly unstable. If so, then how can we understand the above observations of gap solitons in the C_1 and C_3 families? First of all, the photorefractive nonlinearities in the above two experiments were saturable rather than cubic. In addition, these nonlinearities were anisotropic rather than isotropic. Under these saturable and anisotropic nonlinearities (see Zozulya and Anderson (1995)), some gap solitons of the C_1 and C_3 families might be stable. Even if all solitons in those families are still unstable with saturable and anisotropic nonlinearity, if the initial condition is sufficiently close to the exact soliton profile, and if the propagation distance is not long enough, then the instability will not have a chance to develop, and hence observation of such unstable solitons would still be possible.

In addition to the above observations of gap solitons, many other observations of various gap-soliton states in periodic media have also been reported. For instance, the multisite vortex solitons as shown in Fig. 6.18 under self-focusing nonlinearity have been observed by Neshev et al. (2004) and Fleischer et al. (2004). Similar multisite vortex solitons in the first bandgap under self-defocusing nonlinearity have been observed by Song et al. (2008). Two-dimensional dipole solitons under focusing and defocusing nonlinearities have been observed by Yang et al. (2004a) and Tang et al. (2007). Two-dimensional vector lattice solitons have been observed by Chen et al. (2004a). Necklace-like lattice solitons have been observed by Yang et al. (2005). Gap-soliton trains under focusing and defocusing nonlinearities have been observed by Chen et al. (2004b) and Lou et al. (2007). Even soliton trains residing inside the Bloch bands (i.e., embedded soliton trains) have been observed by Wang et al. (2007). All these observations have significantly enhanced our understanding of nonlinear wave phenomena in periodic media in nature.

Chapter 7

Numerical Methods for Nonlinear Wave Equations

Numerical computations play an important role in the study of nonlinear waves. Many important phenomena, such as elastic collision of KdV solitons and fractal scattering in solitary wave interactions, were discovered first by numerical computations (see Zabusky and Kruskal (1965) and Chapter 5). Such numerical discoveries motivated further analytical investigations, which then led to deeper understanding of those phenomena. Numerical computations are especially important for the study of nonintegrable equations. Indeed, in Chapters 5 and 6 on nonintegrable equations, many numerical results were presented. A dynamic interplay between numerics and analysis was a distinctive feature for the topics in those chapters. In recent years, great advances have been made on numerical methods for nonlinear wave equations, and many highly accurate and efficient numerical techniques have been developed. In addition, some convenient commercial software for scientific computing (such as MATLAB) have also become available, which greatly facilitated the coding and implementation of those numerical algorithms. Due to these advances, nontrivial numerical computations of nonlinear wave equations can often be done with several-inch-long MATLAB codes and hence are accessible to anybody with or without prior programming experience.

In this chapter, we describe numerical methods for various aspects of nonlinear wave computations such as evolution simulation, computation of solitary wave solutions, and computation of linear-stability spectra of solitary waves. Fundamental properties of these methods, such as their accuracy, numerical stability, convergence conditions, and convergence speeds, will be elaborated. We will show that all these methods we present have spectral spatial accuracy; i.e., the spatial error decays exponentially with the grid spacing (for smooth functions). Thus these methods yield highly accurate numerical results. In addition, for almost all these numerical algorithms, short MATLAB codes for nontrivial problems will be displayed, so that the reader can quickly grasp these algorithms and apply them to other situations. The reader may have noticed that the previous six chapters contain a wide variety of numerical results. Almost all those numerical results were obtained by the numerical methods to be described in this chapter.

7.1 Numerical Methods for Evolution Simulations

Evolution simulation of a nonlinear wave equation starting from a given initial condition is certainly important for the understanding of the nonlinear wave system. A celebrated example is the numerical simulation of the KdV equation by Zabusky and Kruskal (1965), which led to the discovery of solitons and the invention of the inverse scattering transform method (Gardner et al. (1967)). Early evolution simulations of nonlinear wave equations primarily used finite difference methods (such as in Zabusky and Kruskal (1965)), which have low spatial accuracy. Starting from the 1970s, various types of spectral methods were developed. A common feature of these spectral methods is that their spatial accuracy is spectral, which is much more accurate than finite difference methods. Then when coupled with high-order time-stepping schemes (such as Runge–Kutta methods), high overall accuracy can be achieved. All these methods are very easy to implement, especially using software packages such as MATLAB. This makes evolution simulation of nonlinear wave equations very affordable and convenient now. Many types of spectral methods have been invented. In this section, we describe three of them: the pseudospectral method, the split-step method, and the integrating-factor method. For each method, we determine its order of accuracy and condition for numerical stability. To demonstrate their performances, we apply them to the familiar NLS equation. The corresponding MATLAB codes will also be presented. It must be stressed that our choice of the NLS equation as the example is purely a matter of convenience. These numerical methods have nothing to do with the integrability property of the NLS equation and thus can be equally applied to other nonintegrable equations in arbitrary spatial dimensions. Modifications of our sample MATLAB codes to those general evolution equations are often straightforward.

7.1.1 Pseudospectral Method

One of the first spectral methods developed for wave equations is the pseudospectral method (see Orszag (1971) and Fornberg and Whitham (1978) for its early applications, and Gottlieb and Orszag (1977), Fornberg (1998), Trefethen (2000), and Boyd (2001) for further explanations). To describe this method, let us take the NLS equation

$$iu_t + u_{xx} + 2|u|^2 u = 0 \tag{7.1}$$

as an example. After this equation is discretized in space, it becomes

$$u_{n,t} = i(u_{n,xx} + 2|u_n|^2 u_n), \tag{7.2}$$

where u_n represents the solution on a grid point x_n. The basic idea of the pseudospectral method is to use the discrete Fourier transform to evaluate the spatial derivative $u_{n,xx}$, and use an appropriate time-stepping scheme (Runge–Kutta, Adams, Leapfrog, etc.) to advance in time. The computation of $u_{n,xx}$ by the discrete Fourier transform is

$$u_{n,xx} = \mathcal{F}^{-1}[(ik)^2 \mathcal{F}(u_n)], \tag{7.3}$$

where $\mathcal{F}$ and $\mathcal{F}^{-1}$ represent the discrete Fourier and inverse Fourier transforms, and k is the wavenumber. For a smooth function $u(x,t)$, the accuracy of this computation is spectral; i.e., the error is smaller than any power of the spatial spacing Δx (Trefethen (2000), Boyd (2001)). The reason is that when a continuous function is discretized, the discretization error is caused by aliasing of high wavenumbers to low wavenumbers (the discretized function has a finite spectral bandwidth of $[-\pi/\Delta x, \pi/\Delta x]$, while the continuous function has an infinite spectral bandwidth). However, for smooth functions, energy in high wavenumbers is smaller than any inverse power of the wavenumber, which can be proved by repeated use of integration by parts to the Fourier transform formula. Thus the discretization error is smaller than any power of Δx, which gives the spectral accuracy of $u_{n,xx}$ by formula (7.3). If the function $u(x)$ can be extended to an analytic function in a horizontal strip of the complex x plane, then this discretization error decays exponentially with Δx due to the Paley–Wiener theorem (Trefethen (2000)).

After $u_{n,xx}$ is obtained, the spatially discretized equation (7.2) can then be advanced in time by a time-stepping scheme such as the Runge–Kutta method. Here we choose the classical fourth-order Runge–Kutta method for time stepping, hence the temporal accuracy is fourth order. The MATLAB code of the resulting scheme for the NLS equation (7.1) is displayed below. This code, as well as all the other MATLAB codes in this chapter, can also be downloaded directly from the SIAM Web page, www.siam.org/books/mm16, or from the author's Web page, www.cems.uvm.edu/~jyang.

Program 2

```
% p2.m: the pseudospectral method for solving the NLS equation
% iu_t+u_{xx}+2|u|^2u=0.

  L=80; N=256; dt=0.02; tmax=20; nmax=round(tmax/dt);
  dx=L/N; x=[-L/2:dx:L/2-dx]'; k=[0:N/2-1 -N/2:-1]'*2*pi/L; k2=k.^2;
  u=1.2*sech(1.2*(x+20)).*exp(i*x)+0.8*sech(0.8*x);
  udata=u; tdata=0;
```

```
for nn=1:nmax                                   % integration begins
  du1=i*(ifft(-k2.*fft(u))+2*u.*u.*conj(u));  v=u+0.5*du1*dt;
  du2=i*(ifft(-k2.*fft(v))+2*v.*v.*conj(v));  v=u+0.5*du2*dt;
  du3=i*(ifft(-k2.*fft(v))+2*v.*v.*conj(v));  v=u+    du3*dt;
  du4=i*(ifft(-k2.*fft(v))+2*v.*v.*conj(v));
  u=u+(du1+2*du2+2*du3+du4)*dt/6;
  if mod(nn,round(nmax/25)) == 0
     udata=[udata u]; tdata=[tdata nn*dt];
  end
end                                             % integration ends
waterfall(x, tdata, abs(udata'));               % solution plotting
colormap([0 0 0]); view(10, 60)
text(-2,  -6, 'x', 'fontsize', 15)
text(50, 5, 't', 'fontsize', 15)
zlabel('|u|', 'fontsize', 15)
axis([-L/2 L/2 0 tmax 0 2]); grid off
set(gca, 'xtick', [-40 -20 0 20 40])
set(gca, 'ytick', [0 10 20])
set(gca, 'ztick', [0 1 2])
```

The initial condition in this code is taken as two single solitons of the NLS equation which are moving toward each other. Running this code, we get Fig. 7.1(a), which shows that the collision between these two solitons is elastic, and the only effect of the collision is a position shift (a phase shift also exists, but it is not visible in the $|u|$ plot of Fig. 7.1(a)). This numerical result agrees with the inverse scattering theory of the NLS equation in Chapter 2.

We have explained that the pseudospectral method is spectrally accurate in space. To verify this fact, we take a very small time-step value $\Delta t = 0.0005$ in the above code, so that the temporal error is negligible. We also take the initial condition to be a single moving soliton

$$u(x,0) = \operatorname{sech}(x+10)\, e^{ix}. \tag{7.4}$$

This initial condition is convenient as its exact solution at later times is simply $u(x,t) = \operatorname{sech}(x-2t+10)\, e^{ix}$, which makes it easy for us to calculate the error of the numerical solution at later times for different grid spacings. Running Program 2 using various spacings Δx, the error of the solution versus $1/\Delta x$ at time $t = 10$ is plotted in Fig. 7.1(b). Here the error is calculated as the ∞-norm of the difference between the numerical solution and the exact solution. For comparison, a benchmark exponentially decaying function $e^{-5.5/\Delta x}$ is also plotted. We see that the error indeed decreases exponentially with $1/\Delta x$, confirming the spectral spatial accuracy of the pseudospectral method. One can notice that an error

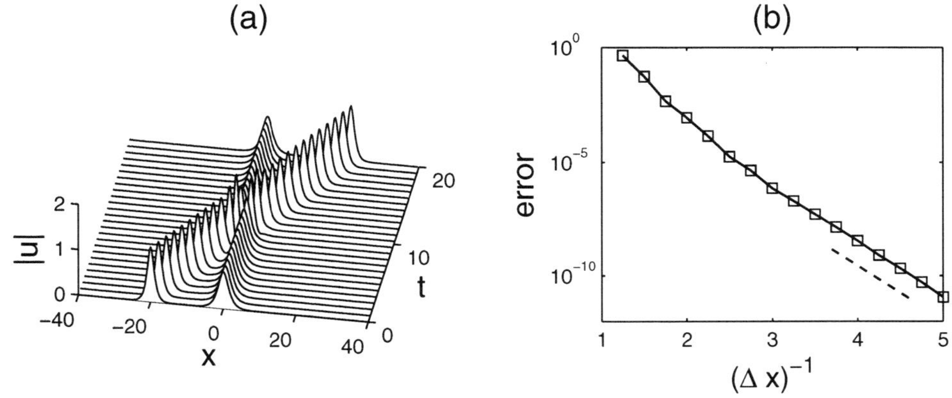

Figure 7.1. (a) *Output of Program 2 (pseudospectral method) for the simulation of two-soliton collision in the NLS equation* (7.1)*;* (b) *error of the pseudospectral method versus* $1/\Delta x$*, demonstrating spectral spatial accuracy of the method. Here the initial condition is a single soliton* (7.4)*,* $\Delta t = 0.0005$*, and* $t = 10$ *in Program* 2*. The dashed line is a benchmark exponentially decaying function* $e^{-5.5/\Delta x}$.

below 10^{-10} is already achieved for spacing as large as $\Delta x = 0.2$, which is unimaginable for finite-difference-type methods.

In Program 2 (with either two-soliton or single-soliton initial condition), if we take a larger time-step value, say $\Delta t = 0.03$, then the numerical solution becomes unstable, and this numerical instability develops very rapidly (within one time unit). This indicates that the pseudospectral method has a numerical stability restriction on the time-step size Δt. The reason for this stability condition is that for the numerical solution to be stable, eigenvalues of the linearized spatial discretization operator, multiplied by Δt, must lie in the stability region of the time-stepping scheme (Trefethen (2000, page 103)). For small grid spacing Δx, the largest eigenvalues of the linearized spatial discretization operator (7.2) for the NLS equation are $\pm i\pi^2/\Delta x^2$, and they are induced by the spatial derivative term $i\partial_{xx}$ in the NLS equation and the highest wavenumbers $\pm\pi/\Delta x$ admitted on the discrete spatial grid. These largest eigenvalues set the limit on the time-step size for numerical stability. The stability region of the fourth-order Runge–Kutta method is well known. The boundary points of this stability region on the imaginary axis are $\pm 2\sqrt{2}i$. Thus, in order for these largest eigenvalues (scaled by Δt) to lie in this stability region, we must have

$$\frac{\Delta t}{\Delta x^2} \leq \frac{2\sqrt{2}}{\pi^2}, \tag{7.5}$$

which is the necessary and sufficient stability condition for this pseudospectral method on the NLS equation (7.1). This stability condition can be understood more easily as follows.

It is known that for pseudospectral methods, the stability condition is generally set by the highest Fourier mode on the spatial grid. Let us consider the time evolution of this highest Fourier mode $c(t)e^{i\pi x/\Delta x}$, where $c(t)$ is the coefficient of this mode. Substituting this Fourier mode into the NLS equation (7.1), we see that when Δx is small, the equation for the Fourier coefficient $c(t)$ is approximately $c'(t) = i(\pi^2/\Delta x^2)c(t)$. In order for the fourth-order Runge–Kutta method to be stable on this ODE, we get the same stability condition as (7.5). In Program 2, $\Delta x = L/N = 0.3125$, and thus the stability condition (7.5) gives the time-step restriction $\Delta t \leq 0.028$. This explains why Program 2 is stable when $\Delta t = 0.02$ but unstable when $\Delta t = 0.03$. If other explicit time-stepping schemes are used for the NLS equation, their stability conditions would be similar to (7.5), except that the constant on the right-hand side of (7.5) would be different. If the pseudospectral method is applied to the KdV equation, the stability condition would be $\Delta t/\Delta x^3 \leq C$, where the constant C depends on the particular time-stepping scheme which is used. When the fourth-order Runge–Kutta method is taken, this constant would be $C = 2\sqrt{2}/\pi^3$. For other nonlinear wave equations, stability conditions of the pseudospectral method can be similarly derived.

The pseudospectral method is highly accurate and easy to implement for any wave equation in any dimension. Its main drawback is its stability condition (such as (7.5)), which forces the user to take small time-step values. If the time-step size violates this stability condition, numerical instability would rapidly develop and render numerical results useless. This stability constraint is particularly severe if the order of spatial derivatives in the wave equation is high (such as in the fifth-order KdV equation). In such cases, the pseudospectral method is not competitive, and other numerical methods should be used. In the next two subsections, we will describe two alternative methods which have milder stability constraints.

7.1.2 Split-Step Method—Accuracy and Numerical Stability

Another evolution method we introduce is the split-step method. In this method, the evolution equation is split into several pieces which are integrated separately. These separate integrations are then combined together to form the numerical solution at the next time step. The idea of the split-step method appeared long time ago (see Strang (1968) and Yanenko (1971), for instance). This method was further developed in later years, and high-order schemes for linear equations were constructed (see Yoshida (1990)). In the past twenty years, this method became popular and was widely used in the physical community (see Boyd (2001) and Agrawal (2007), for example). The spatial accuracy of this method is also spectral if spatial derivatives are computed by Fourier transforms. More importantly, this method has milder stability constraints on the size of the time step. In addition, some important properties of the original evolution equation (such as energy conservation) are preserved by this method, and this can be important for long-time simulations.

The temporal accuracy of the split-step method for *linear* equations has been analyzed extensively in the literature. However, its temporal accuracy for *nonlinear* equations is still largely an open question. In this section, the temporal accuracy of the split-step method for both linear and nonlinear equations will be examined. We will show that its accuracy for general nonlinear equations is the same as that for linear equations (at least for schemes up to fourth-order accurate in time). Regarding its numerical stability, a widely held misconception is that this method is numerically stable for all sizes of the time step. This is not true. In other words, the split-step method also has a stability condition on the size of the time step, at least for certain nonlinear equations such as the NLS equation (see Weideman and Herbst (1986)). But its stability condition is less stringent than that for the pseudospectral method. In addition, its numerical instability, when it arises, is much weaker than that in the pseudospectral method. An interesting phenomenon we will reveal on its numerical stability is that, over a wide time-step interval, stable and unstable values of the time step are densely intertwined. Hence a stable time-step size may appear above an unstable time-step size. This phenomenon does not occur in the pseudospectral method, where stable time-step sizes are always less than unstable time-step sizes.

Split-Step Method for Linear Equations

We first introduce the split-step method for a linear system of PDEs with time-independent coefficients. Suppose this system can be written as

$$\mathbf{u}_t = (L+M)\mathbf{u}, \tag{7.6}$$

where $\mathbf{u}(\mathbf{x},t)$ is a vector function, $\mathbf{x}$ is a multidimensional spatial variable, and L, M are two linear time-independent operators. The exact solution of this equation at time h can be formally written as

$$\mathbf{u}(\mathbf{x},h) = e^{h(L+M)}\mathbf{u}(\mathbf{x},0), \tag{7.7}$$

where h is the time-step size. Now we consider two split equations

$$\mathbf{v}_t = L\mathbf{v} \tag{7.8}$$

and

$$\mathbf{v}_t = M\mathbf{v}, \tag{7.9}$$

whose solutions at time h can be formally written as $e^{hL}\mathbf{v}(\mathbf{x},0)$ and $e^{hM}\mathbf{v}(\mathbf{x},0)$. If the original equation (7.6) is not easy to solve but the split equations (7.8) and (7.9) are, then the idea of the split-step method is to approximate $e^{h(L+M)}$ by a sequence of split operators as

$$e^{h(L+M)} \approx e^{\beta_n hM} e^{\alpha_n hL} \ldots e^{\beta_1 hM} e^{\alpha_1 hL}, \tag{7.10}$$

where coefficients α_j and β_j are constants. The error of the above approximation can be calculated by expanding its left and right sides as power series in h and then comparing their difference. For instance, the expansion for its left side is

$$e^{h(L+M)} = 1 + h(L+M) + \frac{1}{2}h^2(L+M)^2 + \cdots, \tag{7.11}$$

and so on. If the difference between the two sides of (7.10) is $O(h^{n+1})$, then the split-step scheme (7.10) will be nth order accurate in time. The simplest split-step scheme is when we take $\alpha_1 = \beta_1 = 1$ and other coefficients to be zero. In this case, we get

$$S_1(h) \equiv e^{hM} e^{hL} = e^{h(L+M)} + O(h^2), \tag{7.12}$$

and thus this splitting scheme is first-order accurate in time (for the generic case where L and M do not commute). If we take $\alpha_1 = \alpha_2 = \frac{1}{2}$, $\beta_1 = 1$, and other coefficients to be zero, then we get

$$S_2(h) \equiv e^{\frac{1}{2}hL} e^{hM} e^{\frac{1}{2}hL} = e^{h(L+M)} + O(h^3), \tag{7.13}$$

thus this symmetric splitting method is second-order accurate in time. This splitting was proposed by Strang (1968) and is sometimes called *Strang splitting*. In the implementation of this scheme, since two operators $e^{\frac{1}{2}hL}$ between successive time steps can be combined as one operator e^{hL}, the actual implementation then becomes

$$\mathbf{u}(\mathbf{x},t) \approx e^{\frac{1}{2}hL} e^{hM} e^{hL} e^{hM} \cdots e^{hL} e^{hM} e^{\frac{1}{2}hL} \mathbf{u}(\mathbf{x},0). \tag{7.14}$$

Thus the computational effort of this second-order Strang splitting is actually the same as that of the first-order splitting (7.12). This is analogous to quadrature, where the second-order trapezoidal rule involves the same amount of computations as the first-order Riemann summation. Higher-order split-step schemes can be constructed by taking more terms in (7.10) and selecting coefficients α_j and β_j appropriately. If this is done by power series expansions to Eq. (7.10), the algebra will quickly become complicated. A more efficient way was proposed by Yoshida (1990). The idea is to symmetrically assemble a few lower-order schemes together to obtain a higher-order one. Specifically, to obtain a fourth-order split-step scheme $S_4(h)$, we express $S_4(h)$ as a symmetric product of three second-order schemes:

$$S_4(h) = S_2(c_1 h)\, S_2(c_0 h)\, S_2(c_1 h), \tag{7.15}$$

where c_0 and c_1 are real constants. To determine c_0 and c_1, we use a standard formula

$$e^X e^Y e^X = e^{2X+Y+\frac{1}{6}[Y,Y,X]-\frac{1}{6}[X,X,Y]+\cdots}, \tag{7.16}$$

where X and Y are any operators, $[X,Y] \equiv XY - YX$, $[X,X,Y] \equiv [X,[X,Y]]$, etc. This formula can be obtained by repeated application of the Baker–Campbell–Hausdorff formula for the operator $e^X e^Y$:

$$e^X e^Y = e^{X+Y+\frac{1}{2}[X,Y]+\frac{1}{12}([X,X,Y]+[Y,Y,X])+\frac{1}{24}[X,Y,Y,X]+\cdots}. \tag{7.17}$$

Using this formula (7.16), it is easy to find that $S_2(h)$ can be expressed as

$$S_2(h) = e^{h(L+M)+h^3 A_3+h^5 A_5+\cdots}, \tag{7.18}$$

where

$$A_3 = \frac{1}{12}[M,M,L] - \frac{1}{24}[L,L,M], \tag{7.19}$$

and so on. It is important to notice that in the exponent of (7.18), only odd powers of h are present. Substituting (7.18) into (7.15) and utilizing the formula (7.16), we find that

$$S_4(h) = e^{(c_0+2c_1)h(L+M)+(c_0^3+2c_1^3)h^3 A_3+O(h^5)}. \tag{7.20}$$

Thus if we choose (c_0, c_1) such that

$$c_0 + 2c_1 = 1, \quad c_0^3 + 2c_1^3 = 0, \tag{7.21}$$

i.e.,

$$c_1 = \frac{1}{2-2^{1/3}}, \quad c_0 = 1 - 2c_1, \tag{7.22}$$

then the split-step scheme (7.15) would be fourth-order accurate. If this scheme is unfolded into a product of individual split operators of the form (7.10), then this fourth-order scheme is such that $n = 4$, and

$$\begin{aligned} &\alpha_1 = \frac{1}{2}c_1, \quad \alpha_2 = \frac{1}{2}(1-c_1), \quad \alpha_3 = \frac{1}{2}(1-c_1), \quad \alpha_4 = \frac{1}{2}c_1, \\ &\beta_1 = c_1, \qquad \beta_2 = 1-2c_1, \qquad \beta_3 = c_1, \qquad \beta_4 = 0. \end{aligned} \tag{7.23}$$

To obtain a sixth-order split-step scheme, we express it as a symmetric product of fourth-order schemes S_4 and repeat the above process. In view of the exponential form (7.18) of $S_2(h)$ as well as the formula (7.16), we see that only odd powers of h are present in the exponent (7.20) of $S_4(h)$ as well. Thus repetition of the above procedure will produce a sixth-order scheme. Along this line, split-step methods of any even order can be constructed. It should be remembered, however, that very high order split-step methods are very complicated and are not necessarily better. Extensions of these schemes to time-dependent operators L and M can be found in Bandrauk and Shen (1994). It was shown that when the time level in the integration of each split operator is chosen properly, the order of accuracy of the above split-step schemes will be maintained.

Split-Step Method for Nonlinear Equations

Next, we consider split-step methods for a general system of nonlinear wave equations

$$\mathbf{u}_t = M(\mathbf{u}) + N(\mathbf{u}), \tag{7.24}$$

where at least one of the operators $M(\mathbf{u})$ and $N(\mathbf{u})$ is nonlinear in $\mathbf{u}$ and M, N do not depend explicitly on time. Note that if M and N depend on the time t explicitly, the time can be added to the variable $\mathbf{u}$ to form an expanded system. This expanded system is then in the form of (7.24). Thus Eq. (7.24) covers all nonlinear wave equations with or without explicit time dependence.

Split-step schemes constructed above for linear equations with time-independent coefficients have been widely used to compute nonlinear equations (7.24) in the scientific community; see Kassam and Trefethen (2005), and Agrawal (2007). In this case, the split equations to be solved are

$$\mathbf{v}_t = M(\mathbf{v}) \tag{7.25}$$

and

$$\mathbf{v}_t = N(\mathbf{v}). \tag{7.26}$$

To solve nonlinear equations by split-step methods, a common practice is to formally extend the previous split-step schemes (developed for linear equations) directly to nonlinear equations. For instance, the extension of the Strang splitting scheme (7.13) to the nonlinear equation (7.24) is that, starting from the initial condition $\mathbf{u}(\mathbf{x},0)$, we first integrate the M-operator equation (7.25) by half a step $h/2$ to get an intermediate solution $\mathbf{v}_1(\mathbf{x})$. Then starting from $\mathbf{v}_1(\mathbf{x})$, we integrate the N-operator equation (7.26) by one step h to get another intermediate solution $\mathbf{v}_2(\mathbf{x})$. Lastly, starting from $\mathbf{v}_2(\mathbf{x})$, we integrate the M-operator equation (7.25) again by half a step $h/2$. The resulting function is then our numerical approximation $S_2(h)\mathbf{u}(\mathbf{x},0)$ for the nonlinear solution $\mathbf{u}(\mathbf{x},h)$. Extension of other linear splitting schemes (such as $S_4(h)$) to the nonlinear equation (7.24) is similar.

The order of accuracy of these extended schemes to nonlinear equations (7.24) is a nontrivial issue. This accuracy cannot be inferred from linear equations, because the formal exponential-operator manipulations we have used to determine their accuracy for linear equations are not valid for nonlinear equations. Javanainen and Ruostekoski (2006) employed symbolic calculations and directly verified that, for the Gross–Pitaevskii equation (1.62), the above nonlinear extensions of linear split-step schemes (up to S_4) have the same order of accuracy as for linear equations. Below, we will prove that for general nonlinear equations (7.24), the extended split-step schemes S_1, S_2, and S_4 also have first, second, and fourth order of accuracy in time, just as for linear equations. The proof will be based on direct verification, and it holds for both Hamiltonian and non-Hamiltonian equations (7.24). Since the proof for S_1 is trivial, we consider only S_2 and S_4 below. The proof follows Yang (2009b).

Our proof will use the method of lines. The idea is to discretize the wave equation (7.24) onto a spatial grid, so that it becomes a system of ODEs, with $\mathbf{u}$ being the vector of solution values on the spatial grid. If we can determine the order of accuracy of split-step schemes for this nonlinear ODE system with the length of the vector $\mathbf{u}$ arbitrary, then the same order of accuracy will hold for the nonlinear wave system (7.24). Thus Eq. (7.24) will be treated as a nonlinear ODE system below.

First we consider the Strang splitting scheme $S_2(h)$. Its extension to the nonlinear ODEs (7.24) is that, starting from the initial condition $\mathbf{u}(0)$, we integrate the M-operator equation (7.25) by half a step $h/2$ and obtain the vector $\mathbf{v}_1$. Then starting from $\mathbf{v}_1$, we integrate the N-operator equation (7.26) by one step h and obtain the vector $\mathbf{v}_2$. Lastly, starting from $\mathbf{v}_2$, we integrate the M-operator equation (7.25) by half a step $h/2$. The resulting vector is then our numerical solution $S_2(h)\mathbf{u}(0)$ at time $t = h$. Note that $S_2(h)\mathbf{u}(0)$ depends on $\mathbf{u}(0)$ nonlinearly for the nonlinear equation (7.24). For convenience, we assume below that integrations of the split equations (7.25)–(7.26) are exact. If not, as long as the accuracy of these split-equation integrations is second order or higher, then the order of accuracy of the numerical solution $S_2(h)\mathbf{u}(0)$ will not be affected.

In order to determine the accuracy of the numerical solution $S_2(h)\mathbf{u}(0)$, we first derive the h-expansion of the exact solution $\mathbf{u}(h)$. Expanding $\mathbf{u}(h)$ into Taylor series and utilizing (7.24), we get

$$\mathbf{u}(h) = \mathbf{u}(0) + (M+N)h + \frac{1}{2}\nabla(M+N)\cdot(M+N)h^2 + O(h^3), \tag{7.27}$$

where $\mathbf{u}(0)$ is the initial solution, M, N are evaluated at the initial solution $\mathbf{u}(0)$, and ∇ is the gradient respect to the vector $\mathbf{u}$.

Next we derive the h-expansion of the numerical solution $S_2(h)\mathbf{u}(0)$. Notice that $\mathbf{v}_1$ is the solution of the M-operator equation (7.25) at time $t = h/2$ starting from the initial condition $\mathbf{u}(0)$. Thus by Taylor expansion, we get

$$\mathbf{v}_1 = \mathbf{u}(0) + \frac{1}{2}Mh + \frac{1}{8}\nabla M\cdot Mh^2 + O(h^3), \tag{7.28}$$

where M is evaluated at the initial condition $\mathbf{u}(0)$. The vector $\mathbf{v}_2$ is the solution of the N-operator equation (7.26) at time $t = h$ starting from the initial condition $\mathbf{v}_1$. Thus by Taylor expansion, we get the expression for $\mathbf{v}_2$ as

$$\mathbf{v}_2 = \mathbf{v}_1 + N(\mathbf{v}_1)h + \frac{1}{2}\nabla N(\mathbf{v}_1)\cdot N(\mathbf{v}_1)h^2 + O(h^3), \tag{7.29}$$

where $N(\mathbf{v}_1)$ is the operator N evaluated at $\mathbf{v}_1$. Inserting (7.28) into (7.29) and expanding around the initial condition $\mathbf{u}(0)$, the expression (7.29) for $\mathbf{v}_2$ then becomes

$$\mathbf{v}_2 = \mathbf{u}(0) + \left(\frac{1}{2}M + N\right)h + \frac{1}{2}\left[\frac{1}{4}\nabla M\cdot M + \nabla N\cdot(M+N))\right]h^2 + O(h^3), \tag{7.30}$$

where M and N are evaluated at $\mathbf{u}(0)$. Continuing this calculation for $S_2(h)\mathbf{u}(0)$, which is the solution of the M-operator equation (7.25) at time $t = h/2$ starting from the initial condition $\mathbf{v}_2$, we find that

$$S_2(h)\mathbf{u}(0) = \mathbf{u}(0) + (M+N)h + \frac{1}{2}\nabla(M+N)\cdot(M+N)h^2 + O(h^3), \tag{7.31}$$

where M and N are evaluated at $\mathbf{u}(0)$. Comparing this result with (7.27), we see that

$$S_2(h)\mathbf{u}(0) = \mathbf{u}(h) + O(h^3), \tag{7.32}$$

thus the Strang-splitting scheme $S_2(h)$ is also second-order accurate for nonlinear equations (7.24).

The error term of $S_2(h)\mathbf{u}(0)$ turns out to be important for the accuracy analysis of $S_4(h)$ on nonlinear equations (7.24). Following (7.32), let us write the h-expansion of $S_2(h)\mathbf{u}(0)$ as

$$S_2(h)\mathbf{u}(0) = \mathbf{u}(h) + Fh^3 + Gh^4 + O(h^5), \tag{7.33}$$

where the error terms F and G are functions of the initial condition $\mathbf{u}(0)$. For general nonlinear ODEs (7.24), a key relation between these two error terms is

$$G = \frac{1}{2}\left[\nabla(M+N)\cdot F + \nabla F\cdot(M+N)\right], \tag{7.34}$$

where ∇ is the gradient with respect to the vector $\mathbf{u}$ (or $\mathbf{u}(0)$) as before, and M, N are evaluated at the initial condition $\mathbf{u}(0)$. This relation will play a critical role in the error analysis of $S_4(h)$ on nonlinear equations (7.24). For linear equations (7.6), by utilizing (7.18) and the Zassenhaus formula

$$e^{X+Y} = e^X e^Y e^{-\frac{1}{2}[X,Y]} e^{\frac{1}{6}(2[Y,X,Y]+[X,X,Y])}\cdots, \tag{7.35}$$

one can easily find that

$$F = A_3\mathbf{u}(0), \quad G = \frac{1}{2}[(L+M)A_3 + A_3(L+M)]\mathbf{u}(0), \tag{7.36}$$

which are linear functions of $\mathbf{u}(0)$. Here A_3 is given in Eq. (7.19). Thus the relation (7.34) is satisfied when one notices that for linear equations (7.6), $M(\mathbf{u})$ and $N(\mathbf{u})$ in (7.34) are $L\mathbf{u}$ and $M\mathbf{u}$, respectively. For nonlinear equations (7.24), expressions for F and G are more complicated. To obtain them, one needs to carry out Taylor expansions (7.27)–(7.31) to the fourth order in h. In this process, vector notations as in (7.27)–(7.31) are no longer sufficient since higher-order tensors start to appear. Because of this, one has to switch to index notations for performing Taylor expansions and vector manipulations. After straightforward but tedious calculations, we have proved the relation (7.34), and the details are omitted.

Now we use relation (7.34) to prove the fourth-order accuracy of S_4 for nonlinear equations (7.24). According to (7.15) and (7.22), $S_4(h)\mathbf{u}(0)$ is

$$S_4(h)\mathbf{u}(0) = S_2(c_1h)\, S_2(c_0h)\, S_2(c_1h)\,\mathbf{u}(0). \tag{7.37}$$

Let us denote

$$\mathbf{u}_1 \equiv S_2(c_1h)\mathbf{u}(0), \quad \mathbf{u}_2 \equiv S_2(c_0h)\mathbf{u}_1. \tag{7.38}$$

Then in view of the $S_2(h)\mathbf{u}(0)$ formula (7.33), we see that

$$\mathbf{u}_1 = \mathbf{u}(c_1h) + F[\mathbf{u}(0)]c_1^3h^3 + G[\mathbf{u}(0)]c_1^4h^4 + O(h^5), \tag{7.39}$$

$$\mathbf{u}_2 = \widehat{\mathbf{u}}(c_0h)|_{\mathbf{u}_0=\mathbf{u}_1} + F(\mathbf{u}_1)c_0^3h^3 + G(\mathbf{u}_1)c_0^4h^4 + O(h^5), \tag{7.40}$$

and

$$\begin{aligned} S_4(h)\mathbf{u}(0) &= S_2(c_1h)\mathbf{u}_2 \\ &= \widehat{\mathbf{u}}(c_1h)|_{\mathbf{u}_0=\mathbf{u}_2} + F(\mathbf{u}_2)c_1^3h^3 + G(\mathbf{u}_2)c_1^4h^4 + O(h^5). \end{aligned} \tag{7.41}$$

Here $\widehat{\mathbf{u}}(c_0h)|_{\mathbf{u}_0=\mathbf{u}_1}$ is the exact solution of Eq. (7.24) after time c_0h when the initial condition is $\mathbf{u} = \mathbf{u}_1$, and $\widehat{\mathbf{u}}(c_1h)|_{\mathbf{u}_0=\mathbf{u}_2}$ is the exact solution of Eq. (7.24) after time c_1h when the initial condition is $\mathbf{u} = \mathbf{u}_2$. This first term in (7.40) can be calculated from (7.27) and (7.39) as

$$\begin{aligned} \widehat{\mathbf{u}}(c_0h)|_{\mathbf{u}_0=\mathbf{u}_1} &= \widehat{\mathbf{u}}(c_0h)|_{\mathbf{u}_0=\mathbf{u}(c_1h)} \\ &\quad + \left[1 + \nabla(M+N)|_{\mathbf{u}(c_1h)} \cdot c_0h\right][\mathbf{u}_1 - \mathbf{u}(c_1h)] + O(h^5), \end{aligned} \tag{7.42}$$

where $\nabla(M+N)|_{\mathbf{u}(c_1h)}$ is the gradient of $M+N$ evaluated at $\mathbf{u} = \mathbf{u}(c_1h)$. Notice that the first term in (7.42) is simply $\mathbf{u}[(c_1+c_0)h]$. Thus,

$$\begin{aligned} \mathbf{u}_2 &= \mathbf{u}[(c_1+c_0)h] + \left[1 + \nabla(M+N)|_{\mathbf{u}(c_1h)} \cdot c_0h\right][\mathbf{u}_1 - \mathbf{u}(c_1h)] \\ &\quad + F(\mathbf{u}_1)c_0^3h^3 + G(\mathbf{u}_1)c_0^4h^4 + O(h^5). \end{aligned} \tag{7.43}$$

Similarly, we have

$$\begin{aligned} S_4(h)\mathbf{u}(0) &= \mathbf{u}(h) + \left[1 + \nabla(M+N)|_{\mathbf{u}[(c_1+c_0)h]} \cdot c_1h\right]\{\mathbf{u}_2 - \mathbf{u}[(c_1+c_0)h]\} \\ &\quad + F(\mathbf{u}_2)c_1^3h^3 + G(\mathbf{u}_2)c_1^4h^4 + O(h^5). \end{aligned} \tag{7.44}$$

Now we expand the right-hand side of the above equation as well as $\mathbf{u}_2$ around the initial condition $\mathbf{u}(0)$. Utilizing (7.39), (7.43), and the relation (7.21), we find that

$$S_4(h)\mathbf{u}(0) = \mathbf{u}(h) + c_1^3(1-3c_1)[\nabla(M+N)\cdot F + \nabla F\cdot(M+N) - 2G]h^4 + O(h^5), \tag{7.45}$$

where functions M, N, F, and G are evaluated at the initial condition $\mathbf{u}(0)$. Then recalling the relation (7.34), we see that the split-step scheme $S_4(h)$ is fourth-order accurate for nonlinear equations (7.24) as well.

A natural question one may ask now is whether other higher-order split-step schemes (such as $S_6(h)$) have the same order of accuracy for both linear and nonlinear equations. This question requires further investigation.

Now we apply the split-step scheme S_4 to the NLS equation (7.1) and demonstrate its fourth-order accuracy. This NLS equation can be rewritten as

$$u_t = iu_{xx} + 2i|u|^2u. \tag{7.46}$$

Thus we take its split operators as $M(u) = iu_{xx}$ and $N(u) = 2i|u|^2u$. Here the split equation $v_t = M(v)$ can be solved by the discrete Fourier transform, and the split equation $v_t = N(v)$ has the exact solution formula $v(x,t) = v(x,0)e^{2i|v(x,0)|^2t}$. The corresponding MATLAB code for the S_4 scheme is presented below.

Program 3

```
% p3.m: fourth-order split-step method for solving the NLS equation
% iu_t+u_{xx}+2|u|^2u=0.

  L=4*pi; N=128; dt=0.01; tmax=2; nmax=round(tmax/dt);
  dx=L/N; x=[-L/2:dx:L/2-dx]'; k=[0:N/2-1 -N/2:-1]'*2*pi/L;
  u=sin(x).^2; udata=u; tdata=0;
  c=1/(2-2^(1/3));                           % scheme coefficients
  a1=c/2; a2=(1-c)/2; a3=a2; a4=c/2;
  b1=c; b2=1-2*c; b3=c;
  E1=exp(-a1*dt*i*k.^2);
  E2=exp(-a2*dt*i*k.^2);
  E3=exp(-a3*dt*i*k.^2);
  E4=exp(-a4*dt*i*k.^2);
  for nn=1:nmax                              % integration begins
    v=ifft(fft(u).*E1);
    v=v.*exp(b1*dt*i*2*v.*conj(v));
    v=ifft(fft(v).*E2);
    v=v.*exp(b2*dt*i*2*v.*conj(v));
    v=ifft(fft(v).*E3);
    v=v.*exp(b3*dt*i*2*v.*conj(v));
    u=ifft(fft(v).*E4);
    if mod(nn,round(nmax/25)) == 0
       udata=[udata u]; tdata=[tdata nn*dt];
```

```
  end
end                                         % integration ends
waterfall(x, tdata, abs(udata'));           % solution plotting
colormap([0 0 0]); view(5, 60)
text(-0.4,  -0.4, 'x', 'fontsize', 15)
text(7, 1, 't', 'fontsize', 15)
text(-3.7, -0.3, '-\pi', 'fontsize', 14)
text(2.9, -0.3, '\pi', 'fontsize', 14)
zlabel('|u|', 'fontsize', 15)
axis([-L/2 L/2 0 tmax 0 2]); grid off
set(gca, 'xtick', [-pi 0 pi], 'xticklabel',{'','0',''})
set(gca, 'ytick', [0 1 2], 'yticklabel',{'0','','2'})
set(gca, 'ztick', [0 1 2])
```

In this code, we take the initial condition to be a periodic function $u(x,0) = \sin^2 x$ over the interval $[-2\pi, 2\pi]$, discretized by 128 points. The output of this program is displayed in Fig. 7.2(a), and it shows a recurrence phenomenon in the solution.

The spatial accuracy of this split-step method S_4 is spectral since its evaluation of ∂_{xx} is performed by the discrete Fourier transform. The temporal accuracy of this method is fourth order. To verify this fourth-order temporal accuracy, we rerun Program 3 to time $t = 2$ using various time-step values. The spatial accuracy in these simulations is extremely high, thus the temporal error is dominant. The "exact" solution at $t = 2$ is taken as the numerical solution with a very small time step $\Delta t = 10^{-4}$. At larger time-step values, the numerical solution is compared to this "exact" solution, and the error is calculated as the ∞-norm of their difference. The results of these errors versus time steps are plotted in Fig. 7.2(b). For comparison, a benchmark fourth-order decaying function $0.3\Delta t^4$ is also plotted. It is seen that the error indeed decays in proportion to Δt^4, confirming the fourth-order temporal accuracy of this S_4 scheme.

The split-step methods have several advantages. One advantage is that they preserve some important properties of the original evolution equation. For instance, the power $\int |u|^2 dx$, which is a conserved quantity of the NLS equation (7.1), is also conserved in the split-step methods because each split-operator integration is power invariant. This power conservation can be important for long-time evolution simulations. Another advantage of the split-step methods is that a larger time-step size can often be used without the onset of numerical instability. Indeed, for $L = 4\pi$ and $N = 128$ as in Program 3, the fourth-order pseudospectral method would require $\Delta t < 0.003$ for numerical stability (see Eq. (7.5)), but the time steps in Program 3 can be much larger than 0.003 without signs of numerical instability. The split-step methods are particularly suitable for NLS-type equations, where

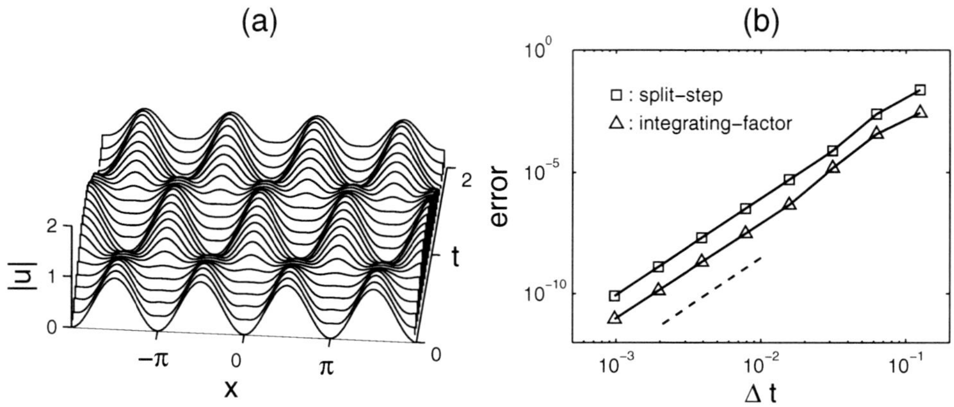

Figure 7.2. (a) *Output of Program* 3 *(the fourth-order split-step method) for the simulation of the NLS equation* (7.1) *with a periodic initial condition;* (b) *errors versus time-step size for the fourth-order split-step method and integrating-factor method at* $t = 2$. *The dashed line is the benchmark function* $0.3\Delta t^4$.

the split equations (7.25) and (7.26) can be solved exactly, hence their integrations take little numerical effort. For KdV-type equations, the split equation containing the nonlinear term does not have a simple solution formula and has to be obtained either by additional time-stepping schemes (such as Runge–Kutta methods) or by interpolation methods along characteristic lines. For such equations, the split-step methods will be less competitive.

Numerical Stability of Split-Step Methods

For computations shown in Fig. 7.2, the split-step method could take a large time step Δt (such as 0.1) without signs of numerical instability. However, this does not mean the split-step method is unconditionally stable. In fact, if we run Program 3 with $\Delta t = 0.01$ for a very long time (such as to $t = 250$), high-wavenumber Fourier modes will become prominent, signifying numerical instability (see Fig. 7.7 later). Weideman and Herbst (1986) analyzed this numerical instability for the first-order split-step method S_1 and obtained its stability condition for the NLS equation with near-constant initial conditions. Below we will extend that analysis to derive stability conditions for both the first-order and higher-order split-step methods on the NLS equation. We will show that when the time-step size is below a certain threshold value, these split-step methods will be numerically stable. We will also reveal that when the time-step size is above this sufficient stability threshold, intervals of stable and unstable time-step sizes will intertwine in an intricate way. In particular, a stable time-step size can be much larger than an unstable time-step size, which is quite surprising.

In addition, we will show that numerical instability in the split-step methods, when it arises, is quite weak (compared to that in the pseudospectral method). Thus this numerical instability often does not contaminate the numerical solution for short-time simulations (as in Program 3). Furthermore, we will show that all these results on numerical instability, which are obtained analytically for near-constant initial conditions, still hold qualitatively for other initial conditions such as solitons and periodic functions. The analysis below follows Yang (2009b).

(1) Numerical Instability in the First-Order Split-Step Method

Let us consider split-step methods as applied to the NLS equation

$$iu_t + u_{xx} + 2|u|^2 u = 0. \tag{7.47}$$

The first-order split-step method S_1 for this equation is

$$u_{n+1} = \mathcal{F}^{-1}\left\{ e^{-ik^2 h}\mathcal{F}[u_n e^{2ih|u_n|^2}]\right\}, \tag{7.48}$$

where h is the time-step size (i.e., Δt), $\mathcal{F}$ and $\mathcal{F}^{-1}$ represent Fourier and inverse Fourier transforms, and k is the Fourier wavenumber. Here the nonlinear step is performed first, followed by the linear step, and the linear step is computed by the Fourier transform (as in Program 3). The NLS equation (7.47) admits an exact x-independent solution

$$u(x,t) = a\, e^{2i|a|^2 t}, \tag{7.49}$$

where a is a constant. Without loss of generality, we take a to be a real number since the NLS equation is phase invariant. When this solution is perturbed to

$$u(x,t) = e^{2ia^2 t}\,[a + \tilde{u}(x,t)], \quad \tilde{u}(x,t) \ll 1, \tag{7.50}$$

we now analyze how the perturbation $\tilde{u}$ evolves under the split-step scheme S_1 in (7.48). Substituting (7.50) into (7.48) and neglecting terms of $O(\tilde{u}^2)$ and higher, we obtain the iteration equation for the perturbation as

$$\Psi_{n+1} = \widehat{S}_1 \Psi_n, \tag{7.51}$$

where $\Psi_n = \mathcal{F}([\tilde{u}_n, \tilde{u}_n^*]^T)$, the iteration matrix $\widehat{S}_1$ is

$$\widehat{S}_1 = \widehat{M}(h)\,\widehat{N}(h), \tag{7.52}$$

and

$$\widehat{M}(h) = \begin{bmatrix} e^{-ik^2 h} & \\ & e^{ik^2 h} \end{bmatrix}, \quad \widehat{N}(h) = \begin{bmatrix} 1 + 2ia^2 h & 2ia^2 h \\ -2ia^2 h & 1 - 2ia^2 h \end{bmatrix}. \tag{7.53}$$

The $\widehat{M}$ matrix gives the perturbation iteration after the linear step of the method, and the $\widehat{N}$ matrix gives the perturbation iteration after the nonlinear step. Eigenvalues of the iteration matrix $\widehat{S}_1$ determine if the perturbations will grow or not. It is easy to see that $\det(\widehat{S}_1) = 1$. Thus eigenvalues of $\widehat{S}_1$ are

$$\lambda = \beta \pm \sqrt{\beta^2 - 1}, \tag{7.54}$$

where

$$\beta = \frac{1}{2}\mathrm{tr}(\widehat{S}_1) = \cos(k^2 h) + 2a^2 h \sin(k^2 h). \tag{7.55}$$

Perturbations will grow if $|\lambda| > 1$, i.e., when $|\beta| > 1$. Rewriting β as

$$\beta = r\cos(k^2 h - \theta), \tag{7.56}$$

where

$$r = \sqrt{1 + 4a^4 h^2}, \quad \tan\theta = 2a^2 h, \quad 0 < \theta < \frac{1}{2}\pi, \tag{7.57}$$

we see that Fourier modes with wavenumbers in the windows

$$0 < \mathrm{mod}(k^2 h, \pi) < 2\theta \tag{7.58}$$

are unstable. The largest growth occurs at wavenumbers where $\mathrm{mod}(k^2 h, \pi) = \theta$, and the largest growth rate is

$$|\lambda|_{\max} = \sqrt{1 + 4a^4 h^2} + 2a^2 h. \tag{7.59}$$

To illustrate, the graph of $|\lambda|$ versus $k^2 h$ at $2a^2 h = 0.1$ is plotted in Fig. 7.3. Here $|\lambda|$ is the larger of the two eigenvalues in (7.54). From this graph, windows of unstable Fourier modes and their growth rates can be clearly seen. As $2a^2 h \to 0$, $\theta \to 0$, and thus all these unstable windows shrink toward their left edge points $0, \pi, 2\pi, \ldots$. At the same time, the heights of these unstable windows decrease according to the formula (7.59).

It is important to recognize that, among these windows of unstable modes, the first window

$$0 < k^2 h < 2\theta \tag{7.60}$$

is *not* induced by numerical instability. Rather it is caused by modulation instability of the constant solution (7.49) in the NLS equation (7.47). But all the other windows

$$k^2 h \in (\pi, \pi + 2\theta),\ (2\pi, 2\pi + 2\theta), \ldots \tag{7.61}$$

are true windows of numerical instability. In the numerical implementation of the split-step method S_1 in (7.48), the Fourier wavenumbers involved are discrete:

$$k = 0,\ \pm k_0,\ \pm 2k_0, \ldots,\ \pm\frac{1}{2}N k_0, \tag{7.62}$$

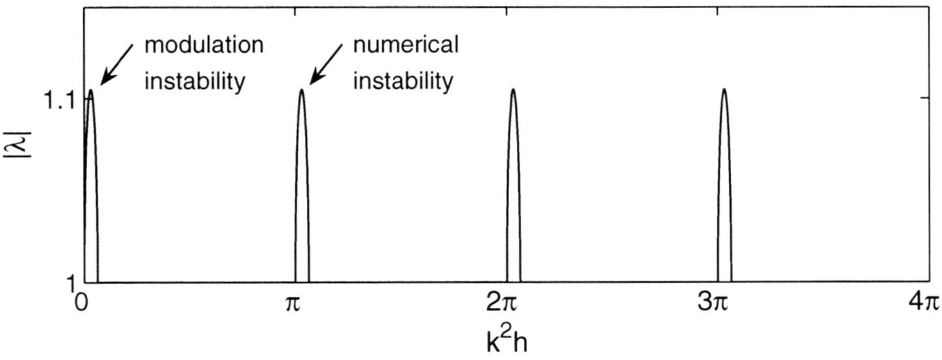

Figure 7.3. *Regions of unstable Fourier modes and their growth rates in the first-order split-step method S_1 as applied to the NLS equation* (7.47) *with near-constant initial conditions* (7.50) *and* $2a^2h = 0.1$.

where $k_0 = 2\pi/L$ is the wavenumber separation, L is the length of the computational interval, and N is the number of spatial grid points. Thus the necessary and sufficient condition for numerical stability is that none of these discrete Fourier wavenumbers (7.62) falls in the numerical-instability windows (7.61). Note that the largest discrete Fourier wavenumber in (7.62) is

$$|k|_{\max} = \frac{1}{2}Nk_0 = \frac{\pi}{\Delta x}; \tag{7.63}$$

thus a sufficient condition for numerical stability is $|k|^2_{\max}h < \pi$, i.e.,

$$\frac{\Delta t}{\Delta x^2} < \frac{1}{\pi}. \tag{7.64}$$

This sufficient stability condition is quantitatively close to the necessary and sufficient stability condition (7.5) of the pseudospectral method. If the time-step condition (7.64) is not met, discrete Fourier wavenumbers (7.62) may fall in the unstable windows (7.61), causing numerical instability. The above numerical-stability conditions are the ones which were derived by Weideman and Herbst (1986).

The numerical instability in the split-step method S_1 may be viewed as a parametric resonance. In the split-step simulations, repeated alternations between integrations of the two split operators with each time step h induce an oscillation with frequency $2\pi/h$. This frequency may be regarded as the parametric frequency. At the same time, a Fourier mode with wavenumber k in the split linear Schrödinger equation of (7.47) oscillates at its natural frequency k^2. When this natural frequency k^2 is a multiple of one half of the parametric frequency $2\pi/h$, i.e.,

$$k^2 \approx \frac{n\pi}{h}, \quad n = 1,2,\ldots, \tag{7.65}$$

parametric resonance would occur, hence the Fourier mode with this resonant wavenumber k would be excited. The resonance condition (7.65) is roughly the same as the analytical instability condition (7.61) when θ is small (i.e., when $a^2h \ll 1$). This parametric resonance resembles sideband instability induced by periodically spaced amplifiers in long-distance fiber communications (Matera et al. (1993), Ablowitz et al. (1996)) and excitation of radiation modes in optical fibers with periodic dispersion variation (Pelinovsky and Yang (2004)). It also resembles generation of sideband frequencies in the spectrum of a soliton pulse in certain mode-locked lasers due to periodic disturbance in the laser resonator (this sideband is called a Kelly sideband in optics, after the work by Kelly (1992)). However, this interpretation of numerical instability in the S_1 method as a parametric resonance is only a heuristic argument. This numerical instability would not arise when the S_1 method is applied to a *linear* Schrödinger equation with a potential term, even though one might also expect such instability based on the above parametric-resonance argument.

It is important to realize that the condition (7.64) is sufficient, but not necessary, for numerical stability. Indeed, in actual numerical computations, one can often take time steps much larger than that required by (7.64) without triggering numerical instability. To illustrate, let us take the length of the computational interval as $L = 4\pi$, and the amplitude of the unperturbed solution (7.49) as $a = 1$. At various numbers of grid points N and time steps Δt (i.e., h), we can determine if the computation is numerically stable or not by checking if any of the discrete Fourier wavenumbers (7.62) falls in the numerical-instability windows (7.61). The collection of numerically stable pairs $(N, \Delta t)$ is then plotted in Fig. 7.4. The curve of sufficient numerical-stability condition (7.64) is also displayed for comparison. One can see from this figure that at each value of the grid points N, there are many Δt values larger than the time-step size set by the sufficient condition (7.64), where the computation is still numerically stable. These stable Δt values form a certain number of disconnected segments, and these stable segments are intertwined with unstable segments of Δt. Due to this intertwined structure, many stable time-step sizes are actually larger than many unstable time-step sizes. This is a distinctive feature of the split-step method which does not appear in the pseudospectral method (in the latter method, stable time-step sizes are always less than unstable time-step sizes; see (7.5)). The reason for this intertwining of stable and unstable segments of time steps above the sufficient stability threshold (7.64) is that when the time step Δt (i.e., h) increases, discrete quantities k_n^2h, with discrete wavenumber k_n given in (7.62), will move along the k^2h-axis in Fig. 7.3. With a continuous increase of Δt above the threshold (7.64), these discrete quantities k_n^2h sometimes fall in the unstable windows (7.61), then fall outside these unstable windows, and then fall in these windows again, thus creating intertwining segments of stable and unstable time steps.

Even though stable time steps exist above the sufficient stability threshold (7.64), it should be pointed out that unstable time steps do appear right above this threshold, as can

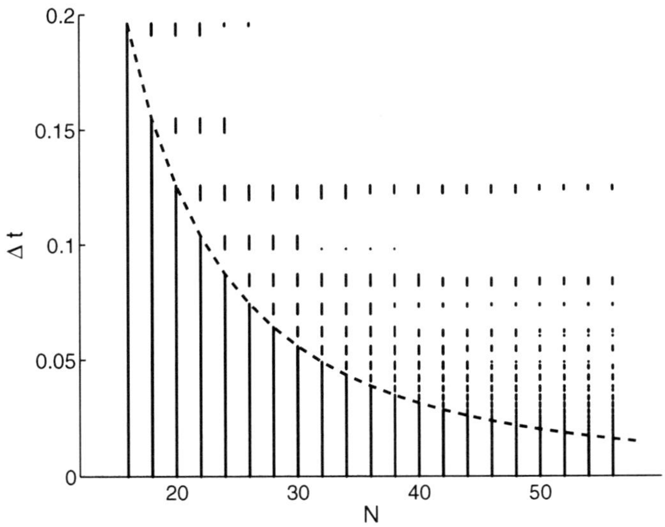

Figure 7.4. *Regions of stable time steps* Δt *at various numbers of grid points* N *in the split-step method* S_1 *applied to the NLS equation* (7.47) *with near-constant initial conditions* (7.50). *Here* $L = 4\pi$ *and* $a = 1$.

be seen in Fig. 7.4. The reason is that at this threshold $h = \Delta t_c$, $|k|^2_{\max}\Delta t_c = \pi$, which is at the left edge of the first numerical-instability window in Fig. 7.3. Thus for the time step h slightly above Δt_c, $|k|^2_{\max} h$ would fall into the first numerical-instability window, and hence numerical instability would arise. This means that the sufficient stability threshold (7.64) is indeed a sharp condition to guarantee numerical stability. Immediately above this threshold, unstable and stable time steps will start to mix, as Fig. 7.4 depicts.

When the time step Δt does not fall in the stable regions (see Fig. 7.4), the split-step method will be numerically unstable. Small disturbances will grow by a factor of $|\lambda|$ with each step. The largest growth rate $|\lambda|_{\max}$ is given by the formula (7.59). When $a^2 h \ll 1$,

$$|\lambda|_{\max} = 1 + 2a^2 h + O(a^4 h^2); \tag{7.66}$$

thus at a given simulation time t, perturbations will grow by a maximal factor of

$$|\lambda|^{t/h}_{\max} \approx (1 + 2a^2 h)^{t/h} \approx e^{2a^2 t}. \tag{7.67}$$

This maximal growth factor holds for all values of grid points N and step size Δt. It is important to notice that this factor is bounded by the simulation time t. Thus for short-time simulations, perturbation growth will be limited, and hence numerical instability will not be very serious. For example, if $t = 5$ and $a = 1$, then perturbations will grow by a maximal factor of $e^{10} = O(10^4)$. Assuming that perturbations are due to numerical roundoff errors, which are below the level of 10^{-14} in MATLAB, then perturbations at time $t = 5$ will rise only to the level below 10^{-9}, which is still very small. For long-time simulations, however, these numerical instabilities can be much worse and should be suppressed.

(2) Numerical Instability in the Fourth-Order Split-Step Methods

The above stability analysis was performed for the first-order split-step method S_1. Now we consider numerical stabilities in higher-order split-step methods. For the second-order method S_2, since its execution scheme (7.14) is the same as S_1 (except the first and last steps), its numerical stability property is thus identical to that of the first-order method S_1. Regarding the fourth-order method S_4, when it is applied to the NLS equation (7.47), there are two different versions which differ by the choice of M either as the linear (dispersive) operator or as the nonlinear operator of the NLS equation. In other words, these two versions of S_4 differ by whether first performing the linear operator or the nonlinear operator. Let us denote the former version as $S_4^{(a)}$ and the latter version as $S_4^{(b)}$ (the method used in Program 3 is $S_4^{(a)}$). When these two versions are applied to simulate the NLS equation (7.47) with near-constant initial conditions (7.50), by extending the previous analysis, we can derive iteration equations for perturbations in these two versions as

$$\Psi_{n+1} = \widehat{S}_4^{(a)} \Psi_n \tag{7.68}$$

and

$$\Psi_{n+1} = \widehat{S}_4^{(b)} \Psi_n, \tag{7.69}$$

where iteration matrices $\widehat{S}_4^{(a,b)}$ are

$$\widehat{S}_4^{(a)} = \widehat{M}(\alpha_4 h)\widehat{N}(\beta_3 h)\widehat{M}(\alpha_3 h)\widehat{N}(\beta_2 h)\widehat{M}(\alpha_2 h)\widehat{N}(\beta_1 h)\widehat{M}(\alpha_1 h) \tag{7.70}$$

and

$$\widehat{S}_4^{(b)} = \widehat{N}(\alpha_4 h)\widehat{M}(\beta_3 h)\widehat{N}(\alpha_3 h)\widehat{M}(\beta_2 h)\widehat{N}(\alpha_2 h)\widehat{M}(\beta_1 h)\widehat{N}(\alpha_1 h). \tag{7.71}$$

Here matrices $\widehat{M}$ and $\widehat{N}$ have been defined in Eq. (7.53), and coefficients $\{\alpha_j, \beta_j\}$ are given in Eq. (7.23). The determinants of these two iteration matrices are both equal to one because so are the determinants of matrices $\widehat{M}$ and $\widehat{N}$. Thus eigenvalues of these iteration matrices are

$$\lambda^{(a,b)} = \beta^{(a,b)} \pm \sqrt{\beta^{(a,b)2} - 1}, \tag{7.72}$$

where

$$\beta^{(a,b)} = \frac{1}{2}\mathrm{tr}(\widehat{S}_4^{(a,b)}). \tag{7.73}$$

The traces of matrices $\widehat{S}_4^{(a,b)}$ are real, and so $|\lambda| > 1$ when $|\beta^{(a,b)}| > 1$. Like the S_1 case, eigenvalues $\lambda^{(a,b)}$ are functions of k^2h and $2a^2h$, i.e.,

$$\lambda^{(a,b)} = \lambda^{(a,b)}(k^2h, 2a^2h). \tag{7.74}$$

However, since the coefficients $\{\alpha_j, \beta_j\}$ are irrational, it is easy to see that $\lambda^{(a,b)}$ are not periodic functions in k^2h anymore. To illustrate, the larger of the two $|\lambda|$ values in (7.72) versus k^2h at $2a^2h = 0.1$ is displayed in Fig. 7.5(a, b) for $S_4^{(a,b)}$, respectively. These graphs are qualitatively similar to that of the S_1 method in Fig. 7.3. In these graphs, window sequences of unstable Fourier modes appear near points $k^2h = 0, \pi, 2\pi, \ldots$. The first unstable window is due to modulation instability of constant solutions in the NLS equation (7.47), while all the other windows are induced by numerical instability. As $2a^2h \to 0$, all these windows shrink toward points $0, \pi, 2\pi, \ldots$, and their peaks decrease in a roughly linear manner. Notice that for both $S_4^{(a)}$ and $S_4^{(b)}$, numerically unstable Fourier modes appear when $k^2h \gtrsim \pi$. Since the largest wavenumber in numerical computations is $|k|_{\max} = \pi/\Delta x$ (see (7.63)), a sufficient condition for numerical stability in these S_4 methods is then

$$\frac{\Delta t}{\Delta x^2} \lesssim \frac{1}{\pi}, \tag{7.75}$$

which is about the same as the sufficient condition (7.64) for the S_1 method. If Δt is above the threshold set by (7.75), segments of stable and unstable time steps will mix with each other just like in the S_1 method (see Fig. 7.4). Beside these common qualitative features, notable quantitative differences also exist on numerical instabilities between the $S_4^{(a)}$ and $S_4^{(b)}$ methods. For instance, one can see from Fig. 7.5 that the instability growth rate in $S_4^{(a)}$ can be several times larger than that in $S_4^{(b)}$. This means that at time steps Δt where numerical instabilities appear in both methods, this instability can become visible much earlier in $S_4^{(a)}$ than in $S_4^{(b)}$. Thus, different choices of operators in the S_4 method do make a difference on numerical instabilities.

(3) Numerical Instability for General Initial Conditions

The above stability analysis on split-step methods was performed for simulating the NLS equation with near-constant initial conditions. In practical computations, the initial conditions are rarely of that type, and thus one may wonder whether the above stability results still apply. It turns out that many of these results still hold for arbitrary initial conditions, at least qualitatively. For instance, a sufficient numerical stability condition for the S_1, S_2, and S_4 methods with arbitrary initial conditions is still (7.75). When the time step is above this sufficient stability threshold, segments of stable and unstable time steps will mix as in Fig. 7.4. To illustrate, let us consider two simulations of the NLS equation (7.47) by

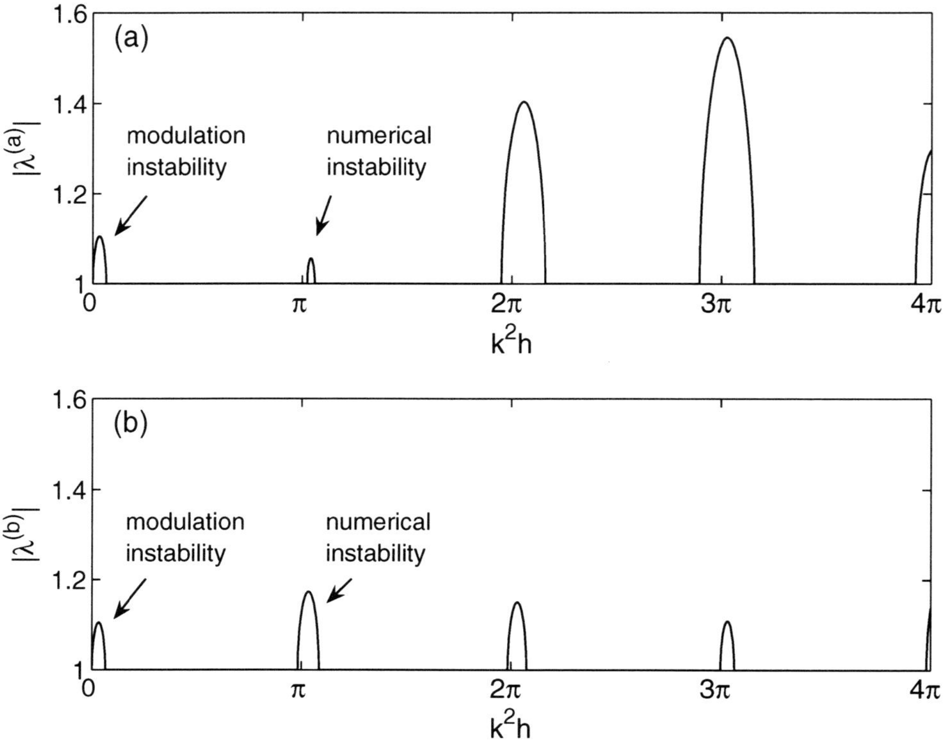

Figure 7.5. *Regions of unstable Fourier modes and their growth rates in the two versions of the fourth-order split-step methods* $S_4^{(a)}$ *and* $S_4^{(b)}$ *as applied to the NLS equation* (7.47) *with near-constant initial conditions* (7.50) *and* $2a^2h = 0.1$.

the $S_4^{(a)}$ method. One is the simulation of a stationary soliton, where $L = 32\pi, N = 1024$, and the initial condition is $u(x,0) = \text{sech}(x)$. The other one is the recurrence simulation in Program 3, where $L = 4\pi, N = 128$, and the initial condition is a periodic function $u(x,0) = \sin^2 x$. At each time-step size Δt, we perform the simulation to a very long time and determine whether the simulation is numerically stable or not (by monitoring the Fourier spectrum of the solution for signs of high-wavenumber instability). The stable regions of time steps thus obtained are displayed in Fig. 7.6(a, b) for the two simulations, respectively. One can see that above the sufficient stability threshold (7.75), which is $\Delta t_c \approx 0.003$ for the discretizations of both simulations, there are many segments of stable Δt values as well, just like in simulations of near-constant initial conditions analyzed above (see Fig. 7.4). Thus this interweaving of stable and unstable time steps above the sufficient stability threshold (7.75) is a generic feature of split-step methods in NLS simulations. Note that in both of

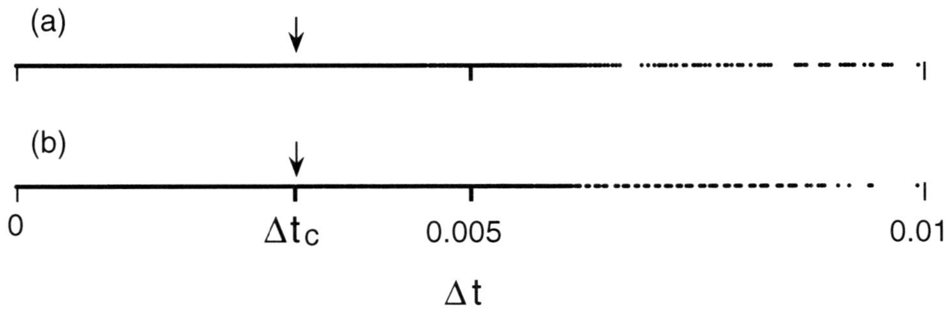

Figure 7.6. *Stable time steps Δt for the fourth-order split-step method $S_4^{(a)}$ applied to two simulations of the NLS equation:* (a) *soliton simulation with $L = 32\pi$, $N = 1024$, and $u(x,0) = \text{sech}\,(x)$;* (b) *recurrence simulation with $L = 4\pi$, $N = 128$, and $u(x,0) = \sin^2 x$ (see Program* 3*). The time step Δt_c marked by arrows is the sufficient stability threshold given by* (7.75).

these two non-near-constant simulations, we have found unstable time steps right above the sufficient stability threshold Δt_c, just like in near-constant simulations (see Fig. 7.4). This means that the threshold (7.75) is also a sharp condition to guarantee numerical stability for non-near-constant simulations. In Fig. 7.6, these unstable time steps near Δt_c are rare, so they are not visible in the figure. The same occurs for the near-constant initial condition in Fig. 7.4 when N is large.

While numerical-stability results obtained from near-constant initial conditions often hold for non-near-constant initial conditions, two main differences do exist. One difference is that in near-constant simulations, the interweaving of stable and unstable time steps above the sufficient stability threshold would disappear when the computational interval is wide enough (because the wavenumber separation $k_0 = 2\pi/L$ then becomes very small, and so some discrete wavenumbers can always hit unstable windows and create instability). But in non-near-constant simulations, this interweaving may persist even for wide computational intervals. For example, in the soliton simulation above, the computational interval is very wide ($L = 32\pi$), but this interweaving phenomenon persists; see Fig. 7.6(a). The reason for this difference is that in near-constant simulations, each Fourier component in the perturbations evolves independently, and each growing mode contains a single Fourier component; see Eq. (7.54) and Fig. 7.3. But in soliton simulations, different Fourier components in the perturbations couple with each other. This can be seen when one linearizes the split-step method around the soliton solution, similar to what was done for the near-constant case earlier in this subsection. Because of this, when perturbations grow, each growing mode has to contain many coupled Fourier components. Thus even if

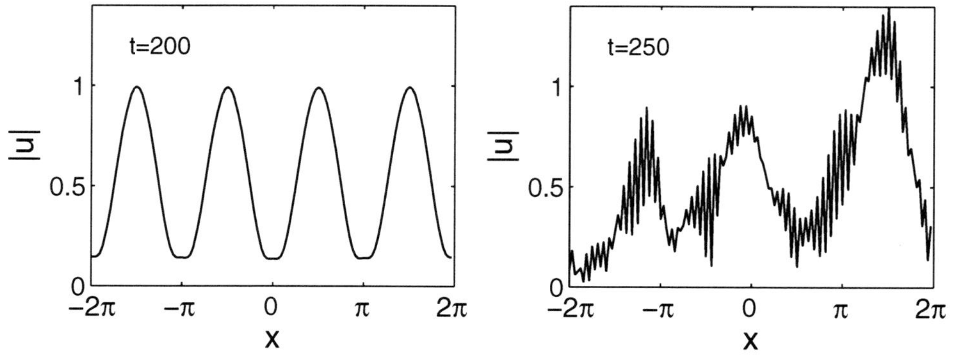

Figure 7.7. *Numerical instability in long-time simulations of the NLS equation by the fourth-order split-step method* $S_4^{(a)}$ *with* $L = 4\pi, N = 128$, $\Delta t = 0.01$, *and a periodic initial condition* $u(x,0) = \sin^2 x$ *(see Program 3). Shown are solutions at times* $t = 200$ *and* 250.

the computational interval is wide (i.e., the wavenumber separation is small), some of these Fourier wavenumbers in the growing mode may still be absent (if they do not fall on the discrete set of Fourier wavenumbers in the computations), and then numerical instability can still be avoided.

The other difference between near-constant and non-near-constant simulations is that in the latter case, numerical instability (if it exists) often develops much more slowly than in the former case. To illustrate, let us consider the recurrence simulation of the NLS equation in Program 3 by the $S_4^{(a)}$ method. At the time step $\Delta t = 0.01$ as used in that program, numerical instability does arise (see Fig. 7.6(b)). For instance, the numerical solution at time $t = 250$ is plotted in Fig. 7.7, where numerical instability (characterized by significant presence of high-wavenumber disturbances) is clearly seen. But this numerical instability, even at time $t = 200$, is still very weak and invisible, as one can see from Fig. 7.7. Thus, numerical instabilities in split-step methods (at unstable values of time steps) are often not a concern except in long-time computations. This explains why, in Program 3, where the simulation ran only to $t = 2$ (see Fig. 7.2(a)), the numerical instability was not an issue at all even though the time step used was unstable. This contrasts the pseudospectral method of the previous subsection, where the numerical instability (for time steps above the stability threshold (7.5)) develops very quickly and is disastrous even for short-time simulations.

In this subsection, numerical instability of the split-step methods was analyzed only for the NLS equation. Similar analysis can also be developed for more general classes of equations such as the generalized NLS equations (with arbitrary forms of nonlinearity) and the coupled NLS equations, and qualitatively similar results would be obtained.

7.1.3 Integrating-Factor Method and Its Numerical Stability

Another spectral method which has less restrictive stability conditions on the time step is the integrating-factor method (Milewski and Tabak (1999), Trefethen (2000), Boyd (2001)). This method is based on the observation that, in the Fourier space, the linear differential part of the equation admits simple analytical solutions. Then one can use this linear solution as an integrating factor and solve the transformed equation which involves only nonlinear terms. Since the linear terms are treated exactly, the resulting nonlinear equation is no longer stiff (or less stiff), and hence larger time steps may be used without triggering numerical instability. For the NLS equation (7.1), its Fourier transform is

$$\hat{u}_t + ik^2\hat{u} - 2i\mathcal{F}(|u|^2u) = 0, \tag{7.76}$$

where $\hat{u} = \mathcal{F}(u)$, $\mathcal{F}$ represents the Fourier transform, and k is the Fourier wavenumber. Multiplying this equation by an integrating factor e^{ik^2t}, we find that (7.76) reduces to a strictly nonlinear equation for $\hat{v} = \hat{u}\, e^{ik^2t}$:

$$\hat{v}_t = 2i\, e^{ik^2t}\mathcal{F}(|u|^2u), \tag{7.77}$$

where $u = \mathcal{F}^{-1}(\hat{v}e^{-ik^2t})$. This nonlinear equation for $\hat{v}$ can then be integrated one time step by a time-stepping scheme of the user's choice. When the solution $\hat{v}$ after one time step is obtained, the corresponding physical solution u is then calculated, and the integration repeats again. Here we choose the fourth-order Runge–Kutta method for time stepping. The MATLAB code of the resulting integrating-factor method is shown below.

Program 4

```
% p4.m: the 4th-order integrating-factor method for solving the
% NLS equation:  iu_t+u_{xx}+2|u|^2u=0.

 L=4*pi; N=128; dt=0.01; tmax=2; nmax=round(tmax/dt);
 dx=L/N; x=[-L/2:dx:L/2-dx]';
 k=[0:N/2-1 -N/2:-1]'*2*pi/L; E=exp(i*k.^2*dt/2); E2 = E.*E;
 u=sin(x).^2; udata=u; tdata=0;
 for nn = 1:nmax                           % integration begins
   v=fft(u);                    dv1=2*i*    fft(u.*u.*conj(u));
   w=ifft((v+dv1*dt/2)./E );    dv2=2*i*E.* fft(w.*w.*conj(w));
   w=ifft((v+dv2*dt/2)./E );    dv3=2*i*E.* fft(w.*w.*conj(w));
   w=ifft((v+dv3*dt   )./E2);   dv4=2*i*E2.*fft(w.*w.*conj(w));
   v=v+(dv1+2*dv2+2*dv3+dv4)*dt/6;   u=ifft(v./E2);
   if mod(nn,round(nmax/25)) == 0
```

```
      udata=[udata u]; tdata=[tdata nn*dt];
    end
  end                                          % integration ends
  waterfall(x, tdata, abs(udata'));            % solution plotting
  colormap([0 0 0]); view(5, 60)
  xlabel('x', 'fontsize', 15)
  ylabel('t', 'fontsize', 15)
  zlabel('|u|', 'fontsize', 15)
  axis([-L/2 L/2 0 tmax 0 2]); grid off
```

The output of this program is the same as Fig. 7.2(a) (except for the axis labels).

For a general wave system in arbitrary spatial dimensions,

$$\mathbf{u}_t = L(\partial_{\mathbf{x}})\mathbf{u} + N(\mathbf{u}, \partial_{\mathbf{x}}), \tag{7.78}$$

where $\mathbf{u}(\mathbf{x}, t)$ is a vector function, L is a linear differential operator with constant coefficients, and N is a nonlinear function of $\mathbf{u}$ and its spatial derivatives, the integrating-factor method is to integrate

$$\widehat{\mathbf{v}}_t = e^{-L(i\mathbf{k})t}\mathcal{F}[N(\mathbf{u}, \partial_{\mathbf{x}})], \tag{7.79}$$

where $\widehat{\mathbf{v}} = e^{-L(i\mathbf{k})t}\mathcal{F}(\mathbf{u})$, $\mathcal{F}(\cdot)$ represents the Fourier transform, and $\mathbf{k}$ is the wavenumber.

Regarding the accuracy of the integrating-factor method, its spatial accuracy is spectral if spatial differentiation is performed by the Fourier transform. The temporal accuracy is determined by the time-stepping scheme the user chooses to integrate Eq. (7.79). In Program 4, the fourth-order Runge–Kutta method was used, and thus its temporal accuracy is fourth order. To illustrate, the errors of the solution from Program 4 with different time steps Δt are displayed in Fig. 7.2(b). It is seen that the error decreases in proportion to Δt^4, confirming the fourth-order temporal accuracy. This figure also shows that at the same time-step value, the integrating-factor method gives more accurate results than the split-step method.

Like the split-step method, the integrating-factor method also suffers numerical instability for certain time-step sizes. For instance, in Program 4, if we take $\Delta t = 0.015$ and $t = 80$ (with other parameters the same), we will observe numerical instability. Below we analyze this numerical instability in the integrating-factor method.

Numerical Stability of the Integrating-Factor Method

Now we study the numerical stability in the integrating-factor method. We will show that for NLS-type equations, their numerical-stability properties qualitatively resemble those of the split-step method. In other words, when the time-step size is below a certain threshold, this method is numerically stable. When the time-step size is above this threshold, intervals of stable and unstable time steps will intertwine in a complicated way. In addition, the strength of numerical instability in this method is comparable to that in the split-step method and is much weaker than that in the pseudospectral method. The procedure of our analysis follows that in Sec. 7.1.2 for the split-step method.

Below, we examine numerical instability of the integrating-factor method on the NLS equation. Here the fourth-order Runge–Kutta time-stepping scheme will be used as in Program 4. To analytically investigate this question, we consider computation of the NLS equation (7.47) with near-constant initial conditions (7.50). Inserting the perturbed solution (7.50) into the integrating-factor method (7.77) with fourth-order Runge–Kutta time stepping, and linearizing it around the constant solution, we obtain the iteration equation for the perturbation as

$$\Psi_{n+1} = \hat{I}_4 \Psi_n. \tag{7.80}$$

Here $\Psi_n = \mathcal{F}([\tilde{u}_n, \tilde{u}_n^*]^T)$, $\mathcal{F}(\cdot)$ represents the Fourier transform, the iteration matrix $\hat{I}_4$ is

$$\hat{I}_4 = \widehat{M}(h)\left[1 + \frac{1}{6}h(K_1 + 2K_2 + 2K_3 + K_4)\right], \tag{7.81}$$

$$\widehat{M}(h) = \mathrm{diag}\,(e^{-ik^2h}, e^{ik^2h}), \tag{7.82}$$

$$K_1 = 2ia^2 Q(0), \tag{7.83}$$

$$K_2 = 2ia^2 Q\left(\frac{1}{2}h\right)\left(1 + \frac{1}{2}K_1 h\right), \tag{7.84}$$

$$K_3 = 2ia^2 Q\left(\frac{1}{2}h\right)\left(1 + \frac{1}{2}K_2 h\right), \tag{7.85}$$

$$K_4 = 2ia^2 Q(h)(1 + K_3 h), \tag{7.86}$$

$$Q(t) \equiv \begin{bmatrix} 1 & e^{2ik^2t} \\ -e^{-2ik^2t} & -1 \end{bmatrix}, \tag{7.87}$$

k is the Fourier wavenumber, and h is the time-step size. It is easy to see that the iteration matrix $\hat{I}_4$ is a function of k^2h and $2a^2h$, i.e.,

$$\hat{I}_4 = \hat{I}_4(k^2h, 2a^2h). \tag{7.88}$$

In addition, $\hat{I}_4$ is 2π-periodic in k^2h. Thus eigenvalues of $\hat{I}_4$ are also functions of k^2h and $2a^2h$, and 2π-periodic in k^2h. To illustrate these eigenvalues, their graph versus k^2h is displayed in Fig. 7.8 with $2a^2h = 0.1$. As in the split-step methods (see Figs. 7.3 and 7.5), this graph exhibits a sequence of unstable windows at the right side of points $k^2h = n\pi$, $n = 0, 1, 2, \ldots$. The first unstable window (with $n = 0$) is due to modulation instability, while the other windows are due to numerical instability. Thus a sufficient condition for numerical stability in this fourth-order integrating-factor method is that $|k|^2_{\max} h < \pi$, i.e.,

$$\frac{\Delta t}{\Delta x^2} < \frac{1}{\pi}, \tag{7.89}$$

which is the same as that in the split-step methods S_1, S_2, and S_4. The necessary and sufficient condition for numerical stability is that none of the discrete Fourier wavenumbers

$$k = 0,\ \pm k_0,\ \pm 2k_0, \ldots,\ \pm\frac{1}{2}Nk_0 \tag{7.90}$$

arising in numerical computations falls in these numerical-instability windows. Here $k_0 = 2\pi/L$, and L is the length of the computational domain. For time steps Δt above this sufficient stability threshold (7.89), since these discrete Fourier wavenumbers (7.90) may miss the numerical-instability windows (see Fig. 7.8), there would exist stable segments of time steps which are intertwined with unstable segments of time steps, just like in the split-step methods (see Fig. 7.4). This mixing of stable and unstable time steps above the sufficient stability threshold (7.89) holds not only for simulations of near-constant initial conditions, but also for simulations of non-near-constant initial conditions. To illustrate, we consider the same two simulations of the NLS equation as in Fig. 7.6. One is the simulation of a stationary soliton, with $L = 32\pi, N = 1024$, and an initial condition $u(x,0) = \mathrm{sech}(x)$, and the other one is the recurrence simulation in Program 4, with $L = 4\pi, N = 128$, and an initial condition $u(x,0) = \sin^2 x$. Stable time steps for these two simulations are displayed in Fig. 7.9(a, b). One can see that in both cases many stable segments of time steps appear above the sufficient stability threshold (7.75), just as we expected. However, in both simulations, unstable time steps appear right above the sufficient stability threshold (7.89), meaning that the threshold (7.89) is a sharp condition to guarantee numerical stability in this integrating-factor method. All these numerical-stability behaviors are qualitatively very similar to those of the split-step methods in the previous subsection. Some quantitative differences do exist between the two methods, though. For example, when Fig. 7.9 is compared to Fig. 7.6, one can see that the stable ranges of time steps in this fourth-order integrating-factor method are wider than those in $S_4^{(a)}$, meaning that this integrating-factor method has better numerical-stability properties than the split-step method $S_4^{(a)}$ (and $S_4^{(b)}$ as well). This better stability behavior of the integrating-factor method can be understood by comparing the unstable windows of Fourier modes in Fig. 7.8 with those in Fig. 7.5 for the split-step methods. One can see that the unstable windows in the integrating-factor method are narrower than their counterparts in the split-step methods; thus stable ranges of time steps Δt in the integrating-factor method are wider.

In this subsection, numerical instability of the integrating-factor method was analyzed for the NLS equation. Similar analysis can be developed for general NLS-type equations,

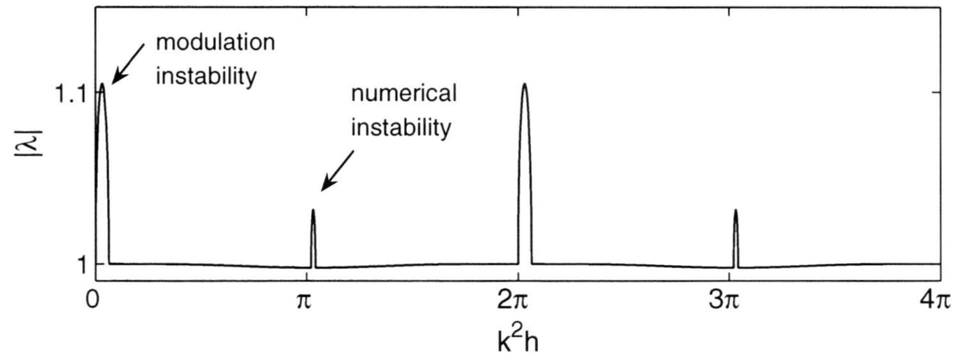

Figure 7.8. *Regions of unstable Fourier modes and their growth rates in the fourth-order integrating-factor method as applied to the NLS equation* (7.47) *with near-constant initial conditions* (7.50) *and* $2a^2h = 0.1$.

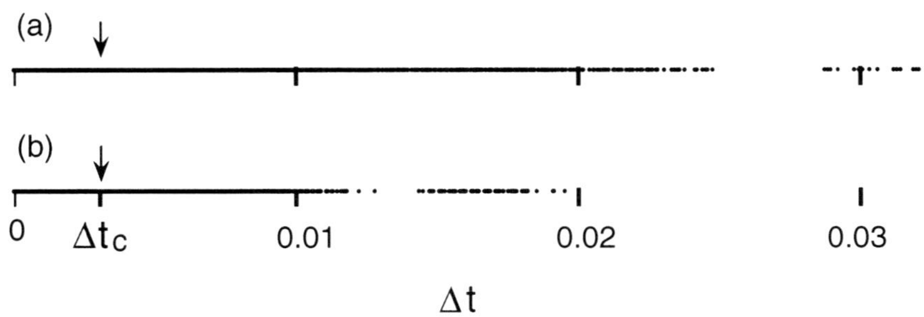

Figure 7.9. *Stable time steps* Δt *for the fourth-order integrating-factor method applied to two simulations of the NLS equation:* (a) *soliton simulation with* $L = 32\pi, N = 1024$, *and* $u(x,0) = \text{sech}(x)$*;* (b) *recurrence simulation with* $L = 4\pi, N = 128$, *and* $u(x,0) = \sin^2 x$ *(see Program* 4*). The time step* Δt_c *marked by arrows is the sufficient stability threshold given by* (7.89).

and qualitatively similar results can be obtained. This analysis can also be developed for other types of equations (such as KdV-type equations), but this has not been done yet.

Before concluding this section on evolution simulations, we should mention damping boundary conditions for simulating wave evolution in an infinite spatial interval. Recall that in all our sample MATLAB codes for the simulation of the NLS equation, the Fourier transform was used. This means that we assumed the solution to be spatially periodic, with the period being the length of the numerical box (i.e., parameter L in the codes). However, in many initial-value simulations, the spatial interval is infinite. If one starts with a localized initial condition, the solution generally emits energy radiation which escapes to far distances. In such cases, if we truncate the infinite domain to a finite numerical interval and assume the periodic boundary condition, the energy radiation which is supposed to escape to far distances will reenter the computational box and corrupt the true solution inside. To overcome this problem, a common practice is to impose a damping boundary condition, so that when the wave travels to the edge of the computational box, it is damped out. Since the damped solution becomes zero (or approximately zero) near the boundaries, the periodic boundary condition can still be used, and thus the sample MATLAB codes above will still work. The simplest way to implement the damping boundary condition is that, after each time step of simulation, the solution is multiplied by a factor $e^{-D(\mathbf{x})h}$, where $D(\mathbf{x})$ is a nonnegative damping function, and h is the time step. The damping function $D(\mathbf{x})$ is generally chosen so that its value is zero in the interior of the numerical box (hence not affecting the interior solution), but increases smoothly from zero to a certain positive value in a buffer zone near edges of the numerical box. The solution in the buffer zone then is damped out. The damping strength is tuned by the height of the damping function $D(\mathbf{x})$ and should be set properly according to the underlying problem. One common mistake in choosing the damping function is that this function be made to rise too rapidly inside the buffer zone (out of a desire to damp out the solution quickly in a narrow buffer region). When radiation hits such a rapidly rising damping function, a significant portion of the waves will be reflected back to the interior, hence corrupting the accuracy of the interior solution. Thus the damping function should rise slowly in the buffer zone instead. In this case, the reflected radiation is believed to be exponentially small in ϵ, where ϵ^{-1} is the length scale of the damping function (see Boyd (2001, page 341)). But a rigorous justification of this exponentially small reflection by a slowly varying damping function seems to be still lacking to the author's knowledge.

7.2 Numerical Methods for Computations of Solitary Waves

In the study of nonlinear wave phenomena, solitary waves play an important role. These waves are special localized solutions of nonlinear wave equations which maintain their shapes upon propagation. If these waves are stable, they can become a nonvanishing part in the long-time solution and may even dominate the long-time solution (solitons in the NLS and KdV equations are examples). Even if they are unstable, they may still play important roles in the solution dynamics. For instance, the ground-state solitary waves in the 2D NLS equation are important in the collapse theory of that equation (see Sec. 5.10). So far, a number of numerical methods have been developed for the computation of solitary waves.

One is the shooting method. This method is efficient for 1D problems, but does not apply in higher dimensions. The Newton's method is also available (see Boyd (2001), for instance), but its traditional implementation takes much effort in higher dimensions, and may also run into difficulties at times (Boyd (2007)). In this section, we describe several other iteration methods which are efficient and also simple to implement in any dimensions.

7.2.1 Petviashvili Method

The first method we present is the Petviashvili method. This method was proposed for constant-coefficient equations with power-law nonlinearity by Petviashvili (1976). It is based on the fixed-point iteration idea, but with the key improvement of introducing a stabilizing factor. Convergence conditions of this method for constant-coefficient equations with power-law nonlinearity were obtained by Pelinovsky and Stepanyants (2004). Extensions of this method to more general wave equations were given by Musslimani and Yang (2004), Ablowitz and Musslimani (2005), and Lakoba and Yang (2007a). These Petviashvili-type methods often converge fast and are easy to implement. However, they can only converge to the ground states of nonlinear wave equations and would diverge for excited states. For vector equations, they may even diverge for the ground states.

Below we use the multidimensional NLS equation with power-law nonlinearity

$$iU_t + \nabla^2 U + |U|^{p-1} U = 0 \tag{7.91}$$

to present this method. Here $p > 1$, and ∇^2 is the multidimensional spatial Laplacian operator. This equation admits a ground state solution

$$U(\mathbf{x}, t) = u(\mathbf{x}) e^{i\mu t}, \tag{7.92}$$

where $u(\mathbf{x}) > 0$, and $\mu > 0$ is the propagation constant. This ground state satisfies the equation

$$\nabla^2 u + u^p = \mu u, \tag{7.93}$$

or

$$Mu = u^p, \tag{7.94}$$

where $M = \mu - \nabla^2$. This equation can be rewritten as

$$u = M^{-1} u^p. \tag{7.95}$$

To determine this ground state, if one directly iterates Eq. (7.95), the iterative solution will diverge to zero or infinity. The basic idea of the Petviashvili method is to introduce a stabilizing factor into this fixed-point iteration. The purpose of this stabilizing factor is to pump up the iteration function when it decays, and to suppress it when it grows. Then the Petviashvili method for Eq. (7.93) is

$$u_{n+1} = S_n^\gamma M^{-1} u_n^p. \tag{7.96}$$

Here S_n^γ is the stabilizing factor, where

$$S_n = \frac{\langle u_n, M u_n \rangle}{\langle u_n, u_n^p \rangle}, \tag{7.97}$$

the inner product $\langle \cdot, \cdot \rangle$ is the standard one in the space of square-integrable functions,

$$\langle \mathbf{f}, \mathbf{g} \rangle = \int_{-\infty}^{\infty} \mathbf{f}^\dagger(\mathbf{x})\, \mathbf{g}(\mathbf{x})\, d\mathbf{x}, \tag{7.98}$$

the superscript "†" represents the Hermitian, and γ is a constant. This constant γ can be chosen heuristically as follows. Suppose the exact solution $u(\mathbf{x}) = O(1)$. After some iterations, if the iteration function becomes $u_n(\mathbf{x}) = O(\epsilon)$, where $\epsilon \ll 1$ or $\epsilon \gg 1$, then we see from (7.97) that $S_n = O(\epsilon^{1-p})$, and thus $u_{n+1} = O(\epsilon^{(1-p)\gamma+p})$ from (7.96). In order for the iteration function u_{n+1} to become $O(1)$, we require

$$(1-p)\gamma + p = 0. \tag{7.99}$$

Thus γ should be taken as

$$\gamma = \frac{p}{p-1}. \tag{7.100}$$

We will show below that this γ value is indeed optimal.

Regarding the convergence of this Petviashvili method (7.96) for Eq. (7.93), we have the following theorem.

Theorem 7.1. *For the ground state $u(\mathbf{x})$ in Eq.* (7.93), *if the initial condition is sufficiently close to the exact solution $u(\mathbf{x})$, then the Petviashvili method* (7.96) *converges if and only if $1 < \gamma < \frac{p+1}{p-1}$. The fastest convergence occurs when $\gamma = \gamma_{opt} \equiv \frac{p}{p-1}$.*

This theorem is a special case of that considered in Pelinovsky and Stepanyants (2004). We will prove it by a method which is a modification of that used in Pelinovsky and Stepanyants (2004).

Proof. By assumption, the initial condition is close to the exact solution, thus we can use the linearization method to prove this theorem. Let

$$u_n(\mathbf{x}) = u(\mathbf{x}) + \tilde{u}_n(\mathbf{x}), \qquad \tilde{u}_n \ll u, \tag{7.101}$$

where $\tilde{u}_n$ is the error term. Inserting it into the Petviashvili method (7.96) and neglecting terms of $O(\tilde{u}_n^2)$ and higher, we obtain the linear iteration equation for the error $\tilde{u}_n$ as

$$\tilde{u}_{n+1} = (1+\mathcal{L})\tilde{u}_n, \tag{7.102}$$

where the iteration operator $\mathcal{L}$ is

$$\mathcal{L}\Psi = M^{-1}L_1\Psi - \gamma\frac{\langle u, L_1\Psi\rangle}{\langle u, Mu\rangle}u, \tag{7.103}$$

and

$$L_1 = -M + p\,u^{p-1} = \nabla^2 - \mu + p\,u^{p-1} \tag{7.104}$$

is the linearization operator of the stationary ground-state equation (7.93). This L_1 operator is the counterpart of operator (5.37) for the generalized NLS equations (5.31). The error $\tilde{u}_n$ from iterations (7.102) would decay (hence the Petviashvili method (7.96) would converge) if and only if

1. the kernel of $\mathcal{L}$ contains only translation-invariance eigenfunctions u_{x_j}; these eigenmodes cause a position shift to the solitary wave and do not affect convergence of the iterations;
2. all nonzero eigenvalues λ of $\mathcal{L}$ satisfy the condition $|1+\lambda| < 1$, i.e., $-2 < \lambda < 0$.

In order to determine whether these conditions are satisfied, we need to examine the eigenvalues of $\mathcal{L}$. We will show that eigenvalues of $\mathcal{L}$ are the same as eigenvalues of $M^{-1}L_1$, except that the single positive eigenvalue of $M^{-1}L_1$ is shifted in $\mathcal{L}$. Thus we first analyze eigenvalues of $M^{-1}L_1$.

First of all, $M^{-1}L_1$ has no continuous spectrum because $M^{-1}L_1 \to -1$ when $|\mathbf{x}| \to \infty$. Second, notice that M is positive definite, and L_1 is Hermitian. According to the *Sylvester inertia law* (see Horn and Johnson (1991, Theorem 7.6.3)), if A is a positive-definite matrix, and B a Hermitian matrix, then all eigenvalues of the product matrix AB are real, and all eigenvectors of AB form a complete set. In addition, the matrix AB has the same number of positive, negative, and zero eigenvalues as B. Since linear operators can be treated as matrices (through discretization for instance), the Sylvester inertia law then applies to linear operators as well. In view of this law, all eigenvalues of $M^{-1}L_1$ are real, and the number of its positive eigenvalues is the same as that of L_1. Let us list the eigenvalues of $M^{-1}L_1$ in decreasing order:

$$\Lambda_1 > \Lambda_2 > \Lambda_3 > \cdots. \tag{7.105}$$

Notice that for the ground state $u(\mathbf{x}) > 0$, the Schrödinger operator L_1 has a unique positive eigenvalue (Weinstein (1986, Proposition 4.2)). Thus $M^{-1}L_1$ also has a single positive eigenvalue $\Lambda_1 > 0$. Indeed, from Eqs. (7.94) and (7.104), we see that

$$M^{-1}L_1 u = (p-1)u. \tag{7.106}$$

Thus the largest eigenvalue Λ_1 of $M^{-1}L_1$ is

$$\Lambda_1 = p - 1 > 0. \tag{7.107}$$

For operator L_1, its second largest eigenvalue is zero, since $L_1 u_{x_j} = 0$ when differentiating Eq. (7.93) with respect to x_j. Thus the second largest eigenvalue of $M^{-1}L_1$ is also zero, i.e., $\Lambda_2 = 0$. All the other eigenvalues $\Lambda_3, \Lambda_4, \ldots$ are negative. These negative eigenvalues have a lower bound. To see this, suppose the eigenfunction of eigenvalue Λ_k is ψ_k ($k = 1, 2, \ldots$), i.e.,

$$M^{-1}L_1\psi_k = \Lambda_k \psi_k. \tag{7.108}$$

This eigenvalue equation can be rewritten as

$$M\psi_k = \frac{p\,u^{p-1}}{1+\Lambda_k}\psi_k. \tag{7.109}$$

Taking the inner product of this equation with ψ_k, we get

$$\langle \psi_k, M\psi_k\rangle = \frac{p}{1+\Lambda_k}\langle \psi_k, u^{p-1}\psi_k\rangle. \tag{7.110}$$

Since M is positive definite and $u(\mathbf{x}) > 0$, we see from this equation that

$$\Lambda_k > -1 \tag{7.111}$$

for all eigenvalues of $M^{-1}L_1$. Thus $\Lambda_{\min} = -1$ is the lower bound of $M^{-1}L_1$'s eigenvalues.

Now we determine eigenvalues of the iteration operator $\mathcal{L}$. This operator does not have continuous spectrum either since $\mathcal{L} \to -1$ when $|\mathbf{x}| \to \infty$. Below we show that $\mathcal{L}$'s discrete eigenvalues are simply

$$\lambda = (1-\gamma)\Lambda_1,\ \Lambda_2,\ \Lambda_3,\ \Lambda_4, \ldots, \tag{7.112}$$

where $\Lambda_1, \Lambda_2, \ldots$ are $M^{-1}L_1$'s eigenvalues (see the illustration in Fig. 7.10). To prove this, we notice that any two eigenfunctions ψ_j and ψ_k of $M^{-1}L_1$ satisfy the orthogonality relation

$$\langle M\psi_j,\ \psi_k\rangle = \delta_{jk}, \tag{7.113}$$

where δ_{jk} is 1 or 0, depending on whether or not $j = k$. This orthogonality relation can be easily proved by writing the eigenvalue equations of $M^{-1}L_1$ for ψ_j and ψ_k as $L_1\psi_j = \Lambda_j M\psi_j$ and $L_1\psi_k = \Lambda_k M\psi_k$, then taking the inner products of these equations with ψ_k and ψ_j, respectively, and utilizing the Hermitian property of L_1. Due to these orthogonality

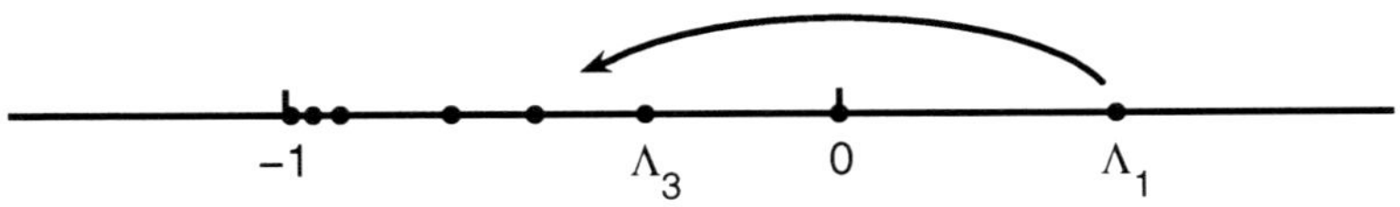

Figure 7.10. *Schematic illustration on spectra of operators $M^{-1}L_1$ and $\mathcal{L}$. The spectrum of $\mathcal{L}$ is the same as that of $M^{-1}L_1$, except that the positive eigenvalue Λ_1 of $M^{-1}L_1$ is shifted to $(1-\gamma)\Lambda_1$ in $\mathcal{L}$.*

relations and the fact that the eigenfunction of $M^{-1}L_1$ for eigenvalue Λ_1 is $\psi_1 = u(\mathbf{x})$, see (7.106), we get

$$\langle Mu, \psi_k \rangle = 0, \quad k = 2, 3, \ldots. \tag{7.114}$$

Utilizing this relation as well as

$$\langle u, L_1 \psi \rangle = \langle Mu,\, M^{-1} L_1 \psi \rangle, \tag{7.115}$$

we find that

$$\mathcal{L}\psi_k = \Lambda_k \psi_k, \quad k = 2, 3, \ldots. \tag{7.116}$$

In addition, from Eq. (7.106), we directly obtain

$$\mathcal{L}u = (1-\gamma)\Lambda_1 u. \tag{7.117}$$

Thus eigenvalues of $\mathcal{L}$ are given in (7.112). According to the Sylvester inertia law, eigenfunctions of $M^{-1}L_1$ form a complete set, and hence all $\mathcal{L}$'s eigenvalues are contained in (7.112).

From the above analysis, we also see that the kernel of $\mathcal{L}$ is the same as the kernel of $M^{-1}L_1$, which in turn is the same as the kernel of L_1. It is known that the kernel of L_1 contains only translation-invariance modes u_{x_j}; thus the kernel of $\mathcal{L}$ also contains only these invariance modes.

Summarizing the above results, we see that the Petviashvili method (7.96) will converge if and only if

$$-2 < (1-\gamma)\Lambda_1 < 0, \tag{7.118}$$

i.e.,

$$1 < \gamma < \frac{p+1}{p-1}. \tag{7.119}$$

When it converges, the error $\tilde{u}_n$ decays geometrically as R^n, where the convergence factor R is given by

$$R = \max\{|1 + (1-\gamma)\Lambda_1|,\ 1 + \Lambda_3\}. \tag{7.120}$$

When $\gamma = \gamma_{opt} = p/(p-1)$, $R = R_{\min} = 1 + \Lambda_3$ is minimal, thus the convergence is the fastest. This ends the proof of Theorem 7.1. □

We would like to point out that, for the Petviashvili method, the minimum of R in Eq. (7.120) is actually reached when γ falls anywhere in the interval

$$1 - \frac{\Lambda_3}{p-1} \leq \gamma \leq 1 + \frac{\Lambda_3 + 2}{p-1} \tag{7.121}$$

with $\gamma_{opt} = p/(p-1)$ lying in its center. This means that optimal convergence of the Petviashvili method occurs for γ anywhere in the above interval.

One attractive feature of the Petviashvili method is that it is easy to implement in arbitrary spatial dimensions. To illustrate, let us consider the NLS equation (7.93) in two dimensions with $p = 3$:

$$u_{xx} + u_{yy} + u^3 = \mu u. \tag{7.122}$$

The MATLAB code of the Petviashvili method for the ground state in this equation with $\mu = 1$ is displayed below.

Program 5

```
% p5.m: Petviashvili method for computing the ground state in the
% 2D NLS equation: u_{xx}+u_{yy}+u^3=u.

 Lx=30; Ly=30; Nx=128; Ny=128; maxerror=1e-10; nmax=300;
 x=-Lx/2:Lx/Nx:(Lx/2-Lx/Nx); y=-Ly/2:Ly/Ny:(Ly/2-Ly/Ny);
 kx=[0:Nx/2-1 -Nx/2:-1]*2*pi/Lx; ky=[0:Ny/2-1  -Ny/2:-1]*2*pi/Ly;
 [X,Y]=meshgrid(x,y); [KX,KY]=meshgrid(kx,ky); K2=KX.^2+KY.^2;
 p=3; gamma=p/(p-1);
 mu=1; u=2.2*exp(-(X.^2+Y.^2));
 for nn=1:nmax
    Mu=ifft2((mu+K2).*fft2(u));
    u3=u.*u.*u;
    L0u=u3-Mu;
    uerror(nn)=max(max(abs(L0u)));
    if uerror(nn) < maxerror
        break
    end
    S=sum(sum(u.*Mu))/sum(sum(u.*u3));
    u=S^gamma*ifft2(fft2(u3)./(mu+K2));
 end
 subplot(221)
```

```
mm=36:94; mesh(x(mm), y(mm), u(mm,mm));
axis([x(mm(1)) x(mm(end)) y(mm(1)) y(mm(end)) 0 2]);
xlabel('x'); ylabel('y'); zlabel('|u|'); title('(a)');
subplot(222)
semilogy(0:nn-1, uerror, 'linewidth', 2);
axis([0 50 1e-11 10]);
xlabel('number of iterations'); ylabel('error'); title('(b)')
```

The output of this program is shown in Fig. 7.11. It is seen that the error of the solution drops below 10^{-10} after only 46 iterations. The program takes about 0.5 seconds on a PC with a 2 GHz processor.

The above Petviashvili method and its convergence conditions were formulated for the NLS equation with power nonlinearity (7.93). This method and its convergence conditions can be readily extended to a more general class of solitary wave equations with power nonlinearity,

$$Mu = u^p, \tag{7.123}$$

where M is a linear self-adjoint and positive-definite differential operator with constant coefficients (when $M = \mu - \nabla^2$, (7.123) reduces to (7.93)). For this more general equation, the Petviashvili method is the same as (7.96). Its linearization operator L_1 is still (7.104), i.e., $L_1 = -M + p\, u^{p-1}$. Assuming that the kernel of L_1 contains only the translation-invariance eigenmodes u_{x_j}, and the initial condition is sufficiently close to the exact solution, then the

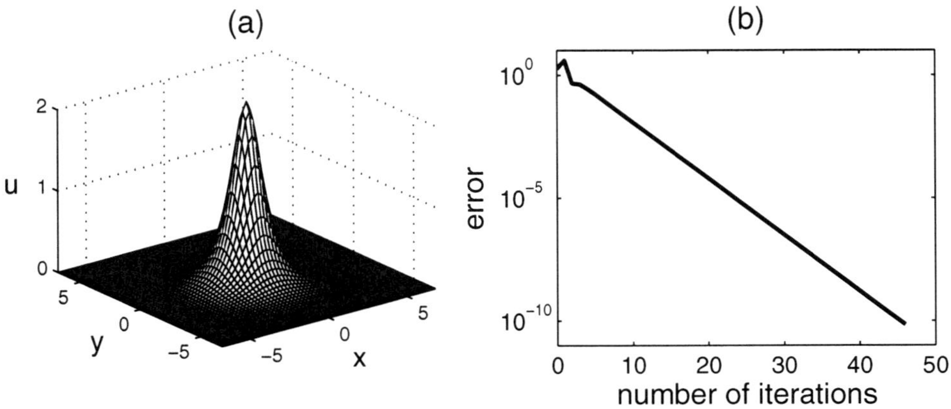

Figure 7.11. *Output of Program 5 on computation of the ground state in the 2D NLS equation (with $\mu = 1$) by the Petviashvili method:* (a) *profile of the ground state;* (b) *error diagram of Petviashvili iterations.*

Petviashvili method (7.96) converges if and only if the following three conditions are all met (Pelinovsky and Stepanyants (2004)):

1. the linearization operator L_1 contains a single positive eigenvalue;
2. $1 < \gamma < \frac{p+1}{p-1}$;
3. either $u^{p-1}(\mathbf{x}) \geq 0$ for all $\mathbf{x}$, or eigenvalues of $M^{-1}L_1$ are all larger than -2.

The first condition usually means that the solitary wave $u(\mathbf{x})$ should be a ground state. The second condition ensures that the single positive eigenvalue of $M^{-1}L_1$ is shifted to the interval $(-2,0)$ in the iteration operator $\mathcal{L}$. The third condition ensures that all eigenvalues of the iteration operator $\mathcal{L}$ are larger than -2. These convergence conditions for the general equation (7.123) can be easily proved by minor modifications to the proof of Theorem 7.1 and are thus omitted. For the ground states in the NLS equation (7.93), the first and third conditions are already satisfied; thus the Petviashvili method converges if and only if the second condition is met (see Theorem 7.1).

The Petviashvili method has been further extended to equations with a more general type of nonlinearity and potential as well as to vector equations (Musslimani and Yang (2004), Ablowitz and Musslimani (2005), Lakoba and Yang (2007a)). These extensions are less trivial. The extended methods are often found to converge to the ground states of the wave system, and diverge for the excited states, similar to the original Petviashvili method. But the precise convergence conditions of these extended methods are not known.

7.2.2 Accelerated Imaginary-Time Evolution Method

Another method to compute solitary waves is the imaginary-time evolution method. In this method, one seeks solitary waves by turning the solitary wave equation into a fictitious time evolution equation of diffusive type, and normalizing the solution after each step of time integration to maintain a fixed power. If the solution evolves into a steady state, then a solitary wave would be obtained. The idea of this method as applied to linear equations is well known (see Koonin (1986, Sec. 7.4), for instance). In the past decade, this idea has also been applied to nonlinear equations in the physics community (see Chiofalo et al. (2000), for instance). This method, in its original form, is quite slow. Recently, an acceleration technique was introduced into this method (García-Ripoll and Pérez-García (2001), Yang and Lakoba (2008)). The accelerated imaginary-time evolution method is much faster. This method also has an interesting property, which is that its convergence is directly connected to the linear stability of the solitary wave if the solitary wave is sign definite (i.e., does not cross zeros); see Yang and Lakoba (2008). Thus both the solitary wave and its linear stability can be determined simultaneously by these same imaginary-time iterations.

For the ease of presentation, we consider multidimensional NLS equations with a general type of nonlinearity and potential

$$iU_t + \nabla^2 U + F(|U|^2, \mathbf{x})U = 0, \tag{7.124}$$

where $F(\cdot,\cdot)$ is a real-valued function. Solitary waves of Eq. (7.124) are sought in the form

$$U(\mathbf{x},t) = u(\mathbf{x})e^{i\mu t}, \tag{7.125}$$

where $u(\mathbf{x})$ is a real-valued localized function, and μ is the propagation constant. The function $u(\mathbf{x})$ satisfies the equation

$$L_{00}u = \mu u, \tag{7.126}$$

where

$$L_{00} \equiv \nabla^2 + F(u^2, \mathbf{x}). \tag{7.127}$$

In the original imaginary-time evolution method (ITEM), one numerically integrates the diffusive equation

$$u_t = L_{00}u, \tag{7.128}$$

which is obtained from the dispersive equation (7.124) by replacing t with $-it$ (hence the name "imaginary time"), and then normalizes the solution after each step of time integration to maintain a fixed power. The power P of the solitary wave $u(\mathbf{x})$ is defined as

$$P(\mu) = \int_{-\infty}^{\infty} u^2(\mathbf{x};\mu)d\mathbf{x}. \tag{7.129}$$

The simplest implementation of numerical time integration on Eq. (7.128) is to use the Euler method, whereby the ITEM scheme is

$$u_{n+1} = \left[\frac{P}{\langle \hat{u}_{n+1}, \hat{u}_{n+1}\rangle}\right]^{\frac{1}{2}} \hat{u}_{n+1} \tag{7.130}$$

and

$$\hat{u}_{n+1} = u_n + [L_{00}u]_{u=u_n}\Delta t. \tag{7.131}$$

Here u_n is the solution after n iterations, and Δt is the step size of time integration which is an auxiliary parameter of the scheme. If iterations (7.130)–(7.131) converge to a solitary wave $u(\mathbf{x})$, then it is easy to see that this $u(\mathbf{x})$ would satisfy Eq. (7.126), with its power being P and the propagation constant μ being equal to

$$\mu = \frac{1}{P}\langle u, L_{00}u\rangle. \tag{7.132}$$

This original ITEM scheme is very slow, because the time integration of the diffusive equation (7.128) by the Euler method suffers stringent stability restrictions on the step size Δt. To overcome this difficulty, one idea is to use implicit time-stepping methods (such as the backward Euler method) to integrate the imaginary-time equation (7.128). Another idea is to introduce a preconditioning operator to Eq. (7.128), which results in an accelerated but still explicit scheme (Yang and Lakoba (2008)). This scheme will be described below.

In our accelerated ITEM, instead of evolving the original imaginary-time equation (7.128), we evolve the following preconditioned imaginary-time equation

$$u_t = M^{-1}[L_{00}u - \mu u], \tag{7.133}$$

where M is a positive-definite and self-adjoint preconditioning (or acceleration) operator. The stationary solution of this equation is still $u(\mathbf{x})$. Applying the Euler method to this new equation, the accelerated imaginary-time evolution method (AITEM) is

$$u_{n+1} = \left[\frac{P}{\langle \hat{u}_{n+1}, \hat{u}_{n+1}\rangle}\right]^{\frac{1}{2}} \hat{u}_{n+1}, \tag{7.134}$$

$$\hat{u}_{n+1} = u_n + M^{-1}(L_{00}u - \mu u)_{u=u_n,\ \mu=\mu_n}\Delta t, \tag{7.135}$$

and

$$\mu_n = \left.\frac{\langle M^{-1}u,\ L_{00}u\rangle}{\langle M^{-1}u,\ u\rangle}\right|_{u=u_n}. \tag{7.136}$$

Here P is the power defined in (7.129), which is prefixed. Notice that our updating formula (7.136) for μ_n is different from the usual formula (7.132). This special updating formula enables us to derive the convergence conditions of the above AITEM. If M is the identity operator, then the above scheme is closely related to the original ITEM (7.130)–(7.131), and both exhibit very slow convergence. But if M takes other sensible choices, convergence of the AITEM will be much faster.

To demonstrate the convergence speed of the AITEM, we consider the 2D NLS equation with a periodic potential

$$iU_t + U_{xx} + U_{yy} - V_0\left(\sin^2 x + \sin^2 y\right)U + |U|^2 U = 0, \tag{7.137}$$

which models nonlinear light propagation as well as Bose–Einstein condensate dynamics in optical lattices under self-focusing nonlinearity (see Chapter 6). This model admits a family of on-site lattice solitons in the form $U(x,y,t) = u(x,y)e^{i\mu t}$, where $u(x,y) > 0$ and satisfies the equation

$$u_{xx} + u_{yy} - V_0\left(\sin^2 x + \sin^2 y\right)u + u^3 = \mu u, \tag{7.138}$$

and μ is the propagation constant which lies in the semi-infinite gap of the underlying linear system. Note that the propagation constant μ here is related to ω of Chapter 6 by $\mu = -\omega$; see Eqs. (6.5) and (6.40). When $V_0 = 6$, the power curve of this family is shown in Fig. 7.12(a). At $\mu = -3.7$, the corresponding soliton is displayed in Fig. 7.12(b), and its power value is $P = 1.9092$. This solution can be computed by the AITEM. To do so, we take

$$M = c - \nabla^2, \tag{7.139}$$

where c is a positive constant. With parameter choices $c = 2$ and $\Delta t = 0.9$, the MATLAB code of the AITEM is displayed below.

Program 6

```
% p6.m: AITEM for computing lattice solitons in the 2D NLS equation
% u_{xx}+u_{yy}-6(sin(x)^2+sin(y)^2)u+u^3=mu*u

 Lx=10*pi; Ly=10*pi; Nx=128; Ny=128; maxerror=1e-10; nmax=500;
 dx=Lx/Nx; x=-Lx/2:dx:Lx/2-dx; kx=[0:Nx/2-1  -Nx/2:-1]*2*pi/Lx;
 dy=Ly/Ny; y=-Ly/2:dy:Ly/2-dy; ky=[0:Ny/2-1  -Ny/2:-1]*2*pi/Ly;
 [X,Y]=meshgrid(x, y); [KX,KY]=meshgrid(kx, ky); K2=KX.^2+KY.^2;
 V=6*(sin(X).^2+sin(Y).^2);
 P=1.9092; c=2; DT=0.9;
 u=sech(2*sqrt(X.^2+Y.^2));                        % initial condition
 u=u*sqrt(P/(sum(sum(abs(u.*u)))*dx*dy));          % power normalization
 for nn=1:nmax                                     % iteration starts
    L00u=ifft2(-K2.*fft2(u))+(u.*u-V).*u;
    Minvu=ifft2(fft2(u)./(c+K2));
    mu=sum(sum(Minvu.*L00u))/sum(sum(Minvu.*u));
    L0u=L00u-mu*u;
    uerror(nn)=max(max(abs(L0u)));
    if uerror(nn) < maxerror
        break
    end
    u=u+ifft2(fft2(L0u)./(c+K2))*DT;
    u=u*sqrt(P/(sum(sum(abs(u.*u)))*dx*dy));
 end                                               % iteration ends
 mu
 subplot(221); mm=35:94; mesh(x(mm), y(mm), u(mm,mm));
 axis([x(mm(1)) x(mm(end)) y(mm(1)) y(mm(end)) 0 1.3]);
 subplot(222); semilogy(0:nn-1, uerror, 'linewidth', 2);
 axis([0 170 1e-11 10])
```

Running this code, one gets the soliton profile as well as the error diagram as shown in Fig. 7.12(b, c). The error of the solution drops below 10^{-10} in 162 iterations, and the CPU time used is about 2.5 seconds (on a PC with a 2GHz processor).

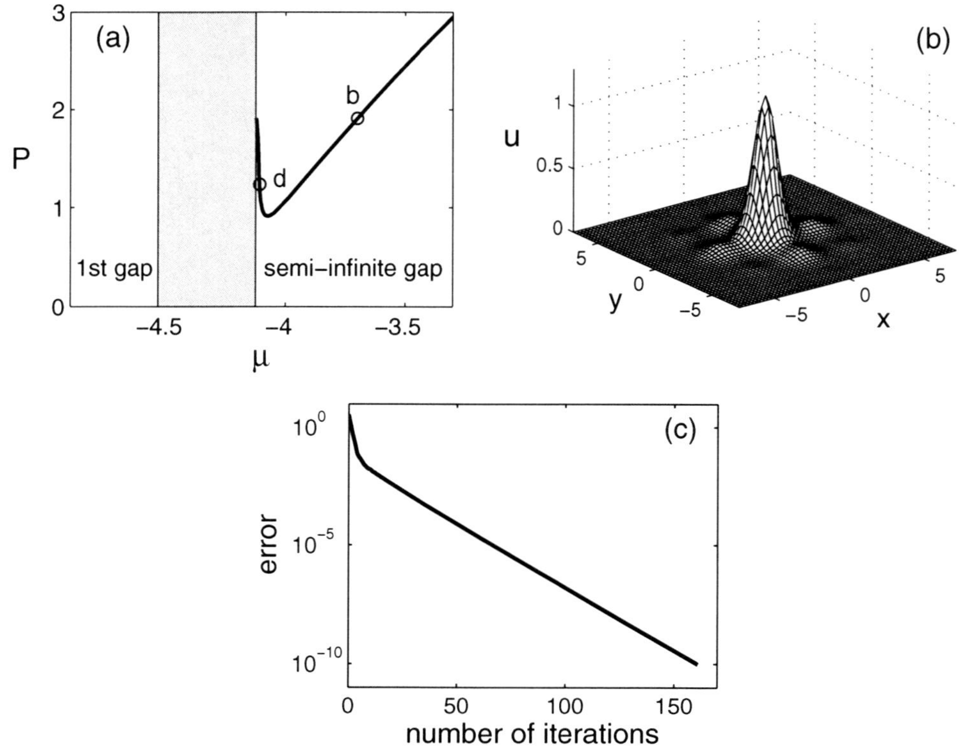

Figure 7.12. *Output of Program* 6 *on the computation of on-site positive solitons in the* 2*D NLS equation with a periodic potential* (7.138) *by the AITEM:* (a) *power curve for this family of solitons (this curve is the same as that in the semi-infinite gap of Fig.* 6.12 *in view of* $\mu = -\omega$*);* (b) *soliton profile at* $\mu = -3.7$ *(this* μ *is marked by letter* "b" *in* (a)*);* (c) *error diagram of AITEM iterations for the soliton in* (b).

Convergence Conditions of AITEM

Applying the AITEM to other on-site positive solitons in Eq. (7.138), we find that when $P'(\mu) > 0$ (see Fig. 7.12(a)), the AITEM always converges when Δt is below a certain threshold. But when $P'(\mu) < 0$, the AITEM always diverges for any $\Delta t \neq 0$. This brings up the issue of convergence conditions of the AITEM. This question was resolved by Yang and Lakoba (2008). Let us consider the solitary wave equation (7.126) for the multidimensional NLS equation (7.124) with a general type of nonlinearity and potential. The linearization operator of this solitary wave equation is defined as

$$L_1 = \nabla^2 - \mu + F(u^2, \mathbf{x}) + 2u^2 F_{u^2}(u^2, \mathbf{x}), \tag{7.140}$$

where F_{u^2} is the partial derivative of F with respect to u^2. Note that this operator is the same as (5.70) in the stability theory of solitary waves in Eq. (5.64) (i.e., Eq. (7.124) above). We also denote $n_+(L)$ as the number of positive eigenvalues in an operator L. Then the precise convergence conditions of the AITEM for the solitary wave equation (7.126) is given by the following theorem (see Yang and Lakoba (2008)).

Theorem 7.2. *Assume that when Eq. (7.126) is translation invariant (i.e., the function F is $\mathbf{x}$-independent), the kernel of L_1 contains only translation-induced eigenmodes u_{x_j}; and when Eq. (7.126) is not translation invariant, the kernel of L_1 is empty. Assume also that the initial condition is sufficiently close to the exact solution $u(\mathbf{x})$. In addition, define $\Delta t_{max} = -2/\lambda_{min}$, where λ_{min} is the minimum (negative) eigenvalue of the following operator $\mathcal{L}$:*

$$\mathcal{L}\psi = M^{-1}\left(L_1\psi - \frac{\langle M^{-1}u, L_1\psi\rangle}{\langle M^{-1}u, u\rangle}u\right). \tag{7.141}$$

Then if $\Delta t > \Delta t_{max}$ or $\Delta t < 0$, the AITEM (7.134)–(7.136) diverges. If $0 < \Delta t < \Delta t_{max}$, the following convergence statements on this AITEM hold.

1. *If $n_+(L_1) = 0$ and $P'(\mu) \neq 0$, then the AITEM converges.*

2. *If $n_+(L_1) = 1$, then $n_+(M^{-1}L_1) = 1$. In this case, denoting the eigenfunction of this single positive eigenvalue of $M^{-1}L_1$ as $\varphi_1(\mathbf{x})$, then when $\langle\varphi_1, u\rangle \neq 0$, the AITEM converges for $P'(\mu) > 0$ and diverges for $P'(\mu) < 0$. But when $\langle\varphi_1, u\rangle = 0$, the AITEM diverges.*

3. *If $n_+(L_1) > 1$, then the AITEM diverges.*

Proof. Here we sketch the proof of this theorem. Since the initial condition is close to the exact solution, we can use the linearization method to analyze the AITEM. Let

$$u_n(\mathbf{x}) = u(\mathbf{x}) + \tilde{u}_n(\mathbf{x}), \qquad |\tilde{u}_n| \ll |u|, \tag{7.142}$$

where $\tilde{u}_n(\mathbf{x})$ is the error. Substituting this equation into (7.135)–(7.136) and neglecting terms of $O(\tilde{u}_n^2)$ and higher, we find that the error satisfies the following linear iteration equation:

$$\tilde{u}_{n+1} = (1 + \Delta t\, \mathcal{L})\tilde{u}_n, \tag{7.143}$$

where the iteration operator $\mathcal{L}$ is as given in Eq. (7.141). Substituting (7.142) into the power-normalization step (7.134) and linearizing, one finds that the error is orthogonal to $u(\mathbf{x})$, i.e., $\langle\tilde{u}_n, u\rangle = 0$, for all index n. Thus convergence of the AITEM is determined by eigenvalues of the following constrained eigenvalue problem on the iteration operator $\mathcal{L}$:

$$\mathcal{L}\psi = \lambda\psi, \qquad \psi \in S, \tag{7.144}$$

where

$$S = \{\psi(\mathbf{x}) : \langle \psi, u \rangle = 0\}. \tag{7.145}$$

Specifically, the necessary and sufficient conditions for the AITEM to converge are as follows:

1. if Eq. (7.126) is translation invariant, the kernel of $\mathcal{L}$ contains only translation-induced eigenmodes u_{x_j} (which lead to a spatially translated solitary wave of Eq. (7.126) and do not affect the convergence of iterations); if Eq. (7.126) is not translation-invariant, the kernel of $\mathcal{L}$ is empty;

2. for all nonzero eigenvalues λ of $\mathcal{L}$, $|1+\lambda\,\Delta t| < 1$, i.e.,

$$-2 < \lambda\,\Delta t < 0. \tag{7.146}$$

On the first condition, noticing the relation $L_1 u_\mu = u$ by differentiating (7.126) with respect to μ, we see that the solution ψ_0 to the zero-eigenvalue equation $\mathcal{L}\psi_0 = 0$ is $\alpha\, u_\mu$, plus functions in the kernel of L_1. Here α is a constant. Due to our assumption on the kernel of L_1, we see that if $P'(\mu) \neq 0$, then in order for $\psi_0 \in S$, i.e., $\langle \psi_0, u\rangle = 0$, we must have $\alpha = 0$; hence ψ_0 is in the kernel of L_1, and the first condition above is satisfied.

Now we consider the second condition. If $\Delta t > \Delta t_{max}$, with Δt_{max} given in Theorem 7.2, then the left inequality in (7.146) is not met, and thus the AITEM diverges. If $\Delta t < 0$, since $\mathcal{L}$ always has negative eigenvalues, the right inequality in (7.146) then is not met, and thus the AITEM diverges. If $0 < \Delta t < \Delta t_{max}$, the left inequality in (7.146) is satisfied, and so we only need to consider the right inequality in (7.146). This will be done below.

The right inequality in (7.146) is simply $\lambda < 0$ (since $\Delta t > 0$ here). Now we analyze when this condition is met. To facilitate the analysis, we introduce the following new variables and operators:

$$\widehat{\psi} = M^{1/2}\psi, \quad \widehat{u} = M^{-1/2}u, \quad \widehat{L}_1 = M^{-1/2}L_1 M^{-1/2}. \tag{7.147}$$

Under this variable transform, the eigenvalue problem (7.144) becomes

$$\widehat{L}_1\widehat{\psi} - \lambda\widehat{\psi} = \widehat{H}\widehat{u}, \tag{7.148}$$

with the constraint

$$\langle \widehat{\psi}, \widehat{u}\rangle = 0. \tag{7.149}$$

Here $\widehat{H} = \langle \widehat{u}, \widehat{L}_1\widehat{\psi}\rangle / \langle \widehat{u}, \widehat{u}\rangle$. Since $\widehat{L}_1$ is similar to $M^{-1}L_1$, then in view of the Sylvester inertia law (see Sec. 7.2.1), $n_+(\widehat{L}_1) = n_+(M^{-1}L_1) = n_+(L_1)$.

The eigenvalue problem (7.148)–(7.149) is very similar to the stationary-point problem (5.42) and (5.49) of the Lagrange functional (5.48) in the linear stability analysis of positive solitary waves in Sec. 5.3. In particular, the eigenvalue λ here corresponds to the parameter κ in that stability analysis. By expanding $\widehat{\psi}$ and $\widehat{u}$ into the complete set of eigenfunctions of $\widehat{L}_1$ and repeating the same calculation as in Sec. 5.3, we readily see that when $n_+(\widehat{L}_1) = n_+(L_1) > 1$, a positive eigenvalue λ exists, and hence the AITEM diverges. When $n_+(\widehat{L}_1) = n_+(L_1) = 0$, there are no positive eigenvalues. Thus if $P'(\mu) \neq 0$, where the first convergence condition is also satisfied (see above), the AITEM then converges. When $n_+(\widehat{L}_1) = n_+(L_1) = 1$, if $\langle \varphi_1, u \rangle = 0$, where φ_1 is the eigenfunction of the single positive eigenvalue Λ_1 of $M^{-1}L_1$, then $\widehat{\psi} = M^{1/2}\varphi_1$ and $\lambda = \Lambda_1 > 0$ satisfy the eigenvalue problem (7.148)–(7.149), and hence the AITEM diverges. When $n_+(\widehat{L}_1) = n_+(L_1) = 1$ and $\langle \varphi_1, u \rangle \neq 0$, following the calculations in Sec. 5.3 and using our assumption on the kernel of L_1, we find that the sign of the largest eigenvalue λ_{max} in (7.148) depends on the sign of $Q(0) = \langle \widehat{u}, \widehat{L}_1^{-1}\widehat{u} \rangle = \langle u, L_1^{-1}u \rangle$, which is equal to $P'(\mu)/2$. When $P'(\mu) > 0$, $\lambda_{max} < 0$, and hence the AITEM converges. When $P'(\mu) < 0$, $\lambda_{max} > 0$, and hence the AITEM diverges. This ends the proof of this theorem. □

For on-site positive solitons in the semi-infinite gap in the 2D NLS equation with a periodic potential (7.138), the kernel of L_1 is empty. In addition, $n_+(L_1) = 1$. Furthermore, $u(\mathbf{x}) > 0$ and $\varphi_1(\mathbf{x}) > 0$ for all $\mathbf{x}$ values, so $\langle \varphi_1, u \rangle \neq 0$. Thus according to Theorem 7.2, the AITEM converges when $P'(\mu) > 0$ and $0 < \Delta t < \Delta t_{max}$, and diverges when $P'(\mu) < 0$, which is precisely what we numerically observed (see Fig. 7.12).

Now we make some remarks on the convergence conditions in Theorem 7.2. For NLS-type equations (7.124) with general forms of nonlinearity and potential, the ground states (which we loosely define as positive and single-humped solitary waves) often satisfy $n_+(L_1) = 1$, and the excited states often satisfy $n_+(L_1) > 1$. In such common cases, Theorem 7.2 shows that the AITEM can converge to ground states, but always diverges for excited states. This behavior is analogous to the Petviashvili method. Exceptions do exist though; see Yang and Lakoba (2008).

Connection between AITEM's Convergence and Linear Stability of Positive Solitary Waves

The three convergence statements in Theorem 7.2 are almost identical to the linear-stability conditions in Theorem 5.2 for positive solitary waves in the generalized NLS equations (7.124); see Sec. 5.3. This indicates that the AITEM converges to a positive solitary wave (for $0 < \Delta t < \Delta t_{max}$) if and only if this wave is linearly stable. The only minor difference between the convergence and stability results is in the case of $n_+(L_1) = 1$, where the convergence theorem has a condition on $\langle \varphi_1, u \rangle$, which is absent in the stability theorem. However, for positive solitary waves $u(\mathbf{x})$ with $n_+(L_1) = 1$, the condition $\langle \varphi_1, u \rangle \neq 0$ in Theorem 7.2 is met in generic cases (when M is the identity operator or in the form $c - \nabla^2$, condition $\langle \varphi_1, u \rangle \neq 0$ can be shown to strictly hold; see Yang and Lakoba (2008)); thus linear stability implies convergence of the AITEM in this case as well. This connection between convergence and linear stability is a unique property of the AITEM. Because of it, if a positive solitary wave can be obtained by AITEM iterations in Eq. (7.124), then this solitary wave is also linearly stable. For example, the soliton of Eq. (7.137) as obtained by AITEM in Fig. 7.12(b) is linearly stable.

Optimal Acceleration of AITEM

The choice of the preconditioning operator M in the AITEM is an important question. For practical purposes, M should be chosen so that it is easily invertible. For the NLS-type equations (7.126), a convenient choice of M is

$$M = c - \nabla^2, \quad c > 0, \tag{7.150}$$

which can be easily inverted by the Fourier transform (see Program 6). For this form of M, what value of c should we take so that the AITEM converges the fastest? If the potential in Eq. (7.126) is localized or absent, then the optimal c value is $c_{opt} = \mu$. Specifically, we have the following theorem (Yang and Lakoba (2008)).

Theorem 7.3. *For Eq.* (7.126) *with* $\lim_{|\mathbf{x}|\to\infty} F(0,\mathbf{x}) = 0$, *if the function*

$$\mathcal{V}(\mathbf{x}) \equiv F(u^2,\mathbf{x}) + 2u^2 F_{u^2}(u^2,\mathbf{x})$$

does not change sign, then $c_{opt} = \mu$ *in the AITEM* (7.134)–(7.136).

This theorem applies to localized potentials. For Eq. (7.138), the potential is not localized, and so the above theorem does not apply.

Most of the results for NLS-type equations (7.124) in this section can be extended to a wider class of equations where ∇^2 is replaced by a general linear self-adjoint and negative-definite differential operator with constant coefficients; see Yang and Lakoba (2008).

In the end of this section, we would like to mention another variation of the AITEM which was proposed by Yang and Lakoba (2008). In this variation, instead of normalizing the solution at each step by its power, one normalizes the solution by its amplitude. This amplitude-normalized AITEM converges faster than the power-normalized AITEM. In addition, it can converge even if $P'(\mu) \le 0$. But the connection between convergence and linear stability is lost for this amplitude-normalized AITEM.

7.2.3 Squared-Operator Iteration Methods

Both the Petviashvili method and the AITEM generally only converge to the ground states of a nonlinear wave system and would diverge for excited states. In this section, we present two iteration methods which can converge for both the ground states and the excited states. These methods are based on the idea of integrating a "squared-operator" evolution equation.

Squared-Operator Iteration Method

Consider solitary waves in a general real-valued vector nonlinear wave system in arbitrary spatial dimensions, which can be written in the following form:

$$\mathbf{L}_0\mathbf{u}(\mathbf{x}) = 0. \tag{7.151}$$

Here $\mathbf{x}$ is the multidimensional spatial variable, and $\mathbf{u}(\mathbf{x})$ is a real-valued vector solitary wave, i.e., $\mathbf{u} \to 0$ as $|\mathbf{x}| \to \infty$. For complex-valued solitary waves, the equation can still be written in the above form with $\mathbf{u}$ containing the real and imaginary parts of the complex solution. In Eq. (7.151), the propagation constant of the solitary wave is lumped into the operator $\mathbf{L}_0$. Let $\mathbf{L}_1$ denote the linearized operator of Eq. (7.151) around the solution $\mathbf{u}$, i.e.,

$$\mathbf{L}_0(\mathbf{u}+\widetilde{\mathbf{u}}) = \mathbf{L}_1\widetilde{\mathbf{u}} + O(\widetilde{\mathbf{u}}^2), \qquad \widetilde{\mathbf{u}} \ll 1. \tag{7.152}$$

To obtain the solitary wave $\mathbf{u}(\mathbf{x})$ in Eq. (7.151), the key idea here is to numerically integrate the "squared-operator" evolution equation

$$\mathbf{u}_t = -\mathbf{M}^{-1}\mathbf{L}_1^{\dagger}\mathbf{M}^{-1}\mathbf{L}_0\mathbf{u}, \tag{7.153}$$

where $(\cdot)^{\dagger}$ denotes the Hermitian of the underlying quantity, and $\mathbf{M}$ is a positive-definite and Hermitian operator. The reason is to guarantee the convergence of numerical iterations, as we will demonstrate below. The operator $\mathbf{M}$ is introduced to speed up the convergence, in the same spirit as the preconditioning technique in the AITEM. Using the simplest time-stepping method for Eq. (7.153)—the forward Euler method, the resulting iteration method for computing the solitary wave $\mathbf{u}(\mathbf{x})$ in Eq. (7.151) is (see Yang and Lakoba (2007))

$$\mathbf{u}_{n+1} = \mathbf{u}_n - \left[\mathbf{M}^{-1}\mathbf{L}_1^{\dagger}\mathbf{M}^{-1}\mathbf{L}_0\mathbf{u}\right]_{\mathbf{u}=\mathbf{u}_n}\Delta t, \tag{7.154}$$

where Δt is an auxiliary time-step parameter. This method is called the squared-operator iteration method (or SOM for short).

We make three remarks here. First, unlike the AITEM in the previous subsection, the above SOM does not need power normalization. In this SOM, the propagation constant is prefixed, not the power. The reason for it is that the iteration operator of the SOM does not have any positive eigenvalues (see below), and thus direct SOM iterations can converge. Second, other forms of "squared-operator" equations other than (7.153) can also be used. An example is $\mathbf{u}_t = -\mathbf{M}^{-2}\mathbf{L}_1^{\dagger}\mathbf{L}_0\mathbf{u}$. These different "squared-operator" equations lead to iteration schemes whose performances are comparable to the above SOM (Yang and Lakoba (2007)). Third, other types of time-integration schemes (such as explicit higher-order Runge–Kutta methods and implicit methods) can also be used to integrate Eq. (7.153). But these other integration schemes are generally less efficient than the forward Euler method (7.154), because the extra computations in them outweigh the benefits they may bring.

The most important property of the SOM (7.154) is that it converges for general solitary waves as long as the time step Δt is below a certain threshold. Specifically, we have the following convergence result on the SOM (Yang and Lakoba (2007)).

Theorem 7.4. *Assume that when Eq.* (7.151) *is translation invariant, the kernel of* $\boldsymbol{L}_1$ *contains only translation-induced eigenmodes* $\boldsymbol{u}_{x_j}$; *and when Eq.* (7.151) *is not translation-invariant, the kernel of* $\boldsymbol{L}_1$ *is empty. Assume also that the initial condition is sufficiently close to the exact solution* $\boldsymbol{u}(\mathbf{x})$. *In addition, define* $\Delta t_{max} = -2/\Lambda_{min}$, *where* Λ_{min} *is the minimum (negative) eigenvalue of the iteration operator*

$$\mathcal{L} = -\boldsymbol{M}^{-1}\boldsymbol{L}_1^{\dagger}\boldsymbol{M}^{-1}\boldsymbol{L}_1. \tag{7.155}$$

Then the SOM (7.154) *converges to the solitary wave* $\boldsymbol{u}(\mathbf{x})$ *of Eq.* (7.151) *if* $0 < \Delta t < \Delta t_{max}$, *and diverges otherwise.*

Proof. We use the linearization technique to analyze the SOM (7.154) and prove the theorem. Let

$$\mathbf{u}_n = \mathbf{u} + \widetilde{\mathbf{u}}_n, \tag{7.156}$$

where $\widetilde{\mathbf{u}}_n(\mathbf{x}) \ll 1$ is the error. When Eq. (7.156) is substituted into the SOM (7.154) and only terms of $O(\widetilde{\mathbf{u}}_n)$ are retained, we find that the error satisfies the following linear equation:

$$\widetilde{\mathbf{u}}_{n+1} = (1 + \Delta t\, \mathcal{L})\, \widetilde{\mathbf{u}}_n, \tag{7.157}$$

where $\mathcal{L}$ is the iteration operator defined in (7.155). Note that $\mathcal{L}$ is similar to a Hermitian and semi-negative-definite operator

$$\mathcal{L}_h = -\left(\mathbf{M}^{-1/2}\mathbf{L}_1\mathbf{M}^{-1/2}\right)^{\dagger}\left(\mathbf{M}^{-1/2}\mathbf{L}_1\mathbf{M}^{-1/2}\right) \tag{7.158}$$

since $\mathcal{L} = \mathbf{M}^{-1/2}\mathcal{L}_h\mathbf{M}^{1/2}$. Thus all eigenvalues of $\mathcal{L}$ are real and nonpositive. In addition, all eigenfunctions of $\mathcal{L}$ form a complete set in the square-integrable functional space, and hence the error $\widetilde{\mathbf{u}}_n$ can be expanded over $\mathcal{L}$'s eigenfunctions. If $0 < \Delta t < \Delta t_{min}$, then no eigenmode in the error can grow since $|1 + \Lambda\Delta t| \le 1$ for all these eigenmodes. In fact, all eigenmodes will decay with iterations except those in the kernel of $\mathcal{L}$. It is easy to see that the kernel of $\mathcal{L}$ is the same as the kernel of $\mathbf{L}_1$. Then according to the assumption, the kernel of $\mathcal{L}$ either is empty or contains only translation-induced eigenmodes $\mathbf{u}_{x_j}$

which only lead to a spatially shifted solution and do not affect the convergence of iterations. Thus the SOM (7.154) converges. On the other hand, if $\Delta t > \Delta t_{min}$, the eigenmode with eigenvalue Λ_{min} in the error will grow exponentially because $1+\Lambda_{min}\Delta t < -1$, and hence the SOM diverges. If $\Delta t < 0$, the eigenmode with negative eigenvalues in the error will grow exponentially, and hence the SOM diverges as well. Thus Theorem 7.4 is proved. □

Modified Squared-Operator Iteration Method

The SOM (7.154) contains a preconditioning operator $\mathbf{M}$ for the purpose of accelerating its convergence. Despite that, the SOM can still be very slow in many situations. The reason is that due to this squaring of the operator, the condition number is also squared. It is well known that large condition numbers lead to slow convergence of iterations. Below, we incorporate another acceleration technique—mode elimination into the SOM (Lakoba and Yang (2007b)). The resulting method, which is called the modified squared-operator method (or MSOM for short), will converge faster than the SOM.

The MSOM proposed by Yang and Lakoba (2007) for computing solitary waves in Eq. (7.151) is

$$\mathbf{u}_{n+1} = \mathbf{u}_n - \left[\mathbf{M}^{-1}\mathbf{L}_1^{\dagger}\mathbf{M}^{-1}\mathbf{L}_0\mathbf{u} - \alpha_n\langle \mathbf{G}_n,\, \mathbf{L}_1^{\dagger}\mathbf{M}^{-1}\mathbf{L}_0\mathbf{u}\rangle \mathbf{G}_n\right]_{\mathbf{u}=\mathbf{u}_n}\Delta t, \tag{7.159}$$

where

$$\alpha_n = \frac{1}{\langle \mathbf{M}\mathbf{G}_n, \mathbf{G}_n\rangle} - \frac{1}{\langle \mathbf{L}_1\mathbf{G}_n,\, \mathbf{M}^{-1}\mathbf{L}_1\mathbf{G}_n\rangle\,\Delta t} \tag{7.160}$$

and

$$\mathbf{G}_n = \mathbf{u}_n - \mathbf{u}_{n-1}. \tag{7.161}$$

The motivation for this MSOM can be explained briefly as follows. Consider the SOM (7.154), and denote the slowest-decaying eigenmode of $\mathcal{L}$ in the error as $\mathbf{G}(\mathbf{x})$ with eigenvalue Λ_s. Our idea is to construct a modified linear iteration operator for the error so that the eigenmode $\mathbf{G}(\mathbf{x})$ decays quickly, while the decay rates of the other eigenmodes of $\mathcal{L}$ remain the same. If so, then this modified iteration scheme would converge faster than the original SOM. For this purpose, consider the modified linear iteration operator

$$\mathcal{L}_M\Psi = \mathcal{L}\Psi - \alpha\langle \mathbf{M}\mathbf{G},\, \mathcal{L}\Psi\rangle \mathbf{G}, \tag{7.162}$$

where α is a constant. Since $\mathbf{G}(\mathbf{x})$ is an eigenfunction of $\mathcal{L}$, and recalling that eigenfunctions of this $\mathcal{L}$ are orthogonal to each other under the $\mathbf{M}$-weighted inner product, we readily see that this modified iteration operator and the original one have identical eigenfunctions. Their eigenvalues are identical too except the one for eigenmode $\mathbf{G}(\mathbf{x})$. The eigenvalue of this

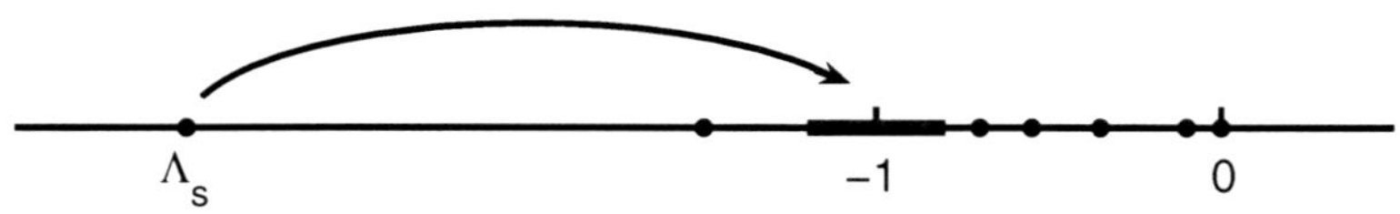

Figure 7.13. *Schematic illustration on spectra of SOM's iteration operator $\mathcal{L}$ and MSOM's iteration operator $\mathcal{L}_M$. The spectrum of $\mathcal{L}_M$ is the same as that of $\mathcal{L}$, except that the slow-mode eigenvalue Λ_s of SOM is shifted to -1.*

eigenmode shifts from Λ_s of $\mathcal{L}$ to $(1-\alpha\langle \mathbf{MG}, \mathbf{G}\rangle)\Lambda_s$ of $\mathcal{L}_M$. This eigenvalue relation between $\mathcal{L}$ and $\mathcal{L}_M$ is illustrated in Fig. 7.13. Then if we choose α so that the decay factor of this eigenmode is zero, i.e.,

$$1+[1-\alpha\langle \mathbf{MG}, \mathbf{G}\rangle]\Lambda_s \Delta t = 0, \tag{7.163}$$

or equivalently,

$$\alpha = \frac{1}{\langle \mathbf{MG}, \mathbf{G}\rangle}\left(1+\frac{1}{\Lambda_s \Delta t}\right), \tag{7.164}$$

then eigenmode $\mathbf{G}(\mathbf{x})$, which decays the slowest in SOM, starts decaying the fastest, while decay rates of the other eigenmodes in the error are unaffected. Thus convergence is improved. In practical situations, the slowest-decaying mode $\mathbf{G}(\mathbf{x})$ and its eigenvalue Λ_s in SOM are not known. A reasonable approximation for $\mathbf{G}(\mathbf{x})$ can be taken as the difference between successive iterations $\mathbf{u}_n - \mathbf{u}_{n-1}$, which motivates Eq. (7.161) in the MSOM. The corresponding eigenvalue Λ_s can be approximated by the Rayleigh quotient as $\Lambda_s \approx \langle \mathbf{MG}_n,\ \mathcal{L}\mathbf{G}_n\rangle/\langle \mathbf{MG}_n, \mathbf{G}_n\rangle$. Substituting this approximation into (7.164), we get α as given by formula (7.160). Corresponding to the modified linear iteration operator (7.162), the modified iteration method is then the MSOM (7.159).

Below we demonstrate the performances of the SOM and the MSOM in an example. The example we choose is the computation of gap solitons in the 2D NLS equation with periodic potentials and self-defocusing nonlinearity. The corresponding model equation is

$$u_{xx}+u_{yy}-V_0\left(\sin^2 x+\sin^2 y\right)u-u^3=\mu u. \tag{7.165}$$

This equation admits a family of solitons in the first bandgap, which have one main intensity peak at a potential minimum, and solutions at neighboring potential minima are out of phase. At $V_0=6$, the power curve of this solution family is displayed in Fig. 7.14(a). When $\mu=-5$, the profile of this gap soliton is shown in Fig. 7.14(b). Note that both the Petviashvili-type method and the AITEM diverge for this soliton because the linearization operator $\mathbf{L}_1$ has a band of positive eigenvalues. Thus we will compute it by the SOM and the MSOM below. In both methods, we take the preconditioning operator $\mathbf{M}$ as $c-\nabla^2$. Note also that $\mathbf{L}_1^\dagger = \mathbf{L}_1$ here. The MATLAB code of the MSOM for this equation is displayed below.

Program 7

```
% p7.m: the MSOM for computing gap solitons in the 2D NLS equation
% with a periodic potential and self-defocusing nonlinearity:
% u_{xx}+u_{yy}-6(sin(x)^2+sin(y)^2)u-u^3=mu*u

 Lx=10*pi; Ly=10*pi; N=128; nmax=1000; maxerror=1e-10;
 x=-Lx/2:Lx/N:Lx/2-Lx/N; dx=Lx/N; kx=[0:N/2-1  -N/2:-1]*2*pi/Lx;
 y=-Ly/2:Ly/N:Ly/2-Ly/N; dy=Ly/N; ky=[0:N/2-1  -N/2:-1]*2*pi/Ly;
 [X,Y]=meshgrid(x, y); [KX, KY]=meshgrid(kx, ky); K2=KX.^2+KY.^2;
 V=6*(sin(X).^2+sin(Y).^2);
 mu=-5; u=1.15*sech(sqrt(X.^2+Y.^2)).*cos(X).*cos(Y);
 c=2.9; DT=1.7;
 for nn=1:nmax                              % MSOM iterations start
    uold=u;
    F0=-V-u.*u-mu; F1=-V-3*u.*u-mu;
    L0u=ifft2(-K2.*fft2(u))+F0.*u;
    uerror(nn)=max(max(abs(L0u))); uerror(nn)
    if uerror(nn) < maxerror
        break
    end
    MinvL0u=ifft2(fft2(L0u)./(c+K2));
    L1MinvL0u=ifft2(-K2.*fft2(MinvL0u))+F1.*MinvL0u;
    MinvL1MinvL0u=ifft2(fft2(L1MinvL0u)./(c+K2));
    if nn == 1
        u=u-MinvL1MinvL0u*DT;
    else
        L1G=ifft2(-K2.*fft2(G))+F1.*G;
        MinvL1G=ifft2(fft2(L1G)./(c+K2));
        MG=ifft2(fft2(G).*(c+K2));
        alpha=1/sum(sum(MG.*G))-1/(DT*sum(sum(L1G.*MinvL1G)));
        u=u-(MinvL1MinvL0u-alpha*sum(sum(G.*L1MinvL0u))*G)*DT;
    end
    G=u-uold;
 end                                        % MSOM iterations end
 subplot(221); mm=35:94; mesh(x(mm), y(mm), u(mm,mm));
 axis([x(35) x(94) y(35) y(94) 0 1.3])
 subplot(222); semilogy(0:nn-1, uerror, 'linewidth', 2);
 axis([0 350 1e-11 10])
```

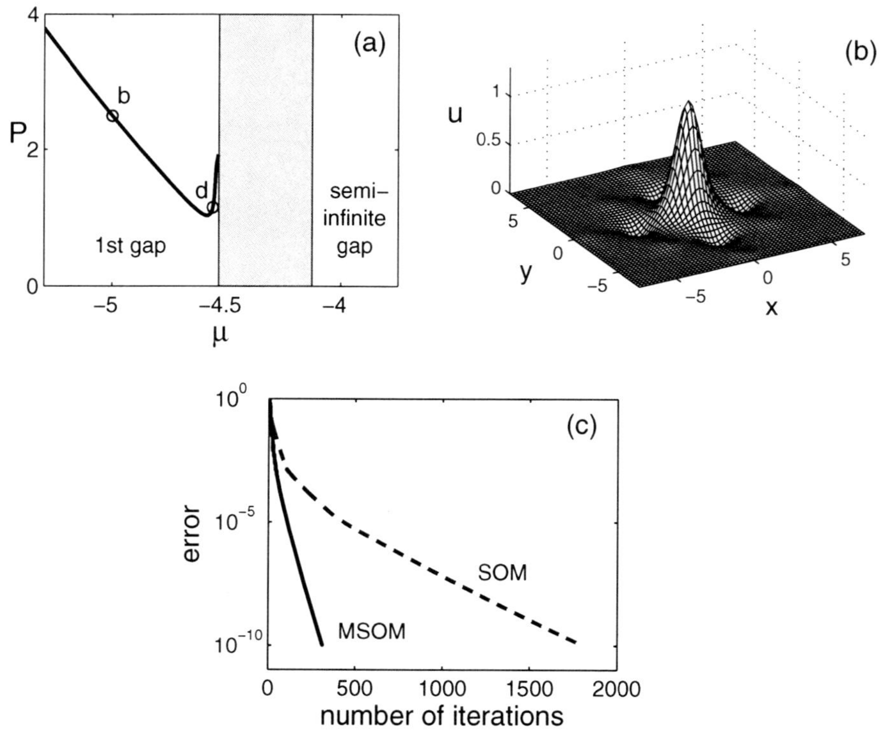

Figure 7.14. *Output of Program 7 on computation of gap solitons in the 2D NLS equation with a periodic potential and self-defocusing nonlinearity* (7.165) *by SOM and MSOM methods:* (a) *power curve for this soliton family (this curve is the same as that in the first gap of Fig.* 6.12 *under* $\mu = -\omega$*);* (b) *profile of the gap soliton at* $\mu = -5$ *(marked by letter* "b" *in* (a)*);* (c) *error diagrams of SOM and MSOM for the soliton in* (b).

In this code, we took $c = 2.9$ and $\Delta t = 1.7$, which are near optimal. The output of this code is displayed in Fig. 7.14(b, c). We see that the error drops below 10^{-10} after 313 MSOM iterations, and the CPU time used is about 11 seconds (on a PC with a 2GHz processor). For the SOM, the MATLAB code is much shorter. At the near-optimal parameters $c = 1.5$ and $\Delta t = 0.51$, the error diagram of the SOM is shown in Fig. 7.14(c) as well. It is seen that the error drops below 10^{-10} after 1810 SOM iterations, or 37 seconds. Thus the MSOM is about 3 times as fast as the SOM on this example.

In both the SOM and MSOM above, the propagation constant μ is fixed. In some situations, solutions with certain fixed powers are desired. In such cases, power-conserving versions of the SOM can be used; see Yang and Lakoba (2007) for details.

7.2.4 Newton Conjugate-Gradient Methods

The conjugate-gradient method and its generalizations are the most prominent iterative methods for solving large systems of linear equations (Hestenes and Stiefel (1952), Golub and Van Loan (1996), Ryabenkii and Tsynkov (2007)). In this section, we adopt them for computing solitary waves. The basic idea is to use Newton iterations, coupled with conjugate-gradient iterations to solve the resulting linear Newton-correction equation. The resulting methods will be called Newton conjugate-gradient (Newton-CG) methods. These methods can converge for both the ground states and excited states of a nonlinear wave system. More importantly, they converge much faster than the other existing iteration methods, often by orders of magnitude. In addition, they are easy to implement for general equations in arbitrary dimensions as well. These methods were proposed by Yang (2009a) and will be described below.

Let us consider solitary waves in a general real-valued vector nonlinear wave system in arbitrary spatial dimensions,

$$\mathbf{L}_0\mathbf{u}(\mathbf{x}) = 0, \tag{7.166}$$

where $\mathbf{x}$ is the multidimensional spatial variable, and $\mathbf{u}(\mathbf{x})$ is a real-valued vector solitary wave. Suppose we have an approximate solution $\mathbf{u}_n(\mathbf{x})$ which is close to the exact solution $\mathbf{u}(\mathbf{x})$. To obtain the next iteration solution $\mathbf{u}_{n+1}(\mathbf{x})$, we express the exact solution $\mathbf{u}(\mathbf{x})$ as

$$\mathbf{u}(\mathbf{x}) = \mathbf{u}_n(\mathbf{x}) + \mathbf{e}_n(\mathbf{x}), \tag{7.167}$$

where $\mathbf{e}_n(\mathbf{x}) \ll 1$ is the error term. Then we substitute this expression into Eq. (7.166) and expand it around $\mathbf{u}_n(\mathbf{x})$, which gives

$$\mathbf{L}_0\mathbf{u}_n + \mathbf{L}_{1n}\mathbf{e}_n = O(\mathbf{e}_n^2). \tag{7.168}$$

Here $\mathbf{L}_{1n}$ is the linearization operator $\mathbf{L}_1$ of the solitary wave equation (7.166) evaluated at the approximate solution $\mathbf{u}_n(\mathbf{x})$. If we neglect the higher-order term on the right-hand side of Eq. (7.168), the remaining equation becomes a linear inhomogeneous equation for the error $\mathbf{e}_n$. This suggests that we update the approximate solution as

$$\mathbf{u}_{n+1}(\mathbf{x}) = \mathbf{u}_n(\mathbf{x}) + \Delta\mathbf{u}_n(\mathbf{x}), \tag{7.169}$$

where the updated amount $\Delta\mathbf{u}_n$ is computed from the linear inhomogeneous equation for $\mathbf{e}_n$, which is rearranged as

$$\mathbf{L}_{1n}\Delta\mathbf{u}_n = -\mathbf{L}_0\mathbf{u}_n. \tag{7.170}$$

The above idea is identical to the traditional Newton method (Boyd (2001)). As such, if the linear correction equation (7.170) is solved exactly (or to very high accuracy), then iterations (7.169) will converge to the exact solution $\mathbf{u}(\mathbf{x})$ quadratically. These nonlinear Newton iterations (7.169) form the outer iterations of the Newton-CG method.

To solve the linear operator equation (7.170), instead of discretizing it into a matrix equation and then using direct methods such as LU or QR factorization (as people normally do in the traditional Newton method), we will use conjugate-gradient-type methods to solve it. These CG iterations form the inner iterations of the Newton-CG method. Below we will describe the CG method for solving Eq. (7.170) when the linearization operator $\mathbf{L}_1$ is self-adjoint, which is often the case in conservative wave systems. The case of non-self-adjoint operators $\mathbf{L}_1$ will be discussed in the end of this section.

When $\mathbf{L}_1$ is self-adjoint, Eq. (7.170) corresponds to a linear matrix equation where the matrix is real symmetric. The idea of the CG method for solving a symmetric-matrix equation is to first pick a set of search directions which are conjugate to each other (i.e., orthogonal to each other under the matrix-weighted inner product), and then take a step along each search direction so that the residue is minimized. Rigorous convergence results for this method can be obtained only when the matrix is symmetric and positive definite. In such cases, this method yields the exact solution in n steps, where n is the size of the matrix. When the matrix size is large, this method often gives the solution to the required accuracy in much fewer than n steps if a suitable preconditioning matrix is introduced. Indeed the preconditioned CG method is generally recognized as the most efficient iteration method for solving linear equations with symmetric positive-definite matrices. If the matrix is not positive definite, the CG method may break down due to division by zero during iterations. For Eq. (7.170) in solitary wave computations, the operator $\mathbf{L}_1$ is always non-positive-definite or non-negative-definite since it has both positive and negative eigenvalues; thus the application of the CG method on (7.170) carries the risk of breakdown. However, our extensive testing shows that this breakdown rarely occurs (in fact it was never encountered), and the preconditioned CG iterations always quickly converge. On this basis, we propose using the preconditioned CG method to solve the linear Newton-correction equation (7.170).

Now we describe the preconditioned CG method as applied to the linear operator equation (7.170). To simplify notations, we drop the subscripts "n" in Eq. (7.170). In addition, we always take the initial guess $\Delta\mathbf{u}^{(0)}$ to be zero for simplicity. Then the preconditioned CG method for Eq. (7.170) is

$$\Delta\mathbf{u}^{(0)} = 0,$$

$$\mathbf{R}^{(0)} = -\mathbf{L}_0\mathbf{u},$$

$$\mathbf{D}^{(0)} = \mathbf{M}^{-1}\mathbf{R}^{(0)},$$

$$a^{(i)} = \frac{\langle \mathbf{R}^{(i)}, \mathbf{M}^{-1}\mathbf{R}^{(i)}\rangle}{\langle \mathbf{D}^{(i)}, \mathbf{L}_1\mathbf{D}^{(i)}\rangle},$$

$$\Delta \mathbf{u}^{(i+1)} = \Delta \mathbf{u}^{(i)} + a^{(i)} \mathbf{D}^{(i)},$$

$$\mathbf{R}^{(i+1)} = \mathbf{R}^{(i)} - a^{(i)} \mathbf{L}_1 \mathbf{D}^{(i)},$$

$$b^{(i+1)} = \frac{\langle \mathbf{R}^{(i+1)}, \mathbf{M}^{-1} \mathbf{R}^{(i+1)} \rangle}{\langle \mathbf{R}^{(i)}, \mathbf{M}^{-1} \mathbf{R}^{(i)} \rangle},$$

$$\mathbf{D}^{(i+1)} = \mathbf{M}^{-1} \mathbf{R}^{(i+1)} + b^{(i+1)} \mathbf{D}^{(i)}.$$

Here $i = 0, 1, 2, \ldots$ is the index of CG iterations, the inner product is the standard one (7.98) in the square-integrable functional space, and the operator $\mathbf{M}$ is a self-adjoint and positive-definite preconditioning operator. The operator $\mathbf{M}$ should be chosen to be easily invertible. In practice, it is often taken as the linear differential part of the operator $\mathbf{L}_0$. These CG iterations are embedded inside the Newton iterations (7.169), and the resulting method is the Newton-CG method for computing solitary waves.

The CG iterations above are terminated when the approximate solution $\Delta \mathbf{u}_n^{(i)}$ to the linear Newton-correction equation (7.170) has reached a certain accuracy. The error of this solution can be measured by the function $\mathbf{R}^{(i)}$ in the CG iterations, which is the residue of the linear equation (7.170), $-\mathbf{L}_0 \mathbf{u}_n - \mathbf{L}_{1n} \Delta \mathbf{u}_n^{(i)}$. Specifically, this error can be measured by the $\mathbf{M}^{-1}$ weighted 2-norm of $\mathbf{R}^{(i)}$, $\| \mathbf{R}^{(i)} \|_M \equiv \langle \mathbf{R}^{(i)}, \mathbf{M}^{-1} \mathbf{R}^{(i)} \rangle^{1/2}$, which appears in the CG iterations. The accuracy with which the linear Newton-correction equation (7.170) is solved is an important parameter in the Newton-CG method. Remember from Eq. (7.169) that the approximate soliton solution $\mathbf{u}_{n+1}$ is updated by $\mathbf{u}_n + \Delta \mathbf{u}_n$. Thus in order to avoid waste in CG iterations (due to oversolving Eq. (7.170)), a sensible choice for the stopping criterion of CG iterations is that the error of $\Delta \mathbf{u}_n^{(i)}$ falls below a certain fraction of the error of the solution $\mathbf{u}_n$ itself (Yang (2009a)). The accuracy of $\mathbf{u}_n$ can be measured by $\| \mathbf{L}_0 \mathbf{u}_n \|_M$, which is the $\mathbf{M}^{-1}$ weighted 2-norm of the residue $\mathbf{L}_0 \mathbf{u}_n$ of the nonlinear wave equation (7.166). Then we take the stopping criterion of CG iterations to be

$$\| \mathbf{R}^{(i)} \|_M < \epsilon_{cg} \| \mathbf{L}_0 \mathbf{u}_n \|_M . \tag{7.171}$$

Here ϵ_{cg} is a small positive error reduction parameter for CG iterations. Notice that the residue $\mathbf{L}_0 \mathbf{u}_n$ is the inhomogeneous term of the linear equation (7.170). Also notice that due to our choice of the zero initial condition for CG iterations, $\mathbf{R}^{(0)} = -\mathbf{L}_0 \mathbf{u}_n$. Thus the stopping criterion (7.171) for CG iterations is simply

$$\| \mathbf{R}^{(i)} \|_M < \epsilon_{cg} \| \mathbf{R}^{(0)} \|_M . \tag{7.172}$$

Regarding the choice of the error reduction parameter ϵ_{cg}, if it is set too small, this leads to oversolving of the linear equation (7.170), which is wasteful. On the other hand, if ϵ_{cg} is set too large, which means that Eq. (7.170) is solved too inaccurately, then the Newton iterations (7.169) may not converge at all. Thus the optimal ϵ_{cg} value should be neither too small nor too large. Our numerical testings show that the optimal ϵ_{cg} is generally in the range between 10^{-1} and 10^{-3}. In the numerical examples below, we set $\epsilon_{cg} = 10^{-2}$. At this ϵ_{cg} value, the error of the numerical solution $\mathbf{u}_n$ roughly drops by a factor of $\epsilon_{cg}^{-1} = 10^2$ after each Newton's iteration. If the soliton's accuracy is set at 10^{-10}, then the number of Newton iterations in the Newton-CG method typically ranges from 5 to 8.

The above Newton-CG method has been used to compute solitary waves in a wide variety of physical systems (Yang (2009a)). It is found to converge for both the ground states and excited states. More importantly, its convergence is almost always faster than the other iteration methods described earlier. This is especially so when the propagation constant lies near the continuous spectrum, where the Newton-CG method can be orders of magnitude faster than the other methods. This fast convergence of the Newton-CG method will be illustrated by two examples below.

The first example is the computation of on-site positive lattice solitons in the 2D NLS equation with a periodic potential and self-focusing nonlinearity (7.138), i.e.,

$$u_{xx} + u_{yy} - V_0\left(\sin^2 x + \sin^2 y\right)u + u^3 = \mu u. \tag{7.173}$$

These solitons exist in the semi-infinite gap of this system. The power curve of these solitons at $V_0 = 6$ is as shown in Fig. 7.12(a). At the propagation constant $\mu = -4.11$ near the right edge of the first Bloch band (see letter "d" on the power curve), the profile of this soliton is shown in Fig. 7.15(a). This solution has low amplitude and is quite broad, and its computation is quite challenging. For this soliton, the AITEM (with power normalization) does not converge since $P'(\mu) < 0$ here. A generalized Petviashvili method as proposed by Musslimani and Yang (2004) can work though, and it will be compared with the Newton-CG method. In the Newton-CG method, the preconditioning operator is taken as $M = c - \nabla^2$ with $c = 3$, and the parameter ϵ_{cg} is taken as 10^{-2}. In the generalized Petviashvili method of Musslimani and Yang (2004), the parameter c is taken as 1.9 which is near optimal. In both methods, the computational domain is taken as a square of $-10\pi < x, y < 10\pi$, discretized by 256 points along each dimension. The initial condition is taken as

$$u_0(x, y) = 0.49\,\mathrm{sech}\, 2\sqrt{x^2 + y^2}.$$

Error diagrams of both methods are displayed in Fig. 7.15(b) (the error is defined as the maximum of $|\mathbf{L}_0\mathbf{u}_n|$). It is seen that for the error to drop below 10^{-10}, the Newton-CG method takes seven Newton iterations, which consist of a total of 257 CG iterations. This number of CG iterations is 12 times less than the number of generalized Petviashvili iterations. In terms of the CPU time, the Newton-CG method takes only 22 seconds in MATLAB (on a PC with a 2GHz processor), which is also 12 times less than the generalized Petviashvili

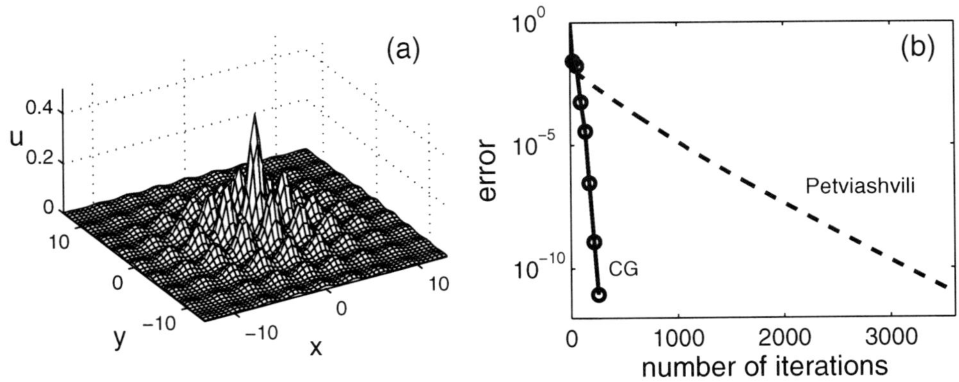

Figure 7.15. *Computation of on-site positive solitons in the 2D NLS equation with a periodic potential* (7.173) *by the Newton-CG method and the generalized Petviashvili method ($V_0 = 6$):* (a) *solution profile at $\mu = -4.11$ (marked by letter "d" on the power curve of Fig.* 7.12(a)*);* (b) *error diagrams of the two methods versus the number of CG or Petviashvili iterations (each circle represents a Newton's iteration point in the Newton-CG method). (After Yang* (2009a).*)*

method. Thus the Newton-CG method is more than an order of magnitude faster than the generalized Petviashvili method on this example.

The second example is the computation of on-site gap solitons in the 2D NLS equation with a periodic potential and self-defocusing nonlinearity (7.165), i.e.,

$$u_{xx} + u_{yy} - V_0\left(\sin^2 x + \sin^2 y\right)u - u^3 = \mu u. \tag{7.174}$$

These solitons exist in the first bandgap of the system, and their power curve at $V_0 = 6$ is as shown in Fig. 7.14(a). At the propagation constant $\mu = -4.56$ near the left edge of the first Bloch band (see letter "d" on the power curve), the profile of this gap soliton is as shown in Fig. 7.16(a). This soliton is challenging to compute as well not only because its propagation constant lies near the Bloch band, but also because it is sign changing. For this gap soliton, neither AITEM nor the generalized Petviashvili method can converge, and thus it will be computed by the MSOM and Newton-CG methods. In the Newton-CG method, the preconditioning operator is taken as $M = c - \nabla^2$ with $c = 3$, and the parameter ϵ_{cg} is taken as 10^{-2}. The computational domain is taken as a square of $-10\pi < x, y < 10\pi$, discretized by 256 points along each dimension. The initial condition is taken as

$$u_0(x, y) = 0.56\,\text{sech}\sqrt{x^2 + y^2}\,\cos x\,\cos y.$$

The corresponding MATLAB code of the Newton-CG method is displayed below.

Program 8

```
% p8.m: the Newton-CG method for computing gap solitons in the 2D NLS
% equation with a periodic potential and self-defocusing nonlinearity
% u_{xx}+u_{yy}-6(sin(x)^2+sin(y)^2)u-u^3=mu*u

 Lx=20*pi; Ly=20*pi; N=256; maxerror=1e-10; errorCG=1e-2;
 x=-Lx/2:Lx/N:Lx/2-Lx/N; kx=[0:N/2-1  -N/2:-1]*2*pi/Lx;
 y=-Ly/2:Ly/N:Ly/2-Ly/N; ky=[0:N/2-1  -N/2:-1]*2*pi/Ly;
 [X,Y]=meshgrid(x,y); [KX,KY]=meshgrid(kx,ky); K2=KX.^2+KY.^2;
 V=6*(sin(X).^2+sin(Y).^2); c=3; mu=-4.56;
 u=0.56*sech(0.5*sqrt(X.^2+Y.^2)).*cos(X).*cos(Y);
 nnt=0;                  % nnt: number of Newton iterations
 ncg=0;                  % ncg: total number of CG iterations
 while 1                 % Newton iterations begin
    nnt=nnt+1;
    L0u=ifft2(-K2.*fft2(u))-(V+u.*u+mu).*u;
    uerror(nnt)=max(max(abs(L0u))); uerror(nnt)
    numcg(nnt)=ncg;
    if uerror(nnt) < maxerror
        break
    end
    du=0*u;                          % CG iterations begin
    R=-L0u; MinvR=ifft2(fft2(R)./(c+K2));
    R2=sum(sum(R.*MinvR)); R20=R2;
    D=MinvR;
    while (R2 > R20*errorCG^2)
        L1D=ifft2(-K2.*fft2(D))-(V+3*u.*u+mu).*D;
        a=R2/sum(sum(D.*L1D));
        du=du+a*D;
        R=R-a*L1D; MinvR=ifft2(fft2(R)./(c+K2));
        R2old=R2;
        R2=sum(sum(R.*MinvR));
        b=R2/R2old;
        D=MinvR+b*D;
        ncg=ncg+1;
    end                              % CG iterations end
    u=u+du;
 end                     % Newton iterations end
 subplot(221);           % solution plotting
```

```
nr=70:1:188; mesh(x(nr),y(nr),u(nr, nr));
axis([x(70) x(188) x(70) x(188) -0.3 0.5]); view([-20 20]);
xlabel('x'); ylabel('y'); zlabel('u', 'rotation',0);
subplot(222);
semilogy(numcg, uerror, numcg, uerror, 'o');
axis([0 300 1e-12 1]);
xlabel('number of iterations'); ylabel('error')
```

One can see that the Newton-CG method is quite easy to implement. Running this code, one gets the gap soliton in Fig. 7.16(a) and the error diagram in Fig. 7.16(b). For the MSOM, the preconditioning operator is also taken as $M = c - \nabla^2$, with parameters c and Δt taken as their optimal values $c_{opt} = 2.5$ and $\Delta t_{opt} = 0.7$. The other conditions of the MSOM are identical to those in the Newton-CG method. The error diagram of the corresponding MSOM is also displayed in Fig. 7.16(b). It is seen that for the error to drop below 10^{-10}, the Newton-CG method takes six Newton iterations consisting of a total of 252 CG iterations. This number of CG iterations is 40 times less than the number of MSOM iterations. In terms of the CPU time, the Newton-CG method takes about 21 seconds (on a PC with a 2GHz processor), which is 120 times faster than the MSOM. Thus on this example, the Newton-CG method is over two orders of magnitude faster than the MSOM.

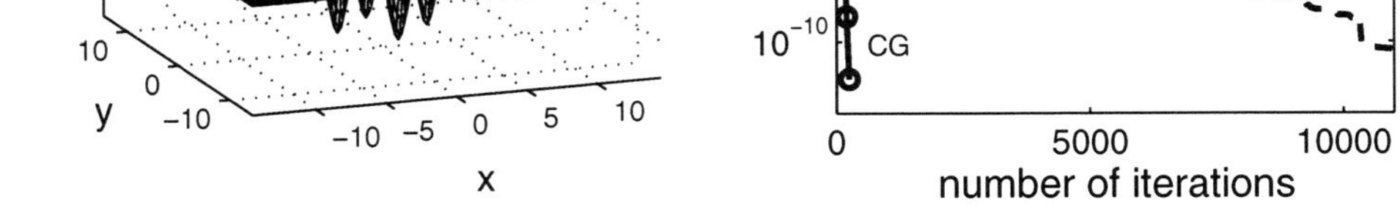

Figure 7.16. *Output of Program* 8 *on computation of gap solitons in the* 2D *NLS equation with a periodic potential and self-defocusing nonlinearity* (7.174) *by the Newton-CG method and MSOM (*$V_0 = 6$*):* (a) *solution profile at* $\mu = -4.56$ *(marked by letter "d" on the power curve of Fig.* 7.14(a)*);* (b) *error diagrams of the two methods versus the number of CG or MSOM iterations (each circle represents a Newton's iteration point in the Newton-CG method). (After Yang* (2009a).*)*

The fast convergence of the Newton-CG method can be heuristically understood as follows. For a matrix equation $Ax = b$, if the matrix A is symmetric and positive definite, then the error of the preconditioned CG method (by a preconditioning matrix M) decays at least by a factor of $R_{cg} = (\sqrt{\kappa} - 1)/(\sqrt{\kappa} + 1)$ with each iteration (Golub and Van Loan (1996)). Here κ is the spectral condition number of the matrix $M^{-1}A$, i.e., $\kappa = |\lambda_{max}/\lambda_{min}|$, where λ_{max} and λ_{min} are the largest and smallest eigenvalues of $M^{-1}A$. From this relation, we see that the number of CG iterations needed to reach a certain relative reduction in the error is roughly proportional to $\sqrt{\kappa}$. If we extrapolate this result to CG iterations on Eq. (7.170) in the Newton-CG method, then we can expect that the number of these CG iterations needed to reach a certain error reduction is roughly proportional to $\sqrt{\kappa}$, where κ is the spectral condition number of the operator $\mathbf{M}^{-1}\mathbf{L}_1$. For the Petviashvili method and the AITEM, however, the errors decay by a factor of $R = (\kappa - 1)/(\kappa + 1)$; see Pelinovsky and Stepanyants (2004), Yang and Lakoba (2008). Thus the number of Petviashvili iterations and the number of AITEM iterations needed to reach a certain error reduction are roughly proportional to κ. For the SOM and the MSOM, the errors decay roughly at the rate of $R = (\kappa^2 - 1)/(\kappa^2 + 1)$, since the spectral condition number of the squared operator in the SOM and the MSOM is roughly κ^2; see Yang and Lakoba (2007). Thus the number of SOM iterations and the number of MSOM iterations needed to reach a certain error reduction are roughly proportional to κ^2. In most solitary wave computations, κ is moderate or large, and thus the Newton-CG method converges faster or much faster than the Petviashvili method, AITEM, SOM, and MSOM.

In addition to the fast convergence, the Newton-CG method has some other advantages as well. For instance, this method can converge regardless of the kernel structure of the linearization operator $\mathbf{L}_1$. In other words, this method can converge even if the kernel of $\mathbf{L}_1$ contains eigenfunctions which are not translation-induced eigenmodes (Yang (2009a)). This contrasts the other iteration methods described earlier, which would not converge under such a circumstance. This indicates that the Newton-CG method can exhibit fast convergence in general situations.

The above Newton-CG method applies when the linearization operator $\mathbf{L}_1$ is self-adjoint, which is often the case in conservative wave systems. If $\mathbf{L}_1$ is non-self-adjoint, the CG iterations on the linear Newton-correction equation (7.170) generally do not converge. In such cases, one may solve Eq. (7.170) by the biconjugate-gradient method (Barret et al. (1994), Golub and Van Loan (1996), Yang (2009a)) or other extended CG methods (such as the generalized minimum residual method in Saad and Schultz (1986)). Such extended CG methods on non-self-adjoint equations do not provide guaranteed convergence in general. If one seeks a CG-type method with a guaranteed convergence, one can multiply the linear equation (7.170) by the adjoint operator $\mathbf{L}_{1n}^{\dagger}$ and turn it into a normal equation. The linear operator $\mathbf{L}_{1n}^{\dagger}\mathbf{L}_{1n}$ of this normal equation is self-adjoint and semi-positive-definite, and thus it can be solved by preconditioned CG iterations with guaranteed convergence. But the convergence speed of CG iterations on the normal equation will be slower since the condition number of the squared operator $\mathbf{L}_{1n}^{\dagger}\mathbf{L}_{1n}$ is much larger.

7.3 Numerical Methods for Linear-Stability Eigenvalues of Solitary Waves

When a solitary wave is obtained, an important problem is to determine its linear-stability spectrum, which consists of eigenvalues of the linear-stability operator of the solitary wave. This spectrum contains valuable information on the solitary wave. For instance, if this spectrum contains eigenvalues with positive real parts, then the solitary wave is linearly unstable. In this case, the largest real part of the eigenvalues gives the maximal growth rate of perturbations. If the spectrum contains purely imaginary discrete eigenvalues, these eigenvalues are the internal modes which cause long-lasting shape oscillations to the solitary wave. In this section, we describe numerical methods to compute this linear-stability spectrum for solitary waves.

Below we use the multidimensional NLS equation with general types of nonlinearities and potentials,

$$iU_t + \nabla^2 U + F(|U|^2, \mathbf{x})U = 0, \tag{7.175}$$

as an example to describe the numerical methods for stability spectrum. Here $F(\cdot,\cdot)$ is a real-valued function. This equation admits solitary waves of the form $U(\mathbf{x},t) = u(\mathbf{x})e^{i\mu t}$, where μ is the propagation constant, and $u(\mathbf{x})$ is treated as a general complex-valued solitary wave in this section. To analyze the linear stability of this solitary wave, it is perturbed by normal modes as

$$U(\mathbf{x},t) = \left\{u(\mathbf{x}) + [v(\mathbf{x}) + w(\mathbf{x})]e^{\lambda t} + [v^*(\mathbf{x}) - w^*(\mathbf{x})]e^{\lambda^* t}\right\} e^{i\mu t}, \tag{7.176}$$

where $v(\mathbf{x}), w(\mathbf{x}) \ll 1$ are normal-mode perturbations, and λ is the eigenvalue of this normal mode. Inserting this perturbed solution in (7.175) and linearizing, we obtain the following linear-stability eigenvalue problem:

$$\mathbf{L}\Psi = \lambda\Psi, \tag{7.177}$$

where

$$\mathbf{L} = i \begin{pmatrix} G_0 & \nabla^2 + G_1 \\ \nabla^2 + G_2 & -G_0 \end{pmatrix}, \quad \Psi = \begin{pmatrix} v \\ w \end{pmatrix}, \tag{7.178}$$

$$G_0 = \frac{1}{2}(u^2 - u^{*2})F_{|u|^2}(|u|^2,\mathbf{x}),$$

$$G_1 = -\mu + F(|u|^2,\mathbf{x}) + \left[|u|^2 - \frac{1}{2}(u^2 + u^{*2})\right] F_{|u|^2}(|u|^2,\mathbf{x}),$$

$$G_2 = -\mu + F(|u|^2,\mathbf{x}) + \left[|u|^2 + \frac{1}{2}(u^2 + u^{*2})\right] F_{|u|^2}(|u|^2,\mathbf{x}),$$

and $F_{|u|^2}$ is the partial derivative of $F(|u|^2,\mathbf{x})$ with respect to $|u|^2$. This eigenvalue problem (7.177) is an extension of a similar eigenvalue problem (5.19) from real-valued solitary waves to complex-valued solitary waves. Below we present numerical methods to compute eigenvalues of the linear-stability operator $\mathbf{L}$.

7.3.1 Fourier Collocation Method for the Whole Spectrum

To obtain the whole spectrum of the linear-stability operator $\mathbf{L}$, we will use the Fourier collocation method. In this method, one expands the eigenfunction Ψ into a Fourier series and turns Eq. (7.177) into a matrix eigenvalue problem for the Fourier coefficients of the eigenfunction Ψ. The structure of the resulting matrix depends heavily on the spatial dimension of Eq. (7.175). We will treat the 1D and 2D cases separately below.

Fourier Collocation Method in One Dimension

In one dimension, we first truncate the infinite x-axis into a finite interval $[-L/2, L/2]$, where L is the length of the interval. On this interval, we expand the eigenfunction $\Psi = [v,w]^T$ as well as the functions G_0, G_1, and G_2 in (7.178) into Fourier series:

$$v(x) = \sum_n a_n e^{ink_0x}, \quad w(x) = \sum_n b_n e^{ink_0x}, \tag{7.179}$$

$$G_j = \sum_n c_n^{(j)} e^{ink_0x}, \quad j = 0,1,2, \tag{7.180}$$

where $k_0 = 2\pi/L$. Substituting these expansions into the eigenvalue problem (7.177) and equating the coefficients of the same Fourier modes, the following eigenvalue system for the coefficients $\{a_j, b_j\}$ will be obtained:

$$\sum_n c_n^{(0)} a_{j-n} - (k_0 j)^2 b_j + \sum_n c_n^{(1)} b_{j-n} = -i\lambda a_j, \tag{7.181}$$

$$-(k_0 j)^2 a_j + \sum_n c_n^{(2)} a_{j-n} - \sum_n c_n^{(0)} b_{j-n} = -i\lambda b_j, \tag{7.182}$$

where $-\infty < j < \infty$. Truncating the number of Fourier modes to $-N \le j \le N$, this infinite-dimensional eigenvalue problem then reduces to the following finite-dimensional one:

$$i\begin{bmatrix} \mathcal{C}_0 & \mathcal{D}+\mathcal{C}_1 \\ \mathcal{D}+\mathcal{C}_2 & -\mathcal{C}_0 \end{bmatrix}\begin{bmatrix} A \\ B \end{bmatrix} = \lambda \begin{bmatrix} A \\ B \end{bmatrix}, \tag{7.183}$$

where

$$\mathcal{D} = (ik_0)^2 \operatorname{diag}(-N, -N+1, \dots, N-1, N)^2,$$

matrices $\mathcal{C}_0, \mathcal{C}_1$, and $\mathcal{C}_2$ are of the form

$$\mathcal{C} = \begin{pmatrix} c_0 & c_{-1} & \cdots & c_{-N} & & & \\ c_1 & c_0 & c_{-1} & \ddots & \ddots & & \\ \vdots & c_1 & c_0 & \ddots & \ddots & \ddots & \\ c_N & \ddots & \ddots & \ddots & \ddots & \ddots & c_{-N} \\ & c_N & \ddots & \ddots & \ddots & \ddots & \vdots \\ & & \ddots & \ddots & \ddots & \ddots & c_{-1} \\ & & & c_N & \cdots & c_1 & c_0 \end{pmatrix}$$

with coefficients c_n taking values $c_n^{(j)}$ for $\mathcal{C}_j$, and

$$A = (a_{-N}, a_{-N+1}, \dots, a_N)^T, \quad B = (b_{-N}, b_{-N+1}, \dots, b_N)^T. \tag{7.184}$$

The matrix eigenvalue problem (7.183) can be solved by either the QR algorithm (Golub and Van Loan (1996)) (when all eigenvalues are desired) or the Arnoldi algorithm (Arnoldi (1951), Sorensen (1992)) (when a portion of the eigenvalues are needed).

Below we illustrate the above algorithm by an example. Let us consider the following 1D cubic-quintic NLS equation:

$$iU_t + U_{xx} - |U|^2U + |U|^4U = 0. \tag{7.185}$$

Here the cubic nonlinearity is of self-defocusing type, and the quintic nonlinearity is of self-focusing type. This equation admits the following solitary waves (see Sec. 5.1):

$$U(x,t) = u(x)\,e^{i\mu t}, \tag{7.186}$$

where

$$u(x) = \left(\frac{4|B|\mu}{B+\cosh(2\sqrt{\mu}\,x)}\right)^{1/2}, \quad B = -\left(1+\frac{16}{3}\mu\right)^{-1/2}. \tag{7.187}$$

For this real solitary wave,

$$G_0 = 0, \quad G_1 = -\mu - u^2 + u^4, \quad G_2 = -\mu - 3u^2 + 5u^4.$$

The MATLAB code to compute the whole spectrum of this wave is displayed below.

Program 9

```
% p9.m: Fourier collocation method to compute the whole stability
% spectrum of solitary waves in equation: iU_t+U_xx-|U|^2U+|U|^4U=0

  N=150; L=40; Nx=4*N; dx=L/Nx; x=-L/2:dx:L/2-dx; k0=2*pi/L;
  mu=1; B=-1/sqrt(1+16*mu/3);
  u=sqrt(-4*B*mu./(B+cosh(2*sqrt(mu)*x)));
  G1=-mu-u.^2+u.^4; G2=-mu-3*u.^2+5*u.^4;
  for n=-N:1:N
      c1(n+N+1)=dx*sum(G1.*exp(-i*k0*n*x))/L;
      c2(n+N+1)=dx*sum(G2.*exp(-i*k0*n*x))/L;
  end
  D=(i*k0*diag([-N:N])).^2;
  C0=zeros(2*N+1);
  C1=toeplitz([c1(N+1:2*N+1) zeros(1,N)],[c1(N+1:-1:1) zeros(1,N)]);
  C2=toeplitz([c2(N+1:2*N+1) zeros(1,N)],[c2(N+1:-1:1) zeros(1,N)]);
  M=[ C0     D+C1
      D+C2    -C0 ];
  eigvalues=eig(i*M);
  plot(eigvalues, '.'); axis([-5 5 -5 5]);
```

On a 2GHz PC, this code takes about 7 seconds. The output of this code is displayed in Fig. 7.17(a). It is seen that the spectrum contains a pair of real simple eigenvalues $\pm\lambda_u \approx \pm 2.1612$, and a zero eigenvalue of multiplicity four. The rest of the spectrum is continuous eigenvalues. This solitary wave is linearly unstable due to the positive eigenvalue λ_u. This instability is of Vakhitov–Kolokolov type since $P'(\mu) < 0$ here. The accuracy of the two simple discrete eigenvalues $\pm\lambda_u$ from this code is about 10^{-7}, and the accuracy of the quadruple-fold zero eigenvalue is about 10^{-4}.

The Fourier collocation method presented above is spectrally accurate, i.e., the error of discrete eigenvalues in the spectrum decays exponentially with the number of Fourier modes. To confirm this spectral accuracy, we vary the number of Fourier modes in Program 9. The error of the unstable eigenvalue λ_u versus the number of Fourier modes $(2N+1)$ is displayed in Fig. 7.17(b). For comparison, the benchmark exponentially decaying function $e^{-0.08(2N+1)}$ is also shown. One can see that this error indeed decays exponentially, thus

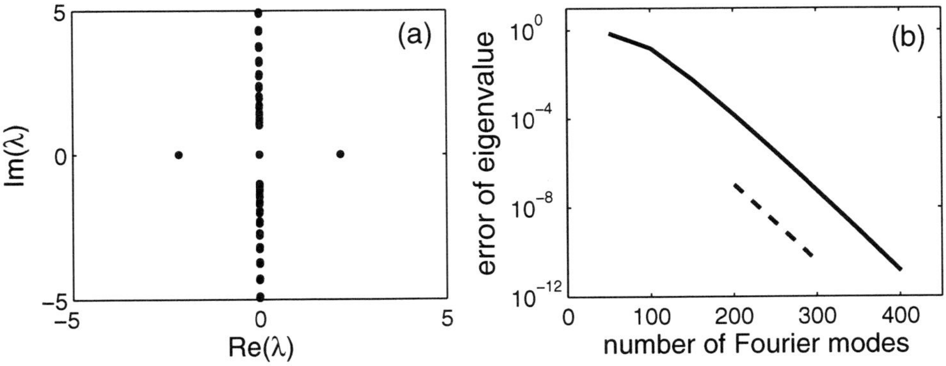

Figure 7.17. *Computation of the whole stability spectrum for the solitary wave* (7.187) *in the* 1*D cubic-quintic NLS equation* (7.185) *with* $\mu = 1$*:* (a) *output of Program* 9 *which gives the whole stability spectrum;* (b) *error of the unstable eigenvalue* λ_u *versus the number of Fourier modes* $(2N+1)$*. The dashed line is the benchmark function* $e^{-0.08(2N+1)}$ *for comparison.*

the spectral accuracy is confirmed. Note that in Program 9, the Fourier coefficients $c_n^{(1)}$ and $c_n^{(2)}$ of functions G_1 and G_2 are computed by the simple Riemann summation. As we have explained in Sec. 2.4.3 (see also Trefethen (2000) and Boyd (2001)), for periodic or exponentially decaying integrand functions (as is the case here), this Riemann summation for the integrals is spectrally accurate rather than first-order accurate in grid spacing Δx. Thus it does not compromise the spectral accuracy of the above Fourier collocation method, as Fig. 7.17(b) clearly confirms.

Fourier Collocation Method in Two Dimensions

In two dimensions, the stability eigenvalue problem (7.177) can be similarly solved by the Fourier collocation method, but the building of the reduced matrix is different. In this case, we first truncate the infinite (x,y) plane into a finite region of $-L_x/2 \le x \le L_x/2, -L_y/2 \le y \le L_y/2$, where L_x and L_y are the lengths along the x and y directions. In this finite region, we expand the eigenfunction $\Psi = [v,w]^T$ as well as the functions G_0, G_1, and G_2 into Fourier series:

$$v(x,y) = \sum_{m,n} a_{m,n}\, e^{imk_{0x}x+ink_{0y}y}, \quad w(x,y) = \sum_{m,n} b_{m,n}\, e^{imk_{0x}x+ink_{0y}y}, \tag{7.188}$$

$$G_j = \sum_{m,n} c_{m,n}^{(j)}\, e^{imk_{0x}x+ink_{0y}y}, \quad j = 0,1,2, \tag{7.189}$$

where $k_{0x} = 2\pi/L_x$ and $k_{0y} = 2\pi/L_y$. Substituting these expansions into the eigenvalue problem (7.177) and equating coefficients of the same Fourier modes, the following eigenvalue system for the coefficients $\{a_{j,k}, b_{j,k}\}$ will be obtained:

$$\sum_{m,n} c^{(0)}_{m,n} a_{j-m,k-n} - [(k_{0x}j)^2 + (k_{0y}k)^2] b_{j,k} + \sum_{m,n} c^{(1)}_{m,n} b_{j-m,k-n} = -i\lambda a_{j,k},$$

$$-[(k_{0x}j)^2 + (k_{0y}k)^2] a_{j,k} + \sum_{m,n} c^{(2)}_{m,n} a_{j-m,k-n} - \sum_{m,n} c^{(0)}_{m,n} b_{j-m,k-n} = -i\lambda b_{j,k},$$

where $-\infty < j,k < \infty$. Truncating the number of Fourier modes to $-N \le j,k \le N$, this eigenvalue problem then becomes similar to (7.183), except that matrices $\mathcal{D}, \mathcal{C}_0, \mathcal{C}_1$, and $\mathcal{C}_2$ are different. This matrix eigenvalue problem can be solved by either the QR or Arnoldi algorithm (Arnoldi (1951), Golub and Van Loan (1996)).

To illustrate this 2D algorithm, we consider the linear stability of vortex solitons in the 2D NLS equation

$$iU_t + U_{xx} + U_{yy} + |U|^2U = 0. \tag{7.190}$$

These vortex solitons are of the form $U(x,y,t) = u(r)e^{i\theta + i\mu t}$, where (r,θ) are the polar coordinates of the (x,y) plane, and μ is the propagation constant. These solitons can be computed by either the SOM, MSOM, or Newton-CG method. At $\mu = 1$, the amplitude profile of this vortex is shown in Fig. 7.18(a). The full stability spectrum of this vortex soliton is computed by the above Fourier collocation method. The MATLAB code for the computation of both the vortex soliton and its stability spectrum is displayed below.

Program 10

```
% p10.m: Fourier collocation method for computing the whole stabilit
% spectrum of vortex solitons in 2D equation: iU_t+U_xx+U_yy+|U|^2U=(

 N=20; Lx=20; Ly=20; Nx=4*N; Ny=4*N; maxuerror=1e-7;
 dx=Lx/Nx;x=-Lx/2:dx:Lx/2-dx; kx0=2*pi/Lx;kx=[0:Nx/2-1 -Nx/2:-1]*kx0
 dy=Ly/Ny;y=-Ly/2:dy:Ly/2-dy; ky0=2*pi/Ly;ky=[0:Ny/2-1 -Ny/2:-1]*ky0
 [X, Y]=meshgrid(x, y); [TH R]=cart2pol(X,Y);
 [KX, KY]=meshgrid(kx, ky); K2=KX.^2+KY.^2;
 mu=1; u=2.5*R.*sech(R).*exp(i*TH); c=2; DT=0.95; uerror=1;
 while uerror > maxuerror             % SOM for vortex soliton starts
    F=u.*conj(u)-mu;
    L0u=ifft2(-K2.*fft2(u))+F.*u;
    uerror=max(max(abs(L0u)))
    MinvL0u=ifft2(fft2(L0u)./(c+K2));
```

```
    L1MinvL0u=ifft2(-K2.*fft2(MinvL0u))+F.*MinvL0u    ...
              +2*u.*real(conj(u).*MinvL0u);
    MinvL1MinvL0u=ifft2(fft2(L1MinvL0u)./(c+K2));
    u=u-MinvL1MinvL0u*DT;
end                                    % SOM for vortex soliton ends
G0=(u.*u-conj(u).*conj(u))/2;
G1=-mu+2*u.*conj(u)-(u.*u+conj(u).*conj(u))/2;
G2=-mu+2*u.*conj(u)+(u.*u+conj(u).*conj(u))/2;
for j=1:2*N+1
  for k=1:2*N+1
    efactor=exp(-i*((j-N-1)*kx0*X+(k-N-1)*ky0*Y))/(Lx*Ly);
    c0(j,k)=dx*dy*sum(sum(G0.*efactor));
    c1(j,k)=dx*dy*sum(sum(G1.*efactor));
    c2(j,k)=dx*dy*sum(sum(G2.*efactor));
  end
end
D=zeros((2*N+1)^2); C0=D; C1=D; C2=D;
for jj=1:(2*N+1)^2
  j=-N+floor((jj-1)/(2*N+1));  k=-N+mod(jj-1, 2*N+1);
  for kk=1:(2*N+1)^2
    m=-N+floor((kk-1)/(2*N+1));  n=-N+mod(kk-1, 2*N+1);
    ind1=j-m+N+1; ind2=k-n+N+1;
    if ind1 > 0 & ind1 < 2*N+2 & ind2 > 0 & ind2 < 2*N+2
       C0(jj, kk)=c0(ind1, ind2);
       C1(jj, kk)=c1(ind1, ind2);
       C2(jj, kk)=c2(ind1, ind2);
    end
  end
  D(jj, jj)=-((j*kx0)^2+(k*ky0)^2);
end
M=[ C0    D+C1
    D+C2   -C0 ];
M=sparse(M); eigvalues=eigs(i*M, 200, 'SM');
subplot(221)
mesh(x, y, abs(u)); view([-30, 50]);
subplot(222)
plot(eigvalues, '.', 'markersize', 10); axis([-3 3 -3 3])
```

In this code, the vortex soliton is computed by the SOM. The algorithm is formulated for the real and imaginary parts of the vortex solution. In its execution, the real and imaginary

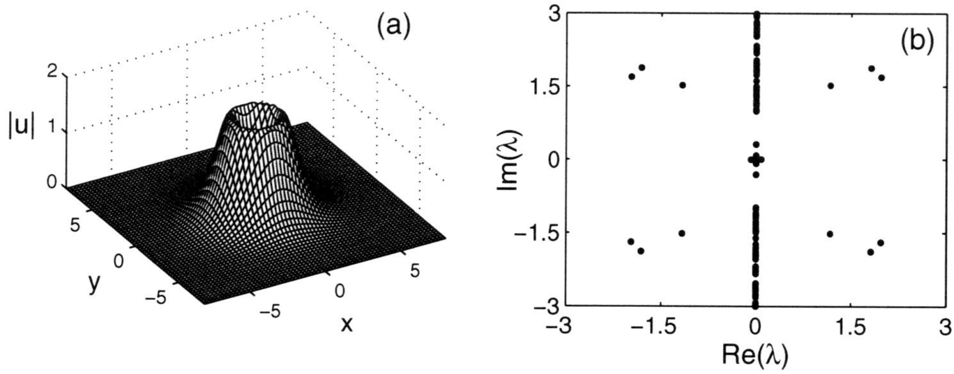

Figure 7.18. *Output of Program* 10 *on computation of the whole stability spectrum for the vortex soliton in the 2D NLS equation* (7.190) *with* $\mu = 1$*:* (a) *amplitude profile of the vortex soliton;* (b) *whole stability spectrum of this soliton.*

parts are recombined, which doubles the speed of the algorithm. The choice of the SOM is solely because its MATLAB code is the shortest. If faster speed is desired, this SOM can be replaced by the MSOM or the Newton-CG method. Eigenvalues of the resulting matrix are obtained by the Arnoldi algorithm through the use of MATLAB command "eigs." The spatial box in this code is taken as $-10 \leq x, y \leq 10$, and 41 Fourier modes ($N = 20$) are used along each dimension. On a 2GHz PC, this code takes about 2.6 minutes to finish. The output of this program is shown in Fig. 7.18. It is seen that the discrete spectrum of this vortex soliton contains three quadruples of complex eigenvalues, a pair of purely imaginary eigenvalues, and a zero eigenvalue of multiplicity eight. The complex eigenvalues reveal that this vortex soliton is linearly unstable due to oscillatory instabilities. The relative errors of complex eigenvalues from this code are about 0.01%, the relative errors of the purely imaginary eigenvalues are about 0.1%, and the maximum errors of the multifold zero eigenvalue are about 0.08.

It is noted that the Fourier collocation method above can produce not only the eigenvalue spectrum, but also the corresponding eigenfunctions. For example, in Program 9, if the MATLAB command "[eigvectors, eigvalues]=eig(i*M)" is used instead of "eigvalues=eig(i*M)," then both the eigenvectors and eigenvalues of the matrix iM would be obtained. Recall that these eigenvectors are the Fourier coefficients of eigenfunctions $v(x)$ and $w(x)$; see (7.179) and (7.184). Thus by inserting these Fourier coefficients into Fourier expansions (7.179), the corresponding eigenfunctions (v, w) would be obtained.

In this Fourier collocation method for the stability spectra of 2D solitary waves, if the number of Fourier modes needs to increase (for higher accuracy or when a larger spatial domain is taken), this method can quickly run into difficulty due to memory and speed constraints. Notice that the size of the underlying matrix is $2(2N+1)^2$ by $2(2N+1)^2$, which goes up very quickly with larger N. For example, if N increases to 30, the resulting matrix will be 7442×7442. Moreover, this matrix is dense. This large dense matrix puts a high demand on the computer's memory. Indeed, MATLAB can easily run out of memory with even a modest increase in N. In addition, eigenvalue computations for large-size matrices slow down quickly as well. In the next subsection, we will present iterative methods for the computation of individual eigenvalues and their eigenfunctions. These iterative methods use very little memory and can yield individual eigenvalues very quickly.

7.3.2 Newton Conjugate-Gradient Method for Individual Eigenvalues

Several iteration methods have been proposed by Yang (2008) for computing individual discrete eigenvalues in the linear-stability spectrum. These methods were based on iterating certain time evolution equations associated with the linear stability eigenvalue problem, which is in the same spirit as the ITEM and the SOMs for solitary wave computations. Below we describe another method which is based on Newton and CG iterations. The basic idea is that we treat the linear eigenvalue equation as a nonlinear "wave" equation for the (localized) discrete eigenfunction when the eigenvalue is expressed in terms of the eigenfunction through the Rayleigh quotient. Then this nonlinear "wave" equation for the discrete eigenfunction can be solved by the Newton-CG-type methods in Sec. 7.2.4. Under this treatment, the Newton-CG method for eigenvalue computations is an extension of the previous Newton-CG method for solitary wave computations. The main difference in the Newton-CG method for eigenvalue computations is that, since the underlying linearization operator here is non-self-adjoint, the Newton-correction equation cannot be solved by CG iterations or bi-CG iterations in general. Thus we will turn this Newton-correction equation into a normal equation first, and then use preconditioned CG iterations to solve it. This Newton-CG method for eigenvalues is more efficient than the iteration methods proposed in Yang (2008) and will be described below.

Let us consider a general linear eigenvalue problem

$$\mathbf{L}\Psi = \lambda\Psi, \tag{7.191}$$

where $\mathbf{L}$ is a linear operator, Ψ is its eigenfunction, and λ is its eigenvalue. Suppose this linear operator has a discrete eigenfunction Ψ. Then by taking the inner product of (7.191) with Ψ, the corresponding eigenvalue λ can be expressed as the Rayleigh quotient

$$\lambda = \frac{\langle \Psi, \mathbf{L}\Psi \rangle}{\langle \Psi, \Psi \rangle}. \tag{7.192}$$

Here the inner product is the standard one (7.98) in the square-integrable functional space. Inserting this λ formula into (7.191), this linear eigenvalue equation then becomes a nonlinear "solitary wave" equation for the discrete eigenfunction Ψ:

$$\mathcal{L}_0\Psi = 0, \tag{7.193}$$

where

$$\mathcal{L}_0\Psi = \mathbf{L}\Psi - \frac{\langle\Psi, \mathbf{L}\Psi\rangle}{\langle\Psi,\Psi\rangle}\Psi. \tag{7.194}$$

To solve this nonlinear equation, we use Newton's method first (see Sec. 7.2.4) and iteratively update the approximation for Ψ as

$$\Psi_{n+1} = \Psi_n + \Delta\Psi_n, \tag{7.195}$$

where the updated amount $\Delta\Psi_n$ is computed from the linear Newton-correction equation

$$\mathcal{L}_{1n}\Delta\Psi_n = -\mathcal{L}_0\Psi_n. \tag{7.196}$$

Here $\mathcal{L}_1$ is the linearization operator of the nonlinear equation (7.193):

$$\mathcal{L}_1\Phi = \mathbf{L}_1\Phi - \frac{\langle\Psi, \mathbf{L}_1\Phi\rangle}{\langle\Psi,\Psi\rangle}\Psi, \tag{7.197}$$

$$\mathbf{L}_1 = \mathbf{L} - \lambda, \tag{7.198}$$

λ is given by the formula (7.192), and $\mathcal{L}_{1n}$ is the operator $\mathcal{L}_1$ evaluated at the function Ψ_n. For linear-stability operators of solitary waves (see (7.178)), the operator $\mathcal{L}_{1n}$ is non-self-adjoint in general, and thus the Newton-correction equation (7.196) cannot be solved by CG iterations. Bi-CG iterations generally do not converge either, especially when the eigenvalue λ is complex valued. Other generalized CG methods such as the generalized minimum residual method might work, but such methods involve more effort in implementation. Our plan for solving Eq. (7.196) is to first turn it into a "normal equation," so that it becomes solvable by CG iterations. However, instead of multiplying (7.196) by the Hermitian $\mathcal{L}_{1n}^{\dagger}$ of $\mathcal{L}_{1n}$, as one normally does to get a standard normal equation, we multiply (7.196) by $\mathbf{L}_{1n}^{\dagger}$ and get the following simpler "normal equation":

$$\mathbf{L}_{1n}^{\dagger}\mathcal{L}_{1n}\Delta\Psi_n = -\mathbf{L}_{1n}^{\dagger}\mathcal{L}_0\Psi_n. \tag{7.199}$$

Here $\mathbf{L}_1^{\dagger} = \mathbf{L}^{\dagger} - \lambda^*$ is the Hermitian of $\mathbf{L}_1$, and $\mathbf{L}_{1n}^{\dagger}$ is $\mathbf{L}_1^{\dagger}$ evaluated at Ψ_n. It is easy to check that the operator $\mathbf{L}_{1n}^{\dagger}\mathcal{L}_{1n}$ in this simpler "normal equation" is also self-adjoint and semi-positive-definite, and thus it will be solved by preconditioned CG iterations. The algorithm for preconditioned CG iterations on Eq. (7.199) is the same as that described in Sec. 7.2.4, except that $\mathbf{L}_0$ is replaced by $\mathbf{L}_{1n}^{\dagger}\mathcal{L}_0$, $\mathbf{L}_1$ replaced by $\mathbf{L}_{1n}^{\dagger}\mathcal{L}_{1n}$, and $\mathbf{u}$ replaced by Ψ. The outer Newton iterations (7.195), coupled with inner preconditioned CG iterations on the "normal equation" (7.199), form the Newton-CG method we propose for the eigenvalue problem (7.191).

Now we use an example to illustrate the above Newton-CG method. Let us consider eigenvalues of the vortex soliton in Fig. 7.18 for the 2D NLS equation (7.190). The eigenvalue spectrum of this vortex soliton is shown in Fig. 7.19(a) (it can be compared to the numerically obtained spectrum in Fig. 7.18(b)). A most unstable eigenvalue in this spectrum is $\lambda_u = 1.9718 + 1.6927i$, which is indicated by an arrow in Fig. 7.19(a). To compute this eigenvalue by the Newton-CG method, we take the spatial domain as $-15 \leq x, y, \leq 15$, discretized by 128 points along each dimension. The linear-stability operator $\mathbf{L}$ is given in Eq. (7.178), with $u(x, y)$ being the vortex soliton in Fig. 7.18. The preconditioning operator is taken as

$$\mathbf{M} = (c - \nabla^2)^2 \text{diag}(1, 1), \quad c = 4,$$

and the initial condition is taken as

$$\Psi_0 = \begin{pmatrix} \tanh 0.5r \left[\text{sech}^2(0.5 - 0.2i)r\, e^{-i\theta} + \text{sech}^2(0.5 + 0.5i)r\, e^{-3i\theta} \right] \\ \tanh 0.5r \left[\text{sech}^2(0.5 - 0.2i)r\, e^{-i\theta} - \text{sech}^2(0.5 + 0.5i)r\, e^{-3i\theta} \right] \end{pmatrix},$$

where (r, θ) is the polar coordinate of the (x, y) plane. The corresponding MATLAB code of this Newton-CG method for λ_u (including computation of the vortex soliton itself) is displayed below.

Program 11

```
% p11.m:  Newton-conjugate-gradient method for computing eigenvalues
% of vortex solitons in the 2D NLS equation: iU_t+U_xx+U_yy+|U|^2U=0

 Lx=30; Ly=30; Nx=128; Ny=128; maxerror=1e-10; errorCG=1e-2;
 x=-Lx/2:Lx/Nx:Lx/2-Lx/Nx; kx=[0:Nx/2-1  -Nx/2:-1]*2*pi/Lx;
 y=-Ly/2:Ly/Ny:Ly/2-Ly/Ny; ky=[0:Ny/2-1  -Ny/2:-1]*2*pi/Ly;
 [X, Y]=meshgrid(x, y); [TH R]=cart2pol(X,Y);
 [KX, KY]=meshgrid(kx, ky); K2=KX.^2+KY.^2;
 mu=1; u=2.5*R.*sech(R).*exp(i*TH); c=4; DT=1.95; uerror=1;
 while uerror > maxerror            % SOM for vortex soliton starts
    F=u.*conj(u)-mu;
    L0u=ifft2(-K2.*fft2(u))+F.*u;
    uerror=max(max(abs(L0u)))
    MinvL0u=ifft2(fft2(L0u)./(c+K2));
    L1MinvL0u=ifft2(-K2.*fft2(MinvL0u))+F.*MinvL0u    ...
```

```
            +2*u.*real(conj(u).*MinvL0u);
   MinvL1MinvL0u=ifft2(fft2(L1MinvL0u)./(c+K2));
   u=u-MinvL1MinvL0u*DT;
end                                % SOM for vortex soliton ends
G0=(u.*u-conj(u).*conj(u))/2;
G1=-mu+2*u.*conj(u)-(u.*u+conj(u).*conj(u))/2;
G2=-mu+2*u.*conj(u)+(u.*u+conj(u).*conj(u))/2;
f1=tanh(0.5*R).*sech((0.5-0.2i)*R).^2;
f2=tanh(0.5*R).*sech((0.5+0.5i)*R).^2;
v=f1.*exp(-i*TH)+f2.*exp(-3*i*TH);           % initial conditions
w=f1.*exp(-i*TH)-f2.*exp(-3*i*TH);
ncg=0; nnt=0; fftM=(c+K2).^2;
while 1                     % Newton-CG method for eigenvalue starts
   nnt=nnt+1;
   v2w2=sum(sum(conj(v).*v+conj(w).*w));
   Lv=i*( G0.*v-ifft2(K2.*fft2(w))+G1.*w);
   Lw=i*(-G0.*w-ifft2(K2.*fft2(v))+G2.*v);
   lambda=sum(sum(conj(v).*Lv+conj(w).*Lw))/v2w2;
   curlL0v=Lv-lambda*v;
   curlL0w=Lw-lambda*w;
   vwerror(nnt)=max(max(abs(curlL0v)+abs(curlL0w))); vwerror(nnt)
   numcg(nnt)=ncg;
   if vwerror(nnt) < maxerror
       break
   end
   L1HcurlL0v=-i*( conj(G0).*curlL0v-ifft2(K2.*fft2(curlL0w)) ...
                  +conj(G2).*curlL0w)-conj(lambda)*curlL0v;
   L1HcurlL0w=-i*(-conj(G0).*curlL0w-ifft2(K2.*fft2(curlL0v)) ...
                  +conj(G1).*curlL0v)-conj(lambda)*curlL0w;
   dv=0*v;  dw=0*w;                         % CG iterations start
   Rv=-L1HcurlL0v;  Rw=-L1HcurlL0w;
   MinvRv=ifft2(fft2(Rv)./fftM);  MinvRw=ifft2(fft2(Rw)./fftM);
   R2=sum(sum(conj(Rv).*MinvRv+conj(Rw).*MinvRw));  R20=R2;
   Dv=MinvRv;  Dw=MinvRw;
   while (R2 > R20*errorCG^2)
       L1Dv=i*( G0.*Dv-ifft2(K2.*fft2(Dw))+G1.*Dw)-lambda*Dv;
       L1Dw=i*(-G0.*Dw-ifft2(K2.*fft2(Dv))+G2.*Dv)-lambda*Dw;
       gamma=sum(sum(conj(v).*L1Dv+conj(w).*L1Dw))/v2w2;
       curlL1Dv=L1Dv-gamma*v;
       curlL1Dw=L1Dw-gamma*w;
```

```
        L1HcurlL1Dv=-i*( conj(G0).*curlL1Dv-ifft2(K2.*      ...
          fft2(curlL1Dw))+conj(G2).*curlL1Dw)-conj(lambda)*curlL1Dv;
        L1HcurlL1Dw=-i*(-conj(G0).*curlL1Dw-ifft2(K2.*      ...
          fft2(curlL1Dv))+conj(G1).*curlL1Dv)-conj(lambda)*curlL1Dw;
        a=R2/sum(sum(conj(Dv).*L1HcurlL1Dv+conj(Dw).*L1HcurlL1Dw));
        dv=dv+a*Dv;  dw=dw+a*Dw;
        Rv=Rv-a*L1HcurlL1Dv;  Rw=Rw-a*L1HcurlL1Dw;
        MinvRv=ifft2(fft2(Rv)./fftM);  MinvRw=ifft2(fft2(Rw)./fftM);
        R2old=R2;
        R2=sum(sum(conj(Rv).*MinvRv+conj(Rw).*MinvRw));
        b=R2/R2old;
        Dv=MinvRv+b*Dv;  Dw=MinvRw+b*Dw;
        ncg=ncg+1;
    end                                  % CG iterations end
    v=v+dv;  w=w+dw;
end                        % Newton-CG method for eigenvalue ends
lambda
subplot(221); imagesc(x,y,abs(u)); axis xy
subplot(222); semilogy(numcg, vwerror, 'o', numcg, vwerror);
xlabel('number of CG iterations'); ylabel('error');
axis([0 90 1e-12 100]);
subplot(223); imagesc(x, y, abs(v)); axis xy
subplot(224); imagesc(x, y, angle(v)); axis xy
```

In this program, the vortex soliton is computed by the SOM for its shorter MATLAB coding. The rest of the program is the Newton-CG method for eigenvalue computations of this vortex soliton. The output of this program is shown in Fig. 7.19. It is seen that for the error of the eigenfunction, measured as the maximum of the residue in Eq. (7.191), to drop below 10^{-10}, only a total of 83 CG iterations (in seven Newton iterations) are needed. Regarding the CPU time, the Newton-CG part of Program 11 takes less than 5 seconds to reach accuracy 10^{-10}. This kind of high accuracy and speed is very difficult to reach by the Fourier collocation method of the previous subsection. The lower row of Fig. 7.19 shows the amplitude and phase of the eigenfunction $v(x, y)$. The eigenfunction $w(x, y)$ is similar and thus not shown. Using this Newton-CG method, the other discrete eigenvalues in the stability spectrum of this vortex soliton can be similarly computed. For the other eigenvalues, the MATLAB code is exactly the same, except that the initial condition for the eigenfunction (v, w) is different.

It is noted that when the solitary wave $u(\mathbf{x})$ in Eq. (7.175) is real, and the eigenvalue λ is purely real or purely imaginary (i.e., λ^2 is real), then a more efficient CG-type method

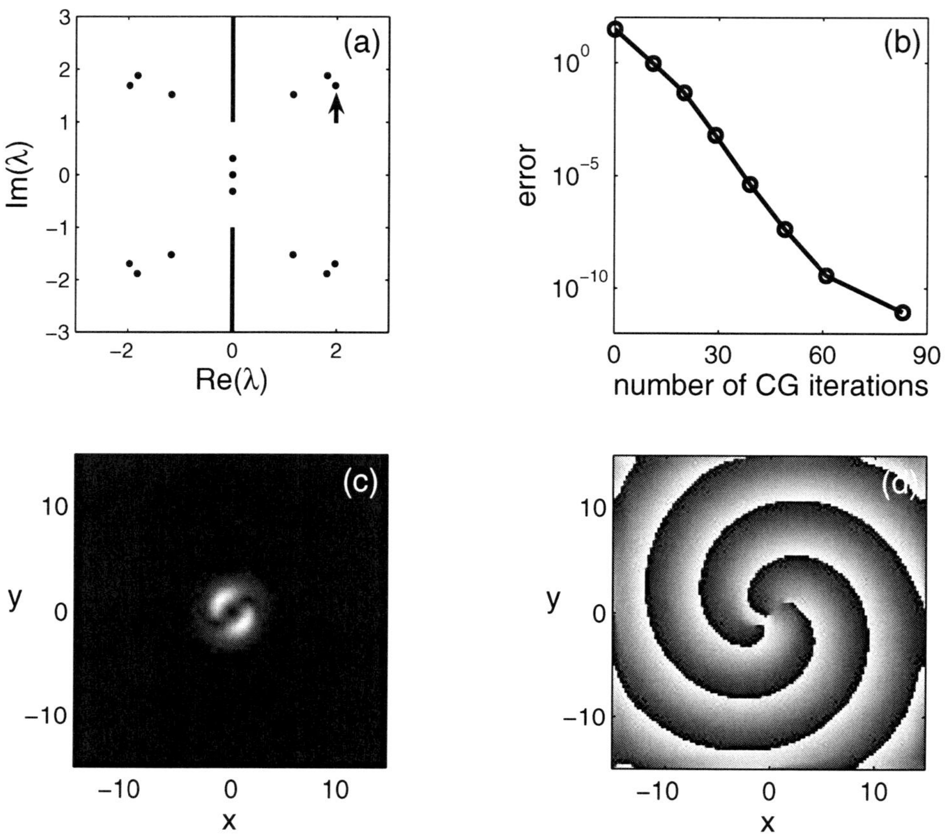

Figure 7.19. *Output of Program* 11 *on computation of an unstable eigenvalue for the vortex soliton in Fig.* 7.18(a) *by the Newton-CG method:* (a) *the whole stability spectrum of this soliton, where the unstable eigenvalue* λ_u *to be computed is indicated by an arrow;* (b) *error diagram of the Newton-CG method versus the number of CG iterations (each circle represents a Newton's iteration);* (c, d) *amplitude and phase of the eigenfunction* $v(x, y)$.

can be used to compute the eigenmode. In this case, the eigenvalue problem (7.177) can be rewritten as

$$L_0 L_1 v = -\lambda^2 v, \tag{7.200}$$

where

$$L_0 = \nabla^2 - \mu + F(u^2, \mathbf{x}), \tag{7.201}$$

$$L_1 = \nabla^2 - \mu + F(u^2, \mathbf{x}) + 2u^2 F_{u^2}(u^2, \mathbf{x}), \tag{7.202}$$

and v is a real eigenfunction. When λ^2 is obtained from the Rayleigh quotient

$$\lambda^2 = -\frac{\langle v, L_0 L_1 v \rangle}{\langle v, v \rangle} \tag{7.203}$$

and inserted into the eigenvalue equation (7.200), this equation becomes a real-valued nonlinear "solitary wave" equation for the discrete eigenfunction $v(\mathbf{x})$. This equation can be readily solved by the Newton-bi-CG method described in Yang (2009a), where the Newton-correction equation is computed directly by bi-CG iterations. In this method, since the Newton-correction equation is not turned into a "normal equation" (which has a higher spectral condition number), it is thus more efficient than the Newton-CG method described earlier in this subsection.

The Newton-CG/bi-CG methods above use very little memory and converge to eigenvalues and eigenfunctions very quickly; thus they can handle situations where large spatial domains and high accuracy are needed. They are ideal for tracing individual eigenvalues when the soliton's propagation constant (or other system parameter) continuously varies. The only potential difficulty in their application is that they require an initial guess for the eigenfunction being sought. In some cases, the basic information on the underlying eigenfunction is known; hence a reasonable initial guess is easy to provide. But there are cases where no prior information is available on the eigenvalues and eigenfunctions to be computed. In such cases, one may first use a low-accuracy Fourier collocation method to obtain an approximate stability spectrum and their corresponding eigenfunctions. These approximate eigenfunctions then can be used as initial guesses for the Newton-CG/bi-CG methods. In certain cases, the eigenfunctions at some special system parameters are known. If so, then one can start the Newton-CG/bi-CG methods from such special parameters and eigenfunctions and use the continuation technique to obtain eigenvalues at the desired system parameters.

Bibliography

M. Ablowitz, I. Bakirtas, and B. Ilan (2005), "Wave collapse in a class of nonlocal nonlinear Schrödinger equation," Physica D 207, 230–253.

M.J. Ablowitz, G. Biondini, S. Chakravarty, R.B. Jenkins, and J.R. Sauer (1996), "Four-wave mixing in wavelength-division-multiplexed soliton systems: Damping and amplification," Opt. Lett. 21, 1646–1648.

M.J. Ablowitz and P. Clarkson (1992), *Solitons, Nonlinear Evolution Equations and Inverse Scattering* (Cambridge University Press, Cambridge, UK).

M.J. Ablowitz, D.J. Kaup, A.C. Newell, and H. Segur (1974), "The inverse scattering transform—Fourier analysis for nonlinear problems," Stud. Appl. Math. 53, 249–315.

M.J. Ablowitz, M.D. Kruskal, and J.F. Ladik (1979), "Solitary wave collisions," SIAM J. Appl. Math. 36, 428–437.

M.J. Ablowitz and Z.H. Musslimani (2005), "Spectral renormalization method for computing self-localized solutions to nonlinear systems," Opt. Lett. 30, 2140–2142.

M.J. Ablowitz and H. Segur (1979), "On the evolution of packets of water waves," J. Fluid Mech. 92, 691–715.

M.J. Ablowitz and H. Segur (1981), *Solitons and the Inverse Scattering Transform* (SIAM, Philadelphia).

G.P. Agrawal (2007), *Nonlinear Fiber Optics*, 4th edition (Academic Press, Boston).

B. Akers and P.A. Milewski (2008), "A stability result for solitary waves in nonlinear dispersive equations," Comm. Math. Sci. 6, 791–797.

T.R. Akylas and Y. Cho (2008), "On the stability of lumps and wave collapse in water waves," Philos. Trans. R. Soc. Lond. A Math. Phys. Eng. Sci. 366, 2761–2774.

T.R. Akylas and R.H.J. Grimshaw (1992), "Solitary internal waves with oscillatory tails," J. Fluid Mech. 242, 279–298.

T.R. Akylas and T.J. Kung (1990), "On nonlinear wave envelopes of permanent form near a caustic," J. Fluid Mech. 214, 489–502.

T.R. Akylas and T.S. Yang (1995), "On short-scale oscillatory tails of long-wave disturbances," Stud. Appl. Math. 94, 1–20.

T.J. Alexander, E.A. Ostrovskaya, and Yu.S. Kivshar (2006), "Self-trapped nonlinear matter waves in periodic potentials," Phys. Rev. Lett. 96, 040401.

D. Anderson (1983), "Variational approach to nonlinear pulse propagation in optical fibers," Phys. Rev. A 27, 3135–3145.

D. Anderson, A. Bonderson, and M. Lisak (1979), "Stability analysis of lower-hybrid cones," Physica Scripta 20, 343–345.

M.H. Anderson, J.R. Ensher, M.R. Matthews, C.E. Wieman, and E.A. Cornell (1995), "Observation of Bose-Einstein condensation in a dilute atomic vapor," Science 269, 198–201.

T. Anker, M. Albiez, R. Gati, S. Hunsmann, B. Eiermann, A. Trombettoni, and M.K. Oberthaler (2005), "Nonlinear self-trapping of matter waves in periodic potentials," Phys. Rev. Lett. 94, 020403.

P. Anninos, S. Oliveira, and R.A. Matzner (1991), "Fractal structure in the scalar $\lambda(\phi^2 - 1)^2$ theory," Phys. Rev. D 44, 1147–1160.

A. Argyros, T.A. Birks, S.G. Leon-Saval, C.M.B. Cordeiro, F. Luan, and P.St.J. Russell (2005), "Photonic bandgap with an index step of one percent," Opt. Express 13, 309–314.

W.E. Arnoldi (1951), "The principle of minimized iterations in the solution of the matrix eigenvalue problem," Quart. Appl. Math. 9, 17–29.

B.B. Baizakov, V.V. Konotop, and M. Salerno (2002), "Regular spatial structures in arrays of Bose-Einstein condensates induced by modulational instability," J. Phys. B 35, 5105–5119.

B.B. Baizakov, B.A. Malomed, and M. Salerno (2003), "Multidimensional solitons in periodic potentials," Europhys. Lett. 63, 642–648.

A.D. Bandrauk and H. Shen (1994), "Higher-order split-step exponential methods for solving coupled nonlinear Schrödinger equations," J. Phys. A 27, 7147–7155.

I.V. Barashenkov, D.E. Pelinovsky, and E.V. Zemlyanaya (1998), "Vibrations and oscillatory instabilities of gap solitons," Phys. Rev. Lett. 80, 5117–5120.

R. Barret, M.W. Berry, T.F. Chan, J. Demmel, J.M. Donato, J. Dongarra, V. Eijihout, R. Pozo, C. Romine, and H. van der Vorst (1994), *Templates for the Solution of Linear Systems: Building Blocks for Iterative Methods* (SIAM, Philadelphia).

G. Bartal, O. Manela, O. Cohen, J.W. Fleischer, and M. Segev (2005), "Observation of second-band vortex solitons in 2D photonic lattices," Phys. Rev. Lett. 95, 053904.

J.T. Beale (1991), "Exact solitary water waves with capillary ripples at infinity," Comm. Pure Appl. Math. 44, 211–257.

D.J. Benney and A.C. Newell (1967), "Nonlinear wave envelopes," J. Math. Phys. 46, 133–139.

D.J. Benney and G.J. Roskes (1969), "Wave instabilities," Stud. Appl. Math. 48, 377–385.

J.L. Bona, P.E. Souganidis, and W.A. Strauss (1987), "Stability and instability of solitary waves of Korteweg-de Vries type," Proc. Roy. Soc. London Ser. A 411, 395–412.

J.P. Boyd (1998), *Weakly Nonlinear Solitary Waves and Beyond-All-Orders Asymptotics* (Kluwer, Boston).

J.P. Boyd (2001), *Chebyshev and Fourier Spectral Methods*, 2nd edition (Dover, Mineola, NY).

J.P. Boyd (2007), "Why Newton's method is hard for travelling waves: Small denominators, KAM theory, Arnold's linear Fourier problem, non-uniqueness, constraints and erratic failure," Math. Comput. Simulation 74, 72–81.

J.P. Boyd and G. Chen (2001), "Weakly nonlinear wavepackets in the Korteweg de Vries equation: The KdV/NLS connection," Math. Comput. Simulation 55, 317–328.

C.C. Bradley, C.A. Sackett, J.J. Tollett, and R. Hulet (1995), "Evidence of Bose-Einstein condensation in an atomic gas with attractive interactions," Phys. Rev. Lett. 75, 1687–1690.

V.A. Brazhnyi, V.V. Konotop, and V. Kuzmiak (2006), "Nature of the intrinsic relation between Bloch-band tunneling and modulational instability," Phys. Rev. Lett. 96, 150402.

V.A. Brazhnyi, V.V. Konotop, V. Kuzmiak, and V.S. Shchesnovich (2007), "Nonlinear tunneling in two-dimensional lattices," Phys. Rev. A 76, 023608.

J. Bronski (1996), "Semiclassical eigenvalue distribution of the Zakharov-Shabat eigenvalue problem," Physica D 97, 376–397.

A.V. Buryak (1995), "Stationary soliton bound states existing in resonance with linear waves," Phys. Rev. E 52, 1156–1163.

D.C. Calvo and T.R. Akylas (1997), "On the formation of bound states by interacting nonlocal solitary waves," Physica D 101, 270–288.

D.C. Calvo, T.S. Yang, and T.R. Akylas (2000), "On the stability of solitary waves with decaying oscillatory tails," R. Soc. Lond. Ser. A. 456, 469–487.

D.K. Campbell, M. Peyrard, and P. Sodano (1986), "Kink-antikink interactions in the double sine-Gordon equation," Physica D 19, 165–205.

D.K. Campbell, J.S. Schonfeld, and C.A. Wingate (1983), "Resonant structure in kink-antikink interactions in ϕ^4 theory," Physica D 9, 1–32.

A.R. Champneys and M.D. Groves (1997), "A global investigation of solitary wave solutions to a two-parameter model for water waves," J. Fluid Mech. 342, 199–229.

A.R. Champneys and B.A. Malomed (1999a), "Moving embedded solitons," J. Phys. A 32, L547–L553.

A.R. Champneys and B.A. Malomed (1999b), "Embedded solitons in a three-wave system," Phys. Rev. E 61, 886–890.

A.R. Champneys, B.A. Malomed, and M.J. Friedman (1998), "Thirring solitons in the presence of dispersion," Phys. Rev. Lett. 80, 4169–4172.

A.R. Champneys, B.A. Malomed, J. Yang, and D.J. Kaup (2001), "Embedded solitons: Solitary waves in resonance with the linear spectrum," Physica D 152, 340–354.

D.Y. Chen (2006), *Introduction to Soliton Theory* (in Chinese) (Science Press, Beijing).

X. Chen and J. Yang (2002), "A direct perturbation theory for solitons of the derivative nonlinear Schroedinger equation and the modified nonlinear Schroedinger equation," Phys. Rev. E 65, 066608.

Z. Chen, I. Makasyuk, A. Bezryadina, and J. Yang (2004a), "Observation of two-dimensional lattice vector solitons," Opt. Lett. 29, 1656–1658.

Z. Chen, H. Martin, E.D. Eugenieva, J. Xu, and A. Bezryadina (2004b), "Anisotropic enhancement of discrete diffraction and formation of two-dimensional discrete-soliton trains," Phys. Rev. Lett. 92, 143902.

Z. Chen and J. Yang (2007), "Optically-induced reconfigurable photonic lattices for linear and nonlinear control of light," in *Nonlinear Optics and Applications*. H.A. Abdeldayem and D.O. Frazier, eds. (Research Signpost, Kerala, India), Chapter 5, 103–150.

R.Y. Chiao, E. Garmire, and C.H. Townes (1964), "Self-trapping of optical beams," Phys. Rev. Lett. 13, 479–482.

M.L. Chiofalo, S. Succi, and M.P. Tosi (2000), "Ground state of trapped interacting Bose-Einstein condensates by an explicit imaginary-time algorithm," Phys. Rev. E 62, 7438–7444.

B.V. Chirikov (1979), "A universal instability of many-dimensional oscillator systems," Phys. Rep. 52, 263–379.

B.I. Cohen, K.M. Watson, and B.J. West (1976), "Some properties of deep-water solitons," Phys. Fluids 19, 345–354.

S.L. Cornish, N.R. Claussen, J.L. Roberts, E.A. Cornell, and C.E. Wieman (2000), "Stable ^{85}Rb Bose-Einstein condensates with widely tunable interactions," Phys. Rev. Lett. 85, 1795–1798.

F. Dalfovo, S. Giorgini, L.P. Pitaevskii, and S. Stringari (1999), "Theory of Bose-Einstein condensation in trapped gases," Rev. Modern Phys. 71, 463–512.

A. Davey and K. Stewartson (1974), "On three-dimensional packets of surface waves," Proc. Roy. Soc. London Ser. A 338, 101–110.

K.B. Davis, M.-O. Mewes, M.R. Andrews, N.J. van Druten, D.S. Durfee, D.M. Kurn, and W. Ketterle (1995), "Bose-Einstein condensation in a gas of sodium atoms," Phys. Rev. Lett. 75, 3969–3973.

B. Deconinck, D.E. Pelinovsky, and J.D. Carter (2006), "Transverse instabilities of deep-water solitary waves," Proc. Roy. Soc. London Ser. A 462, 2039–2061.

M. Desaix, D. Anderson, L. Helczynski, and M. Lisak (2003), "Eigenvalues of the Zakharov-Shabat scattering problem for real symmetric pulses," Phys. Rev. Lett. 90, 013901.

S.V. Dmitriev, Yu.S. Kivshar, and T. Shigenari (2001), "Fractal structures and multiparticle effects in soliton scattering," Phys. Rev. E 64, 056613.

S.V. Dmitriev and T. Shigenari (2002), "Short-lived two-soliton bound states in weakly perturbed nonlinear Schrödinger equation," Chaos 12, 324–331.

N.K. Efremidis and D.N. Christodoulides (2003), "Lattice solitons in Bose-Einstein condensates," Phys. Rev. A 67, 063608.

N.K. Efremidis, J. Hudock, D.N. Christodoulides, J.W. Fleischer, O. Cohen, and M. Segev (2003), "Two-dimensional optical lattice solitons," Phys. Rev. Lett. 91, 213906.

N.K. Efremidis, S. Sears, D.N. Christodoulides, J.W. Fleischer, and M. Segev (2002), "Discrete solitons in photorefractive optically induced photonic lattices," Phys. Rev. E 66, 046602.

B. Eiermann, Th. Anker, M. Albiez, M. Taglieber, P. Treutlein, K.P. Marzlin, and M.K. Oberthaler (2004), "Bright Bose-Einstein gap solitons of atoms with repulsive interaction," Phys. Rev. Lett. 92, 230401.

H. Eisenberg, Y. Silberberg, R. Morandotti, A. Boyd, and J. Aitchison (1998), "Discrete spatial optical solitons in waveguide arrays," Phys. Rev. Lett. 81, 3383–3386.

A.E. Elbert (2003), "Solutions of the extended nonlinear Schrödinger equation that oscillate at infinity," Proc. Steklov Inst. Math. 1, S54–S67.

C. Etrich, U. Peschel, F. Lederer, B.A. Malomed, and Y.S. Kivshar (1996), "Origin of the persistent oscillations of solitary waves in nonlinear quadratic media," Phys. Rev. E 54, 4321–4324.

S.G. Evangelides, L.F. Mollenauer, J.P. Gordon, and N.S. Bergano (1992), "Polarization multiplexing with solitons," J. Lightwave Technol. 10, 28–35.

M. Facao and D.F. Parker (2003), "Stability of screening solitons in photorefractive media," Phys. Rev. E 68, 016610.

L.D. Faddeev and L.A. Takhtadjan (1987), *Hamiltonian Methods in the Theory of Solitons* (Springer-Verlag, Berlin).

Z. Fei, Y.S. Kivshar, and L. Vázquez (1992a), "Resonant kink-impurity interactions in the sine-Gordon model," Phys. Rev. A 45, 6019–6030.

Z. Fei, Y.S. Kivshar, and L. Vázquez (1992b), "Resonant kink-impurity interactions in the ϕ^4 model", Phys. Rev. A 46, 5214–5220.

R. Fischer, D. Trager, D.N. Neshev, A.A. Sukhorukov, W. Krolikowski, C. Denz, and Y.S. Kivshar (2006), "Reduced-symmetry two-dimensional solitons in photonic lattices," Phys. Rev. Lett. 96, 023905.

J.W. Fleischer, G. Bartal, O. Cohen, O. Manela, M. Segev, J. Hudock, and D.N. Christodoulides (2004), "Observation of vortex-ring discrete solitons in 2D photonic lattices," Phys. Rev. Lett. 92, 123904.

J.W. Fleischer, T. Carmon, M. Segev, N.K. Efremidis, and D.N. Christodoulides (2003a), "Observation of discrete solitons in optically induced real time waveguide arrays," Phys. Rev. Lett. 90, 023902.

J.W. Fleischer, M. Segev, N.K. Efremidis, and D.N. Christodoulides (2003b), "Observation of two-dimensional discrete solitons in optically induced nonlinear photonic lattices," Nature 422, 147–150.

M.B. Fogel, S.E. Trullinger, A.R. Bishop, and J.A. Krumhansl (1977), "Dynamics of sine-Gordon solitons in the presence of perturbations," Phys. Rev. B 15, 1578–1592.

A.S. Fokas (1980), "A symmetry approach to exactly solvable evolution equations," J. Math. Phys. 21, 1318–1325.

A.S. Fokas and B. Fuchssteiner (1981), "The hierarchy of the Benjamin-Ono equation," Phys. Lett. A 86, 341–345.

B. Fornberg (1998), *A Practical Guide to Pseudospectral Methods* (Cambridge University Press, Cambridge, UK).

B. Fornberg and G.B. Whitham (1978), "A numerical and theoretical study of certain nonlinear wave phenomena," Philos. Trans. Roy. Soc. London Ser. A 289, 373–404.

J. Fujioka and A. Espinosa (1997), "Soliton-like solution of an extended NLS equation existing in resonance with linear dispersive waves," J. Phys. Soc. Japan 66, 2601–2607.

L. Gagnon and N. Stiévenart (1994), "N-soliton interaction in optical fibers: The multiple-pole case," Opt. Lett. 19, 619–621.

J.J. García-Ripoll and V.M. Pérez-García (2001), "Optimizing Schrödinger functionals using Sobolev gradients: Applications to quantum mechanics and nonlinear optics," SIAM J. Sci. Comput. 23, 1316–1334.

C.S. Gardner, J.M. Greene, M.D. Kruskal, and R.M. Miura (1967), "Method for solving the Korteweg-de Vries equation," Phys. Rev. Lett. 19, 1095–1097.

V.S. Gerdjikov (2005), "Basic aspects of soliton theory," in *Geometry, Integrability and Quantization. Proceedings of the Sixth International Conference*, I.M. Mladenov and A.C. Hirshfeld, eds. (Softex, Sofia, Bulgaria), 78–122.

V.S. Gerdjikov, E.V. Doktorov, and J. Yang (2001), "Adiabatic interaction of N-ultrashort solitons: Universality of the complex Toda chain model," Phys. Rev. E 64, 056617.

V.S. Gerdjikov, D.J. Kaup, I.M. Uzunov, and E.G. Evstatiev (1996), "Asymptotic behavior of N-soliton trains of the nonlinear Schrödinger equation," Phys. Rev. Lett. 77, 3943–3946.

V.S. Gerdjikov and E.Kh. Khristov (1980), "On the evolution equations solvable through the inverse scattering method, I. Spectral theory," Bulgarian J. Phys. 7, 28–41.

G. Golub and C. Van Loan (1996), *Matrix Computations*, 3rd edition (The Johns Hopkins University Press, Baltimore).

S. Gonzalez-Perez-Sandia, J. Fujioka, and B.A. Malomed (2004), "Embedded solitons in dynamical lattices," Physica D 197, 86–100.

R.H. Goodman (2008), "Chaotic scattering in solitary wave interactions: A singular iterated-map description," Chaos, 18, 023113.

R.H. Goodman and R. Haberman (2004), "Interaction of sine-Gordon kinks with defects: The two-bounce resonance," Physica D 195, 303–323.

R.H. Goodman and R. Haberman (2005a), "Vector soliton interactions in birefringent optical fibers," Phys. Rev. E 71, 056605.

R.H. Goodman and R. Haberman (2005b), "Kink-antikink collisions in the ϕ^4 equation: The n-bounce resonance and the separatrix map," SIAM J. Appl. Dyn. Sys. 4, 1195–1228.

R.H. Goodman and R. Haberman (2007), "Chaotic scattering and the n-bounce resonance in solitary-wave interactions," Phys. Rev. Lett. 98, 104103.

J. P. Gordon (1986), "Theory of the soliton self-frequency shift," Opt. Lett. 11, 662–664.

K.A. Gorshkov and L.A. Ostrovsky (1981), "Interactions of solitons in nonintegrable systems: Direct perturbation method and applications," Physica D 3, 428–438.

D. Gottlieb and S.A. Orszag (1977), *Numerical Analysis of Spectral Methods: Theory and Applications* (SIAM, Philadelphia).

M. Grillakis (1988), "Linear instability for nonlinear Schrödinger and Klein-Gordon equations," Comm. Pure Appl. Math. 41, 747–774.

R. Grimshaw (1995), "Weakly nonlocal solitary waves in a singularly perturbed nonlinear Schrödinger equation," Stud. Appl. Math. 94, 257–270.

R. Grimshaw and P. Cook (1996), "Solitary waves with oscillatory tails," in *Proceedings of the Second International Conference on Hydrodynamics*, A.T. Chwang, J.H.W. Lee, and D.Y.C. Leung, eds. (Balkema, Rotterdam), 327–336.

R. Grimshaw and N. Joshi (1995), "Weakly nonlocal solitary waves in a singularly perturbed Korteweg–de Vries equation," SIAM J. Appl. Math. 55, 124–135.

R. Grimshaw and H. Mitsudera (1993), "Slowly-varying solitary wave solutions of the perturbed Korteweg-de Vries equation revisited," Stud. Appl. Math. 90, 75–86.

R. Grimshaw, E. Pelinovsky, T. Talipova, M. Ruderman, and R. Erdelyi (2005), "Short-lived large-amplitude pulses in the nonlinear long-wave model described by the modified Korteweg-de Vries equation," Stud. Appl. Math. 114, 189–210.

E.M. Gromov, L.V. Piskunova, and V.V. Tyutin (1999), "Dynamics of wave packets in the frame of third-order nonlinear Schrödinger equation," Phys. Lett. A 256, 153–158.

E.M. Gromov and V.V. Tyutin (1998), "Stationary waves in a third-order nonlinear Schrödinger equation," Wave Motion 28, 13–24.

E.P. Gross (1961), "Structure of a quantized vortex in boson systems," Il Nuovo Cimento 20, 454–457.

A. Hasegawa and Y. Kodama (1995), *Solitons in Optical Communications* (Clarendon, Oxford).

A. Hasegawa and F. Tappert (1973), "Transmission of stationary nonlinear optical pulses in dispersive dielectric fibers," Appl. Phys. Lett. 23, 142–145.

R. L. Herman (1990), "A direct approach to studying soliton perturbations," J. Phys. A 23, 2327–2362.

M.R. Hestenes and E. Stiefel (1952), "Methods of conjugate gradients for solving linear systems," J. Research Nat. Bur. Standards 49, 409–436.

R. Hirota (2004), *The Direct Method in Soliton Theory* (Cambridge University Press, Cambridge, UK).

R. Horn and C. Johnson (1991), *Matrix Analysis* (Cambridge University Press, New York).

S.M. Hoseini and T.R. Marchant (2006), "Solitary wave interaction and evolution for a higher-order Hirota equation," Wave Motion 44, 92–106.

S.M. Hoseini and T.R. Marchant (2009), "Soliton perturbation theory for a higher order Hirota equation," Math. Comput. Simulation 80, 770–778.

H.C. Hu, B. Tong and S.Y. Lou (2006), "Nonsingular positon and complexiton solutions for the coupled KdV system," Phys. Lett. A 351, 403–412.

N.N. Huang (1996), *Soliton Theory and Perturbation Method* (in Chinese) (Shanghai Science and Education Publisher, Shanghai, China).

G. Hwang, T.R. Akylas, and J. Yang (2010), "Gap solitons and their linear stability in one-dimensional periodic media," preprint.

E. Infeld and T. Lenkowska-Czerwinska (1997), "Analysis of stability of light beams in nonlinear photorefractive media," Phys. Rev. E 55, 6101–6106.

R. Iwanow, R. Schieck, G. Stegeman, T. Pertsch, F. Lederer, Y. Min, and W. Sohler (2004), "Observation of discrete quadratic solitons," Phys. Rev. Lett. 93, 113902.

P.A.E.M. Janssen and J.J. Rasmussen (1983), "Nonlinear evolution of the transverse instability of plane-envelope solitons," Phys. Fluids 26, 1279–1287.

J. Javanainen and J. Ruostekoski (2006), "Symbolic calculation in development of algorithms: Split-step methods for the Gross-Pitaevskii equation," J. Phys. A 39, L179–L184.

C.K.R.T. Jones (1988), "An instability mechanism for radially symmetric standing waves of a nonlinear Schrödinger equation," J. Differential Equations 71, 34–62.

T. Kanna, M. Lakshmanan, P. Tchofo Dinda, and N. Akhmediev (2006), "Soliton collisions with shape change by intensity redistribution in mixed coupled nonlinear Schrödinger equations," Phys. Rev. E 73, 026604.

T. Kano (1989), "Normal form of nonlinear Schrödinger equation," J. Phys. Soc. Japan 58, 4322–4328.

V.I. Karpman and E.M. Maslov (1977), "A perturbation theory for the Korteweg-de Vries equation," Phys. Lett. A 60, 307–308.

V.I. Karpman and E.M. Maslov (1978), "Perturbation theory for solitons," Sov. Phys. JETP 46, 281–291.

V.I. Karpman and V.V. Solovev (1981), "A perturbation theory for soliton systems," Physica D 3, 142–164.

A.-K. Kassam and L.N. Trefethen (2005), "Fourth-order time-stepping for stiff PDEs," SIAM J. Sci. Comput. 26, 1214–1233.

W.L. Kath and N.F. Smyth (1995), "Soliton evolution and radiation loss for the nonlinear Schrödinger equation," Phys. Rev. E 51, 1484–1492.

D.J. Kaup (1976a), "A perturbation expansion for the Zakharov–Shabat inverse scattering transform," SIAM J. Appl. Math. 31, 121–133.

D.J. Kaup (1976b), "Closure of the squared Zakharov-Shabat eigenstates," J. Math. Anal. Appl. 54, 849–864.

D.J. Kaup (1990), "Perturbation theory for solitons in optical fibers," Phys. Rev. A 42, 5689–5694.

D.J. Kaup and T.I. Lakoba (1996), "Squared eigenfunctions of the massive Thirring model in laboratory coordinates," J. Math. Phys. 37, 308–323.

D.J. Kaup, T.I. Lakoba, and Y. Matsuno (1999), "Perturbation theory for the Benjamin-Ono equation," Inverse Problems 15, 215–240.

D.J. Kaup and A.C. Newell (1978a), "Solitons as particles, oscillators, and in slowly changing media: A singular perturbation theory," Proc. R. Soc. London Ser. A 361, 413–446.

D.J. Kaup and A.C. Newell (1978b), "An exact solution for a derivative nonlinear Schrödinger equation," J. Math. Phys. 19, 798–801.

D.J. Kaup and J. Yang (2009), "The inverse scattering transform and squared eigenfunctions for a degenerate 3×3 operator," Inverse Problems 25, 105010.

T. Kawata (1984), "Riemann spectral method for the nonlinear evolution equation", in *Advances in Nonlinear Waves*, Vol. 1, L. Debnath, ed. (Pitman, Boston), 210–225.

J.P. Keener and D.W. Mclaughlin (1977), "Solitons under perturbations," Phys. Rev. A 16, 777–790.

P.L. Kelley (1965), "Self focusing of optical beams," Phys. Rev. Lett. 15, 1005–1008.

S.M. Kelly (1992), "Characteristic sideband instability of periodically amplified average soliton," Electron. Lett. 28, 806–807.

S. Kichenassamy and P.J. Olver (1992), "Existence and nonexistence of solitary wave solutions to higher-order model evolution equations," SIAM J. Math. Anal. 23, 1141–1166.

Y.S. Kivshar and G.P. Agrawal (2003), *Optical Solitons: From Fibers to Photonic Crystals* (Academic Press, San Diego).

Y.S. Kivshar, Z. Fei, and L. Vázquez (1991), "Resonant soliton-impurity interactions," Phys. Rev. Lett. 67, 1177–1180.

Y.S. Kivshar and B. A. Malomed (1989), "Dynamics of solitons in nearly integrable systems," Rev. Modern Phys. 61, 763–915.

Y.S. Kivshar and D.E. Pelinovsky (2000), "Self-focusing and transverse instabilities of solitary waves," Phys. Rep. 331, 117–195.

Y.S. Kivshar, D.E. Pelinovsky, T. Cretegny, and M. Peyrard (1998), "Internal modes of solitary waves," Phys. Rev. Lett. 80, 5032–5035.

M. Klauder, E.W. Laedke, K.H. Spatschek, and S.K. Turitsyn (1993), "Pulse propagation in optical fibers near the zero dispersion point," Phys. Rev. E 47, R3844–R3847.

M. Klaus and J.K. Shaw (2002), "Purely imaginary eigenvalues of Zakharov-Shabat systems," Phys. Rev. E 65, 036607.

M. Klaus and J.K. Shaw (2003), "On the eigenvalues of Zakharov–Shabat systems," SIAM J. Math. Anal. 34, 759–773.

Y. Kodama (1985), "Normal forms for weakly dispersive wave equations," Phys. Lett. A 112, 193–196.

Y. Kodama (1987), "On solitary-wave interaction," Phys. Lett. A 123, 276–282.

Y. Kodama and M.J. Ablowitz (1981), "Perturbations of solitons and solitary waves," Stud. Appl. Math. 64, 225–245.

Y. Kodama and A.V. Mikhailov (2001), "Symmetry and perturbation of the vector nonlinear Schrödinger equation," Physica D 152, 171–177.

S.E. Koonin (1986), *Computational Physics* (Addison-Wesley, Redwood City, CA).

D.J. Korteweg and G. de Vries (1895), "On the change of form of long waves advancing in a rectangular canal and on a new type of long stationary waves," Philosophical Magazine, 5th series, 39, 422–443.

E.A. Kuznetsov, A.V. Mikhailov, and I.A. Shimokhin (1995), "Nonlinear interaction of solitons and radiation," Physica D 87, 201–215.

E.A. Kuznetsov, A.M. Rubenchik, and V.E. Zakharov (1986), "Soliton stability in plasmas and hydrodynamics," Phys. Rep. 142, 103–165.

T.I. Lakoba and D.J. Kaup (1997), "Perturbation theory for the Manakov soliton and its applications to pulse propagation in randomly birefringent fibers," Phys. Rev. E 56, 6147–6165.

T.I. Lakoba and J. Yang (2007a), "A generalized Petviashvili iteration method for scalar and vector Hamiltonian equations with arbitrary form of nonlinearity," J. Comput. Phys. 226, 1668–1692.

T.I. Lakoba and J. Yang (2007b), "A mode elimination technique to improve convergence of iteration methods for finding solitary waves," J. Comput. Phys. 226, 1693–1709.

P. Lax (1968), "Integrals of nonlinear equations of evolution and solitary waves," Commun. Pure Appl. Math. 21, 467–490.

A.J. Lichtenberg and M.A. Lieberman (1992), *Regular and Chaotic Dynamics* (Springer-Verlag, Berlin).

C. Lou, X. Wang, J. Xu, Z. Chen, and J. Yang (2007), "Nonlinear spectrum reshaping and gap-soliton-train trapping in optically induced photonic structures," Phys. Rev. Lett. 98, 213903.

W.X. Ma (2005), "Complexiton solutions to integrable equations," Nonlinear Anal. 63, e2461–e2471.

W.X. Ma, Q. Ding, W.G. Zhang, and B.Q. Lu (1996), "Binary nonlinearization of Lax pairs of Kaup-Newell soliton hierarchy," Il Nuovo Cimento B 111, 1135–1149.

W.X. Ma and R.G. Zhou (1999), "A coupled AKNS-Kaup-Newell soliton hierarchy," J. Math. Phys. 40, 4419–4428.

W.C.K. Mak, B.A. Malomed, and P.L. Chu (2004), "Symmetric and asymmetric solitons in linearly coupled Bragg gratings," Phys. Rev. E 69, 066610.

B.A. Malomed (2002), "Variational methods in nonlinear fiber optics and related fields," Prog. Opt. 43, 71–193.

A.V. Mamaev, M. Saffman, and A.A. Zozulya (1996), "Break-up of two-dimensional bright spatial solitons due to transverse modulation instability," Europhys. Lett. 35, 25–30.

S.V. Manakov (1973), "On the theory of two-dimensional stationary self-focusing of electromagnetic waves," Zh. E' ksp. Teor. Fiz. 65, 1392 [Sov. Phys. JETP 38, 248–253 (1974)].

O. Manela, O. Cohen, G. Bartal, J.W. Fleischer, and M. Segev (2004), "Two-dimensional higher-band vortex lattice solitons," Opt. Lett. 29, 2049–2051.

H. Martin, E.D. Eugenieva, Z. Chen, and D.N. Christodoulides (2004), "Discrete solitons and soliton-induced dislocations in partially coherent photonic lattices," Phys. Rev. Lett. 92, 123902.

F. Matera, A. Mecozzi, M. Romagnoli, and M. Settembre (1993), "Sideband instability induced by periodic power variation in long-distance fiber links," Opt. Lett. 18, 1499–1501.

Y. Matsuno (1995), "Multisoliton perturbation theory for the Benjamin-Ono equation and its application to real physical systems," Phys. Rev. E 51, 1471–1483.

V.B. Matveev (1992), "Generalized Wronskian formula for solutions of the KdV equations: first applications," Phys. Lett. A 166, 205–208.

C.R. Menyuk (1987), "Nonlinear pulse propagation in birefringent optical fibers," IEEE J. Quantum Electron QE-23, 174–176.

R.W. Micallef, V.V. Afanasjev, Y.S. Kivshar, and J.D. Love (1996), "Optical solitons with power-law asymptotics," Phys. Rev. E 54, 2936–2942.

D. Mihalache, L. Torner, F. Moldoveanu, N.-C. Panoiu, and N. Truta (1993a), "Inverse-scattering approach to femtosecond solitons in monomode optical fibers," Phys. Rev. E 48, 4699–4709.

D. Mihalache, N. Truta, N.-C. Panoiu, and D.-M. Baboiu (1993b), "Analytic method for solving the modified nonlinear Schrödinger equation describing soliton propagation along optical fibers," Phys. Rev. A 47, 3190–3194.

P.A. Milewski and E.G. Tabak (1999), "A pseudospectral procedure for the solution of nonlinear wave equations with examples from free-surface flows," SIAM J. Sci. Comput. 21, 1102–1114.

F.M. Mitschke and L.F. Mollenauer (1986), "Discovery of the soliton self-frequency shift," Opt. Lett. 11, 659–661.

K.D. Moll, A.L. Gaeta, and G. Fibich (2003), "Self-similar optical wave collapse: Observation of the Townes profile," Phys. Rev. Lett. 90, 203902.

O. Morsch and M.K. Oberthaler (2006), "Dynamics of Bose-Einstein condensates in optical lattices," Rev. Mod. Phys. 78, 179–215.

J. Moser (1975), *Dynamical Systems, Theory and Applications,* Lecture Notes in Physics, Vol. 38 (Springer-Verlag, Berlin).

Z. Musslimani and J. Yang (2004), "Self-trapping of light in a two-dimensional periodic structure." J. Opt. Soc. Amer. B. Opt. Phys. 21, 973–981.

D.N. Neshev, T.J. Alexander, E.A. Ostrovskaya, Y.S. Kivshar, H. Martin, I. Makasyuk, and Z. Chen (2004), "Observation of discrete vortex solitons in optically induced photonic lattices," Phys. Rev. Lett. 92, 123903.

D. Neshev, E. Ostrovskaya, Y. Kivshar, and W. Krolikowski (2003), "Spatial solitons in optically induced gratings," Opt. Lett. 28, 710–712.

A.C. Newell (1980), "The inverse scattering transform," in *Solitons*, Topics in Current Physics, Vol. 17, R.K. Bullough and P.J. Caudrey, eds. (Springer-Verlag, New York), 177–242.

A.C. Newell (1985), *Solitons in Mathematics and Physics* (SIAM, Philadelphia).

J.J.C. Nimmo (1983), "A bilinear Bäcklund transformation for the nonlinear Schrödinger equation," Phys. Lett. A 99, 279–280.

S.A. Orszag (1971), "Accurate Solution of the Orr-Sommerfeld. Equation," J. Fluid Mech. 50, 689–703.

E.A. Ostrovskaya and Y.S. Kivshar (2004), "Matter-wave gap vortices in optical lattices," Phys. Rev. Lett. 93, 160405.

R.L. Pego and M.I. Weinstein (1992), "Eigenvalues, and instabilities of solitary waves," Philos. Trans. Roy. Soc. London Ser. A 340, 47–94.

D.E. Pelinovsky (2001), "A mysterious threshold for transverse instability of deep-water solitons," Math. Comput. Simulation 55, 585–594.

D.E. Pelinovsky, V.V. Afanasjev, and Y.S. Kivshar (1996), "Nonlinear theory of oscillating, decaying, and collapsing solitons in the generalized nonlinear Schrödinger equation," Phys. Rev. E 53, 1940–1953.

D.E. Pelinovsky, P.G. Kevrekidis, and D. Frantzeskakis (2005), "Stability of discrete solitons in nonlinear Schrodinger lattices," Physica D 212, 1–19.

D.E. Pelinovsky and Y.S. Kivshar (2000), "Stability criterion for multicomponent solitary waves," Phys. Rev. E 62, 8668–8676.

D.E. Pelinovsky, Yu.S. Kivshar, and V.V. Afanasjev (1998), "Internal modes of envelope solitons," Physica D 116, 121–142.

D.E. Pelinovsky and V.M. Rothos (2005), "Bifurcations of travelling breathers in the discrete NLS equations," Physica D 202, 16–36.

D.E. Pelinovsky and Yu.A. Stepanyants (2004), "Convergence of Petviashvili's iteration method for numerical approximation of stationary solutions of nonlinear wave equations," SIAM J. Numer. Anal. 42, 1110–1127.

D.E. Pelinovsky, A.A. Sukhorukov, and Y.S. Kivshar (2004), "Bifurcations and stability of gap solitons in periodic potentials," Phys. Rev. E 70, 036618.

D.E. Pelinovsky and J. Yang (2000), "Internal oscillations and radiation damping of vector solitons," Stud. Appl. Math. 105, 245–267.

D.E. Pelinovsky and J. Yang (2002), "A normal form for nonlinear resonance of embedded solitons," Proc. Roy. Soc. Lond. Ser. A. 458, 1469–1497.

D.E. Pelinovsky and J. Yang (2004), "Parametric resonance and radiative decay of dispersion-managed solitons," SIAM J. Appl. Math. 64, 1360–1382.

D.E. Pelinovsky and J. Yang (2005), "Stability analysis of embedded solitons in the generalized third-order NLS equation," Chaos 15, 037115.

V.I. Petviashvili (1976), "Equation of an extraordinary soliton," Plasma Phys. 2, 469–472.

M. Peyrard and D.K. Campbell (1983), "Kink-antikink interactions in a modified sine-Gordon model," Physica D 9, 33–51.

L.P. Pitaevskii (1961), "Vortex lines in an imperfect Bose gas," Sov. Phys. JETP 13, 451–454.

W. Pogorzelski (1966), *Integral Equations and Their Applications* (Pergamon Press, London).

Y. Pomeau, A. Ramani, and B. Grammaticos (1988), "Structural stability of the Korteweg-de Vries solitons under a singular perturbation," Physica D 31, 127–134.

G.J. Roskes (1976), "Some nonlinear multiphase interactions," Stud. Appl. Math. 55, 231–238.

J.S. Russell (1844), "Report on waves," Report of the Fourteenth Meeting of the British Association for the Advancement of Science, 311–390.

V.S. Ryabenkii and S.V. Tsynkov (2007), *A Theoretical Introduction to Numerical Analysis* (Chapman and Hall/CRC Press, Boca Raton, FL).

Y. Saad and M.H. Schultz (1986), "GMRES: A generalized minimal residual algorithm for solving nonsymmetric linear systems," SIAM J. Sci. Statist. Comput. 7, 856–869.

P.G. Saffman and H.C. Yuen (1978), "Stability of a plane soliton to infinitesimal two-dimensional perturbations," Phys. Fluids 21, 1450–1451.

H. Sakaguchi and B.A. Malomed (2004), "Two-dimensional loosely and tightly bound solitons in optical lattices and inverted traps," J. Phys. B 37, 2225–2239.

S. Sakovich and T. Tsuchida (2000), "Symmetrically coupled higher-order nonlinear Schrödinger equations: Singularity analysis and integrability," J. Phys. A 33, 7217–7226.

J. Satsuma and N. Yajima (1974), "Initial value problems of one-dimensional self-modulation of nonlinear waves in dispersive media," Progr. Theoret. Phys. Suppl. 55, 284–306.

M. Segev, M. Shih, and G.C. Valley (1996), "Photorefractive screening solitons of high and low intensity," J. Opt. Soc. Amer. B Opt. Phys. 13, 706–718.

H. Segur and M.D. Kruskal (1987), "Nonexistence of small-amplitude breather solutions in ϕ^4 theory," Phys. Rev. Lett. 58, 747–750.

A. Sergyeyev and D. Demskoi (2007), "Sasa-Satsuma (complex modified Korteweg-de Vries II) and the complex sine-Gordon II equation revisited: Recursion operators, nonlocal symmetries, and more," J. Math. Phys. 48, 042702.

V.S. Shchesnovich and E.V. Doktorov (1997), "Perturbation theory for solitons of the Manakov system," Phys. Rev. E 55, 7626–7635.

V.S. Shchesnovich and E.V. Doktorov (1999), "Perturbation theory for the modified nonlinear Schrödinger solitons," Physica D 129, 115–129.

V.S. Shchesnovich and J. Yang (2003a), "Higher-order solitons in the N-wave system," Stud. Appl. Math. 110, 297–332.

V.S. Shchesnovich and J. Yang (2003b), "General soliton matrices in the Riemann-Hilbert problem for integrable nonlinear equations," J. Math. Phys. 44, 4604–4639.

Z. Shi, J. Wang, Z. Chen, and J. Yang (2008), "Linear instability of two-dimensional low-amplitude gap solitons near band edges in periodic media," Phys. Rev. A 78, 063812.

Z. Shi and J. Yang (2007), "Solitary waves bifurcated from Bloch-band edges in two-dimensional periodic media," Phys. Rev. E 75, 056602.

Z. Shi and J. Yang (2008), "Stability switching at a power extremum in the generalized NLS equations with external potentials," unpublished.

S.R. Singh, M.I. Carvalho, and D.N. Christodoulides (1995), "Vector interactions of steady-state planar solitons in biased photorefractive media," Opt. Lett. 20, 2177–2179.

M. Skorobogatiy and J. Yang (2009), *Fundamentals of Photonic Crystal Guiding* (Cambridge University Press, Cambridge, UK).

D. Song, C. Lou, L.Tang, X. Wang, W. Li, X. Chen, K. Law, H. Susanto, P. G. Kevrekidis, J. Xu, and Z. Chen (2008), "Self-trapping of optical vortices in waveguide lattices with self-defocusing nonlinearity," Opt. Express, 16, 10110–10116.

D.C. Sorensen (1992), "Implicit application of polynomial filters in a k-step Arnoldi method," SIAM J. Matrix Anal. Appl. 13, 357–385.

G. Strang (1968), "On the construction and comparison of difference schemes," SIAM J. Numer. Anal. 5, 506–517.

M. Struwe (2000), *Variational Methods: Applications to Nonlinear Partial Differential Equations and Hamiltonian Systems,* 3rd edition (Springer-Verlag, Berlin).

C. Sulem and P.L. Sulem (1999), *The Nonlinear Schrödinger Equation* (Springer-Verlag, New York).

S.M. Sun (1991), "Existence of a generalized solitary wave with positive Bond number smaller than 1/3," J. Math. Anal. Appl. 156, 471–504.

A. Szameit, J. Burghoff, T. Pertsch, S. Nolte, A. Tnnermann, and F. Lederer (2006), "Two-dimensional soliton in cubic fs laser written waveguide arrays in fused silica," Opt. Express 14, 6055–6062.

V.I. Talanov (1970), "Focusing of light in cubic media," JETP Lett. 11, 199–201.

Y. Tan and J. Yang (2001a), "Resonance and phase induced window sequences in vector soliton collisions," Phys. Lett. A 288, 309–315.

Y. Tan and J. Yang (2001b), "Complexity and regularity of vector-soliton collisions," Phys. Rev. E 64, 056616.

Y. Tan, J. Yang, and D.E. Pelinovsky (2002), "Semi-stability of embedded solitons in the general fifth-order KdV equation," Wave Motion 36, 241–255.

L. Tang, C. Lou, X. Wang, D. Song, X. Chen, J. Xu, Z. Chen, H. Susanto, K. Law, and P.G. Kevrekidis (2007), "Observation of dipole-like gap solitons in self-defocusing waveguide lattices," Opt. Lett. 32, 3011–3013.

M. Toda (1989), *Theory of Nonlinear Lattices* (Springer-Verlag, Berlin).

L.N. Trefethen (2000), *Spectral Method in MATLAB* (SIAM, Philadelphia).

T. Tsuchida (2004), "N-soliton collision in the Manakov model," Progr. Theoret. Phys. 111, 151–182.

T. Tsuchida and M. Wadati (1998), "The coupled modified Korteweg-de Vries equations," J. Phys. Soc. Japan 67, 1175–1187.

T. Ueda and W.L. Kath (1990), "Dynamics of coupled solitons in nonlinear optical fibers," Phys. Rev. A 42, 563–571.

N.G. Vakhitov and A.A. Kolokolov (1973), "Stationary solutions of the wave equation in a medium with nonlinearity saturation," Izv. Vyssh. Uchebn. Zaved., Radiofiz. 16, 1020–1028 [Radiophys. Quantum Electron. 16, 783–789 (1973)].

J.M. Vanden-Broeck (1991), "Elevation solitary waves with surface tension," Phys. Fluids A 3, 2659–2663.

J.M. Vanden-Broeck and R.E.L. Turner (1992), "Long periodic internal waves," Phys. Fluids A 4, 1929–1935.

S.N. Vlasov, V.A. Petrishchev and V.I. Talanov (1971), "Averaged description of wave beams in linear and nonlinear media (the method of moments)," Izv. Vys. Uchebn. Zaved. Radiofizika 14, 1353–1363 [Radiophys. Quantum Electron. 14, 1062–1070 (1974)].

M. Wadati, K. Konno, and Y.H. Ichikawa (1979), "A generalization of inverse scattering method," J. Phys. Soc. Japan 46, 1965–1966.

M. Wadati, H. Sanuki, and K. Konno (1975), "Relationships among inverse method, Bäcklund transformation and an infinite number of conservation laws," Progr. Theoret. Phys. 53, 419–436.

P.K.A. Wai, H.H. Chen, and Y.C. Lee (1990), "Radiations by solitons at the zero-group-dispersion wavelength of single-mode optical fibers," Phys. Rev. A 41, 426–439.

D. Wang, D. Zhang, and J. Yang (2010), "Integrable properties of the general coupled nonlinear Schrödinger equations," J. Math. Phys. 51, 023510.

H.Y. Wang and X.B. Hu (2008), *Soliton Equations with Self-Consistent Sources* (in Chinese) (Tsinghua University Press, Beijing).

J. Wang and J. Yang (2008), "Families of vortex solitons in periodic media," Phys. Rev. A 77, 033834.

J. Wang, J. Yang, T.J. Alexander, and Y.S. Kivshar (2009), "Truncated-Bloch-wave solitons in optical lattices," Phys. Rev. A 79, 043610.

X. Wang, Z. Chen, J. Wang, and J. Yang (2007), "Observation of in-band lattice solitons," Phys. Rev. Lett. 99, 243901.

J.A.C. Weideman and B.M. Herbst (1986), "Split-Step methods for the solution of the nonlinear Schrödinger equation," SIAM J. Numer. Anal. 23, 485–507.

M.I. Weinstein (1983), "Nonlinear Schrödinger equations and sharp interpolation estimates," Comm. Math. Phys. 87, 567–576.

M. Weinstein (1986), "Lyapunov stability of ground states of nonlinear dispersive evolution equations," Comm. Pure Appl. Math. 39, 51–67.

G.B. Whitham (1974), *Linear and Nonlinear Waves* (Wiley-Interscience, New York).

J. Yan and Y. Tang (1996), "Direct approach to the study of soliton perturbations," Phys. Rev. E 54, 6816–6824.

N.N. Yanenko (1971), *The Method of Fractional Steps* (Springer-Verlag, Berlin).

J. Yang (1997), "Vector solitons and their internal oscillations in birefringent nonlinear optical fibers," Stud. Appl. Math. 98, 61–97.

J. Yang (1999), "Multisoliton perturbation theory for the Manakov equations and its applications to nonlinear optics," Phys. Rev. E 59, 2393–2405.

J. Yang (2000), "Complete eigenfunctions of linearized integrable equations expanded around a soliton solution," J. Math. Phys. 41, 6614–6638.

J. Yang (2001a), "Dynamics of embedded solitons in the extended KdV equations," Stud. Appl. Math. 106, 337–365.

J. Yang (2001b), "Interactions of vector solitons," Phys. Rev. E 64, 026607.

J. Yang (2002a), "Internal oscillations and instability characteristics of $(2+1)$ dimensional solitons in a saturable nonlinear medium," Phys. Rev. E 66, 026601.

J. Yang (2002b), "Eigenfunctions of linearized integrable equations expanded around an arbitrary solution," Stud. Appl. Math. 108, 145–159.

J. Yang (2002c), "Suppression of Manakov-Soliton interference in optical fibers," Phys. Rev. E 65, 036606.

J. Yang (2003), "Stable embedded solitons," Phys. Rev. Lett. 91, 143903.

J. Yang (2008), "Iteration methods for stability spectra of solitary waves," J. Comput. Phys. 227, 6862–6876.

J. Yang (2009a), "Newton-conjugate-gradient methods for solitary wave computations," J. Comput. Phys. 228, 7007–7024.

J. Yang (2009b), "Accuracy and numerical stability of split-step methods for nonlinear wave equations," unpublished.

J. Yang and T.R. Akylas (2003), "Continuous families of embedded solitons in the third-order nonlinear Schrödinger equation," Stud. Appl. Math. 111, 359–375.

J. Yang and Z. Chen (2006), "Defect solitons in photonic lattices," Phys. Rev. E 73, 026609.

J. Yang and D.J. Kaup (2000), "Stability and evolution of solitary waves in perturbed generalized nonlinear Schrödinger equations," SIAM J. Appl. Math. 60, 967–989.

J. Yang and D.J. Kaup (2009), "Squared eigenfunctions for the Sasa-Satsuma equation," J. Math. Phys. 50, 023504.

J. Yang and T.I. Lakoba (2007), "Universally-convergent squared-operator iteration methods for solitary waves in general nonlinear wave equations," Stud. Appl. Math. 118, 153–197.

J. Yang and T.I. Lakoba (2008), "Accelerated imaginary-time evolution methods for the computation of solitary waves," Stud. Appl. Math. 120, 265–292.

J. Yang, I. Makasyuk, A. Bezryadina, and Z. Chen (2004a), "Dipole solitons in optically-induced two-dimensional photonic lattices," Opt. Lett. 29, 1662–1664.

J. Yang, I. Makasyuk, A. Bezryadina, and Z. Chen (2004b), "Dipole and quadrupole solitons in optically-induced two-dimensional photonic lattices: Theory and experiment," Stud. Appl. Math. 113, 389–412.

J. Yang, I. Makasyuk, P.G. Kevrekidis, H. Martin, B.A. Malomed, D.J. Frantzeskakis, and Z. Chen (2005), "Necklace-like solitons in optically induced photonic lattices," Phys. Rev. Lett. 94, 113902.

J. Yang, B.A. Malomed, and D.J. Kaup (1999), "Embedded solitons in second-harmonic-generating systems," Phys. Rev. Lett. 83, 1958–1961.

J. Yang and Z. Musslimani (2003), "Fundamental and vortex solitons in a two-dimensional optical lattice," Opt. Lett. 28, 2094–2096.

J. Yang and Y. Tan (2000), "Fractal structure in the collision of vector solitons," Phys. Rev. Lett. 85, 3624–3627.

J. Yang and Y. Tan (2001), "Fractal dependence of vector-soliton collisions in birefringent fibers," Phys. Lett. A 280, 129–138.

T.S. Yang and T.R. Akylas (1996), "Weakly nonlocal gravity-capillary solitary waves," Phys. Fluids 8, 1506–1514.

A. Yariv (1989), *Quantum Electronics*, 3rd edition (Wiley, New York).

H. Yoshida (1990), "Construction of higher order symplectic integrators," Phys. Lett. A 150, 262–268.

N.J. Zabusky and M.D. Kruskal (1965), "Interaction of "solitons" in a collisionless plasma and the recurrence of initial states," Phys. Rev. Lett. 15, 240–243.

V.E. Zakharov (1968), "Stability of periodic waves of finite amplitude on the surface of a deep fluid," Zh. Prikl. Mekh. Tekh. Fiz 9, 86–94 [Transl. in J. Appl. Mech. Tech. Phys. 9, 190–194].

V.E. Zakharov, S.V. Manakov, S.P. Novikov, and L.P. Pitaevskii (1984), *The Theory of Solitons: The Inverse Scattering Method* (Consultants Bureau, New York).

V.E. Zakharov and A.M. Rubenchik (1974), "Instability of waveguides and solitons in nonlinear media," Sov. Phys. JETP 38, 494–500.

V.E. Zakharov and A.B. Shabat (1971), "Exact theory of two-dimensional self-focusing and one-dimensional self-modulation of waves in nonlinear media," Zh. E'ksp. Teor. Fiz. 61, 118 [Sov. Phys. JETP 34, 62–69 (1972)].

V.E. Zakharov and A.B. Shabat (1979), "Integration of the nonlinear equations of mathematical physics by the method of the inverse scattering problem II," Funk. Anal. Prilozh. 13, 13–22 [Funct. Anal. Appl. 13, 166–174 (1979)].

Y. Zeng, W. Ma, and R. Lin (2000), "Integration of the soliton hierarchy with self-consistent sources," J. Math. Phys. 41, 5453–5489.

Y. Zhang and B. Wu (2009), "Composition relation between gap solitons and Bloch waves in nonlinear periodic systems," Phys. Rev. Lett. 102, 093905.

Y. Zhu, R. Haberman, and J. Yang (2008a), "A universal map for fractal structures in weak interactions of solitary waves," Phys. Rev. Lett. 100, 143901.

Y. Zhu, R. Haberman, and J. Yang (2008b), "A universal separatrix map for weak interactions of solitary waves in generalized nonlinear Schrödinger equations," Physica D 237, 2411–2422.

Y. Zhu, R. Haberman, and J. Yang (2009), "Separatrix map analysis for fractal scatterings in weak interactions of solitary waves," Stud. Appl. Math. 122, 449–483.

Y. Zhu and J. Yang (2007), "Universal fractal structures in the weak interaction of solitary waves in generalized nonlinear Schrödinger equations," Phys. Rev. E 75, 036605.

A.A. Zozulya and D.Z. Anderson (1995), "Propagation of an optical beam in a photorefractive medium in the presence of a photogalvanic nonlinearity or an externally applied electric field," Phys. Rev. A 51, 1520–1531.

Index